ABOUT ISLAND PRESS

Island Press is the only nonprofit organization in the United States whose principal purpose is the publication of books on environmental issues and natural resource management. We provide solutions-oriented information to professionals, public officials, business and community leaders, and concerned citizens who are shaping responses to environmental problems.

In 2006, Island Press celebrates its twenty-second anniversary as the leading provider of timely and practical books that take a multidisciplinary approach to critical environmental concerns. Our growing list of titles reflects our commitment to bringing the best of an expanding body of literature to the environmental community throughout North America and the world.

Support for Island Press is provided by the Agua Fund, The Geraldine R. Dodge Foundation, The Doris Duke Charitable Foundation, The William and Flora Hewlett Foundation, Kendeda Sustainability Fund of the Tides Foundation, The Forrest C. Lattner Foundation, The Henry Luce Foundation, The John D. and Catherine T. MacArthur Foundation, The Marisla Foundation, The Andrew W. Mellon Foundation, The Gordon and Betty Moore Foundation, The Curtis and Edith Munson Foundation, The Oak Foundation, The Overbrook Foundation, The David and Lucile Packard Foundation, The Winslow Foundation, and other generous donors.

The opinions expressed in this book are those of the author(s) and do not necessarily reflect the views of these foundations.

Wildlife—Habitat Relationships

Wildlife—Habitat Relationships

Concepts and Applications

Third Edition

Michael L. Morrison

Bruce G. Marcot

R. William Mannan

ISLANDPRESS

Washington • Covelo • London

Library of Congress Cataloging-in-Publication data.

Morrison, Michael L.
    Wildlife-habitat relationships : concepts and applications / by Michael L. Morrison, Bruce G. Marcot, and R. William Mannan. — 3rd ed.
        p. cm.
    Includes bibliographical references.
    ISBN 1-59726-094-0 (cloth : alk. paper) — ISBN 1-59726-095-9 (pbk. : alk. paper)
1. Habitat (Ecology) 2. Animal ecology.   I. Marcot, Bruce G. II. Mannan, R. William. III. Title.
    QH541.M585 2006
    591.7—dc22

                                                                2006009619

British Cataloguing-in-Publication data available.

Printed on recycled, acid-free paper

Manufactured in the United States of America
10 9 8 7 6 5 4 3 2 1

Contents

Contents

Figures, Tables, and Boxes

Figures, Tables, and Boxes

Tables

Boxes

Preface

When we published our first edition of this book in 1992, the world population stood at a bit over 5.44 billion people and was increasing at an annual growth rate of 1.48%, adding 81,404,054 people to the planet annually, according to the United States Census Bureau. When we published our second edition in 1998, there were over 5.92 billion people, and although the annual growth rate had dropped slightly to 1.31%, it was still adding 78,308,546 people annually. As we completed this, our third edition, in early 2005, the planet was bearing over 6.47 billion of us, with an annual rate of increase of 1.14%, or 74,629,207 people. Concomitantly, just in this 14-year wink of an ecological eye, we have seen striking evidence of continued loss or degradation of the scarcest natural environments, including tropical coral reefs, mangrove swamps, ancient forests, and native grasslands, while urban, suburban, agricultural, and degraded lands, and lands dedicated solely to intensive resource production, continue to spread.

We wish for optimism but cannot ignore the crises in wildlife conservation that seem to confront us everywhere these days. Changes in regional and global climates continue to challenge our understanding but cry for action. There are crises of academia as essential expertise in basic taxonomy and systematics has itself become a moribund species; how crucial these skills are, for if we cannot name and catalog organisms, we cannot hope to document and quantify trends and mobilize action to stem extinctions, local and global. Other writers have despaired of how few conservation biologists these days spend much time in the field, and of how natural history as an empirical science and lifestyle seems to be increasingly forgotten. Perhaps the greatest crisis is what Robert Michael Pyle (1992) wrote of as "the extinction of experience," a growing personal alienation from nature and loss of intimacy with the very environment that sustains us, for, as he wrote, "What is the extinction of the condor to a child who has never known a wren?"

We speak of the "legacies" of ancient forests—large old trees, snags, down logs, and organic material to enrich tomorrow's soils. As well, we need to consider the legacies of our own knowledge and expertise to help others understand and provide wildlife and habitats for tomorrow. Reliable knowledge comes with rigor and scientific study. However, knowledge without action is as fruitless as never evolving from the primordial soup of ignorance in the first place. "Research," as the Oregon political columnist Russell Sadler (1991) once said, "is a race between ignorance and irreversible consequences."

While scientists struggle to understand the relations between human-caused environmental changes, biocomplexity, ecosystem resilience, species viability, and resource sustainability, we cannot lose sight of the astounding rapidity with which all these changes are occurring, nor the accelerated need to educate ourselves and others on the effects of our daily living habits. More than ever, we must all redefine ourselves as perennial students of the planet, whether we are senior managers, researchers, or academic students in the traditional sense. There is far more for us to learn than we ever can, as time for stabilizing or restoring wildlife and their habitats has already run out in portions of our wonderful and crowded world. This is what has led us to dedicate this third edition to wildlife students everywhere, the corollary being that learning and mutual education must never cease.

Let us quickly move beyond alarmism, for that shuts off the lines of listening by those publics, politicians, and purveyors we need to reach. Instead, as a wildlife profession, we can assert a positive vision of wildlife conservation that builds on legacies of knowledge and ecosystems alike. As our numbers grow, we can point to incredibly bold new moves in wildlife conservation that beg respect and emulation. India, which has surpassed 1 billion people since our second edition was published, in an attempt to save the last of its parks and wildlife communities has essentially outlawed clear-felling of forests and most sport hunting, and has given nontimber forest resources great economic and social focus. China has instituted policies of family size constraints and a massive reforestation program in many of their degraded and desertified lands. The Nature Conservancy has successfully run innovative programs of swapping portions of national debt for conserving critical natural areas in some developing countries. Major ecosystem restoration programs have been instituted, such as those in the Everglades of Florida. Top predators—carnivores—have been successfully reintroduced to Yellowstone National Park and elsewhere. There now are more national parks and sanctuaries, and more recovery plans for threatened and endangered species in place, than ever throughout the world.

Such positive steps toward wildlife and habitat conservation include some extreme measures taken literally in the face of collapsing ecosystems and vanishing species, as well as more evolutionary measures designed to better integrate economies with conservation, such as through burgeoning ecotourism and sustainable ecodevelopment programs. Developed countries also can learn much from conservation measures designed to include participation and ownership by local and native peoples, who often are most in need of reliable and sustainable resources and economic growth and stability.

In this way, wildlife conservation should not be viewed as a pastime of the rich but as a plan for the future for us all. A new vision for a near future in which we truly provide for sustainable resources, provide for ecosystem integrity, and foster the health of our biosphere, likely will demand the courage to seek and accept changes in our daily resource use habits and even to shift the very centers of what we value and how we value what we use. We can employ such positive visions, and successes like the ones cited above, as hallmarks and templates to help further such a future. It is not wildlife versus humanity, jobs versus owls, today's food versus tomorrow's inviolate protected area that will foster participation for a sustainable future. Nor will a sustainable future be reached along a gap between the academically educated and the lay public. Only by opening our minds and hearts and all becoming students can we move there together.

The purpose of our first edition was to advance from the point where the many fine, but introductory, texts in wildlife biology left off. Through the second, and now this third, edition, this purpose has not changed. We have further developed this new edition to incorporate the many new ideas that have come our way from several sources. First, we took to heart the independent reviews that appeared in scientific journals. Second, our friends and colleagues showed us their hidden talents as book critics; we also attended to these comments. Finally, we each have tapped into new experiences and studies to present the most current findings, concepts, and visions for future development in research and management.

This book is intended for advanced undergraduates, graduate students, and practicing professionals with a background in general biology, zoology, wildlife biology, conservation biology, and related fields. An understanding of statistics through analysis of variance and regression is helpful, but not essential. Land managers will especially benefit from this book because of its emphasis on the identification of sound research and the interpretation and application of results.

Our approach combines basic field zoology and natural history, evolutionary biology, ecological theory, and quantitative tools. We think that a synthesis of these topics is necessary for a good understanding of ecological processes, and hence good wildlife management. We attempt to draw on the best and recent examples of the topics we discuss, regardless of the species involved or its geographic location. We do concentrate on terrestrial vertebrates from temperate latitudes, with a bias toward North America, because this is where much literature has been developed and where our own experience has occurred. However, because it is the concepts that are important, the specific examples are really of secondary importance. Hence our writing can be used by anyone from any location. We did try, however, to bring in examples from amphibians, reptiles, birds, and mammals (both large and small), and from different ecosystems and locations, to help individuals from different backgrounds better understand the application of concepts to their particular interests.

We emphasize the need for critical evaluation of methodologies and their applications in wildlife research. Management decisions all too often are based on data of unknown reliability— that is, from research conducted using biased methods, low sample sizes, and inappropriate analyses. We understand also that, all too often, managers are faced with making decisions using unreliable or incomplete data. The general dearth of monitoring, validation, and adaptive management research forces a vicious cycle. This does not need to—and should not—persist.

To aid both the student and the professional, we have tried to explain fundamental concepts of ecological theory and assessment so that the use of more advanced technical tools is more acceptable, more often sought, and more appropriately applied. Ultimately, the success of conservation efforts depends on gathering, analyzing, and interpreting reliable information on species composition, communities, and habitat. We hope that this book encourages such rigor in concept and practice.

Literature Cited

Pyle, R. M. 1992. Intimate relations and the extinction of experience. In *Left Bank #2: Extinction*, 61–69. Hillsboro, OR: Blue Heron Publishing.

Sadler, R. 1991. Paper presented at the New Perspectives Conference, USDA Forest Service, Roanoke VA, December 3, 1991.

About the Third Edition

The second edition forms the core of this new work. We have revised much of the text, introduced much new material in each chapter to supplement that previously offered, and updated reference citations throughout.

In Part 1, Chapters 1 through 3 cover *central concepts of wildlife–habitat relationships* and lay the foundation on which the rest of the book is constructed. Chapter 1 discusses the historical background and philosophical attitudes that have shaped the wildlife profession and influenced how research should be approached. Chapter 2 reviews the evolutionary background against which the current distribution, abundance, and habits of animals developed. In this edition, we defer discussion of keystone species to a broader and updated discussion of "key ecological functions" of species in Chapter 11. Chapter 3 discusses habitat relationships from the perspective of vegetation ecology and population biology. In Chapter 3, we have updated and substantially expanded our discussion of the niche as it appeared in the second edition. Specifically, we have developed how the study of multiple limiting factors likely holds the key to advancing our study of habitat relationships. We still address population responses, and have expanded our discussion of population viability, genetics, metapopulation dynamics, and related concepts.

In Part 2, Chapters 4 through 10 form the heart of the book and cover *measurement and modeling of wildlife–habitat relationships*. Chapter 4 discusses fundamental approaches to study design and experimental methodologies, reviewing the philosophy of various ways of gaining reliable knowledge, and the challenges to conducting scientific investigations and having the result be accepted in society. We have added new examples, and new subjects of concern about science are now addressed including information theoretic versus traditional hypothesis testing, and relativism. Chapters 5 and 6 review the many methods that have been used to develop wildlife-habitat relationships, including field methods, data analysis, sampling biases, and data interpretation. We re-organized Chapter 5 to more explicitly encompass analyses across spatial scales, and have updated our discussions of methodologies to include the increasing use of new technologies. Chapter 6 also incorporates discussion of multivariate statistics, which we have updated with additional comments on methods and misuses of the techniques. Although we do not delve heavily into methods of multivariate analyses, we think that we are more effective by emphasizing the concept of multivariate analyses, proper sampling methods, and interpretation of results. We have also added new information on model selection procedures, such as AIC. Chapter 7 covers behavioral sampling and analysis in wildlife research, and has been expanded to include more information on the fundamental causes of an individual's behavior. Chapters 8 and 9 review characterization of patterns of habitat within landscapes, and population responses, respectively, including habitat fragmentation, study of metapopulations, and landscape ecology, topics that continue to be emphasized by researchers and managers alike. Chapter 8 presents the rationale for a landscape-perspective of habitat relationships, definition and classification of landscapes, basics of landscape ecology, concepts of spatial and temporal scales in ecological study and their implications for managing habitat in landscapes, ways to depict and measure habitat heterogeneity including habitat fragmentation, and reviews disturbance ecology and management implications; all material—concepts, summary of studies, and citations—has been brought up to date since the previous edition. Chapter 9 focuses on population response to landscape conditions and patterns, and reviews how researchers and managers have viewed wildlife response to habitat edges, boundary effects, and succession and climate; provides an updated discussion of population viability, metapopulation dynamics, and effects of population isolation; updates discussions of biogeographic implications of habitat isolation and patterns, species–area relations; and discusses implication

for conserving and monitoring wildlife in heterogeneous environments, including utility of habitat corridors. All of this material has been updated since the last edition. Chapter 10 reviews and updates the utility and development of wildlife–habitat relationships models, including discussion of how to select models, depict uncertainty, and implications of prediction errors and model validity for research hypothesis-testing and management decision making. Chapter 10 also updates discussions from the last edition on traditional types of models, and presents a new section on more recent, avant garde wildlife habitat modeling approaches that draw from fields of decision support, Bayesian statistics, and various knowledge-based approaches only recently being developed for wildlife habitat modeling. The chapter also updates a discussion on recent developments in various approaches to modeling land allocations for habitat conservation and on recent results of model validation.

In Part 3, Chapters 11 and 12 cover *management of wildlife–habitat relationships.* Chapter 11 introduces the topic of wildlife and habitat management in the context of ecosystem management. We discuss and illustrate wildlife management goals in an evolutionary and ecological context, and provide all-new material and examples on a broad environmental and functional approach, including an ecographic (mapping) approach to evaluating and managing for key environmental correlates, key ecological functions, and key cultural functions of wildlife. We newly discuss implications for conservation of ecosystem services, thinking beyond wildlife population viability in a community and ecosystem context, and practical approaches to managing for evolutionary potential of wildlife. We also provide an updated discussion of adaptive management and review both failures and successes in this area. Chapter 12 presents a framework for advancing our understanding of wildlife through modified approaches to habitat relationships, raises a call for greater emphasis on the synthetic field of restoration ecology, and makes a plea for improvements to our educational system. We present this material partly as a prescription, and partly as a "null model" on which we can debate the best means of advancing our profession. In this edition we have refined our recommendations on how wildlife and habitat might be studied if we are to improve our understanding of what determines distribution and abundance, and ultimately leads to the recovery and preservation of species.

New to this edition is a brief glossary of key terms that every wildlifer should know. The book concludes, as did the second edition, with an author index and a general subject index.

Developing this latest edition entailed our extensively reviewing a massive amount of recent literature and discussing concepts, findings, and approaches with many researchers and managers. In one sense, little has changed since the early 20th century; habitat is still the crux and essential foundation for wildlife conservation, although there continue to be rapid advances in approaches to conceptualizing, measuring, modeling, and managing habitat. We have tried to keep pace with such advances in this edition and have prioritized new and expanded discussions on topics with the most promise for successfully understanding and conserving wildlife and habitats.

Lastly, in this volume we have again demonstrated the robust, positive growth rate of new editions, despite our wonderful editor's decree for density-dependent limits to growth. When pressed, our answer is simply, "Knowledge should be boundless."

Acknowledgments

First, we thank Barbara Dean, Executive Editor, Island Press, for guiding this third edition through the publication process. And once again we thank Allen Fitchen, former director of the University of Wisconsin Press, for encouraging us to put our thoughts onto paper and shepherding this project through the first and second editions. Many individuals have reviewed chapters through the three editions of this book, including several additional referees for this volume—Roel Lopez, Tom O'Neil, John Marzluff, Paul Krausman, Shawn Smallwood, Luke George, Bob Steidl, and William Matter; we thank you all for your insights. We also thank Joyce VanDeWater for her diligent and excellent help in finalizing the figures for this edition, and Carly Johnson for formatting and editing our penultimate draft.

Numerous individuals helped shape our views of wildlife biology and science in general; we cannot list them all. We especially appreciate the dialogues shared with our graduate students over the years. In addition, discussions with many ecologists and managers, domestically and internationally, helped us identify recent scientific advances and critical management issues. We also thank the numerous authors whom we have cited in this book for their research efforts and insightful analyses.

We dedicated the first edition of our book to Drs. E. Charles Meslow and Jack Ward Thomas, "who taught us that wildlife conservation truly succeeds when practiced with honor, rapport, and rigor." Our second edition was dedicated to the community of field biologists throughout the world "who daily tend to the inheritance of succeeding generations." Both of these dedications bear repeating there, for they are both more valid than ever.

We dedicate this edition to wildlife students of the world, including those who learn in academia and those who continue to educate themselves throughout their careers. Though learning will come informed action and the courage and spirit to educate others, thus ensuring the future of wildlife everywhere.

PART I

Concepts of Wildlife–Habitat Relationships

The Study of Habitat:
A Historical and Philosophical Perspective

It is a good morning exercise for a research scientist to discard a pet hypothesis every day before breakfast. It keeps him young.

KONRAD LORENZ

An animal's habitat is, in the most general sense, the place where it lives. All animals, except humans, can live in an area only if basic resources such as food, water, and cover are present and if the animals have adapted in ways that allow them to cope with the climatic extremes and the competitors and predators they encounter. Humans can live in areas even if these requirements are not met, because we can modify environments to suit our needs or desires and because we potentially have access to resources such as food or building materials from all over the world. For these reasons, humans occupy nearly all terrestrial surfaces of the earth, but other species of animals are restricted to particular kinds of places.

The distribution of animal species among environments and the forces that cause these distributions have frequently been the subjects of human interest, but for different reasons at different times. The primary purpose of this introductory chapter is to review some of the reasons why people study the habitats of animals and to outline how these reasons have changed over

time. We also introduce the major concepts that will be addressed in this book.

Curiosity about Natural History

Throughout recorded history, humans, motivated by their curiosity, have observed and written about the habits of animals. The writings of naturalists were, for centuries, the only recorded sources of information about animal–habitat relationships. Aristotle was among the first and best of the early naturalists. He observed animals and wrote about a wide variety of subjects, including breeding behavior, diets, migration, and hibernation. Aristotle (384–322 BC) also noted where animals lived and occasionally speculated about the reasons why:

> A number of fish also are found in sea-estuaries; such as the saupe, the gilthead, the red mullet, and, in point of fact, the greater part of the gregarious fishes. . . . Fish penetrate into the Euxine [estuary]

for two reasons, and firstly for food. For the feeding is more abundant and better in quality owing to the amount of fresh river-water that discharges into the sea. . . . Furthermore, fish penetrate into this sea for the purpose of breeding; for there are recesses there favorable for spawning, and the fresh and exceptionally sweet water has an invigorating effect on the spawn. (Aristotle 344 BC)

Interest in natural history waned after Aristotle's death. Politics and world conquest were the focus of attention during the growth of the Roman Empire, and interest in religion and metaphysics suppressed creative observation of the natural world during the rise of Christendom (Beebe 1988). As a result, little new information was documented about animals and their habitats for nearly 1700 years after the death of Aristotle. Yet, as Klopfer and Ganzhorn (1985) noted, painters in the medieval and pre-Renaissance periods still showed an appreciation for the association of specific animals with particular features of the environment. "Fanciful renderings aside, peacocks do not appear in drawings of moors nor moorhens in wheatfields" (Klopfer and Ganzhorn 1985, 436). Similar appreciation is seen in artwork from India, China, Japan (e.g., Sumi paintings), and elsewhere during this period. Thus keen observers noticed relationships between animals and their habitats during the Dark Ages, but few of their observations were recorded.

The study of natural history was renewed in the seventeenth and eighteenth centuries. Most naturalists during this period, such as John Ray (1627–1705) and Carl Linnaeus (1707–1778), were interested primarily in naming and classifying organisms in the natural world (Eiseley 1961). Explorers made numerous expeditions into unexplored or unmapped lands during this period, often with the intent of locating new

trade routes or identifying new resources. Naturalists usually accompanied these expeditions or traveled on their own, collecting and recording information about the plants and animals they observed. Many Europeans during this period also collected feathers, eggs, pelts, horns, and other parts of animals for "collection cabinets." Some cabinets were serious scientific efforts, but most were not. Nevertheless, new facts about the existence and distribution of animals worldwide were gathered during this time, and the resulting advances in knowledge generated considerable curiosity about the natural world.

During the nineteenth century, naturalists continued to describe the distribution of newly discovered plants and animals, but they also began to formulate ideas about how the natural world functions. Charles Darwin (1809–1882) was among the most prominent of these naturalists. His observations of the distributions of similar species were one set of facts among many that he marshaled to support his theory of *evolution by natural selection* (Darwin 1859). The work of Darwin is highlighted here, not only because he recorded many new facts about animals, but also (and more importantly) because the theory of evolution by natural selection forms the framework and foundation of the field of ecology.

Curiosity about Ecological Relationships

In the early 1900s, curiosity about how animals interact with their environment provided the impetus for numerous investigations into what are now called *ecological relationships*. Interest in these relationships initially led to detailed descriptions of the distribution of animals along environmental gradients or among plant communities. Merriam (1890), for example, identified the changes that occur in plant and animal species on an elevational gradient, and Adams

(1908) studied changes in bird species that accompany plant succession. Biologists living in this period postulated that climatic conditions and availability of food and sites to breed were the primary factors determining the distributions of animals they observed (see Grinnell 1917a).

Biologists in the early to mid-1900s, however, recognized that the distribution of some animals could not be explained solely on the basis of climate and essential resources. David Lack (1933) was apparently the first to propose that some animals (in this case, birds) recognize features of appropriate environments, and that these features are the triggers that induce animals to select a place to live. Areas without these features, according to Lack, generally will not be inhabited, even though they might contain all the necessary resources for survival. Lack's ideas gave birth to the concept of habitat selection and stimulated considerable research on animal–habitat relationships during the next 60 years.

Svardson (1949) developed a general conceptual model of *habitat selection,* and Hilden (1965) later expressed similar ideas. Their models characterized habitat selection as a two-stage process in which organisms first use general features of the landscape to select broadly from among different environments, and then respond to subtler habitat characteristics to choose a specific place to live. Svardson (1949) also suggested that factors other than those associated with the structure of the environment influence selection. For example, whether an animal stays or leaves a particular place could be influenced by *conspecifics* (Butler 1980), *interspecific competitors* (Werner and Hall 1979), and *predators* (Werner et al. 1983), as well as by features of the environment that are directly or indirectly related to resources needed for survival and reproduction. Habitat selection, therefore, has come to be recognized as a complicated process involving several levels of discrimination and spatial scales and a number of potentially interacting factors. Study of these factors and the behaviors involved in habitat selection has resulted in a wealth of information about why we find animals where we do (see Stauffer [2002] for an overview of the recent history of habitat studies).

The distribution of animals is also intimately tied to the concept of *niche.* This concept has been defined in multiple ways over time (see e.g., Schoener 1989; Griesemer 1992; Pianka 1994 for historical overviews) and continues to be the subject of much discussion. Grinnell (1917b) formally introduced the term when he was attempting to identify the reasons for the distribution of a single species of bird. His assessments included spatial considerations (e.g., reasons for a close association with a vegetation type), dietary dimensions, and constraints placed by the need to avoid predators (Schoener 1989). Thus, in this view, the niche included both positional and functional roles in the community. Elton (1927) later described the niche as the status of an animal in the community and focused on trophic position and diet. Views of the niche articulated by Grinnell and Elton are often contrasted, but Schoener (1989) argued that they had much in common, including the idea that a niche denotes a "place" in the community, dietary considerations, and predator-avoiding traits. Hutchinson (1957) articulated the multivariate nature of the causes of animal distribution in his presentation of the *n*-dimensional niche. In this view, niche dimensions are represented by multiple environmental gradients. A given species (or population) can exist in only a subset of the conditions defined by all the gradients (its potential or fundamental niche) but may be further restricted in distribution (its realized niche) by predators and competitors. Odum (1959) viewed the niche as the position or status of an organism in an ecosystem resulting from its behavioral and morphological adaptations. His idea of the niche was dependent on both where an organism lives and what it does, but he separated, to some degree, habitat from niche with the analogy that an organism's "address" is its habitat and its "profession" is its niche. More recent ideas about the niche (e.g., MacArthur and Levins 1967; Levins 1968; Schoener 1974) consider niche axes as resources (i.e., those important for an animal) and niche as the combination of several "utilization distributions" along those axes. The point of our brief review of the concept of the niche is to illustrate that, although the term can be viewed in a variety of ways, most concepts include elements that are traditionally considered part of habitat. Thus studies designed to describe or define an animal's niche (of which there have been many) almost always elucidate animal–habitat relationships as well.

Hunting Animals for Food and Sport

The earliest humans relied, in part, on killing animals for survival, and they undoubtedly recognized and exploited the patterns of association between the animals they hunted and the kinds of places where these animals were most abundant. Use of fire by Native Americans altered the ecosystems in which they lived (Botkin 1990) and influenced (probably intentionally) the number of animals they hunted. Similarly, people who later made an "economic" living by trapping and hunting, or could afford the luxury of hunting for sport, knew where to find animals and probably speculated accurately about the habitat features that influenced the abundance of game species. Marco Polo reported, for example, that in the Mongol Empire in Asia, Kublai Khan (AD 1215–1294) increased the number of quail and partridge available to him for falconry by planting patches of food, distributing grain during the winter, and controlling cover (Leopold 1933).

This advanced system of habitat management suggests a general understanding of the habitat requirements of target game species, but it is unlikely that the information was obtained through organized studies of habitat use. Also, the men who hunted and trapped for subsistence or sport rarely recorded their knowledge about habitats for posterity.

Not until people began to attempt to apply biology systematically to the management of game as a "crop" in the early 1900s did they realize that "science had accumulated more knowledge of how to distinguish one species from another than of the habits, requirements, and inter-relationships of living population" (Leopold 1933, 20). The absence of information about habitat requirements of most animals and the desire to increase game populations by manipulating the environment stimulated detailed investigations of the habitats and life histories of game species. H. L. Stoddard's work on bobwhite quail (*Colinus virginianus*), published in 1931, and Errington and Hammerstrom's work on pheasants, published in 1937, exemplify early efforts of this kind.

Studies similar to Stoddard's have been conducted on most game animals in North America from 1930 through the present day (e.g. Bellrose 1976; Wallmo 1981; Thomas and Toweill 1982), but many of these studies only summarize general habitat associations and do not identify critical habitat components. Since the early 1980s, the number of hunters has increased while undeveloped land available for managing wild animal populations has decreased. The need to manage populations more intensively is therefore great, and detailed knowledge of habitat requirements is essential for this task. Studies of the habitat requirements of game animals continue to be conducted, as one can easily see by reviewing recent scientific journals on wildlife management.

Public Interest and Environmental Laws

Human activities have dramatically disturbed natural environments in North America and throughout the world. These disturbances have been associated primarily with the rapid increase in the size of the human population and the exploitation of natural resources, including wild animals, for human use. Interest in wild animals by the general public also increased during this period, and concern about the negative effects of human activities on animal populations and other aspects of the natural environment eventually led to the passage of laws in the United States that were designed to aid management of wild animals or reduce environmental degradation. The following summary pertains to U.S. history; it is beyond the scope of the text to review public interest and environmental law in other nations.

Public interest early in the century focused on "game" animals, and some laws passed in the 1930s reflected this interest. The Migratory Bird Hunting Stamp Act of 1934 and the Pittman-Robertson Federal Aid in Wildlife Restoration Act of 1937, for example, primarily taxed sportsmen and provided funds for management of waterfowl and other hunted species (see table 1.1). As noted in the previous section, information needed for management of these species stimulated efforts to describe and quantify their habitats.

An increase in environmental awareness during the 1960s and 1970s broadened the scope of the kinds of animals about which the general public was concerned. Animal species not hunted for sport and without any other apparent economic utility were also perceived as having value. (The ethical rationales underlying these values are discussed in the next section.) Among the laws passed during this period were the National Environmental Policy Act (1969), the En-

dangered Species Conservation Act (1973), the Federal Land Policy and Management Act (1976), and the National Forest Management Act (1976) (Bean 1977; see also table 1.1). Legislators designed these laws, in part, to ensure that all wildlife species and other natural resources were considered in the planning and execution of human activities on public lands. Knowledge of the habitats of animal species is obviously required before the effects of an environmental disturbance can be fully evaluated, before a refuge for an endangered species can be designed, or before animal habitats can be maintained on lands managed under a multiple-use philosophy. Biologists responded to the need for information about habitat requirements by studying, often for the first time, numerous species of "nongame" animals and by developing models to help predict the effects of environmental changes on animal populations (e.g., Verner et al. 1986).

Public interest in the nonconsumptive use of animals has not waned in recent years. In the United States in 2001, 66 million people over 16 years of age spent over $38.4 billion observing, feeding, or photographing wildlife (U.S. Fish and Wildlife Service 2003). The funding mechanisms for managing animals in the United States, however, have not kept pace with the broadening umbrella of public interest. Many state fish and game agencies have developed nongame management programs that emphasize identifying and managing habitats, but these programs are often limited by inadequate funding, and the sources of funds are, with rare exception, not broad based or user related. In Arizona, for example, the nongame program is funded by a fixed percentage of the funds generated by the state lottery.

Efforts to increase the funding base for managing nongame animals and their habitats were initiated in the mid-1990s, when legislation was

Table 1.1.   Important U.S. legislation stimulating the study, preservation, or management of animal habitat

Title	Year	Action
Migratory Bird Treaty Act	1929	Provided for the establishment of wildlife refuges
Migratory Bird Hunting Stamp Act	1934	Required a federal migratory bird hunting license; funds used to purchase lands for refuges
Fish and Wildlife Coordination Act	1934	Authorized conservation measures in federal water projects and required consultation with the U.S. Fish and Wildlife Service and states concerning any water project
Pittman-Robertson Federal Aid in Wildlife Restoration Act	1937	Provided an excise tax on sporting arms and ammunition to finance research on a federal and state basis
Multiple Use–Sustained Yield Act	1960	Stipulated that national forests would be managed for outdoor recreation, range, timber, watershed, and wildlife and fish
Wilderness Act	1964	Gave Congress authority to identify and set aside wilderness areas
Classification and Multiple Use Act	1969	Mandated multiple-use management on lands administered by the Bureau of Land Management
National Environmental Policy Act (NEPA)	1969	Stipulated that environmental impact statements would be prepared for any federal project that affected the quality of the human environment; wildlife habitat considered part of that environment
Endangered Species Conservation Act	1973	Provided protection for species and the habitat of species threatened with extinction.
Sikes Act Extension	1974	Directed secretaries of Agriculture and Interior to cooperate with state game and fish agencies to develop plans for conservation of wildlife, fish, and game
Forest and Rangelands Renewable Resources Planning Act (FRPA)	1974	Called for units of the national forest system to prepare land management plans for the protection and development of national forests.
Federal Land Policy and Management Act	1976	Mandated land use plans on lands administered by the Bureau of Land Management
National Forest Management Act	1976	Stipulated that the plans called for in the FRPA would comply with NEPA and that management would maintain viable populations of existing native vertebrates on national forests

Source: Based in part on Gilbert and Dodds (1987, 17).

written that called for a federal tax on outdoor equipment, such as binoculars and tents, used in activities associated with the nonconsumptive enjoyment of wildlife. Funds generated by this act, like those from the Pittman-Robertson Federal Aid in Wildlife Restoration Act, would have been distributed to the states on a matching basis. This initial effort failed, but similar legislation is currently being promoted (e.g., the Teaming with Wildlife initiative). Legislation of this kind, if made into federal law, would distribute the burden of paying for wildlife management among those who most benefit, and would allow state agencies to manage more thoroughly the habitats of a wide variety of species. It would also stimulate the acquisition of information about those habitats.

Ethical Concerns

Another impetus for studying habitat partly underlies the public interest and environmental laws outlined in the previous section and relates to an ethical concern for the future of wildlife and natural communities (Schmidtz and Willott 2002). This concern is, in part, a humanistic one, insofar as the health of natural systems affects our use and enjoyment of natural resources in the broadest sense. From a utilitarian viewpoint, the world is also our habitat, and its health directly relates to our own. The ethical concern, however, transcends humanism in that wildlife and natural communities are intrinsic to the world in which we have evolved and now live. Writers of legal as well as ethical literature have argued that nonhuman species have, in some sense, their own natural right to exist and grow (e.g., Stone 1974, 1987). The study of wildlife species and their habitats in this context may deepen our appreciation for and ethical responsibility to other species and natural systems.

Why should we be concerned about species and habitats that offer no immediate economic or recreational benefits? Several rather standard philosophical arguments offer complementary and even conflicting rationales. One viewpoint argues for conserving species and their environments because we may someday learn how to exploit them for medical or other benefits (future option values). Another viewpoint argues for preserving species for the unknown (and unknowable) interests of future generations; we cannot speak for the desires of our not-yet-born progeny, who will inherit the results of our management decisions.

Generally, a traditional conflict has pitted ethical humanism against humane moralism. *Ethical humanism,* as championed by Guthrie, Kant, Locke, More, and Aquinas, argues that animals are not "worthy" of equal consideration; animals are not "up to" human levels in that they do not share self-consciousness and personal interests. In effect, this argument allows us to subjugate wildlife and their habitats. Kant argues as much. He advanced his idea on a so-called deontological theme (from the Greek *de-*

ont, "that which is obligatory"). That is, rights—specifically human rights—allow us to view animals as having less value because they are less rational (or are arational); we humans have the duty to manage species and the freedom to subjugate them.

On the other hand, *humane moralism,* as championed in part by Christopher D. Stone, Jeremy Bentham (of the animal liberation movement), and Peter Singer, argues that animals deserve the focus of ethical consideration. According to this argument, humans are moral agents. Animals and, by extension, their habitats require consideration equal to that given humans, even if they do not ultimately receive equal treatment.

There is also a third ethical stance, one that may serve as an impetus for studying and conserving wildlife and their habitats: an ecological ethic. The *ecological ethic,* as proposed by J. Baird Callicott, was most eloquently advanced by Leopold (1949) in his *A Sand County Almanac,* although elements of his philosophy (and much fuller philosophical expositions) can be traced to Henri Berson, Teilhard de Chardin, and John Dewey. The focus of ethical consideration in this view is on both the individual organism and the community in which it resides. Concern for the community is the essence of Leopold's ecological ethic, a holistic ethic that focuses on the relationships of animals with each other and with their environment.

Leopold wrote of soil, water, plants, animals, oceans, and mountains, calling each a natural entity. In his view, animals' functional roles in the community, not solely their utility for humans, provide a measure of their value. By extension, then, to act morally, we must maintain our individual human integrity, our social integrity, and the integrity of the biotic community.

Following such an ecological ethic, a concern for the present and future conditions of wildlife and their habitats motivates the writing of this book. The sad history of massive resource depletion, including extinctions of plant and animal species and the large-scale alteration of terrestrial and aquatic environments, must, in our view, strengthen a commitment to further understanding wildlife and their habitats. Understanding is the necessary prelude to living truly by an ecological ethic.

Concepts Addressed

This book covers both theoretical and applied aspects of wildlife–habitat relationships, with an emphasis on the theoretical framework under which researchers should study such relationships. An appropriate way to begin a preview of the concepts covered in subsequent chapters is to define the term *habitat.* A review of even a few papers concerned with the subject will show that the term is used in a variety of ways. Frequently, *habitat* is used to describe an area supporting a particular type of vegetation or, less commonly, aquatic or lithic (rock) substrates. This use probably grew from the term *habitat type,* coined by Daubenmire (1976, 125) to refer to "land units having approximately the same capacity to produce vegetation."

We, however, view habitat as a concept that is related to a particular species, and sometimes even to a particular population, of plant or animal. Habitat, then, is an area with a combination of resources (like food, cover, water) and environmental conditions (temperature, precipitation, presence or absence of predators and competitors) that promotes occupancy by individuals of a given species (or population) and allows those individuals to survive and reproduce. Habitat of high quality can be defined as those areas that afford conditions necessary for relatively successful survival and reproduction over relatively long periods when compared with

other environments. (We recognize, though, that the habitats of some animals are ephemeral by nature, such as early seral stages or pools of water in the desert after heavy rains.) Conversely, marginal habitat promotes occupancy and supports individuals, but their rates of survival and reproduction are relatively low, or the area is usually suitable for occupancy for relatively short or intermittent periods. Thus quality of habitat is ultimately related to the rates of survival and reproduction of the individuals that live there (Van Horne 1983), to the vitality of their offspring, and to the length of time the site remains suitable for occupancy.

Understanding why a particular population or species occupies only a specific area in a region or why it occupies only a specific continent requires more than just knowledge of the organism's environmental needs and ecological relationships. Explanations for an animal's distribution also require an understanding of its evolutionary history, the climatic history of the area, and even the history of the movements of landmasses. We provide in chapter 2 an overview of the forces, factors, and processes that determine why animals are found where they are and how they came to be there. The information presented emphasizes that both past and present conditions can play significant roles in defining the habitat of an animal. In short, we provide in chapter 2 the evolutionary perspective and conceptual framework we feel are necessary before the study of habitat can proceed successfully.

Important elements of the habitat of an animal are often provided by vegetation. Changes in vegetation can, therefore, alter habitat conditions. Understanding how the structure and composition of vegetation influence the quantity and quality of habitat features is central to understanding the distribution and abundance of animals. We begin chapter 3 with a review of the patterns and processes associated with plant succession and the relationships between animals and vegetative change. We also initiate in chapter 3 a discussion of how the concepts of niche and habitat relate to one another, and we emphasize the importance of focusing investigations of habitat on the resources that allow animals to survive and reproduce and on ecological relationships that may constrain access or use of those resources. We end chapter 3 with a discussion of factors that can influence the dynamics and viability of populations, including how they may be distributed (e.g., the concept of metapopulation), their genetic makeup, movements of animals within and among them, and the influences of other organisms, such as nonnative species and humans.

Studying wildlife–habitat relationships requires knowledge of the scientific method. We review in chapter 4 activities involved in the scientific method and some of the controversial issues associated with its application in ecology and wildlife science. For example, we emphasize in this chapter the need for students and professional biologists to understand the difference between research and statistical hypotheses. We also review weaknesses in traditional "statistical null hypothesis tests" and discuss alternative approaches. We end chapter 4 with an evaluation of the strengths and weaknesses of laboratory and field experiments and offer some general strategies for how to proceed with investigations of wildlife–habitat relationships.

Identifying what constitutes habitat of a population or species is the impetus underlying many activities in wildlife science and management. Designing studies that identify habitat conditions requires considerable thoughtfulness about the needs and perceptual abilities of the species under investigation, the spatial scale at which the study is to be conducted, and the methods for measuring environmental features. We provide in chapter 5 a review and analysis of

what elements of the environment might be measured in studies of habitat and a discussion of the methods commonly used to measure them.

The assessment of what to measure in wildlife habitat and how to measure it, in chapter 5, is followed in chapter 6 with a consideration of when to take the measurements. We focus in chapter 6 on the importance of timing in determining what constitutes habitat. Use of resources by animals can vary on several temporal scales, including time of day, stage of breeding cycle, season of the year, and between years. Deciding which scale or scales to address in a study will obviously influence its design. Evaluation of the numerous factors that can influence whether an animal occupies a given area lends itself to the use of multivariate statistical techniques. We end chapter 6 with a review of the use of these techniques in conceptualizing, analyzing, and understanding wildlife–habitat relationships.

Patterns of resource use detected in animal populations are products of the behaviors of individual animals. We present in chapter 7 the theoretical framework that forms the basis for investigating animal behavior as it relates to habitat. We also review the principal methods used to measure animal behavior. Assessment of diet and foraging behavior is a focus of this chapter because an animal's survival and productivity depend heavily on acquiring food.

Some animals may select habitat through a hierarchical process that begins on "broad spatial scales" or large geographic extents. Furthermore, the distribution of patches of environmental resources (e.g., vegetation types) across the landscape can influence the dynamics of populations and elements of community structure. We review in chapter 8 the basic tenets of landscape ecology, and emphasize that "landscape," like habitat, is best viewed as a species-specific concept. We describe different aspects of "scale," including geographic extent, map resolution, time, biological organization, and administrative hierarchy, and how these aspects can be integrated in landscape ecology. We also discuss in this chapter how animals may respond to different kinds of disturbances and the resulting heterogeneity of resources in patchy landscapes. Associated with this discussion is a review of the management challenges presented by patchy or fragmented environments; this review includes an assessment of the value of retaining remnant patches of natural environments and some level of connectivity between them. In chapter 9, we continue the discussion of wildlife and landscapes but focus on the specific responses of organisms, species, populations, and communities to landscape dynamics.

Models of wildlife–habitat relationships are used for a variety of purposes, including: (1) as descriptions of current levels of understanding; (2) assessing the relative importance of environmental features in the distribution and abundance of organisms; (3) identifying weaknesses in current understanding; and (4) generation of testable hypotheses about animals and systems of interest. In chapter 10, we review the types of models used in the study and management of wildlife and their habitats, and examine how scientific uncertainty affects the use of these models. We also discuss how models can be developed, calibrated, and tested.

The earth and the natural resources on it are changing rapidly, primarily as a result of human use and exploitation. In the future, management of natural resources, including wildlife, will likely require approaches that conceptually force us to think on broader spatial, temporal, and ecological scales. We discuss in chapter 11 the possibility of managing wildlife in an ecosystem context and suggest that the traditional concept of habitat may need to be broadened beyond the basics of food, cover, and water to include ideas such as the ecological roles of other species, abiotic conditions, and natural disturbance re-

gimes. We also suggest that the traditional notion of wildlife may need to be broadened to encompass the full array of biota present in an ecosystem. We propose in this chapter an enhanced approach to depicting, modeling, and predicting the status and condition of wildlife in ecosystems, and advocate the rigorous use of adaptive management as a foundation for land use decisions.

Changes in environmental conditions on Earth, advances in technological devices and analytic methods, and the potential need for new philosophical and conceptual approaches in research and management require that wildlife biologists do their best to keep abreast of new ideas. We discuss in chapter 12 several ideas that may assist us in advancing our understanding of wildlife–habitat relationships. We suggest that the scale on which we conduct research and management be examined rigorously, and explore potential ways that the concept of niche may help focus research in the future. We also call for a more complete integration of the overlapping fields of wildlife ecology and management, conservation biology, and restoration ecology. Changes in the way we educate students and professionals must precede integration of this kind, and we end chapter 12 with some suggestions for how these changes may be initiated.

Our view of the world and how it works is likely to shift over time as current explanations are replaced by better ones. The second edition of this book was motivated by the evolution of ideas presented in the first. Similar shifts in thinking motivated this third edition. However, our ideas about dealing with changes in the world remain the same. We hope that, no matter what changes occur, our readers—current or future conservationists in the broadest sense—remain tied to an ecological land ethic and continue the pursuit of providing vital, productive habitats for wildlife and humans alike.

Literature Cited

Adams, C. C. 1908. The ecological succession of birds. *Auk* 25:109–53.

Aristotle. 344 BC/1862. *Historia animalium.* London: H. G. Bohn.

Bean, M. J. 1977. *The evolution of national wildlife law.* Report to the Council on Environmental Quality. U. S. Government Document, Stock No. 041-011-00033-5.

Beebe, W., ed. 1988. *The book of naturalists.* Princeton, NJ: Princeton Univ. Press.

Bellrose, F. C. 1976. *Ducks, geese and swans of North America.* Harrisburg, PA: Stackpole Books.

Botkin, D. B. 1990. *Discordant harmonies: A new ecology for the twenty-first century.* New York: Oxford Univ. Press.

Butler, R. G. 1980. Population size, social behavior, and dispersal in house mice: A quantitative investigation. *Animal Behavior* 28:78–85.

Darwin, C. 1859/1968. *The origin of species.* New York: Penguin Books.

Daubenmire, R. 1976. The use of vegetation in assessing the productivity of forest lands. *Botanical Review* 42:115–43.

Eiseley, L. 1961. *Darwin's century.* Garden City, NY: Doubleday, Anchor Books.

Elton, C. 1927. *Animal ecology.* London, Sidgwick & Jackson.

Errington, P. L., and F. N. Hamerstrom. 1937. The evaluation of nesting losses and juvenile mortality of the ring-necked pheasant. *Journal of Wildlife Management* 1:3–20.

Gilbert, F. F., and D. G. Dodds. 1987. *The philosophy and practice of wildlife management.* Malabar, FL: Krieger.

Griesemer, J. B. 1992. Niche: Historical perspectives. In *Key-words in evolutionary biology,* ed. E. Fox-Keller and E. A. Lloyd, 231–40. Cambridge, MA: Harvard Univ. Press.

Grinnell, J. 1917a. Field tests of theories concerning distributional control. *American Naturalist* 51:115–28.

Grinnell, J. 1917b. The niche-relations of the California thrasher. *Auk* 34:427–33.

Hilden, O. 1965. Habitat selection in birds. *Annales Zoologici Fennici* 2:53–75.

Hutchinson, G. E. 1957. Concluding remarks. *Cold Spring Harbor Symposium on Quantitative Biology* 22:415–27.

Klopfer, P. H., and J. U. Ganzhorn. 1985. Habitat selection: Behavioral aspects. In *Habitat selection in birds,* ed. M. L. Cody, 435–53. New York: Academic Press.

Lack, D. 1933. Habitat selection in birds with special reference to the effects of afforestation on the Breckland avifauna. *Journal of Animal Ecology* 2:239–62.

Leopold, A. 1933. *Game management.* New York: Scribner's.

Leopold, A. 1949. *A Sand County almanac.* Oxford: Oxford Univ. Press.

Levins, R. 1968. *Evolution in changing environments.* Princeton, NJ: Princeton Univ. Press.

MacArthur, R. H., and R. Levins. 1967. The limiting similarity, convergence and divergence of coexisting species. *American Naturalist* 101:377–85.

Merriam, C. H. 1890. Results of a biological survey of the San Francisco Mountains region and desert of the Little Colorado River in Arizona. USDA Bureau of Biology, *Survey of American Fauna* 3:1–132.

Odum, E. P. 1959. *Fundamentals of ecology.* 2nd ed. Philadelphia: Saunders.

Pianka, E. R. 1994. *Evolutionary ecology.* 5th ed. New York: HarperCollins.

Schmidtz, D., and E. Willott. 2002. *Environmental ethics.* New York: Oxford Univ. Press.

Schoener, T. W. 1974. Resource partitioning in ecological communities. *Science* 185:27–39.

Schoener, T. W. 1989. The ecological niche. In *Ecological concepts,* ed. J. M. Cherrett, 79–113. Oxford: Blackwell Scientific Publications.

Stauffer, D. E. 2002. Linking populations and habitats: Where have we been? Where are we going? In *Predicting species occurrences,* ed. L. M. Scott, P. J. Heglund, M. L. Morrison, J. B. Haufler, M. G. Raphael, W. A. Wall, and F. B. Samson, 53–61. Washington, DC: Island Press.

Stoddard, H. L. 1931. *The bobwhite quail: Its habits, preservation, and increase.* New York: Scribner's.

Stone, C. D. 1974. *Should trees have standing? Toward legal rights for natural objects.* Los Altos, CA: Kaufmann.

Stone, C. D. 1987. *Earth and other ethics.* New York: Harper & Row.

Svardson, G. 1949. Competition and habitat selection in birds. *Oikos* 1:157–74.

Thomas, J. W., and D. E. Toweill, eds. 1982. *Elk of North America.* Harrisburg, PA: Stackpole Books.

U.S. Fish and Wildlife Service. 2003. *2001 national and*

state economic impacts of watching wildlife. Washington, DC: U.S. Fish and Wildlife Service.

Van Horne, B. 1983. Density as a misleading indicator of habitat quality. *Journal of Wildlife Management* 7:893–901.

Verner, J., M. L. Morrison, and C. J. Ralph, eds. 1986. *Wildlife 2000: Modeling habitat relationships of terrestrial vertebrates*. Madison: Univ. of Wisconsin Press.

Wallmo, O. C., ed. 1981. *Mule and black-tailed deer of North America*. Lincoln: Univ. of Nebraska Press.

Werner, E. E., J. F. Gilliam, D. J. Hall, and G. G. Mittelbach. 1983. An experimental test of the effects of predation risk on habitat use in fish. *Ecology* 64:1540–48.

Werner, E. E., and D. J. Hall. 1979. Foraging efficiency and habitat switching in competing sunfishes. *Ecology* 60:256–64.

2 The Evolutionary Perspective

Although we may not fully understand the causes and consequences of many popula-
tional phenomena, we can be confident that all have an evolutionary explanation.
—E. R. PIANKA, *EVOLUTIONARY ECOLOGY*

The chapter opening epigraph simply and eloquently states that evolutionary processes have formed present-day expressions of an animal's ecology. We want the reader to approach each chapter with a clear understanding of the ecological foundation on which this book was written. Our writing centers on the notion that the present distribution and abundance of animals result from adaptations to a host of biotic and abiotic factors and, in some cases, serendipity of species and conditions that happen to come into alignment (e.g., preadaptation). Many of the factors that shaped the form and function of an animal may now be absent or at least of different intensity. The ability of an animal to adjust to future conditions, be they induced by "natural" environmental changes or through human impacts, depends on its preadaptations and its ability through genetic diversity to develop new adaptations to these changes. Thus current conditions and an animal's response to them may not provide the basis for a complete understanding of the reasons for the pattern observed, and could result in inappropriate recommendations for managing long-term evolutionary potential.

The fields of evolutionary ecology and biogeography should thus play a central role in the field of wildlife–habitat analysis. Developing an understanding of why a population is found in a specific area requires knowledge of the organism's ecological relationships: why it is associated with specific combinations of edaphic, aquatic, temperature, and biotic regimes. To understand why it is found only in certain areas requires knowledge of climatic history, which may have resulted in the development of isolated, relict populations. Analysis of the evolutionary history of a population itself and of the geologic history of the landscape is a necessary prerequisite to understanding why the population is distributed as we observe today (see, e.g., Cox and Moore 1993).

Animals evolve by responding to both biotic and abiotic factors. Gene pools can adapt to forces being imposed by the abiotic environment, to forces being applied by other animals, and to changing conditions in the nonanimal biotic environment (e.g., plants). Such adaptations are usually referred to as coevolution, and are certainly of keen interest in how we interpret the ecologies of the animals we study today. Much interest over the past several decades has centered on studying links or webs within groups of animals in an attempt to identify "communities" of organisms. Recently, interest has grown in identifying individual species that play key roles in structuring such communities or at least subsets of species in an area (sensu keystone species).

Competitive interactions may play important roles in shaping morphology and behavior and thus be manifested in distribution and abundance. Under this assumption, many of the interactions between species likely took place in the past, and much of what we observe today is the result of these interactions. Competition often becomes unfalsifiable, the "ghost of competition past" (Connell 1980). Although we cannot solve this dilemma, we intend to use such controversies to show that determination of factors responsible for the animals present in an area and their habitat use are complicated, interrelated, and not completely apparent.

Thus this chapter examines the conceptual framework for viewing wildlife–habitat relationships throughout this book. We argue that, unless theories of biogeography, habitat selection, and community structure are considered when studying and managing for wildlife–habitat relationships, further progress in understanding will be minimal.

The Evolutionary Perspective

The distribution and abundance of animals over the landscape, and the use of habitat, must be considered in light of the geologic and geomorphological events that shaped the area. Everyone has been introduced to the concept of the geo-

logic timetable—the chart that relates geologic developments to the evolution of plants and animals. Here we are most concerned with events shaping our recent past, the Pleistocene to present (Recent epoch).

The Pleistocene

The Pleistocene epoch, which began 2 to 3 million years before the present (BP), is thought to have ended about 10,000 years BP. The Pleistocene was characterized by a series of advances and retreats of the continental ice sheet and glaciers. There were apparently four major advances-retreats in North America: the Nebraskan, Kansan, Illinoian, and Wisconsin. Between these advances of ice were three interglacial periods: the Aftonian, Yarmouth, and Sangamon. Our current Recent epoch is, in fact, probably best viewed as another interglacial period of the Pleistocene (Cox and Moore 1993). These ice sheets obviously had a dramatic impact on the ground cover and the associated environmental conditions; the possible maximum extent of glaciations is shown in figure 2.1. Although vegetation was destroyed, many animals had time to seek and occupy new areas. Of course, such an enormous ice sheet does not just suddenly appear. Further, areas close to but not covered by the ice also underwent dramatic physical and environmental changes because of their proximity to the ice fields. Most of the species currently occupying Earth are survivors of the abiotic and biotic influences of the Pleistocene, especially the last several advance-retreat cycles. In New Mexico, for example, the mid- to late Wisconsin fossil evidence suggests a progression of vegetation types, changing initially from semiarid, moderately warm grasslands or grassy woodlands to cooler, more mesic grassy woodlands, followed by cool, relatively dense sagebrush (*Artemisia*)–grassland-

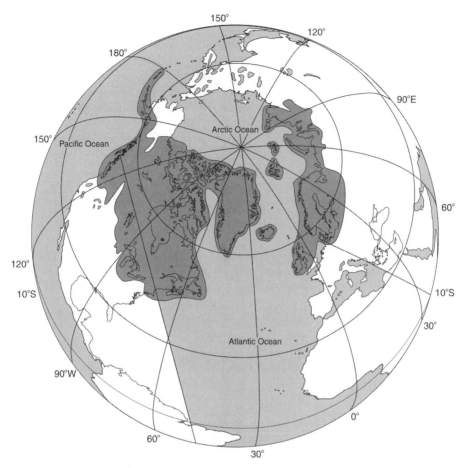

Figure 2.1.   Possible maximum extent of Pleistocene glaciation in the Northern Hemisphere. C = Cordilleran ice; L = Laurentide ice; S = Scandinavian ice; and A = Alpine ice. (From Goudie 1992, fig. 2.9; Reprinted by permission of Oxford University Press.)

woodland with elements from mixed-coniferous forests (Harris 1993).

Charcoal fragments from tropical rainforests in North Queensland, Australia, indicated that *Eucalyptus* woodlands occupied substantial areas of all of the present humid rainforest between about 27,000 BP and 3500 BP. Changes in paleoclimates associated with the most recent glaciations provide a plausible explanation for both the *Eucalyptus* expansion and subsequent rainforest reinvasion (Hopkins et al. 1993). Postglacial succession of vegetation has been reconstructed for many areas (e.g., Ritchie 1976; Amundson and Wright 1979; Elias 1992; Nordt et al. 1994). For example, studies of pollen from lakes in Wisconsin and Michigan have demonstrated a progressively westward movement of hemlock (*Tsuga canadensis*) range limit over the past 6000 years in response to a change in climate toward cooler, wetter conditions (fig. 2.2;

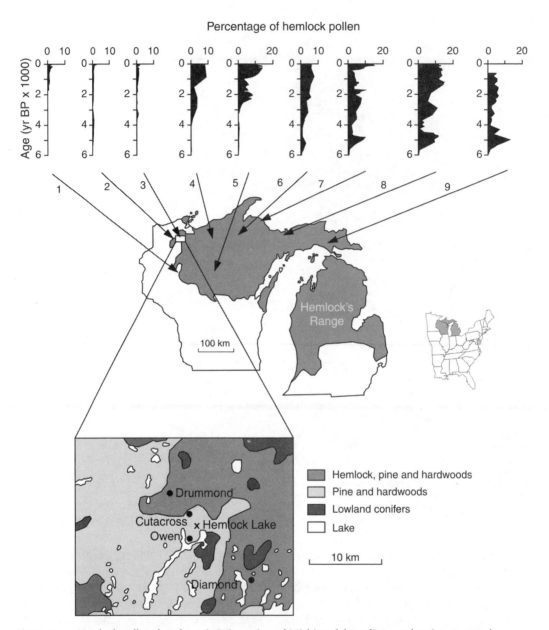

Figure 2.2. Hemlock pollen abundance in Wisconsin and Michigan lake sediments showing westward movement of hemlock range limit over the past 6000 years. The locations of forest hollows and Hemlock Lake are shown in relation to the presettlement distribution of forest types in Wisconsin. (From Parshall 2002, fig. 1; Reprinted by permission of the Ecological Society of America.)

see Parshall 2002 for review). Likewise, O'Brien (2001) summarized literature and data describing the changing climatic and floral-faunal composition in the Midwest during approximately the past 12,000 years. Palynological and sedimentological analyses have allowed paleoclimatologists to isolate postglacial climatic episodes that affected portions of the Midwest. Four of these episodes are depicted in figure 2.3, along with the generalized climatic condition and vegetational changes. By 8050 BC, much of the Midwest harbored modern assemblies of plants and animals, which continued to extend their ranges throughout the Boreal (7550–6550 BC) episode. The slightly cooler and wetter climate during this Boreal episode may have allowed some more northerly plants and animals to return to portions of the Midwest. The climate was, however, generally becoming warmer and drier. Continued drying allowed prairie grasses to return and expand eastward from the Great Plains around 7000–6550 BC. Forests were then retreating into moister locations, such as in valleys and along streams. The height of this dry period occurred around 5000 BC.

Although the areas we now call Alaska and Canada were mostly covered with ice, numerous ice-free pockets existed. These ice-free "refugia" were located primarily along the coasts and nearshore islands, with the ocean acting as a mediating influence on air temperatures (Heusser 1977). The result was a chain of forested refugia that extended south below the ice-covered regions. Palynological evidence—deposits of pollen in sediment layers—has allowed scientists to reconstruct the types of plants inhabiting these refugia. In the Pacific Northwest, for example, vegetation on the Olympic Peninsula through much of the middle and late Wisconsin glacial period resembled the edge of the Pacific Coast forest in Alaska as it appears today (Heusser 1965, 1977). In New Zealand, warm tropical waters of the southwest equatorial Pacific had a

strong ameliorating effect on the mean temperature in the Tasman Sea during the last glaciations. Northern New Zealand had a climate similar to the present during the last glaciation (Wright et al. 1995).

Refugia were also present below the southern extent of the ice sheets. Snow levels in the mountains of the continental United States were much lower than those we see today. During times of glaciation in the Cascade Mountains of the Pacific Northwest, for example, summer temperatures were apparently close to freezing, and precipitation and cloudiness were pervasive (Heusser 1977). In the southwestern United States, Wisconsin glaciation lowered the biotic zones by as much as 1200 meters, so woodlands and forests occurred over much of the ice-free region (Hubbard 1973; Elias 1992). As the glaciers receded northward and upward, bare or nearly bare soil was available for colonization by plants. Plants that had survived the glacial period on refugia were the most available for colonization. In turn, animals inhabiting refugia would likely have been the first colonists.

Distribution of Animals during the Pleistocene

The distribution and abundance of animal species existing currently can be linked to the geologic events of the Pleistocene outlined above. Birds have received the most research effort because of their ability to rapidly travel long distances, thus potentially moving between refugia by flying over unsuitable areas. The retreat of glaciers opened the way for occupation of vast areas by species either preadapted to the newly developing vegetation or able to adjust to the new environmental conditions. The pool of potential colonists is unknown but was probably related to the types of species and their abundances on refugia and the more southern, ice-free areas. A classic example explaining the

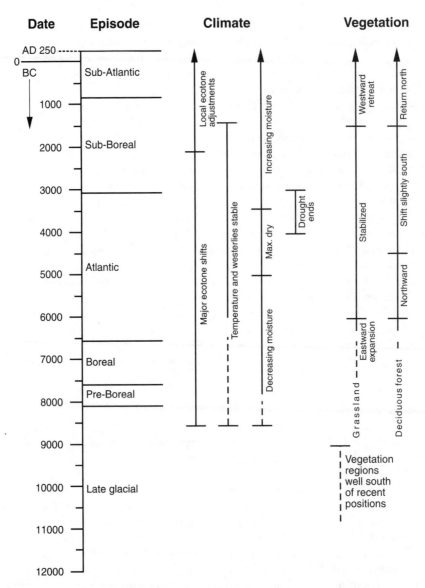

Figure 2.3. Correlation between archaeological episodes and generalized climatic and vegetational changes, 12,000 BC–AD 250. (Modified from O'Brien 2001, fig. 1.2: reprinted by permission of Island Press.)

occupation of North America by wood warblers (Emberizidae: Parulidae) is presented in box 2.1.

During glacial advances, boreal mammals were apparently widespread in lowlands well south of their present ranges. Concurrent with the movements of these mammals northward during interglacial periods were movements of boreal mammals into montane regions. Here, because of the effect of elevation on climate, cool refugia were present. Many of these montane populations have persisted in boreal refugia far south of the northern range of their closest relatives, and the zonation of mammalian distributions on some mountain ranges in the southwestern United States apparently resulted from Pleistocene faunal movements and subsequent retractions from warm, dry lowlands (see, e.g., Brown 1978; Vaughan 1986; Elias 1992). Harris (1990) found that many mammals now occurring in the southern Rocky Mountains ranged south into the Chihuahuan desert in the late Wisconsin. The present ranges of many taxa were probably reached in the early Holocene epoch, but some survived in refugia until the late Holocene (Elias 1992).

Ecological Compromises

Patterns of migration and seasonal changes in habitat use have ramifications for our attempts to determine why animals occupy the areas they do and complicate our attempts to predict the response of animals to environmental changes. The majority of ecological studies have focused on animals during breeding periods. This focus is understandable, given the obvious importance that the production of young plays in survival of the species. However, equally important are nonbreeding areas, and movement to and from geographic areas that serve as seasonal centers of activity. Failure to maintain wintering areas and migratory routes is likely to lead to extinction, or at least substantial population declines. Explor-

ing the causes of seasonal movements and migration is a necessary part of determining means of conserving animals (Brower and Malcolm 1991). Additionally, seasonal shifts in diet and habitat use in the same geographic area are also of likely adaptive and evolutionary significance.

Migration of birds is certainly tied to seasonal changes in climate and food resources, and many migrations resulted from Pleistocene glaciation. In fact, at least eight categories of explanations have been proposed for the evolution of migration (Rappole 1995): ancient environmental changes; availability of resources elsewhere; proximate factors such as photoperiod and temperature changes; climatic changes; seasonal tracking of fruit or nectar; seasonality and interspecific competition; seasonal change in intraspecific dominance interactions; and intense intraspecific competition for breeding sites. Migratory birds breeding in North America spend 5 to 7 months on wintering areas that usually bear little resemblance to their breeding grounds; several months are often spent in migration. Because of a long series of increasing range extensions, the global pattern of seasonal migrations must have been established or reestablished in the 20,000 years or so since the maximum extent of the last glacial advance.

The use of different summer and winter areas, plus environmental conditions within areas along the migration path, set morphological, behavioral, and physiological constraints on the abilities of an animal to efficiently use either the summer or the winter area. For example, Morrison (1983) found that, in the western United States, two populations of Townsend's warblers (*Dendroica townsendi*) apparently exist—one a shorter-winged group that winters in the United States, the other a longer-winged group that migrates to Mexico and Central America (see also Grinnell 1905). Both groups have separate breeding areas. Cody (1985) thought that such morphological adaptations might compromise

Box 2.1   Wood warbler occupation of North America as related to Pleistocene geologic events

Mengel (1964) presented what has become a classic scenario relating the occupation of North America by wood warblers (Emberizidae: Parulinae) to Pleistocene geologic events. By his reasoning, the retreat of glacial ice opened the way for occupation of vast areas by generalist species able to adjust to varying environmental conditions. The pool of potential colonists is unknown but is probably related to the types of species and their abundance in refugia and more southern habitats.

Using a precursor of the black-throated green warbler (*Dendroica virens*) as an example, Mengel postulated that this "pro-*virens*" was able to occupy vast reaches of western North America as the environment warmed (box fig. 2.1). Most warblers apparently evolved in more tropical regions of North, Central, and South America, colonizing western regions of North America by passing through the Caribbean Islands and the southeastern United States. With subsequent advance of the ice sheets, pro-*virens* was isolated to the north on refugia and to the south in warmer regions of the Southwest. During this glacial period, the isolated populations either adapted to their changing surroundings or became extinct. With the warming and concomitant retreat of the glaciers, the populations were able to expand their range as the forest expanded; many were now distinct species (see Mengel 1970). Although the general premise of Mengel is well accepted, many of the specifics are not (Hubbard 1973; Flack 1976). Similar scenarios have been developed for other species of birds (e.g., Hubbard 1973) and mammals (Smith 1981).

Mengel (1964) did not consider winter ranges of these species in his analysis of the speciation process. Rappole (1995), however, evaluated the breeding and wintering ranges of these warblers and found parallels between breeding location, winter range location, and the pattern of species distribution (box fig. 2.2). Rather than assuming that the degree of breeding ground isolation entirely controlled the speciation process, Rappole suggested that changes in both the breeding and wintering grounds were required for speciation to occur, or that perhaps species originated as sedentary populations on the wintering grounds that were isolated by Pleistocene events and evolved a migratory behavior at a later date.

the habitat and foraging site selection of animals during breeding. Finch bills, for example, seem more strongly influenced by their granivorous winter diet than by their mostly insectivorous diet during breeding (Cody 1985). In the Sierra Nevada of California, many small gleaning and hover-gleaning foliage insectivorous songbirds become flakers and probers of bark during winter in response to changes in food abundance and location (Morrison et al. 1985). Some herbivorous mammals make seasonal migrations in concert with seasonal changes in the local availability of high-quality food; and some smaller mammals (e.g., voles, lemmings) change habitat use on a local scale with seasonal changes in food (Batzli 1994).

The fossil record can sometimes be used to reconstruct the former range of species. Changes in distribution can sometimes be related to changes in paleoclimatic factors. For example, Harris

(1993) reconstructed the succession of microtine rodents in New Mexico from the mid- to late Wisconsin. The disappearance of the prairie vole (*Microtus ochrogaster*) and the appearance of the sagebrush vole (*Lemmiscus curtatus*) were interpreted as a result of cooling summer temperatures and a shift in seasonality from warm-season-dominant to cool-season-dominant precipitation (Harris 1993; see box 2.2 for details). Goodwin (1995) reconstructed the Pleistocene distribution of prairie dogs (*Cynomys* spp.), and concluded that fossil evidence supported the historic association of white-tailed prairie dogs (*C. gunnisoni*) with the shrub-steppe, whereas the black-tailed prairie dog (*C. ludovicianus*) had a long history in or near its present range, as indicated by its use of relatively more complex habitats and its more primitive biology.

The fossil record of Quaternary mammals helps us understand the ways communities have

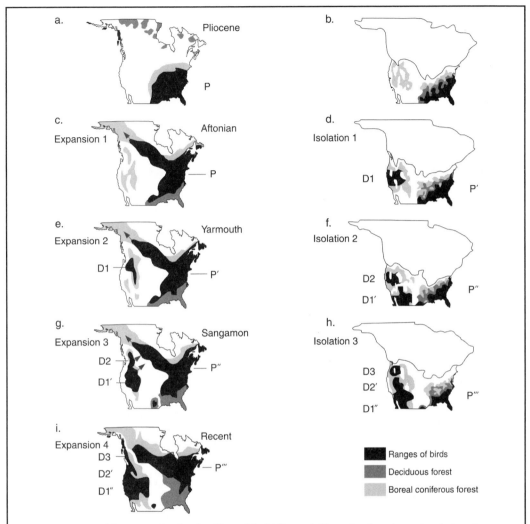

Box Figure 2.1.   Model sequence showing the effects of glacial flow and ebb on the adaptation and evolution of a hypothetical ancestral wood warbler and its descendants. The details of glacial boundaries are approximate. P = parental species; D1, D2, D3 = derivative species; prime marks = number of glacial cycles removed from origin. (From Mengel 1964, fig. 4; courtesy of Cornell Laboratory of Orthinology.)

responded to climatic fluctuations. As reviewed by Graham et al. (1996), during the late Pleistocene, tundra and boreal mammalian species ranged as far south as southern Wisconsin, central Pennsylvania, central Illinois, and northern Alabama. Species that typically inhabit eastern deciduous forests today extended their ranges westward into the Great Plains, and, alternatively, grassland species ranged as far eastward as Tennessee and Virginia. This late

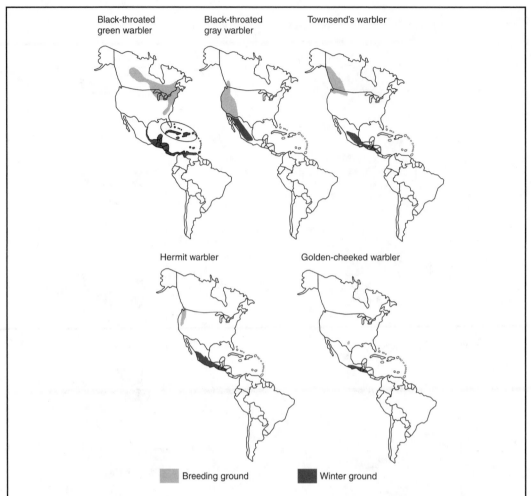

Box Figure 2.2.   Breeding and wintering ground distribution of members of the black-throated green warbler (*Dendroica virens*) superspecies complex. (From Rappole 1995, fig. 6.5; reprinted by permission of the Smithsonian Institute Copyright 1995.)

Pleistocene mammalian fauna also included a diverse megafauna that are now extinct, such as mastodons, camels, and horses.

Graham et al. (1996) outlined two competing models of how mammalian communities might have responded to Pleistocene environmental changes. In the Clementsian model, it is as-sumed that large groups of species were in equi-librium and that the organization of these groups is determined primarily by biological in-teractions, including competition. This model suggests that animal communities can be tracked as basically intact units across tens to hundreds of thousands of years. Alternatively,

Box 2.2   Progression of four species of microtine rodents

Harris (1993) developed the progression, derived from the fossil record of New Mexico, in the relationship between four species of microtine rodents and climatic changes in the mid- to late Wisconsin period (box fig. 2.3). Mid-Wisconsin, semi-arid, moderately warm grasslands were replaced by cooler, more mesic grassy woodlands; these were followed by cool, relatively dense sagebrush (*Artemisia*)–grassland–woodlands with elements from mixed-coniferous forest.

Concomitant with these environmental changes were changes in the microtine rodents occupying the region. Of the four species discovered, only the prairie vole (*Microtus ochrogaster*) was present in the lower (i.e., older) levels of the stratigraphic sequence. In levels 7 to 5, the prairie vole coexisted with the Mexican vole (*M. mexicanus*). There then followed a brief interval (level 4) in which only the Mexican vole was present. Level 3 documents the invasion of the sagebrush vole (*Lemmiscus curtatus*), followed in level 2 by the only occurrence of the long-tailed vole (*M. longicaudus*). Level 2 also shows the peak frequency of the Mexican vole. The record is closed with the sagebrush and Mexican voles being rejoined by the prairie vole. Harris interpreted the disappearance of the prairie vole after level 5 and the appearance of the sagebrush vole in level 3 not only as resulting from cooling summer temperatures but also as indicating a shift in seasonality from warm-season-dominant to cool-season-dominant precipitation.

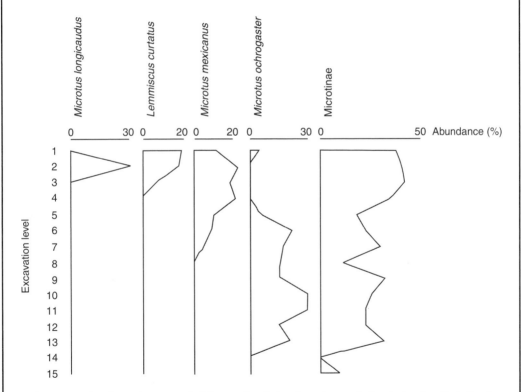

Box Figure 2.3.   Abundances (percentages) of four species of microtine rodents and of total microtines (including those unidentifiable to species) recovered from 15 fossil excavation levels dating from the mid- to late Wisconsin in New Mexico. (From Harris 1993, fig. 4; by permission of Quarterly Research Center.)

the Gleasonian model suggests that species respond in ecological time to environmental change according to individual tolerances, resulting in range shifts that occur at varying species-specific rates. Graham et al. (1996) concluded that species that make up mammal communities responded to late Quaternary environmental changes in a Gleasonian manner. Although these changes were caused by the direct effects of climate change, biological interactions and stochastic events must also have played a role. Thus animal communities are continually emerging in largely unpredictable fashions. Although late Pleistocene communities differed from Holocene ones, they were arranged into similar biogeographic patterns or faunal provinces. These similarities reflect east–west moisture and north–south temperature gradients. Graham et al. (1996) concluded that it is important to consider individualistic shifts in species distributions, nonanalog composition of communities, and changes in environmental heterogeneity in modeling the responses of mammalian communities to past and future environmental changes. Models for future changes must rely increasingly on individual species and their requirements, rather than on species associations.

Sullivan (1988, in Hafner 1993) described a *biogeographic indicator species* as diagnostic of specific environmental conditions through time. Reconstruction of the biogeographic history of such a species provides a reconstruction of the changing distribution of the habitat to which the species is restricted. Thus this reconstruction provides an idea of the historical biogeography of other recent species that are restricted to the same general conditions. Hafner (1993) offered the Nearctic pikas (*Ochotona principes* and *O. collaris*) as biogeographic indicators of cool, mesic, rocky areas. Fossil pika have been found far from extant populations, which is especially evident in Nevada (see fig. 2.4). Hafner provided

the following speculation regarding interpretation of locations with fossil pika, but lacking extant populations: If the overall site reconstruction is xeric and at low elevations, then a suitable rocky microsite must be assumed to have existed nearby. If the overall site reconstruction is one of a cool, mesic area, but fossil remains are rare, they may represent dispersing individuals, but suitable mesic habitat must be assumed to have existed nearby.

Reconstructions such as these have obvious ramifications for modern-day research and management: they enable us to partly understand why species change in distribution as climate and perhaps even interspecific interactions change and, further, allow us to predict changes that might occur with future natural and human-induced changes in the environment. Of course, these ideas assume that habitat specificity is unchanging, an assumption that should be fully explored in any such analysis. For example, fossil evidence of an unchanging morphology does not mean that physiology has remained static through time.

Post-Pleistocene Events

Biogeographic patterns of animal distribution have been explained by both *vicariance* (defined as the distribution that results from the replacement of one member of a species pair by the other, geographically) and *dispersal*. As an example of vicariant distribution, Brown (1978) explained the current distribution of small, nonflying montane mammals in the Great Basin as resulting from post-Pleistocene climatic and vegetational changes that left these species stranded in woodland and forest habitats on mountaintops and isolated by intervening desert scrub. He concluded that the desert vegetation prevented intermountain colonization. In other regions of the Southwest, however, evidence indicates that

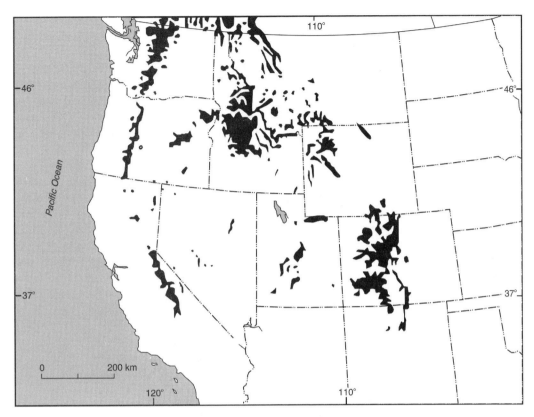

Figure 2.4.   Distribution of extant pikas (*Ochotona princeps*) (shaded area) and late Pleistocene–Holocene fossil records (dots) in western North America. (Modified from Hafner 1993, fig.1; reprinted by permission of Quarterly Research Center.)

post-Pleistocene colonization has dramatically influenced the current distribution of some species (Davis et al. 1988). Davis and his coworkers concluded that both post-Pleistocene dispersal and subsequent colonization as well as vicariant events and subsequent extinction influenced the assemblages of mammals. They stressed that the degree to which each process influences animal distribution should be considered to explain current community composition.

As already discussed, the record of vertebrate biogeographical migrations during the past 125,000 years indicates movements of consider-

able magnitude in response to changing temperature and moisture regimes during glacial advances and retreats. Because climatological changes of different periodicities are occurring continuously, biogeographical migrations continue at present, and records of vertebrate faunal shifts during the last two centuries have been documented for many regions of the world.

Johnson (1994) studied range expansions of 24 species of birds in the contiguous western United States that occurred from the late 1950s to the early 1960s. He concluded that all had enlarged their ranges for reasons apparently unrelated to direct human modifications of the environment. Instead, he proposed that pervasive *climatic changes* over the past several decades were the most likely explanation. Although climatic warming is probably involved, he concluded that increased summer moisture was the overriding factor. Climatic information from the region offered support for wetter and warmer summers in recent decades in the contiguous western United States.

As noted by Johnson, the establishment of substantial numbers of breeding birds well outside their normal breeding ranges must alter the biotic relationships in the new areas; this might be especially evident when a large predator is involved (e.g., the barred owl, *Strix varia*, has been expanding westward to, and southward down, the Pacific Coast).

Lyman and Wolverton (2002) analyzed the number of specimens of bones and teeth per taxon recovered from archaeological sites from the lower Snake River of southeastern Washington. They found that ungulate remains outnumbered those of small mammals during each

2000-year period between 10,000 and 100 BP (fig. 2.5). They speculated that the decrease in ungulates at 6000 BP reflected the aridity that existed between 8000 and 4500 BP, and that the increase in ungulate abundance after 4000 BP reflected the cooler temperatures and increased moisture between 4500 and 2000 BP. Thereafter, the climate was somewhat drier and warmer, approximating modern conditions.

Distributions of vertebrates during the Quaternary indicate distinct north–south changes in their distributions in response to glacial conditions. During the Wisconsin advance, for example, many species shifted southward. The presently disjunct southern populations of relatively northern species reflect southward displacement during Wisconsin glaciation. Such range adjustments, in essence, strand populations of species outside what we now consider their normal ranges. Such populations are often referred to as relicts, and they present difficult management problems because they are usually unable to naturally interchange with other populations of their species. Further, it is difficult to ascertain if

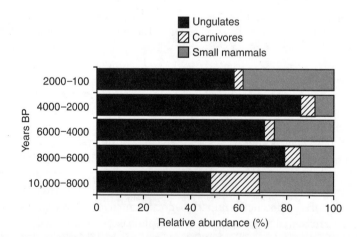

Figure 2.5. Temporal changes in relative abundances of ungulate, carnivore, and small-mammal remains in archaeological collections from the lower Snake River, southeastern Washington. (From Lyman and Wolverton 2002, fig. 3.)

declines in a relict population are due to continuing naturally occurring phenomena, or are being hastened by human-induced changes. These situations are further complicated by human-induced changes in land uses that hinder or even prevent the population from undergoing additional biogeographic shifts in response to environmental changes.

Global Climate Change

Studies conducted on past climate conditions have indicated that drought is likely to be the most devastating consequence of future global change. High-resolution records from lake sediments from the northern Great Plains show pronounced 100- to 130-year drought cycles during the middle Holocene (8000 years BP). During drought phases, grass productivity declined, erosion and forbs increased, and reduction of fuel reduced fires (Clark et al. 2002). During the more humid late Holocene (2800 BP), the climate was less variable and drought was apparently not cyclical. Therefore Clark et al. (2002) concluded that drought severity during past, and possibly future, arid phases cannot be anticipated from the attenuated climate variability evident during contemporary humid phases. Thus there is much uncertainty surrounding the impacts that future climate changes will have on ecosystems.

Root and Schneider (2002) assembled a succinct review of climate change and its implications for the distribution of wildlife. They noted that the ice buildup from about 90,000 BP to 20,000 BP was apparently highly variable and was followed by a 10,000-year transition to the currently relatively stable Holocene period. Ice cores taken from glaciers showed that concentrations of carbon dioxide (CO_2) and methane (CH_4)—both greenhouse gases—and temperature were fairly constant for about the past

10,000 years until nearly AD 1700. This relatively constant relationship ended, coincidentally, at the beginning of the industrial age. Cores of fossil pollen from numerous sites in North America show how boreal coniferous forests, now the predominant forest type in central Canada, were well represented during the last glacial period (15,000–20,000 BP) in what are now the mixed-hardwood and corn belt regions of the United States. Most of Canada was covered by ice during the last glacial period.

These climatic changes certainly influenced animal distributions and could have contributed to the extinctions of many species. It has been difficult, however, to separate the co-occurring influences of climate and humans on the extinctions of megafauna (e.g., mammoths, mastodons, saber-toothed tigers). Recently, evidence indicates that climate change was secondary to human hunting as a cause of extinctions of megafauna over the past 50,000 years (see review by Kerr 2003).

The current global rate of temperature increase exceeds those typical of the sustained average rates experienced over the past 100,000 years or so. As such, it is unlikely that paleoclimatic conditions reconstructed from pollen and the fossil record will be appropriate analogs for global climate changes that are accelerated by human activities (i.e., global warming or cooling). Thus future climates may be very different than one would have predicted had humans not began influencing temperatures through emission of greenhouse gases (Root and Schneider 2002).

The anticipated changes in plant distribution caused by climate change will have substantial influence on the distribution and abundance of animals. As we thoroughly review in this book, animals have strong links to the structure and species composition of plants. As such, the ranges of animals will change as the ranges of plants change. Additionally, changes in local plant species composition and the pattern of succession will have more subtle, but nonetheless substantial, impacts on animal populations. Likewise, changes in average annual temperatures as well as changes in average minimum and maximum temperatures will influence both the north–south and elevation ranges of animals. For example, Root (1988) showed that the eastern phoebe only winters in the United States where average minimum temperatures exceed 4°C. Changes in animal distribution will not be equal across species currently residing in the same region. As such, the behavioral interactions (e.g., predation, competition) typical today will likely vary in a multitude of ways as patterns of temperature and precipitation change (Root and Schneider 2002).

Of course, regardless of the causes, climate change is only one of a multitude of factors that can influence the distribution and abundance of animals. The introduction of exotic species (both plants and animals), changes in the distribution of disease, large-scale changes in land use patterns (e.g., agriculture, suburban development, water drawdown), and pollution all interact with natural changes to create a patchy and highly variable environment. Root and Schneider (2002) concluded that, to document a strong role for climate change in explaining changes in animal and plant populations, increased confidence is achieved using examples that show changes in direction predicted by the physiological tolerances of a particular species. A coherent large-scale pattern exhibited by many species that is consistent with the understanding of the causal mechanisms provides the greatest confidence in the attribution of observed changes in animals of plants to climate change.

Root and Schneider (2002) summarized some of the different types of changes animals are showing in relation to climate change, including changes in abundance, distribution, phenology (e.g., breeding, migration), morphol-

ogy and physiology, and behavior. For example, the northern expansion of the porcupine (*Erethizon dorsatum*) in central Canada has been linked to warming that caused a northward shift in tree line (Payette 1987). An interesting impact on physiology caused by changes in rainfall has been shown in the California quail (*Callipepla californica*). Rainfall affects the chemistry of plants eaten by the quail, which in turn influences reproductive hormones in the birds (Botsford et al. 1988). Similarly, changes in annual temperature can influence the amount of available nesting locations available for northern bobwhite (*Colinus virginianus*) (Guthery et al. 2001). Other studies examining the potential influence of temperature change on the life history of North American animals and plants are summarized in table 2.1.

Bottlenecks

A population may be influenced in a major way by catastrophic events that occur only infrequently. Such catastrophic events are thought to drive a population so low in abundance that it is forced through a "bottleneck" that changes the genotype and its phenotypic expression and possibly its ability to adapt to new environments (Wiens 1977). Genetic diversity is necessary for populations to evolve in response to environmental changes, and heterozygosity levels are linked directly to reduced population fitness through inbreeding depression. This linkage leads to the expectation that levels of heterozygosity and fitness at the population level will be correlated. Finite populations lose genetic variation as a consequence of genetic drift and at the same time become inbred. Inbreeding leads to inbreeding depression in virtually all species studied to date (see Reed and Frankham 2003 for details and review). Thus the amount of genetic variation a population contains is predicted to

Table 2.1.   Studies examining the potential influence of temperature change on the life history of North American animals and plants

Taxa	Number changing	Type of change	Reference
Bird	1	Spring phenology	Brown et al. (1999)
Bird	1	Spring phenology	Dunn and Winkler (1999)
Birds	15	Spring phenology	Bradley et al. (1999)
Mammals	3	Morphology	Post and Stenseth (1999)
Grasses	6	Density	Alward et al. (1999)
Forbs	24	Spring phenology	Bradley et al. (1999)
Tree	1	Morphology	Barber et al. (2000)
Tree	1	Spring phenology	Bradley et al. (1999)

Source: Root and Schneider (2002, table OV.1).

correlate with current fitness and thus with evolutionary potential.

Various studies have identified situations in which genetic difficulties, especially inbreeding resulting from bottlenecks, have been manifested in physiological problems. Probably the most cited study is that on the South African cheetah (*Acinonyx jubatus*) by O'Brien et al. (1983, 1986). They found extreme genetic monomorphism in the two populations they studied and reported low sperm counts, low sperm mobility, a high frequency of abnormally shaped sperm, and concomitant low reproductive output. They thought that these physiological problems and low genetic variability were the result of a severe population bottleneck followed by inbreeding, likely the result of climatic changes in the late Pleistocene. Other examples of species going through apparent population bottlenecks are available (see O'Brien et al. 1983, 1986; Pimm 1991, 156–162). However, Pimm (1991, 159–162) has questioned the conclusions of O'Brien et al. and, in general, the notion that inbreeding will necessarily doom a relatively large population to extinction. It is clear, however, that changes in the distribution and abundance of a species, especially through fragmentation of the species into disjunct populations, will lead to

changes in genetic composition. Human activities producing such fragmented populations can lead to increased genetic differentiation among the populations (e.g., Leberg 1991). The impacts that such changes have on the long-term viability of a species are unknown.

The effective population size necessary to minimize genetic changes has been the subject of much debate (see reviews by Moore et al. 1992; Lacy 1994; Gibbons et al. 1995; Morrison 2002). In captive populations, the "50/500" rule called for a minimum of 50 individuals for short-term breeding programs, but more than 500 individuals for long-term programs. The rationale for the long-term criteria was to allow new mutations to restore heterozygosity and additive genetic variance as rapidly as possible as it is lost to random genetic drift. The concept of effective population size (Wright 1931) has several related meanings, including the number of individuals at which a genetically ideal population (one with random union of gametes) would drift at the rate of the observed population. The rate of genetic drift could be measured as the sampling variance of gene frequencies from parental to offspring generations. Another "rule" proposed that at least 90% of the genetic variation in the source (wild) population be maintained in the captive population (Lacy 1994). Ramey et al. (2000) suggested that it is justifiable to intervene when there has been a severe *genetic bottleneck,* which they defined as an effective population size of less than 10 individuals, and a lack of gene flow with other outbred populations. Some reintroduction and translocation efforts have been successful with as few as around 10 individuals (see Morrison 2002). The 90% retention of genetic diversity guideline referred to expected heterozygosity, and many management programs have used expected heterozygosity as an index of genetic variability. The possibility of a population to adapt at all depends on the presence of sufficient variants, so allelic diversity might be critical to long-term persistence (Lacy 1994).

Coevolution

Our previous discussion centered on the abiotic factors shaping the types and distributions of plants and animals through geologic time, as well as on the near-term effects of human activities, both known and forecasted. But as an organism evolved to meet the demands of a changing environment, it also had to evolve in light of forces exerted on it by other organisms. This joint evolution of two or more taxa that have close ecological relationships but do not exchange genes has been termed *coevolution.* Under coevolution, reciprocal selection pressures operate to make the evolution of either taxon partially dependent on the evolution of the other. Coevolution includes various types of population interactions, including some instances of predation, competition, parasitism, commensalism, and mutualism. Although the classic examples of coevolution involve plant–animal interactions, the term rightfully applies to any interdependent phenomena between taxa (Pianka 1983, 238; for example, competitive displacement of sibling species).

That taxa are interdependent naturally increases our fascination for the living world; consequentially, this interdependence also increases our challenge when trying to understand and then manage these taxa. Nevertheless, it is critical that researchers and managers appreciate the fact that most of our actions will cause some type of reaction by other taxa. Historically, when much of wildlife management centered on featured game or pest species, little consideration was given to the impacts such management had on other organisms (see chapter 1 for a review). Today, however, we understand that professional management, and usually the law, requires some ability to more fully predict the consequences of our management activities. This understanding should begin with careful evaluation of the historic geologic and environmental conditions that shaped the plant and animal communities under study (see box 2.1).

As summarized by Franklin (1988), human activities have been responsible for dramatic changes in the composition and structure of vegetated areas throughout the world. For example, most forests in the temperate zone are now secondary (i.e., previously harvested), and tree species composition has shifted toward commercially valuable species and relatively early successional stages. In addition, standing dead trees (snags), fallen logs, and the litter layers, which are not usually maintained or at a minimum heavily impacted in managed forests, are essential to many organisms and biological processes. Efforts to conserve structural and functional diversity within an area are often linked; for example, by maintaining woody debris, one of the sites of nitrogen fixation is retained, along with the microsite conditions often necessary for numerous organisms (see also chapter 10). Changes in patterns of grazing (by domestic and native herbivores) and fire have similar impacts (e.g., Risser 1988). These changes alter the conditions under which species evolved and are especially critical for coevolved relationships (e.g., the introduction of exotic animals; see chapter 4).

Food Webs

Groups of organisms can be represented by a *food web,* which is simply a diagram of all the trophic relationships among its component species. A food web is generally composed of many food chains, each of which represents a single pathway of the food web (Pianka 1983). Webs thus depict the flows of matter, nutrients, and energy across and within user-defined boundaries (usually referred to arbitrarily as a community). Since first introduced by Elton (1927), the concepts of food passage and nutrient cycling have now achieved a central place in ecological theory and the structuring of communities. Several excellent books and articles explain and document the development of this theory, which often links the structure of food webs to community stability (Cohen 1978; Pimm 1982, 1991; Lawton 1989; Cohen et al. 1990; Hall and Raffaelli 1993; Schmitz et al. 2000; Halaj and Wise 2001).

Pimm (1982) linked food web structure and ecosystem function under three principal categories: resilience, which measures how fast spe-

cies densities return to equilibrium following a perturbation; nutrient cycling, which measures how tightly ecosystems retain the nutrients that make up part of the biomass of their constituent species; and resistance, which measures how productivity or biomass at a particular trophic level changes with changes in feeding rates (e.g., increased or decreased predator pressure). These interrelationships between food web structure, ecosystem functions, and successional state envisioned by Pimm are illustrated in figure 2.6.

The apparent link between food webs and ecosystem functioning was thought by Cohen (1990) to relate to our understanding and subsequent management of the environment in several ways. First, some environmental toxins accumulate along food chains. Second, knowledge of the food web is necessary for anticipating the consequences of species' removals and introductions. Third, an understanding of food webs will help in the design of nature reserves and other land use practices. (Additional details on these topics are found below.)

As already noted, the empirical study of food chains and webs has developed into a prolific branch of ecological theory. However, there remains a fundamental disagreement over whether "bottom-up forces" (e.g., nutrient availability, especially plants) or "top-down forces" (e.g., predators, especially carnivores) predominate in

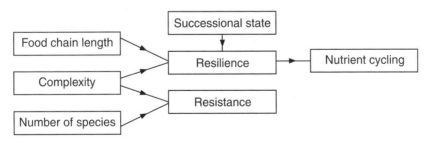

Figure 2.6.   Causal relationships between food web structures, successional state, and ecosystem functions. (Modified from Pimm 1982, fig. 10.2.)

populations and serve to structure communities. Hunter and Price (1992) argued that a synthesis of the top-down and bottom-up forces in terrestrial systems requires a model that incorporates the influences of biotic and abiotic heterogeneity. This view thus incorporates the classic debate between supporters of the relative roles that biotic or abiotic factors played in determining changes in animal abundances (e.g., Andrewartha and Birch 1954; Lack 1954). Hunter and Price (1992) hypothesized that differences among species within a trophic level (rather than assuming that all species on one level are indivisible units), differences in species interactions in a changing environment, and changes in population quality (e.g., reproductive success) with population density are as important determinants of population and community dynamics as are the number of levels in a food web or the position of the system along a resource gradient.

A single bottom-up template has been proposed for viewing the relative contributions of top-down versus bottom-up forces. In figure 2.7a, variability in climate, soil parameters, decomposers, and plant–soil symbionts determines the initial heterogeneity of form exhibited among primary producers. This pattern of heterogeneity represents the template on which the complex interactions among species in real populations and communities are superimposed. Figure 2.7b adds back some level of biological reality by allowing species at each trophic level to affect those in trophic levels below as well as above them, and by including the effects of abiotic heterogeneity at all trophic levels. Hunter and Price thought that the bottom-up template was compelling because plants form a major component of large-scale patterns over landscapes and geographic regions.

Advantages of Hunter and Price's model include that it permits the system to be dominated by species or guilds at any trophic level through feedback loops; it can encompass the mecha-

nisms of interactions among species; and it focuses attention on the extensive heterogeneity in natural systems. The model does not, however, consider changes in community structure nor interactions among animals and their abiotic environment over evolutionary time (Hunter and Price 1992). Rather than trying to dichotomize systems into either bottom-up or top-down categories, we have concluded that a much more fruitful line of research concerns to what extent variation at different trophic levels, or in abiotic factors, can influence the relative strengths of bottom-up and top-down forces. The model proposed by Hunter and Price is a useful means of conceptualizing how groups of organisms might be interacting, and provides a starting point for designing studies that test the influence of heterogeneity at different levels in a food web.

Predators can have direct effects on their prey, and also relatively indirect effects via those prey and other organisms, including down to the level of producers (fig. 2.8; see Schoener et al. 2002 for review). The impacts of a predator can, however, vary between communities for a variety of reasons, including the presence or not of predation on themselves. As such, even the effects of a devastating predator can be reversed if it is under heavy predation pressure. These types of relationships can, therefore, substantially influence food webs and trophic cascades (whereby predators have major effects on lower trophic levels, especially plants).

Additionally, recent work has indicated that changes in behavior, physiology, and other traits of individual organisms may also affect how perturbations are transmitted in food webs. For example, Schoener et al. (2002) noted that reduced prey feeding rate and/or habitat shift motivated by predator-avoidance behavior may affect which and how much food a prey consumes. Monitoring changes in traits suspected to lead to such effects should increase understanding of

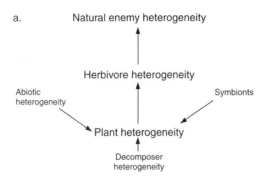

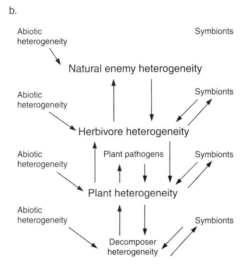

Figure 2.7.   Factors influencing population dynamics and community structure in natural systems. (a) A simple model in which variation among primary producers—determined by climate, soil parameters, and symbionts—cascades up the trophic system to determine heterogeneity among herbivores and their natural enemies. (b) With the addition of feedback loops, organisms at any trophic level can influence heterogeneity at any other level by cascading effects both up and down the system. (From Hunter and Price 1992, fig. 1; reprinted by permission of the Ecological Society of America.)

food web dynamics. Further, the nature and mechanism of trait changes have potential implications for behavioral and evolutionary biology.

Most large terrestrial mammalian predators have been lost from more than 95% of the contiguous United States and Mexico since the arrival of humans. As such, most ecological communities are either missing dominant selective forces or have new ones that are dependent on humans (e.g., hunting). Berger et al. (2001) showed a cascade of ecological events that were triggered by the local extinction of grizzly bears (*Ursus arctos*) and wolves (*Canis lupus*) in the southern Greater Yellowstone Ecosystem, including: (1) the demographic eruption of a large, semiobligate, riparian-dependent

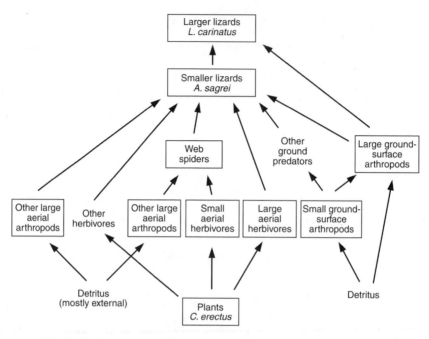

Figure 2.8.   Partial food web depicting relationships between lizards (*Leiocephalus* and *Anolis*) and various web components. Components in rectangles were monitored in the study; arrows indicate direct consumption. (From Schoener et al. 2002, fig. 13; reprinted by permission of the Ecological Society of America.)

herbivore (moose, *Alces alces*) during the past 150 years, (2) the subsequent alteration of riparian vegetation structure and density by ungulate herbivory, and (3) the coincident reduction of avian Neotropical migrants in the degraded riparian communities (fig. 2.9).

Despite the interest among some ecologists in food webs, there are real difficulties in applying food web theories to real-world management. As summarized by Power (1992) and Hall and Raffaelli (1993), there is a need to resolve methodological issues concerning appropriate spatiotemporal scales, to agree on operational definitions for concepts like trophic levels, to evaluate assumptions of the variety of available models of top-down and bottom-up forces, and to develop testable hypotheses that address the dynamic feedbacks between adjacent and non-

adjacent trophic levels (see also Oksanen et al. 1995). The usefulness of food web theory as currently practiced has also been questioned by Peters (1991), who concluded that because food webs represent qualitative, verbal models, they indicate only weakly what we can expect from nature. Because the predictions are not risky (i.e., they do not state specific, quantitative predictions), and because exceptions are explicitly allowed (see, e.g., Pimm 1982; Peters 1991, table 7.3), the models are protected from falsification. Echoing Power's (1992) call for operationalization of definitions, Peters concludes that most tests of food webs are ambiguous, and apparent refutations inconclusive. As discussed in the following section, how we view species interactions can have a substantial impact on how we plan our conservation strategies.

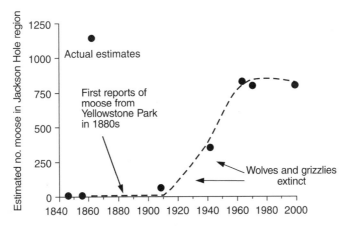

Figure 2.9.   Patterns of colonization and general population trends for moose in relation to loss of grizzly bears and wolves, Jackson Hole region, Greater Yellowstone Ecosystem, 1840s–1990s. (From Berger et al. 2001, fig. 1; reprinted by permission of the Ecological Society of America.)

Hall and Raffaelli (1993) present a sensible discussion of how we might better advance studies of food webs. They stated that if we are to construct models to help us understand and manage the environment, we need to know which structural properties are important and why. Efforts to document the links in webs ever more completely are probably misplaced in view of the alternative avenues available. It is more sensible to focus on specific questions that emerge from food web models using systems that are amenable to study.

Extinctions and Endangered Phenomena

There is also much to be learned regarding conservation of extant species by analyzing the causes of extinctions. Numerous hypotheses have been proposed to explain mammalian extinctions seen in the fossil record (Owen-Smith 1987; see also table 2.2). The *climatic change hypothesis* relates waves of extinction to abrupt changes in climatic conditions. For example, major extinctions of North American mammals

took place at the end of the Pleistocene. The *human predation hypothesis* was developed because of the apparent relationship between the pattern of Pleistocene extinctions of large herbivores and the pattern of human occupation. The *keystone herbivore hypothesis* assumes that the extinction of large herbivores (by whatever event) caused a concomitant change in the distribution and abundance of vegetation, which in turn caused the extinction of animals associated with

Table 2.2.   Dates of major Pleistocene and Holocene mammalian extinctions

Location	Date
North America	11,000 BP
South America	10,000 BP
Northern Eurasia	13,000–11,000 BP
Australia	13,000 BP
West Indies	Mid-postglacial
Madagascar	800 BP
New Zealand	900 BP
Africa and Southeast Asia	40,000–50,000 BP

Source: Martin 1967. Reprinted with permission of Yale University Press.

the now extinct or substantially changed vegetation types. For example, the conversion of the open parklike woodlands and mosaic grasslands typical of North America during the Pleistocene to the more uniform forests and prairie grasslands of today could have been a consequence of the elimination of these herbivores (Owen-Smith 1987). But, like predictions arising from food web theory, conclusions regarding this issue are ambiguous due to the confounding of numerous factors.

Goudie (1992) provided a good summary of the influence of humans and climate change on animal extinctions, concluding that the combination of the two likely influenced the extinction rate of some species. Regardless of the causes of extinction, we need more direct, empirical data on the physical environmental history of the Pleistocene and the biological consequences, with emphasis on species extinctions, of environmental changes. If we can increase our knowledge of the Pleistocene record, then we will be better able to evaluate the consequences of contemporary human activities on the biota. Without consideration of the time perspective available from the geologic record, a complete evaluation of the current rates of extinction will be at best difficult and at worst misleading (Raup 1988). To a certain degree, the fossil record can provide an understanding of how species appear, disappear, and then reappear in an area (see box 2.2).

Comparing contemporary rates of extinction (species originations) with those of the past is only one means of placing the current situation in perspective. On local spatial scales and within relatively recent time periods, we can compare known and supposed species distributions with current conditions to develop plans for habitat restoration and species reintroductions. In southern California, for example, Morrison et al. (1994) used museum records (many from early 1900) to help determine what species

"should" be in a heavily degraded urban park. They could have taken this process a few steps further by hypothesizing which species disappeared from the park due to natural environmental changes, and in contrast, which species were missing historically but have recently expanded their ranges.

In addition to outright extinction of a population or species, natural and human-induced changes in animal populations and their habitats can cause dramatic changes in the behavior of the surviving individuals. Brower and Malcolm (1991) have coined the term *endangered phenomena* to describe the situation in which a spectacular aspect (e.g., massive migrations) of the life history of an organism is endangered (see also chapter 10 on key ecological functions), even though the species itself might not become extinct. They envisioned in the near future an increasing number of species reduced in range and so constrained in numbers that they could no longer exhibit their spectacular life history phenomena. As examples, they include the greatly diminished herds of *Bison* of the North American prairie, and the endangered migrations of the east African wildebeest (*Connochaetus taurinus*), the Atlantic salmon (*Salmo salar*), and numerous North American songbirds. The consequences that such forced changes in life history traits will have on long-term survival of a species are poorly understood but cannot be assumed to be favorable to maintaining natural ecosystems with native communities and species.

Conclusions

Pimm (1991) made the important distinction between short- and long-term studies and the types of management we do. Most management is concerned about short-term changes in diversity; this focus is understandable, given the human time frame. However, over the longer term, we must realize that changes in a species' density will affect the species' immediate predators and prey and, eventually, its competitors. Over even longer periods, the predator and prey of these predators and prey will be affected; a ripple effect often occurs. Thus, over the long term, we must consider both the direct and the indirect, including time-lagged, impacts that our management scenarios will have on entire groups of organisms.

The fossil record is generally credited with allowing only the observation of dynamics in species and communities on time spans of 10^4 years or greater, and only for selected taxa that can be fossilized. However, the paleoecology record can be used to examine patterns and infer dynamic processes on time scales relevant to many debates in contemporary ecology (DiMichele 1994). As summarized by DiMichele, the paleoecological record reveals that ecosystem structure and taxonomic composition commonly persist or repeatedly recur throughout intervals of hundreds of thousands to millions of years. DiMichele has asked: Can the Recent era be a time period of dominantly Gleasonian, individualistic dynamics, not characteristic of most of geological time? Can current theory account for paleontological patterns, and further, can it predict future patterns? Answers to these questions will most certainly help place our current ecological views into a perspective that includes both a short- and a long-term reality.

We know that species diversity has changed over time and is continuing to change for a variety of reasons. Studying the past helps us set contemporary changes in perspective, and assists us in predicting the consequences of future changes in resource use (Pimm 1991). Extant species may have had different geographic distributions, coexisted with different species, and lived under different environmental conditions during the Pleistocene than they do today. Interpretation of the current distribution and habitat affinities of animals should be evaluated in light of these historical biogeographic and ecological data (Goodwin 1995).

Literature Cited

Alward, R. D., J. K. Detling, and D. G. Milchunas. 1999. Grassland vegetation changes and nocturnal global warming. *Science* 283:229–31.

Amundson, D. C., and H. E. Wright Jr. 1979. Forest changes in Minnesota at the end of the Pleistocene. *Ecological Monographs* 49:1–16.

Andrewartha, H. G., and L. C. Birch. 1954. *The distribution and abundance of animals.* Chicago: Univ. of Chicago Press.

Barber, V. A., G. P. Juday, and B. P. Finney. 2000. Reduced growth of Alaskan white spruce in the twentieth century from temperature-induced drought stress. *Nature* 405:668–73.

Batzli, G. O. 1994. Special feature: Mammal–plant interactions. *Journal of Mammalogy* 75:813–15.

Berger, J., P. B. Stacey, L. Bellis, and M. P. Johnson. 2001. A mammalian predator–prey imbalance: Grizzly bear and wolf extinction affect avian neotropical migrants. *Ecological Applications* 11:947–60.

Botsford, L. W., T. C. Wainwright, and J. T. Smith. 1988. Population dynamics of California quail related to meteorological conditions. *Journal of Wildlife Management* 52:469–77.

Bradley, N. L., A. C. Leopold, J. Ross, and W. Huffaker. 1999. Phenological changes reflect climate change in Wisconsin. *Proceedings of the National Academy of Science of the United States of America* 96:9701–4.

Brower, L. P., and S. B. Malcolm. 1991. Animal migrations: Endangered phenomena. *American Zoologist* 31:265–76.

Brown, J. H. 1978. The theory of insular biogeography and the distribution of boreal birds and mammals. *Great Basin Naturalist Memoirs* 2:209–27.

Brown, J. L., S.-H. Li, and N. Bhagabati. 1999. Long-term trend toward earlier breeding in an American bird: A response to global warming? *Proceedings of the National Academy of Science of the United States of America* 96:5565–69.

Clark, J. S., E. C. Grimm, J. J. Donovan, S. C. Fritz, D. R. Engstrom, and J. E. Almendinger. 2002. Drought cycles and landscape responses to past aridity on prairies of the northern Great Plains, USA. *Ecology* 83:595–601.

Cody, M. L. 1985. An introduction to habitat selection in birds. In *Habitat selection in birds,* ed. M. L. Cody, 3–56. New York: Academic Press.

Cohen, J. E. 1978. *Food webs and niche space.* Princeton, NJ: Princeton Univ. Press.

Cohen, J. E. 1990. Food webs and community structure. In *Community food webs: Data and theory,* ed. J. E. Cohen, F. Briand, and C. M. Newman, 1–14. New York: Springer-Verlag.

Cohen, J. E., F. Briand, and C. M. Newman. 1990. *Community food webs: Data and theory.* New York: Springer-Verlag.

Connell, J. H. 1980. Diversity and the coevolution of competitors; or, the ghost of competition past. *Oikos* 35:131–38.

Cox, C. B., and P. D. Moore. 1993. *Biogeography: An ecological and evolutionary approach.* 5th ed. Boston: Blackwell Scientific Publications.

Davis, R., C. Dunford, and M. V. Lomolino. 1988. Montane mammals of the American Southwest: The possible influence of post-Pleistocene colonization. *Journal of Biogeography* 15:841–48.

DiMichele, W. A. 1994. Ecological patterns in time and space. *Paleobiology* 20:89–92.

Dunn, P. O., and D. W. Winkler. 1999. Climate change has affected the breeding date of tree swallows throughout North America. *Proceedings of the Royal Society of London B* 266:2487–90.

Elias, S. A. 1992. Late Quaternary zoogeography of the Chihuahuan desert insect fauna, based on fossil records from packrat middens. *Journal of Biogeography* 19:285–97.

Elton, C. 1927. *Animal ecology.* New York: Macmillan.

Flack, J. A. D. 1976. *Bird populations of aspen forests in western North America.* Ornithological Monographs, no. 19. Lawrence, KS: Allen Press.

Franklin, J. F. 1988. Structural and functional diversity in temperate forests. In *Biodiversity,* ed. E. O. Wilson, 166–75. Washington, DC: National Academy Press.

Gibbons, E. F., Jr., B. S. Durrant, and J. Demarest. 1995. *Conservation of endangered species in captivity: An interdisciplinary approach.* Albany: State Univ. of New York Press.

Goodwin, H. T. 1995. Pliocene-Pleistocene biogeographic history of prairie dogs, genus *Cynomys* (Sciuridae). *Journal of Mammalogy* 76:100–122.

Goudie, A. 1992. *Environmental change.* 3rd. ed. Oxford, UK: Clarendon Press.

Graham, R. W., and 20 coauthors (FAUNMAP Working Group). 1996. Spatial response of mammals to late Quaternary environmental fluctuations. *Science* 272:1601–6.

Grinnell, J. 1905. Status of Townsend's warbler in California. *Condor* 7:52–53.

Guthery, F. S., C. L. Land, and B. W. Hall. 2001. Heat loads on reproducing bobwhites in the semiarid subtropics. *Journal of Wildlife Management* 65:111–17.

Hafner, D. J. 1993. North American pika (*Ochotona principes*) as a late Quaternary biogeographic indicator species. *Quaternary Research* 39:373–80.

Halaj, J., and D. H. Wise. 2001. Terrestrial trophic cascades: How much do they trickle? *American Naturalist* 157:262–81.

Hall, S. J., and D. G. Raffaelli. 1993. Food webs: Theory and reality. *Advances in Ecological Research* 24:187–39.

Harris, A. H. 1990. Fossil evidence bearing on southwestern mammalian biogeography. *Journal of Mammalogy* 71:219–29.

Harris, A. H. 1993. Wisconsinan pre-pleniglacial biotic change in southeastern New Mexico. *Quaternary Research* 40:127–33.

Heusser, C. J. 1965. A Pleistocene phytogeological sketch of the Pacific Northwest and Alaska. In *The Quaternary of the United States,* ed. H. E. Wright Jr., and D. E. Frey, 469–83. Princeton, NJ: Princeton Univ. Press.

Heusser, C. J. 1977. Quaternary palynology on the Pacific slope of Washington. *Quaternary Research* 8:282–306.

Hopkins, M. S., J. Ash, A. W. Graham, J. Head, and R. K. Hewett. 1993. Charcoal evidence of the spatial extent of the *Eucalyptus* woodland expansions and rain forest contractions in North Queensland during the late Pleistocene. *Journal of Biogeography* 20:357–72.

Hubbard, J. P. 1973. Avian evolution in the aridlands of North America. *Living Bird* 12:155–96.

Hunter, M. D., and P. W. Price. 1992. Playing chutes and ladders: Heterogeneity and the relative roles of bottom-up and top-down forces in natural communities. *Ecology* 73:724–32.

Johnson, N. K. 1994. Pioneering and natural expansion of breeding distributions in western North American birds. *Studies in Avian Biology* 15:27–44.

Kerr, R. A. 2003. Megafauna died from big kill, not big chill. *Science* 300:885.

Lack, D. 1954. *The natural regulation of animal numbers.* New York: Oxford Univ. Press.

Lacy, R. C. 1994. Managing genetic diversity in captive

populations of animals. In *Restoration of endangered species: Conceptual issues, planning, and implementation,* ed. M. L. Bowles and C. J. Whelan, 63–89. Cambridge: Cambridge Univ. Press.

Lawton, J. H. 1989. Food webs. In *Ecological concepts,* ed. J. M. Cherrett, 43–78. Oxford, UK: Blackwell Scientific Publications.

Lyman, R. L., and S. Wolverton. 2002. The late prehistoric–early historic game sink in the northwestern United States. *Conservation Biology* 16:73–85.

Leberg, P. L. 1991. Influence of fragmentation and bottlenecks on genetic divergence of wild turkey populations. *Conservation Biology* 5:522–30.

Martin, P. S. 1967. Prehistoric overkill. In *Pleistocene extinctions: The search for a cause,* ed. P. S. Martin and H. E. Wright, 75–120. New Haven, CN: Yale Univ. Press.

Mengel, R. M. 1964. The probable history of species formation in some northern wood warblers (Parulidae). *Living Bird* 3:9–43.

Mengel, R. M. 1970. The North American Central Plains as an isolating agent in bird speciation. In *Pleistocene and Recent environments of the central Great Plains,* ed. W. Dort Jr. and J. K. Jones Jr., 279–340. Dept. of Geology Special Publication no. 3. Lawrence: Univ. of Kansas.

Moore, H. D. M., W. V. Holt, and G. M. Mace. 1992. *Biotechnology and the conservation of genetic diversity.* Symposia of the Zoological Society of London no. 64. Oxford: Oxford Univ. Press.

Morrison, M. L. 1983. Analysis of geographic variation in the Townsend's warbler. *Condor* 85:385–91.

Morrison, M. L. 2002. *Wildlife restoration: Techniques for habitat analysis and animal monitoring.* Washington, DC: Island Press.

Morrison, M. L., T. A. Scott, and T. Tennant. 1994. Wildlife-habitat restoration in an urban park in southern California. *Restoration Ecology* 2:17–30.

Morrison, M. L., I. C. Timossi, K. A. With, and P. N. Manley. 1985. Use of tree species by forest birds during winter and summer. *Journal of Wildlife Management* 49:1098–1102.

Nordt, L. C., T. W. Boutton, C. T. Hallmark, and M. R. Waters. 1994. Late Quaternary vegetation and climate changes in central Texas based on the isotopic composition of organic carbon. *Quaternary Research* 41:109–20.

O'Brien, M. J. 2001. Archaeology, paleoecosystems, and ecological restoration. In *The historical ecology handbook: A restorationist's guide to reference ecosystems,* ed. D. Egan and E. A. Howell, 29–53. Washington, DC: Island Press.

O'Brien, S. J., D. Goldman, C. R. Merril, and M. Bush. 1983. The cheetah is depauperate in genetic variation. *Science* 221:459–62.

O'Brien, S. J., D. E. Wildt, and M. Bush. 1986. The cheetah in genetic peril. *Scientific American* 254:84–92.

Oksanen, T., M. E. Power, and L. Oksanen. 1995. Ideal free habitat selection and consumer-resource dynamics. *American Naturalist* 146:565–85.

Owen-Smith, N. 1987. Pleistocene extinctions: The pivotal role of megaherbivores. *Paleobiology* 13:351–62.

Parshall, T. 2002. Late Holocene stand-scale invasion by hemlock (*Tsuga canadensis*) at its western range limit. *Ecology* 83:1386–98.

Payette, S. 1987. Recent porcupine expansion at tree line: A dendroecological analysis. *Canadian Journal of Zoology* 65:551–57.

Peters, R. H. 1991. *A critique for ecology.* Cambridge: Cambridge Univ. Press.

Pianka, E. R. 1983. *Evolutionary ecology.* 3rd. ed. New York: Harper & Row.

Pimm, S. L. 1982. *Food webs.* New York: Chapman & Hall.

Pimm, S. L. 1991. *The balance of nature? Ecological issues in the conservation of species and communities.* Chicago: Univ. of Chicago Press.

Post, E., and N. C. Stenseth. 1999. Climatic variability, plant phenology, and northern ungulates. *Ecology* 80:1322–39.

Power, M. E. 1992. Top-down and bottom-up forces in food webs: Do plants have primacy? *Ecology* 73:733–46.

Ramey, R. R., II, G. Luikart, and F. J. Singer. 2000. Genetic bottlenecks resulting from restoration efforts: The case of bighorn sheep in Badlands National Park. *Restoration Ecology* 8:85–90.

Rappole, J. H. 1995. *The ecology of migrant birds: A neotropical perspective.* Washington, DC: Smithsonian Institution Press.

Raup, D. M. 1988. Diversity crises in the geological past. In *Biodiversity,* ed. E. O. Wilson, 51–57. Washington, DC: National Academy Press.

Reed, D. H., and R. Frankham. 2003. Correlation between fitness and genetic diversity. *Conservation Biology* 17:230–37.

Risser, P. G. 1988. Diversity in and among grasslands. In *Biodiversity,* ed. E. O. Wilson, 176–80. Washington, DC: National Academy Press.

Ritchie, J. C. 1976. The late-Quaternary vegetational history of the western interior of Canada. *Canadian Journal of Botany* 54:1793–1818.

Root, T. L. 1988. *Atlas of wintering North American birds.* Chicago: Univ. of Chicago Press.

Root, T. L., and S. H. Schneider. 2002. Climate change: Overview and implications for wildlife. In *Wildlife responses to climate change: North American case studies,* ed. S. H. Schneider and T. L. Root, 1–56. Washington, DC: Island Press.

Schmitz, O. J., P. A. Hamback, and A. P. Beckerman. 2000. Trophic cascades in terrestrial systems: A review of the effects of carnivore removals on plants. *American Naturalist* 155:141–53.

Schoener, T. W., D. A. Spiller, and J. B. Losos. 2002. Predation on a common Anolis lizard: Can the food-web effects of a devastating predator be reversed? *Ecological Monographs* 72:383–407.

Smith, C. C. 1981. The indivisible niche of Tamiasciurus: An example of nonpartitioning of resources. *Ecological Monographs* 51:343–63.

Sullivan, R. M. 1988. Biogeography of southwestern montane mammals: An assessment of the historical and environmental predictions. PhD dissertation, Univ. of New Mexico, Albuquerque.

Vaughan, T. A. 1986. *Mammalogy.* 3rd ed. Philadelphia: Saunders College Publishing.

Wiens, J. A. 1977. On competition and variable environments. *American Scientist* 65:590–97.

Wright, I. C., M. S. McGlone, C. S. Nelson, and B. J. Pillans. 1995. An integrated latest Quaternary (stage 3 to present) paleoclimate and paleoceanographic record from offshore northern New Zealand. *Quaternary Research* 44:283–93.

Wright, S. 1931. Evolution in Mendelian populations. *Genetics* 16:97–159.

3 The Habitat, Niche, and Population Perspectives

Euclid would approve with glee
How territoriality
Of bunting perched upon a stone
Is asymptote to hypercone.
— BRUCE G. MARCOT, "ODE TO THE NICHE (AN ECOLOGIST'S NIGHTMARE)"

In this chapter we explore the perspectives of habitat, niche, and populations. Vegetation is a core component of the habitat of terrestrial vertebrates and is thus a central focus of the chapter. We discuss plants as indicators of the physical environment; vegetation succession; and the correlation between animals and vegetation, including animal–plant interactions and relations of wildlife to floristics and vegetation structure. We continue with development of the myriad other environmental features that form the habitat of an animal. Next we look more closely within the habitat of an animal by examining niche relationships and thereby the factors that often drive the behavior, productivity, and survival of an individual. We then turn attention to the relationships of populations and habitats. We focus on spatial and geographic factors that influence habitats and environments, population structure, fitness of organisms, and the ultimate viability of populations.

The Habitat Perspective

In chapter 1 we developed definitions and explored the concept of habitat. Here we look at habitat more closely and discuss the basic factors that constitute habitat, including vegetation and various other factors that often fall under the heading of "special features" in many habitat analyses. Sampling strategies and methods that follow from these earlier concepts are discussed later, in chapters 5 and 6.

The Central Role of Vegetation

That vegetation plays a central role in the life of many animals is self-evident. Indeed, much of this book is devoted to analyzing the distribution and abundance of plants as part of developing wildlife–habitat relationships. Thus it is incumbent on researchers and managers to understand the factors determining the health of plants, how plants are arranged into identifiable units, and how these units change through time. In the following sections we discuss vegetation ecology and relate this ecology to the study of wildlife–habitat relationships.

Plants as Indicators of the Physical Environment

Clements (1920, 3–34) traced the development of the *indicator concept* as applied to plant communities back to the early 1600s and also noted that an understanding of the relationship between soil and plants must have marked the beginnings of agriculture. Many researchers in the mid- to late 1800s, including Hilgard, Chamberlin, Shantz, and Weaver, developed the indicator concept as applied to plants, particularly with application to agricultural sciences (Clements 1920). The concept of using plants as indicators of environmental conditions became ingrained in the scientific literature following the pioneering work of these early scientists.

Vegetation Patterns and Processes of Succession

Humans have long been fascinated with understanding and predicting changes in plant distribution and abundance and often depend on such predictions for survival. As noted by Glenn-Lewin et al. (1992), hunter-gatherer societies desired an understanding of fire-induced changes in vegetation that altered forage for the game animals that were their food source. Modern wildlife conservation depends on our ability to understand the patterns and processes of vegetative change.

Vegetation change as an orderly process is usually referred to as *plant succession.* Glenn-Lewin et al. (1992) reviewed the historical development of the concept of plant and plant community succession. The "Clementsian paradigm" developed from the work of Clements (1904, 1916) on the theory of plant succession, which dominated until the 1940s. His *community-unit theory* held that plant species formed groupings of distinct, clearly bounded types of plant communities, often termed *associations* (Whittaker 1975). Clements viewed succession as an orderly and predictable process in which vegetation change represented the life history of a plant community; the community thus assumed "organism-like" characteristics. These communities were believed to develop to a distinct climax condition, the characteristics of which were controlled by the regional climate (Glenn-Lewin et al. 1992).

The Clementsian views had some early and strong critics. Gleason (1939) and Tansley (1935) argued that plant communities resulted from the overlap in the independent distributions of species with similar environmental needs, and that vegetation change in a region did not necessarily converge toward a similar climax condition. In the 1960s, several researchers attempted to synthesize these disparate views of succession into a unified theory. Margalef (1963) offered the idea that succession was driven from simple ecosystems to more complex systems with more trophic levels and greater species diversity. Similarly, Odum (1969) postulated that ecosystems tended to develop toward greater homeostasis of species composition. As summarized by Glenn-Lewin et al. (1992), this Margalef-Odum synthesis of successional theory was philosophically similar to the Clementsian paradigm.

By the early 1970s ecologists had recognized the inadequacies of the Margalef-Odum synthesis, primarily because of the increased understanding of the need for site-specific data on the proximate causes of vegetation change (Glenn-Lewin et al. 1992; see also Whittaker 1975). Two major trends have occurred in our ideas about succession since the 1970s: first, a shift away from a search for holistic explanations toward a more reductionist and mechanistic approach that emphasizes proximate causes of vegetation changes; and second, a shift away from equilibrium toward nonequilibrium theories (Glenn-Lewin et al. 1992). What has become known as the *individualistic hypothesis,* which contends that plant populations are individually distributed along gradients, and that most communities and their component species intergrade along these gradients rather than forming distinct, separate zones (Whittaker 1975), has grown in popularity. Rather than being a single hypothesis, this individualistic view is actually an umbrella for various views, including hypotheses that consider succession as a gradient in time or resource availability, as the consequence of differential longevity and other population processes, as the result of differences in life history traits, and as a stochastic process (see review by Glenn-Lewin et al. 1992).

In essence, all of these ideas view succession as the combined outcome of populations of plants interacting with a fluctuating environment. The specific pattern observed (i.e., community structure and flora) will vary with location, and with time in the same location. Whittaker (1975) synthesized various ways in which plants could be distributed in relation to one another and to plant communities; we present his summary in box 3.1. Barbour (1996) provided an interesting historic account of the debate over Gleason's original ideas.

Van Hulst (1992) asked, What factors influence future vegetational composition of a site? The answer to this question has obvious ramifications for our study of wildlife–habitat relationships, given the central role that vegetation plays in determining the distribution and abundance of animals. Van Hulst outlined six general categories of factors that influence future vegetation:

- Present vegetation
- Present vegetation in the surrounding area, or immigration of propagules
- Past vegetation (e.g., dormant seeds)
- Present resource levels (e.g., light, humidity, soil minerals)
- Disturbance levels, including herbivory
- Stochastic factors (e.g., climatic variability, supply fluctuations of resources)

Different models are available to help predict the course of succession depending on which of the preceding factors are of primary importance in determining the dynamics of the plant community; Van Hulst (1992) provided details. Making reliable predictions concerning the future course a wildlife population will take often requires reliable predictions of the future course the vegetative community will take. Thus it is important that the influence the factors listed above have on the characteristics of plant communities, including successional patterns, be recognized and understood when developing habitat models.

The Correlation between Animals and Vegetation

Broad-scale relationships between vegetation and animals have long been recognized. The classic *life zone* concept of C. Hart Merriam (1898) used temperature boundaries to describe the limits of major vegetation associations (life zones) and their associated animal species. Holdridge (1947) developed his well-known system using temperature, precipitation, and evapotranspiration to describe major vegetation groupings of the world. In the United States, every state and region has multiple classification systems, many of which are now being revised with the aid of remote-sensing data and geographic information systems (GIS). These

---

Box 3.1   Whittaker's four hypotheses on the distribution of plant species populations

The eminent plant ecologist Robert H. Whittaker wrote an excellent book on plant ecology, including a thorough development of how plant species might come to be associated. He outlined four working hypotheses on ways species populations might be distributed (Whittaker 1975, 113–15):

1. Competing species, including dominant plants, exclude one another along sharp boundaries. Other species evolve toward close association with the dominants and toward adaptations for coexisting with one another. A gradient thus develops with distinct zones; each zone has a characteristic assemblage of plants adapted to one another and gives way at a sharp boundary to another assemblage of species adapted to one another. Such a hypothesis results in relatively discontinuous kinds of plant communities (box fig. 3.1A, panel a).
2. Competing species exclude one another along sharp boundaries but do not become organized into groups with parallel distributions (box fig. 3.1A, panel b).
3. Competition generally does not result in sharp boundaries between assemblages. Evolution of species toward adaptation to one another will, however, result in the appearance of groups of species with similar distributions. These groups characterize different kinds of plant communities, but the communities intergrade continuously (box fig. 3.1A, panel c).
4. Competition does not usually produce sharp boundaries between species populations, and evolution of species in relation to one another does not produce well-defined groups of species with similar distributions. The centers and boundaries of populations of species are scattered along the environmental gradient (box fig. 3.1A, panel d).

Whittaker concluded that when we study the manner in which plant populations increase and decrease in abundance along environmental gradients (e.g., interactions among soil moisture, soil texture, ambient temperature, and other factors), the results support the fourth hypothesis presented above and depicted in box fig. 3.1A, panel d. Although cases of sharp boundaries between species are known, most populations are distributed as presented in figure 3.1B. Thus field observations agree with the "individualistic hypothesis" rather than various "community-unit theories."

---

systems are, in turn, often used to describe the distribution and abundance of wildlife associated with vegetation conditions over broad geographic scales (see chapter 10).

Classically, "wildlife habitat" was described as containing three basic components; namely, cover, food, and water. Dasmann (1964) further divided cover into two components: "habitat requirements" and "escape cover." These two components are not, of course, mutually exclusive. There was a clear recognition that vegetation—cover—provides essential requirements for animals, and that manipulation of this cover could drastically affect wildlife populations.

Of particular interest to wildlife biologists has been the relationships between "edge habitats" and wildlife richness and abundance (see chapter 8). Edge has been described as the joining of two "habitats." We place habitat in quotation marks here because, by the definitions we use in this book, earlier researchers were really referring to the joining of two vegetation associations—popularly referred to as the *edge effect*. Aldo Leopold, in his famous *Game Management* (1933), actually bestowed the edge effect with law status; namely, his *law of interspersion*. This law stated that the edge between two "habitats" will be more favorable as wildlife habitat than either association considered alone. This effect may be a result of the conjoining of separate environments, each with their associated wildlife communities. By sampling across two vegetation associations, including the intersection of the two, one is actually sampling three (or more)

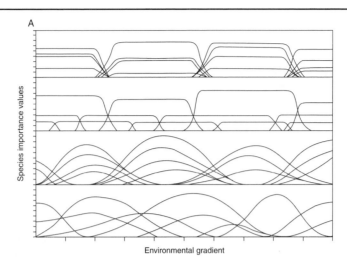

Box Figure 3.1A.   Four hypotheses on plant species distributions along environmental gradients. Each curve represents one species population and the way it might be distributed along the environmental gradient. (From Whittaker 1975. Reprinted by permission of Pearson Education.)

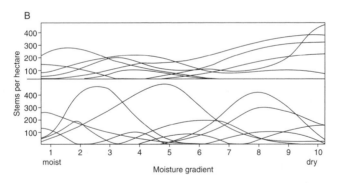

Box Figure 3.1B. Actual distributions of plant species populations along environmental gradients. Species populations are plotted by densities (number of tree stems per hectare). (From Whittaker 1975. Reprinted by permission of Pearson Education.)

vegetation associations. It may not be the edge per se that is relevant; it is the fact that more combinations of vegetation and other habitat factors are sampled, thus usually resulting in a higher species count than if only one environment were sampled.

Edges, often referred to as *ecotones,* are scale-dependent. As shown in table 3.1, within a single area, many edges or ecotones can exist, depending on the spatial scale of interest. These edges are influenced by an increasingly complex array of environmental factors as the scale increases in

Table 3.1.   Ecotone hierarchy for a biome transition area

Ecotone hierarchy	Probable constraints
Biome ecotone	Climate (weather) × topography
Landscape ecotone (mosaic pattern)	Weather × topography × soil characteristics
Patch ecotone	Soil characteristics × biological vectors × species interactions × microtopography × microclimatology
Population ecotone (plant pattern)	Interspecies interactions × intraspecies interactions × physiological controls × population genetics × microtopography × microclimatology
Plant ecotone	Interspecies interactions × intraspecies interactions × physiological controls × plant genetics × microclimatology × soil chemistry × soil fauna × soil microflora, etc.

Source: Gosz 1993, table 1. Reprinted with permission of the Ecological Society of America.

Note: Each level in the ecotone hierarchy has a range of constraints and interactions between the constraints; × symbolizes interactions between constraints. The primary constraints vary with the scale of the ecotone, with an increase in the number of possible constraints at finer scales.

resolution. Such concepts hold extremely important information for researchers hoping to develop wildlife–habitat relationships, and for managers hoping to implement management recommendations. A relatively sedentary amphibian will be most severely affected by ecotones at the patch, population, and even plant levels of table 3.1, whereas relatively mobile large mammals will be most influenced by landscape patterns. Allowing for both groups of animals, as managers are usually called on to do today, greatly increases our information needs and complicates management.

Today the edge concept is no longer viewed as an overriding positive feature of wildlife management. As discussed in chapter 8, increasing edge beyond natural levels leads to fragmented environments, which may cause increased pre-

dation by natural and exotic predators and increased rates of avian nest parasitism (see Paton 1994 for review; also Askins 1995; Mills 1995 [for mammals]; Robinson et al. 1995). This problem is magnified today as fewer and fewer large tracts of undisturbed environments remain. Thus increasing edge can be seen as a measure of increasing environmental degradation for at least some species. Guthery and Bingham's (1992) critique of Leopold's edge concept noted that few modern researchers and managers take note of the fact that Leopold qualified his principle by stating it did not generally apply to mobile species, and that he confined his principle to "edge-obligate species." Guthery and Bingham also noted that animals cannot increase abundance above some maximum density; thus the correlation between animal density and amount of edge has limits. That is, creating edge beyond some optimum amount would not result in increased wildlife density. The increasing number of smaller patches, and the linear or irregularly shaped patches that often result from fragmentation, contribute to an increase in the amount of edge in a landscape (fig. 3.1).

The issues of fragmentation and edge are evident worldwide. For example, large-scale tropical deforestation has led to fragmentation of continuous forests. Through secondary succession, recolonization of deforested lands may restore some of the unique characteristics of these ecosystems. Species composition of restored areas may, however, differ substantially from that of mature forests, in part because of the influence of edge effects. In South Africa, for example, Weiermans and van Aarde (2003) examined a postmining rehabilitation project that had begun 20 years earlier. They found that roads exerted an edge effect, causing a distinction between edge and core assemblages of birds and millipedes, but not of rodents. Distance from the edge apparently affected species richness, diversity, and abundance of individuals and, there-

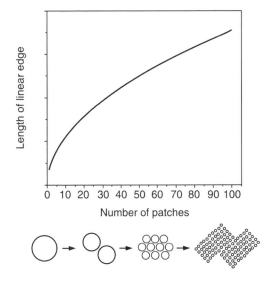

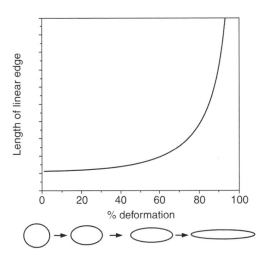

Figure 3.1. The amount of edge proliferates with increasing fragmentation, due to the increased edge per unit area as the number of patches increases (*top*), and as individual patches become, on average, more linear or more irregular in shape (*bottom*). (From Sisk and Battin 2002, fig. 1; by permission of the Cooper Ornithological Society.)

fore, the separation of species into distinct edge and core assemblages.

Thus, as reviewed by Sisk and Battin (2002), perceptions of the relationship between edge and fragmentation are often contradictory. In some cases, edges are thought to benefit certain animal species; but in other cases, edges have a detrimental impact. Part of this confusion likely results from changes in the spatial scale at which species diversity is measured. Historically, biologists have focused on alpha (local) diversity, which can be high in edge situations. As the focus of biologists has expanded to include larger spatial areas, the focus on species diversity has shifted to gamma (regional) diversity, which may be lower in fragmented landscapes due to the loss of edge-avoiding species. Sisk and Battin (2002) advised that biologists adopt a multiscale approach when quantifying and discussing biodiversity. Both local and regional habitat characteristics can influence species richness and community structure. The scale at which communities are studied affects the detection of relationships between habitat characteristics and patterns of habitat selection, species diversity, and species composition, and may obscure observation of differences in how species perceive the scale of environmental variation (Pearman 2002).

Beginning in the 1950s, ecologists began to develop formal relationships between various components of vegetation cover and the kinds of animals in an area and their abundances. Most famous are the foliage height diversity–bird species diversity (FHD-BSD) constructs of MacArthur and MacArthur (1961). In figure 3.2 we see that the diversity of birds rises as vegetation becomes increasingly complex vertically. A plethora of studies followed the early work of the MacArthurs, with most researchers finding significant statistical relationships between FHD and BSD, although there were exceptions (Roth 1976). Karr and Roth (1971) suggested that the

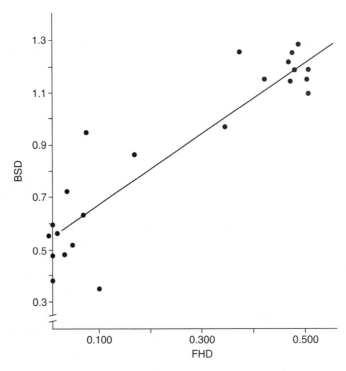

Figure 3.2.   Foliage height diversity (FHD) versus bird species diversity (BSD). The dots represent study sites. (From Willson 1974, fig. 1, by permission of the Ecological Society of America.)

scatter in an FHD-BSD figure likely resulted from important, but unmeasured, factors that influenced birds. Other researchers were able to relate BSD to the heterogeneity, or horizontal "patchiness," of the vegetation (Roth 1976).

Unfortunately, measures such as FHD collapse detailed information on vegetation into a single number. These relationships depict patterns that occur over broad geographic areas. They do indicate, however, that animals respond to measures of complexity in their environment.

As noted by Dasmann (1964) in his introductory wildlife text, "manipulation of plant succession is the principal way of providing the habitat requirements and escape cover required for the support of larger game populations." Thus what theoretical ecologists had "discovered" had already become a principal tool of wildlife management, albeit for a selected group of preferred species. However, as researchers and management agencies became increasingly concerned about the fate of all animal species, these fundamental principles of ecology and wildlife management were applied to a wide array of species. A good indication of the development of researcher and manager interest in this area can be found in the many U.S. Forest Service publications published in the 1970s and thereafter. For example, Thomas et al. (1979) presented many fundamental wildlife–habitat relationships, in-

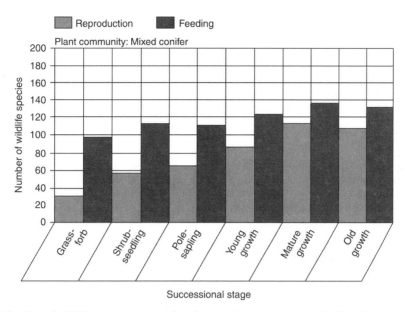

Figure 3.3   Number of wildlife species associated with successional stages in a mixed-conifer community. (From Thomas et al. 1979, fig. 14.)

cluding that shown in figure 3.3. Here we see that a basic relationship exists whereby the number of animal species increases with advancing seral stages.

Although much is known about succession, the application of this knowledge to resource management has not progressed rapidly (Luken 1990). Luken thought that this was because ecologists do not want to work in managed systems, and that most ecological studies are site-specific and do not lead rapidly to new, robust theories. It is clear, however, that wildlife biologists have long understood the basic relationships between management of seral stages and wildlife populations. This interest traditionally centered on a narrow group of game or pest species. Luken continued that the incorporation of succession management into natural resource management decision making was the next step after careful description of succession pathways. The recent

(1990s) move by many federal land management agencies to manage at the multiple-species "ecosystem scale" is an indication of a move in the direction Luken suggested (see also chapter 11).

ANIMAL–PLANT INTERACTIONS

As reviewed by Batzli (1994), responses of animals to plant characteristics first involve behavioral and physiological adjustments of individuals to plants. As a result of such adjustments, individual fitness may change, leading to changes in the distribution and abundance of wildlife populations. The resulting activities of these populations then feed back onto the plants. Further, animals individually affect soils, nutrient cycles, and other environmental factors (fig. 3.4). The ability of many mammals, for example, to modify the environment is well documented; Jones et al. (1996) went so far as to call them

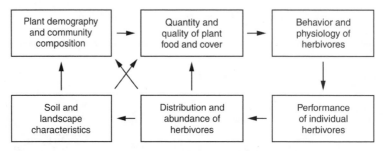

Figure 3.4   Model of interactions between mammalian herbivores and plants. Arrows indicate causal pathways for change, which can be negative or positive. Mammals directly respond to food and cover provided by plants. Mammals influence plants directly (by immediately damaging or facilitating plant success) or indirectly (by affecting substrate or landscape). (From Batzli 1994, fig.1; reprinted by permission of Alliance Communication Group a division of Allen Press Inc.)

"ecosystem engineers." Moose, beavers, prairie dogs, Serengeti ungulates, and pocket gophers are classic examples (Batzli 1994).

The influence of animals on vegetation can be much more subtle than that resulting from beavers or prairie dogs. For example, Snyder (1993) found that ponderosa pine (*Pinus ponderosa*) trees defoliated as a result of bark feeding by Abert's squirrels (*Sciurus aberti*) showed significant reduction in fitness. The squirrels may actually be agents of natural selection in ponderosa pine populations because they are attracted to trees with specific genetically determined traits and because such trees show significant reduction in fitness. Bonser and Reader (1995) tested whether the effects of competition and herbivory on plant growth depended on the above-ground biomass of vegetation. They found that competition and herbivory each have a greater effect on plant growth at sites with higher biomass, and that herbivory had less effect than competition on plant growth at sites with relatively less biomass. The herbivores in their experimental study included voles (*Microtus* spp.), slugs, and snails. Such results show the complex and interactive dynamic between plant biomass, plant competition, and grazing by herbivores (even relatively small ones).

The role of animals in succession management can be viewed in two major ways: (1) the manipulation of seral stages by managers for high and persistent abundances of selected species (the classic function of wildlife biologists); and (2) the control of succession by land managers using grazing animals. The first views animal populations changing as a *result* of succession, while the other views animal populations as *changing* succession itself (Luken 1990, 151–152). Luken went on to make a critical point: when managing wildlife habitat, and thus succession, for animals, a phase of succession is not the ultimate goal, but rather the biotic and abiotic factors in a mix of successional phases that satisfy the resource requirements of the target animal species. Thus Luken acknowledged the species-specific nature of wildlife habitat, as we are emphasizing throughout this book.

Unfortunately for researchers and managers alike, few absolute generalizations can be made about the influence of animals on succession. As illustrated in figure 3.5, numerous pathways exist by which animals can influence plants, and thus succession, including trampling, eating foliage and seeds, and defecation. Animals also function as dispersers of seeds and other plant disseminules. In addition, numerous manage-

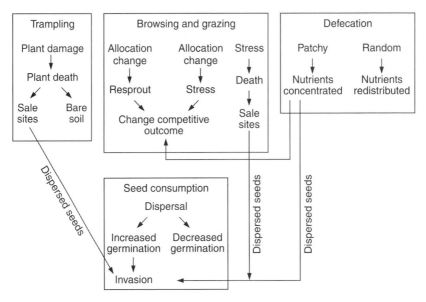

Figure 3.5.   Various activities associated with grazing or browsing animals and their possible effects on plants. (Reproduced from Luken 1990, fig. 7.6, by permission of Springer Science and Business Media..)

ment activities can change the path of succession (table 3.2). The course of succession can be generally predicted when a single management impact is applied. However, multiple impacts render prediction highly difficult (e.g., shrub control, followed by burning, followed by grazing by native and domestic animals). It is this complicated interaction among impacts that prevents managers from making accurate predictions concerning the outcome of planned activities on specific sites. And, when studying animal–succession relationships, or when planning management based on such a study, one must consider the surrounding vegetation associations and seral stages and their possible influence on the area of interest (i.e., the broader landscape-scale perspective).

Domestic animals have had direct and substantial influences on plant succession. A large body of literature has documented the long-term changes caused by overgrazing of cattle. In southeast Arizona and southwest New Mexico, for example, overgrazing has been implicated in changing the historic semidesert grasslands into mesquite-dominated woodlands with little herbaceous cover; fire suppression has exasperated these changes (e.g., see review by Bahre 1991).

Domestic herbivores have also been used, however, to help restore natural successional processes and plant communities. In the Netherlands, for example, cattle, ponies, and sheep have been used in nature reserves to restore ecological processes associated with herbivory (see also Savory 1988). Reserve managers in the United States sometimes use livestock to help reduce the cover of exotic plants as part of attempts to reintroduce native plants. Domestic animals were introduced as "ecological substitutes" for their extinct ancestors. For example, the wild horse (*Equus przewalskii*) disappeared from northern

Table 3.2.   Anticipated effects of management actions on successional state or condition

Management action	Successional stage condition					
Type of control	Grass-forb	Shrub-seedling	Pole-sapling	Young	Mature	Old growth
Shrub control						
Herbicides	>	>	>	<	<	<
Mechanical control	<	<		<	<	<
Controlled burn						
Cold burn	<	<	>	>	<	<
Hot burn	<	<	<	<	<	<
Fertilization	<	>	>	>	>	—
Grazing and browsing (moderate rates)						
Cattle and sheep	<	>	—	—	<	<
Goats	—	<	—	—	<	<
Deer and elk	<	>	—	—	<	<
Planting						
Trees	>	>			<	<
Shrubs	>	<				
Grasses-forbs	>	>		>	>	—
Regeneration cut						
Clearcut				<	<	<
Shelterwood				>	<	<
Seed tree				<	<	<
Salvage			<	<	<	>
Thinning (including single tree selection harvest)		>	>	>	>	—

Source: Thomas et al. 1979, table 3.

Europe in about 2000 BC (Bokdam and de Vries 1992).

FLORISTICS–VEGETATION
STRUCTURE–WILDLIFE RELATIONSHIPS

Two basic and obvious aspects of vegetation can be distinguished: the structure of the plant, or physiognomy; and the taxon of the plant, or floristics. Many authors have initially concluded that vegetation structure and habitat configuration (size, shape, and distribution of vegetation in an area), rather than particular plant taxonomic composition, most determine patterns of habitat occupancy by animals (e.g., see Hilden 1965; Wiens 1969; James 1971; Rotenberry 1985). As Rotenberry observed, however, subsequent studies have shown that plant species composition plays a much greater role in determining occupancy of an area than previously thought. He noted that comparisons of the effects of structure and floristics on animal abundance really involve the spatial scale at which the animal species is being examined. The same species that appears to respond to the physical configuration of an area at a broad (landscape) scale might show little correlation with structure at a more localized scale (see our previous discussion in this chapter of the FHD-BSD relationship). Rotenberry thus reached the same conclusion, albeit independently and from a different direction, as did other researchers interested in the relationship between spatial scale and occupancy of an area (e.g., Johnson 1980; Hutto 1985).

Thus we see that the variables measured, and the spatial resolution of those measurements, should be based on the scale involved and the level of model refinement required. Simple presence–absence studies of animal abundance at broad scales likely do not require analysis of vegetation on a fine taxonomic level. Broad categorizations by vegetative structural class (e.g., sapling, pole, old growth) is probably adequate. Plant taxonomy becomes increasingly important, however, as our studies become more site-specific and call for prediction of wildlife population size or density. Attempts to apply broad, structurally based models of wildlife–habitat relationships to local management situations (e.g., at the stand scale) usually fail (Block et al. 1994).

Nonvegetative Habitat Features

By definition, all features of the environment surrounding an individual animal at any given point in time can be used to describe its habitat. In other words, the most basic way to describe habitat is simply to catalog the features of the environment surrounding the animal. Of course, our goal is to go beyond such a simple cataloging and determine those features that hold special importance to the animal in terms of survival and fitness. Although we discuss the sampling and analytical methods used to identify features of primary importance to animals, here we want to briefly review some of the nonvegetative features of the environment that have been the focus of study.

In designing field studies, biologists have found it difficult for several reasons to identify a small set of factors that might be of relevance to an animal. First, by basic nature, biologists tend to be curious and thus do not want to accidentally leave out any factor that might explain

something about an animal under study, especially some single critical factor. Second, from a standpoint of sampling biases, they do not want to eliminate something of potential importance just because they "think" it has no great merit. This attitude has often been taken to an extreme, with 30 or 40 or 50 or more independent variables being measured to describe, say, the location of each individual under study. Third, biologists are aware of the correlated nature of many environmental features and hope to evaluate data in a multivariate fashion.

Therefore, historically, studies of wildlife habitat have tended to gather data on virtually every feature of the environment of an animal, including rather finely subdivided categories of the same general variable (e.g., tree height, shrub cover). We are not criticizing this approach per se, but rather pointing out that cataloging of environmental features is a common practice. Later we cover the reasons why such a strategy is fraught with sampling and statistical problems. Below we discuss some of the measurements taken to describe habitat and the rationale behind those selections. Again, sampling nonvegetative features of the environment is an essential component of wildlife-habitat analysis.

In addition to the vegetative components discussed above, data collected in terrestrial environments typically includes features of ground cover (sand, gravel, stones, rocks); standing dead or down trees (arguably part of the vegetative component); rocky outcrops; presence and type of water, from ephemeral puddles to rivers; and lichen and moss. Artificial structures, such as poles, fences, houses, trails, roads, and so on, are also recorded. Many or all of these features are recorded by absolute size (e.g., pole height), cover (e.g., percentage of gravel cover), or number (e.g., number of boulders >1 m in diameter).

Burrow et al. (2001) described the habitat of the horned lizard (*Phrynosoma cornutum*) in Texas as including open deserts and grasslands, usually with sparse vegetation. They wanted, however, to provide more specifics on habitat requirements in the hope of elucidating factors important in maintaining populations. Of the habitat descriptors they measured, bare ground, litter, woody stems, and woody canopy varied between lizard and random locations. Thus several nonvegetative features of lizard habitat were important in predicting where the species would occur.

Coarse woody debris (CWD) is a focus of forest management not only because of its value to wildlife but because of its role in the frequency and intensity of fires. Fuel load management often prescribes a reduction in CWD, whereas wildlife management often strives to retain this material for animals. Ucitel et al. (2003) studied the southern red-backed vole (*Clethrionomys gapperi*), a species closely associated with CWD. They showed that CWD coverage was highest in locations used by the species relative to nonused sites (fig. 3.6). Voles were more active in portions of a stand with higher CWD levels than the overall stand average.

It is well known, of course, that most animals need free-standing water to successfully occupy a location. For example, even hibernating bats are thought to arouse during winter to drink (e.g., see Kuenzi and Morrison 2003). The absence of water can lead to abandonment of a hibernaculum or perhaps death. In many amphibians, soil moisture is a key factor in determining animal use of a specific location (e.g., Hyde and Simons 2001).

Thus it should be evident that numerous nonvegetative factors are an essential component of wildlife habitat. Of course, many of the features that make up habitat are interrelated; for example, vegetative canopy cover is partially responsible for soil condition, including water content. Sorting out the key factors that determine occupancy, survival, and fecundity is the subject of much of this book.

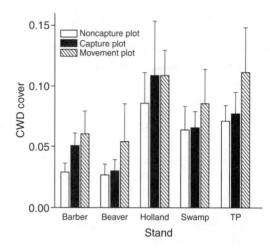

Figure 3.6.   Stand measure of percentage of coarse woody debris (CWD; >7.5 cm diameter) cover calculated separately for capture plots (trap locations where voles were captured); noncapture plots (trap locations chosen randomly out of all locations where voles were not captured); and movement plots (minimum bounding triangle around a vole trail), Montana. (From Ucitel et al. 2003, fig. 2; by permission of the Wildlife Society.)

The Niche Perspective

As developed above and in chapter 1, habitat is a valuable concept that can be used to develop general descriptors of the distribution of animals and lend insight into factors driving survival and fitness. However, we often fail to find commonalities in "habitat" for populations across space and time because we often miss underlying mechanisms that determine occupancy, survival, and fecundity. Habitat per se can only provide part of the explanation of the distribution of an animal. Other factors, including some often related to an animal's niche, must be studied to more fully understand the mechanisms responsible for animal survival and fitness. As we have seen in many wildlife studies (e.g., Collins 1983; Johnson and Igl 2001), what we term *habitat quality* varies in physical attributes for a species across its range because the studies did not measure mechanisms. Rather, our statistical models of habitat are often analyzing surrogates of these mechanisms. Tyre et al. (2001) showed that, even if all important habitat variables could be measured perfectly, and there was no error in surveying the species of interest, demographic stochasticity and the limiting effect of localized dispersal generally prevent an explanation of much more than half of the variation in territory occupancy as a function of habitat quality. They also concluded that habitat models really measure the ability of the species to reach and colonize an area, not birth and death rates.

In chapter 1 we outlined the historic development of the niche concept. Here we will provide additional details on how the niche can be used to advance our understanding of wildlife ecology and management, and interrelate niche and habitat. For example, James et al. (2001) proposed a new model for habitat management based on the concept of the *niche gestalt*. This concept is based on early ideas about the perceptual world of animals, the *Umwelt*, and their species-specific requirements. Of interest in this approach is that subset of the structure of a species' environment that is relevant to its reproduction and survival. In chapters 5 and beyond we discuss how elements of the niche can be measured in wildlife studies. Some of the following material has been synthesized from Morrison (2001) and Morrison and Hall (2002).

As outlined in chapter 1, Joseph Grinnell's (1917) objective was to understand population regulation in terms of resources that limit the distribution and abundance of a single species. Later, MacArthur (1968), Arthur (1987), and others recommended that we describe the choice of resources by animals, and analyze how these choices can be *constrained* by predators, competitors, disease, parasites, disturbance, and other factors.

Huston (2002) concluded that efforts to understand and predict variation in the abundance and distribution of species have been hampered by inadequate theories, mismatches between processes and sampling scales, and inappropriate statistical methods. He argued for a stronger theoretical framework that addresses multiple interacting processes and limiting factors, which he thought would help resolve the patterns underlying the complexity of nature and also contribute to identification of appropriate sampling scales and the development of statistical methods more appropriate for quantifying causal relationships than those that have been traditionally used. Lack of such a framework has been a major impediment to sound resource management and conservation planning.

Likewise, O'Connor (2002) concluded that the conceptual basis of ecology shows that limitations of current concepts are marked and aggravated by generally poor implementation of the available approaches despite a long tradition of informed commentary by the more percipient practitioners. He thought that a central issue is that the scale (extent and grain) at which analysis is conducted is often wrong for the scale of the process of interest, for sparsely distributed sample points may characterize gross distribution but not cast light on local processes. O'Connor (2002) concluded, then, that the key focus should be on factors as limiting agents in species abundance. His argument had two key assumptions. First, that most species can be limited by any of a variety of factors. Second, the influence of any ecologically relevant factor is not additive to the influence of any other factor. Rather, typically only one factor is limiting in any particular situation. Thus he concluded that the widely used approach of correlative analyses is particularly misleading under these assumptions.

The individual organism with its behavior and physiology shaped by natural selection is a fundamental unit of ecological processes. As summarized by Huston (2002), there has been increasing recognition that the individual organism can serve as a fundamental unit that allows integration of processes across dimensions ranging from molecular and genetic to ecosystems and landscapes. Observations and measurements of individual organisms are the basic units of data, which may be aggregated statistically to higher organizational levels, such as populations, communities, landscapes, and biomes. The critical requirements for the success of this individual-based approach are a thorough, quantitative understanding of how individuals of particular species may respond to the range of environmental conditions they may experience, and a quantitative description of the relevant environmental conditions in the specific areas over which we desire to understand and predict ecological patterns at spatial and temporal resolutions appropriate for the organism of interest (Huston 2002).

Huston (2002) developed a strong rationale for how we conceptualize, and thus address, questions concerning the factors determining the distribution and abundance of animals. His approach centers on the fact that the generality of the law of the minimum has held up for over a century and a half and thus provides a conceptual framework that is useful for a wide range of ecological processes. Most ecological processes are, however, influenced by multiple factors, only some of which an organism can control. Although any resource or regulator can potentially occur at levels that limit the rate of a process, it is unlikely that the same single factor will always be limiting because spatial and temporal variation in resources and regulators results in a continual shifting of resources. Such variation results in a continual shifting among potentially key limiting factors. This shifting between limiting factors complicates our understanding and lessens our predictive power. When processes are regulated according to the law of minimum, varia-

tion in a nonlimiting factor has little or no effect on the rate of the process of interest. Process rates will be correlated only with variation in the limiting resource, and only under these conditions will regression analysis reflect the potential effect of the limiting resource on the process. Thus maximum potential response to any specific resource or regulator of a process can only be quantified when all other resources or regulators occur at nonlimiting factors (Huston 2002). Most ecological research, however, is conducted using a mixture of measurements made under both limiting and nonlimiting conditions and thus produce datasets with high variance (and low correlation coefficients) that prevent identification of the actual response of a process to a specific factor. Thus, as noted by Huston (2002), the high variance that is often found in correlations between ecological processes and presumed causal factors may not be sampling error of random "noise," but rather mechanistic consequences of shifts between limiting resources or the effects of other limiting factors such as mortality or dispersal.

These phenomena emphasize the importance of distinguishing between the *physiological (fundamental or potential) niche* and the *ecological (realized or actual) niche* of organisms. This concept makes the distinction between the physiological limitations to species distributions and the biological limitations, and it demonstrates the complex distribution patterns that can result from these interactions (Huston 2002). The observed abundance distribution of a species represents its ecological niche, which usually will not include the full range of conditions under which the species could potentially be found and which, in particular, may not include the physical conditions in which the species actually does best. The observed distribution of a species may differ from its potential distribution (physiological niche) if the species is excluded from the conditions in which it grows best by competition or other negative effects of other species. Depending on the pattern of overlap of the physiological responses of potential competitors, the realized niche of a species that is a poor competitor under optimal environmental conditions may be displaced into a skewed or even a bimodal distribution along the environmental condition.

Heglund (2002) outlined that the foundation of our current modeling efforts lies in the characterization of a species' realized niche rather than simply determining habitat relations. She developed how ecologists search primarily for patterns in resource use for a species, and then attempt to identify specific elements of the environment associated with the occurrence of that species. Although, conceptually, most species should exhibit a unimodal distribution approximating a Gaussian response curve (with a maximum response at some point along an environmental gradient), a number of factors (e.g., competition, predation, disease) place pressure on an organism, and the curve narrows to its realized niche. The realized niche and its optimum may differ from the fundamental niche both in location along a gradient and in the shape of the response. The realized niche can take on a variety of shapes, and failure to recognize this may result in inefficient or incorrect predictive models (Heglund 2002). There are very few cases in which symmetrical bell-shaped species distributions are found in nature.

Although the complex patterns of species distributions in relation to realized ecological conditions make the quantitative prediction of species occurrence difficult, a focus on the fundamental response of species will lead to solutions to this difficulty. Understanding the physiological optima of a species provides critical information about the environmental conditions where it could potentially occur. This information is critical for restoration, conservation, and management of species because these

are the conditions in which, with appropriate management, the species actually does best. Management, such as control of predators or restoration of a different disturbance regime may allow some species to thrive in areas where they are rarely found under present conditions (Huston 2002).

Thus habitat and niche are both potentially valuable but overlapping concepts. A major problem with focusing on habitat (at least as we usually measure it) is that features measured can stay the same while use of important resources (what could be considered niche parameters; see chapter 12) by an animal *within* that habitat can change—for example, changes in the species or size of prey taken by a bird foraging on shrubs. The abundance of many arthropods is closely related to the flowering phenology of plants, which causes shifts in animal foraging behavior (e.g., Hejl and Verner 1990; Keane and Morrison 1999). Differences in behavior between sexes and intersexual competition for food also cause shifts in the portion of a niche axis being used (e.g., Petit et al. 1990; Kelly and Woods 1996; Brennan et al. 2000). Crude differences we identify in habitat studies (e.g., shifts in use of areas under study) are often caused fundamentally by changes in use of specific resources that we fail to see. Habitat, then, if described, for example, only as structural or floristic aspects of vegetation, often fails as a predictor of animal performance because we do not identify constraints on exploitation of critical resources and consideration of critical limiting factors.

Van Horne (2002) recommended that we test the importance of processes thought to drive observed habitat relationships and test for boundary conditions under which the processes may become relatively unimportant (see fig. 3.7). Boundary conditions describe the range of a given variable within which the model is applicable. Van Horne noted that animal ecologists do not usually test boundary conditions

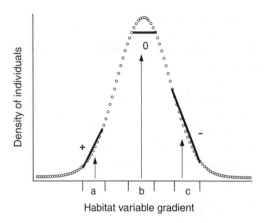

Figure 3.7. Limited sampling along a gradient in a habitat variable may produce a positive (*a*), nonexistent (*b*), or negative (*c*) correlation with a species response variable such as density. (From Van Horne 2002, fig. 24.2; Copyright 2002 Island Press.)

because most work is done within or among discrete vegetation associations, or within some management or other association of artificial "study" area. Although ecologists might arrange sampling locations (e.g., plots, experimental blocks) along an environmental gradient, such sampling is usually too discontinuous to identify boundary conditions. Looking for boundary conditions is most useful in identifying limiting factors.

In her summary of the status of habitat modeling, Van Horne (2002) concluded that habitat models that are very specific will not be useful for making predictions of management effects across time and space unless enormous resources are devoted to their development. They may, however, elucidate processes that can be incorporated into more generalized models that do have predictive power. This process of generalizing requires that boundary conditions for a given habitat model be defined. Models that achieve generality by collecting data at a very broad scale may be useful in the definition of

problems across vegetation associations, in initial inventories, and in identification of large tracts of land for preservation. Because they are so distantly related to the processes that drive population change, however, they will not be useful in identifying the effects of processes-oriented management policies regarding timber harvest, stream flow, grazing, hunting pressure, and the like (Van Horne 2002).

Thus we contend that studies of wildlife–habitat relationships also should encompass factors (not traditionally considered habitat resources) that also influence whether an animal occupies a given site (e.g., presence of predators or competitors). These factors could be considered part of an animal's niche (see chapter 12), and understanding how the niche is expressed across space and time will almost certainly aid in understanding animal–habitat relationships.

The Population Perspective

Why Study Populations?

To conserve wildlife we must ultimately provide for the survival and protection of individual organisms, the populations to which they belong, the communities to which populations appertain, the ecosystems in which communities occur, and the long-term evolutionary potential of species lineages (Soulé 1980). To this end, management of habitat can provide for conditions in which organisms can maximize their *realized fitness*. Realized fitness is measured as the number of viable offspring produced that in turn find mates and suitable environments and successfully reproduce. Fitness is influenced by the dynamics of interactions of individuals within a population, by interactions among populations and species, and by interactions between organisms and their habitats and environments. To study and successfully manage a species' habitat thus requires knowledge of population dynamics and behaviors.

A related reason for studying populations is to better identify the factors in the environment that regulate population trends. This study extends beyond simply recognizing correlates of population trends; we need to know the true environmental causes. Once such linkages are made, we may be able to specify the most efficient means for conservation, or at least predict possible outcomes of our activities.

Populations or Habitats?

If the goal is to provide for conservation of species and populations, then why study population dynamics and trends if general habitat conditions can be provided? The answer is that, whereas habitat is essential to survival of all species, by itself it does not guarantee long-term fitness of individuals and viability of populations.

An example helps to explain. In the Intermountain West of the United States, macrohabitat conditions, measured as vegetation cover associations and structural stages, for the Townsend's big-eared bat (*Corynorhinus townsendii*) is estimated to have increased since the early 1800s by about 3 percent (Marcot 1996). However, populations of this bat likely have *declined* over this period. The reason is that, while the species uses a wide range of macrohabitats, substrates, and roosts, it is particularly vulnerable to human activity. Disturbing females with young adversely affects breeding success, and disturbing winter hibernacula can increase winter mortality (Nagorsen and Brigham 1993). Thus, in this case, the trend in macrohabitats belies the trend in populations, even though providing such habitat is essential to species conservation and restoration. Similarly, the American bison (*Bison bison*) suffered great declines, nearly to extinction, during the 19th century because of overhunting, even though for a long time its

habitat remained suitable and available. Overharvesting or disturbance can cause populations to follow trajectories quite different from those of their habitat. Thus it is often desirable to have knowledge of true population distribution, abundance, and trend, as well as knowledge of habitat; both together tell complementary stories.

Population Concepts Tied to Habitat Analysis and Management

WHAT IS AN OBLIGATE?

Some species can be identified that are *obligate users* of specific environmental features—that is, species that highly specialize in and select for a particular resource, land cover, or vegetation condition. Many assessments of habitat fragmentation, for example, have assumed that wildlife species can be identified as obligate users of forest interiors, forest edges, old-growth forests, or other vegetation or landscape conditions. While in many practical ways this is true for some species, and simple assumptions are most parsimonious, the question is more complicated than may first appear and bears closer inspection.

The response pattern of individual organisms to environmental continua such as temperature or vegetation structure along a sere is typically a bell-shaped curve (fig. 3.8). Assuming that in figure 3.8, "biological response" is something like realized fitness (or some index thereto), the high point of the curve marks the best environmental condition (point e in fig. 3.8) for maximum fitness (BRmax in fig. 3.8), and the two trailing edges of the curve suggest increasingly unsuitable conditions. The maximum potential range of the environmental gradient in the ecosystem can also be depicted (points α and ω in fig. 3.8), even if not used by the organism. Maximum ranges of tolerance by the organism are marked by truncation of the curve at each

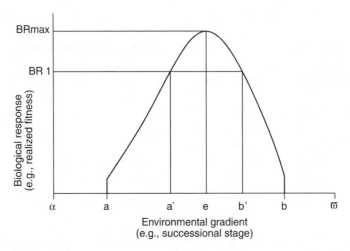

Figure 3.8.   Generalized depiction of a species' biological response to an environmental gradient. See text for description of points.

side (points a and b, but there may be circumstances where a = α and b = ω). In some cases, the truncation may be more abrupt or the curve may be asymmetric, but the basic pattern still holds.

Traditional ecological principles that have used this fundamental precept include Shelford's law of tolerance and Leibig's law of the minimum, as well as measures of Hutchinsonian niche breadth, overlap, and community structure. This curve probably differs in specific shape among different species in the same community, and the same species in different ecosystems. Moreover, the curve presented here represents only a single organism; biological response of all members of a population would be represented by tolerance intervals spanning each side of the curve.

So what is an obligate? The most common view of an obligate species is one defined by narrow (stenotopic) overall use of some resource. That is, the species selects environments only with a specific narrow range (a, b). In one sense, every species is an obligate of some range of con-

ditions along some environmental gradient(s). The term, though, needs quantitative rigor to be applied consistently. Most intended uses of the term refer to a range of tolerance far smaller than the available range of the environmental gradient. "Far smaller" can be defined as a specific proportion of the overall range, such as less than 10% (that is, $[a - b]/[\alpha - \omega]$ <0.10; or, with discretely defined environmental conditions, uses <10% of available or potential conditions). In figure 3.9. Species A has a broad tolerance for an environmental gradient (is *eurytopic*)—in this case, let us say, successional forest stages—whereas Species B has a narrower tolerance (is *stenotopic*) for only one stage. In this example, Species B is an *obligate user* of late successional stages, whereas Species A uses several stages and is thus only a *facultative user* of the late stages. Also, Species B is shown with a taller curve under the precept that at least some habitat specialists might have a greater realized fitness (density being an index to fitness, but this could be misleading) than do generalists in the same environment, although this is not necessarily the rule.

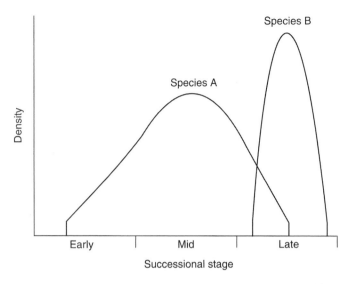

Figure 3.9.   Generalized depiction of two species' biological responses to successional forest stages. Species A is a generalist; Species B is a specialist.

Species B could be naturally rare and always have densities lower than Species A.

The problem is that not many species may demonstrate this simple pattern of absolute adherence to a single environmental condition, such as successional stages, for the range of habitat conditions typically considered in ecological studies. That is, many stenotopic species will also make at least facultative or opportunistic use of other conditions. An example is a "forest interior species" or "old-growth forest species" also using habitat edges or earlier successional stages if such edges or stages also contain some specific elements of older forests (although DeGraaf and Chadwick [1987] strictly defined birds in northeastern U.S. forests as stand condition obligates and forest type obligates using only one condition or type). An example was demonstrated by McConnell (1999), who reported on use of residual trees in seedtree–shelterwood forest stands by red-cockaded woodpeckers (*Picoides borealis*) in Florida. Such habitat conditions may

be less than optimal (biological response <BRmax) but can still provide some degree of habitat connectivity of species across landscapes between nodes of more optimal habitats. In real-world management, such "exceptions" may be the more useful and interesting cases for maintaining demes and metapopulations.

For instance, in a study of forest birds, one of us (Marcot 1985) found that mature forest species such as the brown creeper (*Certhia americana*) and the hermit warbler (*Dendroica occidentalis*)—sometimes thought of as mature-forest obligates—also occurred, but in lower numbers and greater variability, in midsuccessional forest stages where a few (≥7/ha) scattered old Douglas-fir (*Pseudotsuga menzeisii*) trees or snags remained in the overstory (fig. 3.10). In the strict interpretation, neither of these bird species are absolute obligate uses of mature forest, although their greater mean density and lower variability in mature forest strongly suggest such optimal conditions and that the species

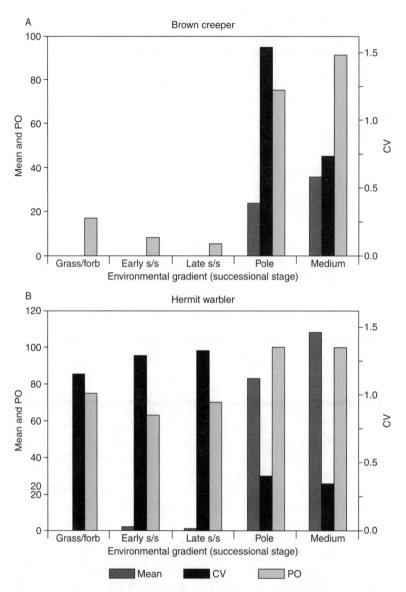

Figure 3.10.   Mean density (*n*/40.5 ha index), coefficient of variation (CV) of density among replicate study plots, and percentage of occurrence (PO, percentage of replicate study plots occupied) of (A) brown creepers (*Certhia americana*) and (B) hermit warblers (*Dendroica occidentalis*) among five successional stages of Douglas-fir (*Pseudotsuga menziesii*) forest in northwestern California. Stages are: grass/forb, early shrub/sapling, late shrub/sapling, pole, and medium tree. (From Marcot 1985.)

should not be called forest habitat generalists. Note too that the pattern of mean density in each of these species roughly matches the general pattern of Species B in figure 3.9.

This example also demonstrates the interplay between three complementary metrics of biological response illustrated in the data shown in figure 3.10: (1) Mean density reaches its greatest value in the best environment (mature forest stage), although it could be nonzero in other, suboptimal stages. (2) The coefficient of variation (CV) of density among replicate study plots is lowest in the best environment. (3) Percent occurrence (PO) of each species among replicate study plots is greatest in the best environment. The data on brown creeper (fig. 3.10A) show these traits well. The data on hermit warbler (fig. 3.10B) also show these patterns, although, if one inspected just the PO patterns alone, one would conclude that pole and medium tree stages were equally suitable, whereas the data on mean density and CV suggest an incrementally better environment in the medium tree stage (greater mean density and lower variation in density than in the pole tree stage).

Presence of these two species in early successional stages of this study, although in lower numbers, may represent some occasional or intermittent use for dispersal, nesting, or foraging, but only more intensive autecological studies on reproductive response would reveal the specific kind of use and influence on fitness and metapopulation structure in the study area. Nonetheless, such early successional stages are used (especially if they contain old-forest components), which could contribute to at least some degree of habitat connectivity for these "mature forest" species (and others like them) across a landscape. Investigations of density, variation in density, and percent occupancy in early successional stages with and without old-forest components can help reveal which elements are important and can help in developing silvicultural guide-

lines for young-forest management (Marcot 1985).

This example also serves to illustrate that degree of specificity of use defines the level of obligate relationship. A more fluid definition of obligate use might allow for some threshold biological response to be identified (point BR 1 in fig. 3.8), for which a specific range of an environmental gradient can be correlated (points a′ and b′). In this case, the threshold could be identified as some percentage of maximum possible response (e.g., BR 1 = 80% of BRmax), and "far smaller" (as mentioned above) now would be defined based on this response range (that is, $[a' - b']/[\alpha - \omega] < 0.10$, or uses <10% of available or potential discrete conditions). In this way, species can be identified that have varying degrees of obligate relation to specific environmental conditions.

A simple example is shown in figure 3.11. Here is plotted the cumulative number of terrestrial vertebrate species that use old-forest structural stages of the interior Columbia River Basin, (based on Marcot et al. 1997) as a function of the percentage (up to 50%) of all other vegetation structural stages used. The figure shows that no species occurs solely in old-forest structural stages. Six species occur in old-forest structural stages, but each also uses about 20% of all other stages as well (that is, two other stages). An additional 17 species use old-forest stages, and about 30% of all other stages too. Thus there is no species that specializes only on old-forest stages, and an increasing number of species generalize in other vegetation stages. The cutoff point for "obligate" old-forest species used by Marcot et al. (1997) was the 20% level, but this was a subjectively derived threshold for comparing degree of specialization of species among other vegetation structural stages and vegetation types.

In summary, use of environmental conditions—be they vegetation types, structural or successional stages, substrates, or other condi-

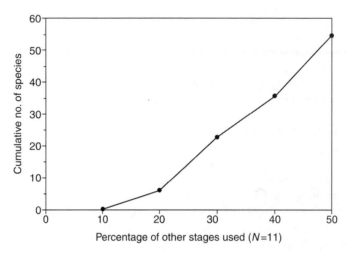

Figure 3.11.   Cumulative number of terrestrial vertebrate species (amphibians, reptiles, birds, and mammals) that use old-forest structural stages as a function of the percentage of all other vegetation structural stages used, in the interior Columbia River basin. For example, 36 species use old-forest structural stages and about 40 percent of all other structural stages (i.e., four other stages). (Based on data in Marcot et al. 1997.)

tions—can be viewed as a gradient of specialization to generalization, and thus as probabilities of organisms using specific spaces and habitats (Marzluff et al. 2004). In identifying obligate users of some condition, the range of tolerance of individual organisms, the mean and variation in use among organisms, and the use of complementary indices such as mean density, coefficient of variation of density, and percent occurrence can all provide useful information (Ruggiero et al. 1988). Recognizing at least facultative use of some habitat conditions can be quite helpful in devising management guidelines for suboptimal "connectivity habitat" to better link optimal habitats or environmental conditions across a landscape.

OF POPULATIONS, DEMES, AND DISTANCES

We introduced the population concept in chapter 1. Briefly, the traditional definition of a *population* is "a group of organisms of common ancestry that are able to reproduce only among themselves and that are usually geographically distinct" (Morris 1992). However, few wild collections of organisms completely interbreed— that is, are completely *panmictic* (i.e., individuals all have equal probability of sharing genes). A local collection of organisms that have a high likelihood of sharing genetic material through interbreeding is called a *deme*. In wild populations, demes can be isolated or can partially interact, and the simple definition of population given here is seldom entirely true. In some literature (e.g., Kindlmann et al. 2005), the term *subpopulation* is used to refer variously to a deme or to a portion of a population in a specific geographic location, or as delineated by nonbiological criteria (for example, administrative or political boundaries).

Several characteristics of habitats and environments can prevent complete panmixia (interbreeding) in wild populations. Factors limiting panmixia include (1) dispersal filters and barriers, which limit distances and locations traveled by dispersing individuals, and (2) patchiness of resources through space. Examples of dispersal filters and barriers that tend to isolate demes and populations are the various mountain ranges ringing the Central Valley of California, which probably inhibit dispersal of kit foxes (*Vulpes macrotis*); and, on an even larger geographic scope, major oceans separating populations of great gray owls (*Strix nebulosa*) between Nearctic (North America) and Palearctic (Eurasia) zoogeographic regions. One example of how patchiness of resources through space could separate demes is the physical separation of aquatic environments that could sever demes of river otters (*Lutra canadensis*) into different river systems and watersheds in North America. Another example is the occurrence of small isolated marshes used as breeding habitats by local colonies and demes of tricolored blackbirds (*Agelaius tricolor*), such as occurs in portions of the Klamath Basin in northern California and southern Oregon.

Behavioral characteristics of individuals and social structures of breeding groups can also act to differentially select for specific breeding individuals and thereby to dissuade complete panmixia. Examples are mate defense in pronghorns (*Antilocapra americana*) and territoriality of iguanid lizards such as the western fence lizard (*Sceloperus occidentalis*). Other factors, including resource competition or predation by other species, natural perturbations of environmental conditions, wide geographic distances as compared to vagility of organisms, and genetic structure of the population, can act to limit the geographic area over which organisms interact and can serve to partially or fully isolate populations genetically (see box 3.2).

Partial panmixia and degrees of isolation among populations can result in *metapopulation*

Box 3.2   How distance can act to genetically isolate organisms and populations

Patterns of spatial distribution of organisms can take many forms (box fig. 3.2). First, consider the simplest case of one contiguous population, as in box figure 3.2, panel C. Even in a contiguously homogeneous habitat, complete panmixia (interbreeding) of individuals within a population can be hindered through isolation-by-distance deme structures of the populations, such as has been found with marmots (*Marmota flaviventris*) across mountaintops in the Great Basin (Floyd et al. 2005). The area over which a deme is effectively panmictic (i.e., individuals all have equal probability of sharing genes) has been called a *neighborhood,* which is the space over which genotypes of offspring are not significantly different from those resulting from randomly selected parents in that area (Chambers 1995). In the case of linearly arrayed habitats, such as shorelines or rivers, a neighborhood, N_L can be defined as

$$N_L = 2\sqrt{\pi\rho\sigma}$$

where ρ = density of breeding individuals (in the case of linear habitat, number of breeding individuals per unit habitat length) and σ = standard deviation of individual dispersal distances, also called root-mean-square dispersal distance. In the case of a two-dimensional habitat area, a neighborhood, N_A can be defined as

$$N_A = 4\pi\rho\sigma^2$$

In this case, ρ = number of breeding individuals per unit habitat area, and σ is defined as above (Lande and Barrowclough 1987).

(box continued on following page)

Box 3.2   Continued

In linear habitats, populations can be considered entirely panmictic if $\rho\sigma^2 > L/10$, where L = the total linear range of the population; in areal habitats, populations are panmictic if $\rho\sigma^2 > 1$ (Maruyama 1977;1 Lande and Barrowclough 1987). Or, panmictic populations must be less than $10\rho\sigma^2$ in linear habitats and $\rho\sigma^2$ in two-dimensional habitats.

Thus a linear neighborhood, N_L, of an entirely panmictic deme of tailed frogs (*Ascaphus truei*) in habitats of montane streams, with $\rho = 1$ adult per m of stream (from Nussbaum et al. 1983, 150, in eastern Washington state) and $\sigma = 50$ m (hypothesized value; no data available), is 177.2 m of stream length. Panmictic populations must be less than $10\rho\sigma^2$, or 25 km of stream length; streams greater than this length necessarily contain more than one neighborhood. (The problem of greater differential dispersal downstream than upstream is not considered here and is left to the student as an exercise. Under this model, how would the size of a panmictic population vary with different densities? With different estimates of root-mean-square dispersal distance?)

Compare the preceding relationship of a panmictic population with a rare species in a linear habitat. In New England northern hardwood forests, a panmictic population, N_L, of the substantially more rare dusky salamander (*Desmognathus fuscus*), also a stream habitat obligate, with $\rho = 0.02$ adults per m of stream (2 per 100 m of stream length; DeGraaf and Rudis 1990, 160) and $\sigma = 50$ m (hypothesized value; no data available), is found in only 3.5 m of stream length. This density is substantially less than the mean density of the population. No panmixia can occur unless σ exceeds 705 m, which is probably unlikely. Thus dusky salamanders in this location may not be able to achieve panmixia, and populations likely would show intergrades of genomes along stream habitat corridors.

Levine (2003) also modeled the ecological consequences of directional dispersal in linear habitats, such as a downstream environment. His simulations showed that population size and species diversity correlated with dispersal distance, increasing downstream, but only when reproduction and mortality rates were such that local neighborhood demes depended on dispersing individuals to maintain them, and only with successful long-distance dispersal. The implications are that, if species are poor dispersers in such environments, then local demes may be at risk, species would not be able to coexist in the presence of intense competition, and species diversity would remain low. Levine and others (e.g., Nathan and Volis 2005) concluded that successful directional dispersal may be a key determinant of species coexistence and diverse communities in severely disperser-limited systems.

Similar examples of the above calculations can also be developed for populations in two-dimensional habitats (also see box 3.4 on N_e). This exercise is left to the student.

Other, more complex models of distance separation of breeding organisms deal with how local concentrations of organisms interact in metapopulations. These are discussed later in this chapter.

For discussions of other isolating mechanisms in animals, see Futuyma (1986).

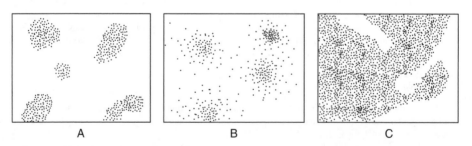

Box Figure 3.2. Some patterns of spatial distribution of organisms. Each dot represents an individual. *A*, Discrete populations, corresponding to the "island model," if mating is more frequent within than between the populations. *B*, Perhaps the most common pattern in nature, ill-defined populations between which density is low. *C*, More or less uniform distribution, corresponding to the simplest "isolation by distance" models, except that regions of unfavorable habitat are acknowledged. (Reproduced from Futuyma 1986, 120, by permission of Sinauer Associates, Inc.)

structures. A metapopulation, or "population of populations," occurs when a species' range "is composed of more or less geographically isolated patches, interconnected through patterns of gene flow, extinction, and recolonization" (Lande and Barrowclough 1987, 106). Both metapopulations and their component populations (or subpopulations) are important foci for conservation (Baguette and Schtickzelle 2003; Baguette and Stevens 2003; for definitions and examples of other key terms related to populations, species, and systems and their relevance for conservation, see table 3.3). Metapopulation structures occur when environmental conditions and species characteristics provide for less than a complete interchange of reproductive individuals and there is greater demographic and reproductive interaction among individuals within than among subpopulations. It is the condition likely most often found in wild animal populations. We draw some conclusions here and in later chapters on managing habitats for viable metapopulations.

POPULATION DYNAMICS AND VIABILITY

The dynamics of wildlife populations are mediated through their relations with their environments. The older German concept of *Umwelt* nicely described the set of biotic and abiotic factors that ultimately influence realized fitness of organisms (e.g., Klopfer 1959) and viability of populations. Today, we would list exogenous and endogenous factors that can influence population viability (table 3.4), including population demography, population genetics, metapopulation dynamics, environmental stochasticity, species biogeography, evolutionary adaptations and selection mechanisms, reproductive ecology and behavior, effects of other species, and effects of human activities.

Population viability is the likelihood of persistence of well-distributed populations to a specified future time period, typically a century or longer. The term *well-distributed population* refers to the need to ensure that individuals can freely interact where "natural" conditions exist. However, all species are limited to some degree in distribution; some species are inherently uncommon, and some are even locally endemic. Thus "well distributed" needs to be interpreted in the context of the ecological distribution of the species in question.

Population viability analysis pertains to estimating likelihoods of persistence or, conversely, extinction. We identify four kinds of extinction (see table 3.5): (1) *local extirpation* of a species in part of its range, particularly from human activity; (2) *global extinction* of a species throughout its entire geographic range; (3) *quasi extinction* of a population or species when it drops below some predetermined or desired abundance; and (4) *phyletic extinction* of a species' lineage over evolutionary time as it evolves into and becomes succeeded by a different species form. Interestingly, in medical surgery jargon, extirpation refers to the excision of an entire organ or tissue, an apt metaphor for the loss of a component of an ecological community. Much of population viability conservation deals with striving to avoid local extirpation and global extinction and uses quasi extinction as one measure (e.g., Otway et al. 2004). Arguments that 90[+]% of all species that have ever lived have gone extinct—thus supporting risky management—fallaciously confuse phyletic (evolutionary) extinction with global (ecological) extinction. Local extirpation is really the most insidious pattern of species loss (Michalski and Peres 2005) and is seldom monitored and quantified in any systematic way (but see Barbraud et al. 2003).

The time span over which population viability is assessed should be scaled according to the species' life history, body size, and longevity, and especially population generation time. We suggest that a rule of thumb might be to use at least 10 generations for gauging lag effects of

Table 3.3. Definitions and examples of key terms related to populations, species, and systems, and their importance for conservation

Term	Definition	Example	Importance for conservation
Population	A (local) collection of interbreeding organisms.	Salamanders in a pond (Wissinger and Whiteman 1992)	A fundamental component of a viable species.
Metapopulation	A set of populations of a species linked by occasional to frequent interbreeding.	Cougars (*Puma concolor*) among the San Andres Mountains, New Mexico (Sweanor et al. 2000)	Links breeding among populations; ensures continued recolonization of habitats.
Subpopulation	One of the population components of a metapopulation; breeding within a subpopulation is more frequent than breeding among subpopulations; focus is on demography.	Local colonies of red-cockaded woodpeckers (*Picoides borealis*) in Florida (McConnell 1999)	Component of a metapopulation; may also be the smallest viable breeding unit, such as a colony.
Deme	A local population or neighborhood of a species viewed as a mostly randomly interbreeding group (Morris 1992); focus is on gene flow.	Local neighborhoods of Galapagos cormorant (*Compsohalieus [Nannopterum] harrisi*) colonies in the Galápagos Islands (Valle 1995).	Describes unit of effectively interbreeding individuals for ensuring genetic diversity within populations.
Race	An interbreeding subgroup of a species whose individuals are geographically, physiologically, or chromosomally distinct from other members of the species (Morris 1992).	Forms of the *Amphibolurus decresii* species complex (agamid lizards) that select for different rock color substrates in Australia (Gibbons and Lillywhite 1981).	Ensures full range of genetic diversity within a species; provides for local adaptation, speciation, and evolutionary potential.
Variety	A botanical taxonomic rank below subspecies, denoting an organism that can interbreed with the subspecies or species but which is distinguished by some minor morphological character; varieties are usually developed in cultivated or horticultural species (Morris 1992).	Often used in reference to plant cultivars but can be more or less synonymous with subspecies in wild or native species.	Maintains genetic variation within the species.
Ecotype	Individuals of a species that are adapted to local climatic, resource, or other conditions and that would be less fit in other parts of the same species' range; often refers to plants.	Woodland caribou (*Rangifer tarandus caribou*) in Idaho and British Columbia (Warren et al. 1996)	Ensures greatest fitness and identifies environmental conditions best suited for local populations.
Morph	Any of two or more phenotypic varieties in a polymorphic population (Morris 1992).	Stewart Island shags (*Phalacrocorax chalconotus*); occur as a dark "bronze" morph and a black and white "pied" morph, often in the same colony.	Provides for genetic diversity and locally adapted forms with greater fitness within a species or superspecies complex.

Table 3.3. Continued

Term	Definition	Example	Importance for conservation
Subspecies	A taxonomic rank immediately below species, indicating a group of organisms that is geographically isolated from, and may display some morphological differences from, other populations of a species, but which is nevertheless able to interbreed with other such groups within the species where their ranges overlap (Morris 1992).	Two subspecies of orchard oriole that are long-distance (*Icterus spurius spurius*) and short-distance (*I. s. fuertesi*) migrants in eastern North America (Baker et al. 2003).	Provides for geographic diversity and evolutionary (speciation) potential.
Species	A collection of potentially interbreeding organisms.	Many examples in taxonomies: e.g., Jones et al. (1992) for mammals; Collins and Taggart (2002) for amphibians and reptiles; American Ornithologists' Union (2003 with supplements) for birds.	Usually the basic unit of conservation of organisms; fundamental components of most of the higher levels of biological organization (e.g., communities).
Assemblage	A set of species, often of the same taxonomic class ("taxocene"), that co-occur.	The set of montane amphibian species in Peñalara Natural Park in Spain (Martinez-Solano et al. 2003).	Contributes to the full diversity of species in an area.
Guild	Traditionally, (bird) species sharing a common foraging mode and substrate (Root 1967); more broadly, any set of species with common resource use patterns with implications of intraguild, interspecific resource competition (Blondel 2003).	Guilds of antbirds associated with forest edges, interiors, and gaps in Nicaragua (Cody 2000),	Provides for the array of species along a resource competition spectrum as part of the full set of evolutionary potential of a community.
Functional group	A set of co-occurring species having the same ecological roles in their ecosystem; generally does not necessarily connote interspecific competition (Blondel 2003).	Sets of mammal species with common ecological roles, in conifer forests of the western United States (Marcot and Aubry 2003)	Provides the set of ecological functions and their redundancy in a system; contributes to resilience of ecosystems to perturbations.
Community	The set of all species in a given area, and their ecological interactions.	Classification of grassland and range plant communities (Ratliff and Pieper 1982).	A basic unit of conservation focus (e.g., rare or native communities).
Ecosystem	The set of all abiotic conditions, and biotic entities and their ecological interactions, in a given area.	Integrated classification of aquatic and terrestrial ecosystems in the Pacific Northwest (Rieman et al. 2000)	A fundamental unit of conservation focus; provides context for other conservation elements.

Table 3.4. Some of the main ecological factors that affect the long-term viability of populations[a]

Population demography
   Population trend, rate of change
   Likelihoods of extinction and quasi extinction
   Vital rates (natality, mortality)
   Sensitivity of population rate of change (lambda) to age- or stage-class-specific vital rates

Population genetics
   Degree of historical and current loss of heterozygosity in genome
   Causes for any loss of heterozygosity (inbreeding, genetic drift, etc.)
   Potential for founder effect to simplify genome in future recolonization and rescue effect dynamics

Metapopulation dynamics
   Number of linked subpopulations
   Number of isolated populations
   Dynamics of future size of populations and numbers of populations
   Likelihood of loss (extinction) or near loss (quasi extinction) of populations
   Locations and conditions of existing and historical key links between subpopulations

Environmental stochasticity
   Types of disturbance events pertinent to persistence of a specific population and species
   Dynamics of a disturbance event: type, frequency, intensity, location, duration, and effect
   Effects of the disturbance event on population isolation, population size (direct mortality), ecology (as mediated through effects on other species such as competitors or predators), and population trend and vital rates
   Descriptions of how stochastic processes have changed, and consequently how their effects have changed throughout the years (e.g., effects on metapopulation dynamics from changes in fire regimes and subsequent changes in forest canopy structure and density; increases in fire frequency in sagebrush communities in Snake River plains and subsequent cheatgrass spread and competition with native grasses)

Biogeography
   Species source pools
   Zones of spread and dispersal
   Dispersal rates

Evolutionary adaptations and selection mechanisms
   Allopatric speciation
   Isolating mechanisms
   Species swamping and hybridization

Reproductive ecology and behavior
   Hybridization
   Dissemination dynamics
   Angiosperm pollination vectors and their status

Effects of other species
   Predation
   Competition
   Parasitism, pathogens, and disease

Human causes of population declines
   Prehistoric (paleoecological), historic, and recent exploitation of populations
   Invasion of exotic species into human-disturbed environments
      burning
      cutting
      grazing
      agricultural conversion
      urban conversion

[a]Such factors should be addressed in a "hard-core" quantitative population viability analysis.

72

Table 3.5.   Types of extinction and their implications for conservation

Term	Definition	Examples	Implications for conservation
Local extirpation	Local loss of members of a species, usually with continued occurrence elsewhere	Local loss and reintroduction of Asiatic wild ass (*Equus hemionus*) in Israel (Saltz and Rubenstein 1995); local loss of wintering birds in North America (Koenig 2001).	Creates gaps in distributions of organisms; isolates demes and subpopulations; reduces local biodiversity (Breininger et al. 1998).
Global extinction	Loss of all members of a species anywhere on Earth	Loss of endemic island avifauna due to predatory snakes (Savidge 1987).	Removes entire species, along with their specific ecological roles, from the ecosystem, thus reducing biodiversity, ecosystem function, and evolutionary potential.
Quasi extinction	Nonzero population levels that drop below a specified or desired number	Levels used to estimate extinction probabilities in population viability modeling (Ginzburg et al. 1982).	Can be used as an early warning threshold for extirpation or extinction.
Phyletic extinction (pseudoextinction)	Evolution in which a species becomes extinct but is replaced by another species that is directly descended from it, so the species lineage continues	Of mass extinctions since the mid-Permian, 75% include pseudoextinctions (Patterson and Smith 1989).	Provides knowledge of when species' lineages have actually continued; can be used to correct estimates of global extinction on monophyletic groups and lineages.

demographic dynamics, and at least 50 generations for genetic dynamics. Longer time spans can be considered if environmental changes can be predicted beyond that time period. Thus, for a population of parrots with a generation time measured perhaps on the order of a decade, population viability should be projected over a century for demographic factors, and over five centuries for genetic factors, whereas for a population of voles—with more rapid reproduction, shorter life spans, and far shorter generation times—population viability may be projected only over a few years.

Logistic Gaussian population models predict smooth (or predictable stepwise) changes (increases or decreases) to some idealized and long-term asymptote representing carrying capacity. But real-world populations seldom exhibit the simple dynamics of Gaussian population models, and animal densities are seldom unimodal within a species range (O'Connor 2002). Rather, populations are complicated by many exogenous (external) and endogenous (internal) factors. One complicating factor is variation in age-specific survivorship and fecundity. Stochastic (random) variations in these vital rates distributed over reproductive individuals within a given generation, such as expressed by a variance in the number of offspring, has been called *demographic stochasticity*. Stochastic variation in average population values of these vital rates over time, such as expressed by the average number of offspring over successive breeding seasons, has been called *environmental stochasticity*. Demographic stochasticity plays a role in affecting population viability where the effective number of breeding individuals is small and where the range of progeny per reproductive adult is large. Environmental stochasticity plays a role in affecting population viability where external factors influencing mortality or fecundity randomly vary over time, such as food levels or suitability of nest sites as affected by weather conditions. Both demographic and environmental stochasticity can contribute to population trajectories being substantially more complex than simple models predict, and both can act to lower likelihoods of population persistence over a given time period. In general, smaller populations can be more subject than larger populations to the adverse effects of both kinds of stochasticity.

Simple Gaussian population models also assume a fixed and measurable carrying capacity and fixed vital rates of fecundity and mortality of organisms in the population. Under such idealized circumstances, even allowing for some degree of demographic and environmental stochasticity, it is easy to identify threshold effects in model populations. In population viability modeling, thresholds are typically conditions of the environment that, when changed slightly past particular values, cause populations to crash (Lande 1987; Oborny and Szabo 2005). (Conversely, some population models refer to explosion thresholds, which are environmental conditions that lead to large increases of populations.) Such threshold conditions have given rise to the concept of *minimum viable populations* (MVPs) (Gilpin and Soulé 1986; Reed 2005). In models, an MVP is the smallest population (typically measured in absolute number of organisms rather than in density or distribution of organisms) that can sustain itself over time, and below which extinction is inevitable. Through modeling stochastic demography, Shaffer (1983) reported the viability and MVP of grizzly bear (*Ursus arctos*) populations. Marshall and Edwards-Jones (1998) addressed MVP levels for reintroducing capercaillie (*Tetrao urogallus*) in Scotland. Many other examples are available in the literature (for example, Reed et al. 2003), but these typically constitute modeling results rather than empirical evidence of threshold effects.

So do actual populations follow such threshold effects? Perhaps they do in rare cases of unchanging environments. But most often, the many factors influencing population viability vary through space and time, rendering identification of specific MVP levels problematic at best and misleading at worst. Reed et al. (2003) defined MVP as a population size having a 99% probability of persistence for 40 generations and used population viability analysis (PVA) on 102 vertebrate species to determine an MVP size of about 7000 adults. Culotta (1995) suggested that the number of individuals needed to ensure a genetically robust population may be as high as 10,000. Such considerations also prompted Thomas (1990) to reexamine simple guidelines for MVP size to ensure long-term viability of populations. Thomas concluded that MVP rules of thumb should be adjusted upward from Soulé's (1987) suggestion of "low thousands" to at least 5500 for undivided (essentially panmictic) populations, and that future work should focus on better understanding the complicating factors in real-world populations. We concur with Thomas and suggest that the term and concept of MVP be abandoned for more realistic models and for empirical studies. We advocate using models of population viability that account for random variations in demographic, environmental, or genetic factors, and that represent location- and case-specific outcomes as likelihoods of population persistence rather than as fixed MVP threshold sizes (see box 3.3).

Lastly, on the topic of population viability analysis, we question whether predictions of viability can be verified or validated. In PVA,

Box 3.3 Population viability modeling and analysis

Population viability has been defined as the likelihood of continued persistence of well-distributed population(s) to a specified future time (Shaffer 1990). A population with "high viability" thus has a high likelihood of continued persistence with a well-distributed population over a long time, say, on the order of a century, or 10 generations. This definition, used in numerous wildlife planning analyses, suggests that viability is a probabilistic event, and that the likelihood of not achieving specific desired viability levels is a measure of risk under any given management plan.

Population viability is affected by many factors intrinsic to a population and to its environment (see table 3.4 in text). Factors that would likely receive attention in a quantitative analysis include population demography, population genetics, metapopulation dynamics, environmental stochasticity, biogeography, evolutionary adaptations and selection mechanisms, reproductive ecology and behavior, rates and causes of population declines, and degree of and reasons for endemism and range restriction. Unfortunately, such empirical information is available for very few species.

So where may one turn for information useful in a population viability analysis? Our experience suggests three main sources: (1) contract reports and rigorously guided panel discussions with leading species experts, both of which can provide information on current population status and suspected reasons for population declines, as well as information on key environmental factors and key ecological functions of species; (2) geographically referenced (GIS) modeling of "key environmental correlates" that most influence the distribution and abundance of selected species; and (3) published empirical studies on species ecology.

Ideally, a population viability analysis (PVA) should combine empirical studies with simulation modeling of demography, genetics, environmental disturbances, and other factors. Such an approach is often referred to as a *quantitative PVA*, which can be provided for only a few species. In the absence of adequate empirical species studies on which to base a quantitative PVA, the three-pronged approach listed above provides at least a starting point for identifying which species might warrant further, more intensive population analyses, and for providing base information for such future analyses. Future analyses can be part of an explicit research, development, and application study plan, and part of the implementation, inventory, and monitoring phases of any habitat or land management plan.

The GIS modeling approach (item 2 above) can be used to produce maps of past, current, and future potentially suitable environments for the species of interest. Each map can be analyzed for how the abundance, distribution, and trend of potentially suitable environments might influence the likelihood of population persistence. The GIS modeling approach is *not* the same as modeling actual or anticipated population sizes, structures, and distributions, and is not a strict replacement for more intensive analysis. This is an important distinction. Results of modeling species' environments must be interpreted as tentative working hypotheses concerning population response. Only follow-up monitoring of populations, environments, or both, would determine the extent to which potentially suitable environments index true population persistence.

By modeling environments instead of *a priori* defined habitat for a specific species, one can provide the basis for assessing conditions for a range of species. Modeling environmental conditions can set the stage for further species-specific PVAs and also for broader assessments of biodiversity. As part of the three-pronged approach—particularly focusing on existing population abundance and distribution, and on patterns and trends of environments—such modeling can be used to draw inferences about potential population viability response by individual species and about potential persistence of species groups, scarce ecological communities, and other aspects of biodiversity.

Coupling modeling of environmental conditions with knowledge of species demographics has the advantage of helping determine population distribution patterns, such as whether a population is potentially large and interbreeding; partially interbreeding (in "metapopulation" patterns); isolated but having relatively large separate populations; or consisting of small and isolated populations. Also useful is determining whether conditions contributing to a particular viability level are due to natural biophysical conditions or are caused by human activities and thus might be ameliorated. But it is important to remember that there is no real substitute for field studies of population and biodiversity conditions.

Validation studies of PVA models have been few. Schiegg et al. (2005) found that spatially explicit, individually based stochastic models predicted population dynamics of red-cockaded woodpeckers (*Picoides borealis*) with high accuracy, although predictions could be improved by modeling how individuals colonize new sites and by improving estimates of sex-specific dispersal behavior. Fleishman et al. (2003) tested model predictions of occurrence of native butterfly species in mountains of Nevada and concluded that it is desirable to separate model predictions of presence from absence. McCarthy

(box continued on next page)

Box 3.3   Continued

et al. (2003) found that PVA model estimates of population extinction of threatened species greatly improved with far longer data sets (100 versus 10 years) and that model results are best presented as rank-order outcomes of potential extinction risk.

Caswell (2001) presented a succinct summary of the strengths and weakness of PVAs. He noted that future population dynamics, including the risk of extinction, can only be predicted through demographic models because they provide the only framework that can integrate birth and death rates, which determine changes in population size. He listed three primary criticisms of PVAs and provided a response to each, which we summarize here:

1. *Models require too much data.* This criticism centers on the fact that vital rates are difficult to measure, yet model results are very sensitive to changes in these rates. Thus uncertainty in vital rates leads directly to uncertainty in model results. Caswell noted, however, that uncertainty is part of the results of a PVA. Further, this criticism really misses the point; namely, that the model (and associated uncertainty) is simply reflecting the fact that vital rates are the critical component of population viability and they are sensitive to relatively small changes in values. That these factors make research difficult is irrelevant.

2. *Models are difficult to validate.* Caswell points out the critical distinction between *validation* and *testing*. The former implies that models can provide a valid, or "true," prediction of the future; they cannot, as they are always approximations. The latter tests model predictions against some other value, such as testing whether the model is true. Caswell further points out, however, that even testing a PVA is a waste of time because the model is never true but is at best an approximation of truth from a set of competing models. Rather, the focus of PVA should be on parameter estimation, model selection, and statistical evidence.

3. *Models are error-prone.* Different models usually lead to different conclusions, which causes users to question their reliability. As summarized by Caswell, models infer consequences from premises; that is, population growth changes as a function of changes in vital rates. Thus different premises result in different conclusions, which is not an error.

In summary, the goal of a PVA is to predict the likely fate of a population. These predictions involve a balance between competing rates of birth, death, immigration, and emigration. Models with relatively low uncertainty require precise estimates of these vital rates. This requirement is not a failure of the modeling approach, but rather a statement on the difficulty involved in studies in population dynamics.

---

viability is calculated as a probability, such as likelihood of extinction or of persistence of a population above some quasi-extinction level for a specified time period (see table 3.5). In the real world, however, a population follows a single deterministic course and outcome. PVAs are usually conducted on species, populations, or metapopulations at risk, where $n = 1$. The only ways that PVA can be empirically tested (thus, aside from comparing outcomes to simulated or modeled populations; e.g., see McCarthy et al. 2003) would be to (1) experimentally manipulate real populations at risk, which is usually an undesirable situation except perhaps for rare circumstances of "experimental populations" (such as deemed by the U.S. Fish and Wildlife Service), or (2) compare PVA predictions with empirical outcomes among multiple populations and species. At best, specific components or functions within a PVA model could be empirically tested—for example, Johnson's (2005) test of whether synchronous extinctions of a Neotropical beetle correlated with local disturbances and patch-size-dependent dispersal. Further, different PVAs may use very different formulae and models, confounding such comparisons (e.g., see box 3.4). For the most part, PVA results expressed as probability distributions do not produce statistically

Box 3.4   Alternative ways of calculating effective population size, N_e

One facet of population viability analysis (PVA) entails calculating effective population size (N_e). Effective population size is the idealized number of fully interbreeding (panmictic) individuals, taking into account certain factors that would serve to change or reduce the degree of panmixia (Kimura and Crow 1963). Just as there are multiple methods and models for conducting PVAs, there are multiple ways to estimate N_e. The user must choose among them according to the purpose of the PVA and the factors that seem to best fit the situation and species. Following are some examples of calculations using $N = 100$:

Situation	Formula	Example calculation	Source
	Basic models of N_e		
Inbreeding effective population size (related to inbreeding rate)	$N_{ei} = \dfrac{(N \cdot k) - 2}{(k - 1) + \dfrac{V_k}{k}}$	$N_{ei} = 49$ if $k = 1$ and $V_k = 2$.	Waples 2002
Variance effective population size (related to changes in allele frequency)	$N_{ev} = \dfrac{(N \cdot k) - k}{1 + \dfrac{V_k}{k}}$	$N_{ev} = 33$ if $k = 1$ and $V_k = 2$.	Waples 2002
	Spatially structured models of N_e		
Variable subpopulation migration rates	$N_{et}W = \dfrac{N_t}{[(1 + P) \cdot (1 - F_{st})] + \dfrac{N \cdot P \cdot F_{st} \cdot n}{(n - 1)}}$	Charcoal fragments from tropical rainforests in North Queensland, Australia, indicated that *Eucalyptus* woodlands occupied substantial areas of all of the present humid rainforest between about 27,000 BP and 3500 BP. Changes in paleoclimates associated with the most recent glaciations provide a plausible explanation for both the *Eucalyptus* expansion and subsequent rainforest reinvasion (Hopkins et al. 1993). $N_{et}W = 91$ if $N_t = 100$, $P = 0.1$, $F_{st} = 0.0001$, and $n = 2$.	Whitlock and Barton 1997
Intergenic genetic drift	$N_{et}N = \dfrac{N_t}{[(1 + F_{is}) \cdot (1 + F_{st})] - (2 \cdot F_{is} \cdot F_{st})}$	$N_{et}N = 99$ if $N_t = 100$, $F_{is} = 0.01$, and $F_{st} = 0.0001$.	Nunney 1999

(box continued on next page)

Box 3.4   Continued

Subpopulation extinction	$N_{et}N2 = \dfrac{n}{4 \cdot (x + e) \cdot F_{st}}$	$N_{et}N2 = 83$ if $n = 2$, $x = 50$, $e = 10$, and $F_{st} = 0.0001$.	Nunney 2000

Various demographic and genetic factors affecting N_e

Sex ratio	$N_e Sex = \dfrac{4 \cdot N_{ef} \cdot N_{em}}{N_{ef} + N_{em}}$	$N_e Sex = 10$, if $N_{ef} = 5$ $N_{em} = 5$	Frankham et al. 2002
Variance in family size	$N_e Fam = \dfrac{(4 \cdot N) - 2}{V_f + 2}$	$N_e Fam = 80$, if $V_f = 3$	Frankham et al. 2002
Variance in family size	$N_e Fam2 = \dfrac{(N \cdot f) - 1}{f - 1 + \left(\dfrac{V_f}{f}\right)}$	$N_e Fam2 = 80$, if $f = 2$ $V_f = 3$	Frankham et al. 2002
Fluctuations in population size over time	$N_d Fluc = \dfrac{t}{\Sigma \left(\dfrac{1}{N_{ei}}\right)}$	$N_e Fluc = 22$, if $N_{ei} = 30, 20,$ 40, 30, 30, 10	Frankham et al. 2002
Inbreeding	$N_e Inb = \dfrac{N}{(1 + F)}$	$N_e Inb = 83$, if $F = 0.2$	Frankham et al. 2002

where:

e = rate of subpopulation extinction
f = mean family size
F = inbreeding rate (0 = none, 1 = completely inbred)
F_{is} = Wright's inbreeding statistic
F_{st} = Wright's measure of genetic differentiation among subpopulations
k = mean number of offspring produced per individual
n = number of subpopulations
N = (censused) population size in the parental generation
N_{ef} = effective number of breeding female parents
N_{ei} = effective population size in the ith generation
N_{em} = effective number of breeding male parents
N_t = total number of individuals in the metapopulation
P = variance in productivity among subpopulations
t = number of generations
v_f = variance in family size
V_k = variance in k across individuals
x = migration rate among subpopulations

A number of other formulations are also available (e.g., Chambers 1995; Rockwell and Barrowclough 1995). Clearly, calculations of N_e vary greatly according to the situation and the formula used, which is one of the troubles—and the art—of PVA: knowing which formula to use. PVAs are also difficult to compare across species or across situations in which researchers used different formulae. That is, it is unclear if differences in predictions of future population sizes result from dif-

---

Box 3.4   Continued

ferences in the species, from the environmental situations (especially among management alternatives), or from the types of formulae and models used. Our advice, as with any modeling exercise, is to (1) understand the basis of the formulae used, including what is excluded as well as included, (2) have an empirical basis and rationale for including specific parameters, and (3) where appropriate, use several formulae or models to help tease out the effects of different model structures from true, expected responses of populations.

---

testable hypotheses. In this way, PVA is more of a planning and management tool, and it is often useful as such. (Also see discussion on model validation in chapter 10.)

POPULATION DEMOGRAPHY

Population demography is the proximate expression of a host of factors that influence individual fitness and population viability (see table 3.4). The reader is directed to other sources on specific techniques for modeling population demography (for example, Caughley 1977; Caswell 2001). How is demography influenced by habitat quality and distribution? Many models of population response assume invariant vital rates of fecundity and mortality, or at least fixed probability distributions around mean values of vital rates. However, vital rates may vary substantially in space and time as a function of many factors, including food quality, weather, imbalance in sex ratios, and other factors. In some cases, populations respond to weather conditions, such as harsh winters, with lag effects measured in seasons or years.

In other cases, the causes of variations in vital rates are unknown. One example is the northern spotted owl (*Strix occidentalis caurina*), which for decades has shown quasi-periodic variations in nesting attempts, successful breeding, and production of young. These periods may last 2 to 4 years and may be synchronized over broad geographic areas and among owl populations that are likely to be largely isolated demographically. Causes of these fluctuations may be attributed to synchronous variations in prey density, although

this is unlikely given the variety of prey taken and the breadth of the geographic area. Other possible causes are lag effects of harsh winters, and some innate mechanism that shuts off breeding investment for one or a few seasons, although the closely related and sympatric barred owl (*Strix varia*) can be a relatively prodigious breeder in the same habitats at the same times. Accurately predicting demographic trends of northern spotted owls through modeling will entail a better understanding of the causes of such variations in vital rates and how these rates are affected by habitat and environmental conditions.

How does population size affect likelihood of decline? We might pose three patterns for a size-decline relation, as depicted in figure 3.12. The first pattern (curve A in fig. 3.12), and really the null model, is that population size does not influence rate of change as measured on a per capita basis. That is, large and small populations alike would tend to experience the same annual rate of per capita change, all other factors being equal. This pattern is unlikely, because many factors that influence population viability (see table 3.4) have greater influence on small populations than on large populations.

For example, the selective contribution of only some genes to successive generations through genetic drift and through inbreeding depression can become a key factor in small populations. Such sampling increases the degree of genetic homozygosity of genotypes (through fixation of deleterious and other alleles) and thereby decreases overall diversity of alleles in gene pools. This decline happens at faster rates

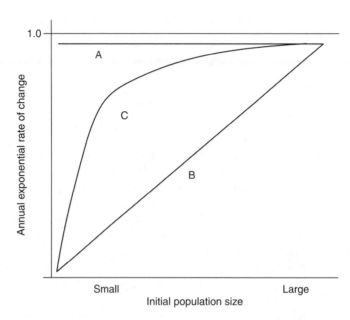

Figure 3.12.   Three patterns of relations between annual exponential rate of change in population and initial size of the population. (A) Populations decline at rates regardless of initial size of the population. (B) Smaller populations decline at a higher rate, and the relation with population size is linear. (C) Smaller populations decline at a higher rate, and the relation with initial population size is nonlinear (e.g., exponential). In this figure, the ordinate is scaled so that declining populations occur at a rate of <1.0. The horizontal line at 1.0 denotes no population change.

in small populations than in large populations. Another reason why small populations might decline differently (potentially faster) than large populations is the Allee effect, in which some degree of social interaction among organisms, afforded by population size or density above some threshold, is necessary to stimulate reproduction (Dennis 2002). Below the threshold, such facilitation is too infrequent or unlikely to support population numbers.

The second pattern for a size-decline relation (curve B in fig. 3.12) is that there is a linear effect of population size on rate of change. Under this pattern, smaller populations are likely to decline faster than larger populations, but the effect is merely a linear increase in the rate of decline with respect to population size. This pattern is

likely to be closer to most real-world cases than the null model of no effect. One argument for a linear relation over no relation has to do more with the simple geometry of increasing difficulty of finding mates in increasingly sparse populations or increasingly fragmented and divided habitats.

However, a linear relation is probably not commonly the case in real-world populations. One major reason is that the effect of small population size on allelic fixation and demographic stochasticity is disproportionately greater than that of larger population sizes. Greater demographic stochasticity means greater variation in the number of progeny among breeding individuals of a given generation and among years or breeding intervals. Mathematically, on a per capita basis, these factors affect smaller populations much more acutely than they do larger populations (Legendre et al. 1999). It is easier for a small population to dip to zero, or to below critical nonzero thresholds, often called quasi-extinction levels (Ginzburg et al. 1982), than it is for a large population to do so, *ceteris paribus.*

The third pattern for a size-decline relation (curve C in fig. 3.12) is a pattern in which smaller populations decline more rapidly than larger ones, and nonlinearly so. Mathematical models of population genetics and demography suggest that this third pattern is often the case. Surprisingly, few real-world examples from the literature are available, but all seem to fit curve C (fig. 3.13). Examples include birds on the Channel Islands of southern California (Jones and Diamond 1976); breeding birds on Bardsey Island off Britain (Diamond 1984); and population data of the plant *Astrocaryum mexicanum* (Menges 1991). This pattern (curve C) may also hold for populations of herpetofauna, mammals, and other taxonomic groups, as suggested by population modeling (for example, Diamond 1984), but needs empirical testing. Under this pattern, small populations are substantially at greater risk of severe decline and extinction than even moderately small populations, and the relation is far worse than linear. This relation has important implications for the timing and funding of species and habitat recovery programs, and for developing realistic expectations for these programs.

POPULATION GENETICS

Many references are available on modeling population genetics (e.g., Young and Clarke 2000; Frankham et al. 2004). However, real-world relations between habitat conditions, population genetics, realized fitness of organisms, and viability of populations are poorly understood at best.

Two aspects of population genetics are of major interest to conservation managers: inbreeding depression and genetic drift, often mistakenly assumed to be the same phenomenon. *Inbreeding depression* refers to fixation and adverse phenotypic expression of deleterious alleles in a population, caused by breeding among often-increasingly related individuals. Deleterious effects can be manufactured through several genetic mechanisms, including increasing the proportion of homozygous recessive genotypes, increasing the proportion of deleterious alleles in overdominance, maintaining the presence of deleterious genetic mutations through successive breeding of the same pedigree line, among others (Keller and Waller 2002). Senner (1980) described three manifestations of inbreeding depression: (1) *individual viability depression,* which is a failure of maturation of young; (2) *fecundity depression,* which is an increase of sterility; and (3) *sex ratio depression,* which is an overabundance of males. Models of inbreeding dynamics have been used in estimating effects on long-term viability of populations (see box 3.4). Perrin and Mazalov (1999) assessed dispersal as an evolved trait that avoids inbreeding. Increases in the level of inbreeding reduce effective population size (fig. 3.14) and can increase the probability of extinction (fig. 3.15).

Sherwin and Moritz (2000) developed a classification of overlapping types of genetically determined variation within and between populations (fig. 3.16). They noted that there has been intense conservation interest in the historical component of genetic diversity (categories 1, 3, 5, and 7 in fig. 3.16)—that is, variants that have accumulated over thousands of generations through the random processes of drift and mutation, and possibly through selection and adaptation. As noted by Sherwin and Moritz, this historical component is irreplaceable because the circumstances that generated the variation can only be hypothesized, and the time scale cannot be replicated in a management program. Variants that affect current adaptation and viability are

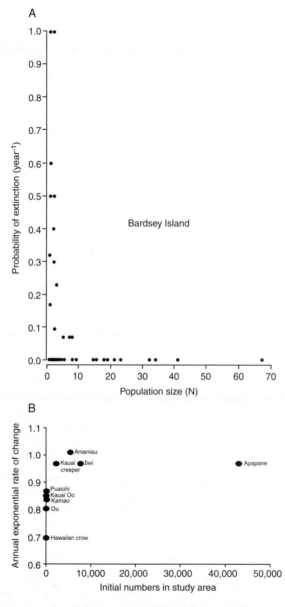

Figure 3.13. Examples of empirical evidence supporting the inverse exponential pattern of relations between changes in population size and initial population size (curve C in fig. 3.12 rates), expressed in two ways. (A) Probability of extinction as a function of population size, for breeding birds of Bardsey Island off Great Britain, 1954–1969 (Diamond 1984,199). (B) Annual exponential rate of change (<1 means population decline) as a function of initial population size, for Hawaiian honeycreepers on Kauai. (Calculated by B. Marcot from data in Scott et al. 1986.)

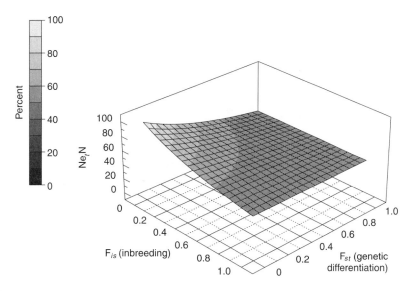

Figure 3.14.   An example of how inbreeding depression influences effective population size, using the model for intergenic genetic drift presented in box 3.4. As degree of inbreeding (F_{is}) increases, effective population size (N_e:N) decreases, potentially overriding any positive influence of genetic differentiation (F_{st}) among subpopulations. Actual effects will vary by species and circumstance.

important in short-term conservation and form a second type of genetic variation (categories 3, 4, 5, and 6 in fig. 3.16). If lost, phenotypes corresponding to these variants, especially categories 4 and 6 in figure 3.16, can potentially be re-created through selection, given that the impact of this selection is not so great as to cause immediate extinction. The third type of genetic variation is that which forms the basis of continued adaptation to changing conditions in the future and factors into relatively long-term conservation efforts (categories 5, 6, 7, and 8 in fig. 3.16). We cannot, however, specify which genetic variants belong to these categories and must rely on inferences about the behaviors of each category.

Variation in categories 4 and 6 is likely to be restored relatively quickly if lost, as may some of the variations in categories 3 and 5. Variation in categories 7 and 8 is vital for long-term viability but is not of any current adaptive significance

and is therefore liable to loss through genetic drift. Once lost, this variation cannot be regained as fast as variation in other categories (Sherwin and Moritz 2000).

Because of the time scale of process of population genetics, much current genetic variation is in the irreplaceable categories (i.e., 1, 3 , 5, and 7), and three of these categories are important to current and future adaptations (i.e., 3, 5, 7). To maintain current population viability, we must also manage variants that are of recent origin and currently of adaptive significance (i.e., categories 4 and 6). We would also strive to manage variants of recent origin that are likely to be of adaptive significance in the future, but which are not so at present (i.e., category 8).

Populations tend to lose genetic variation as they become relatively small and isolated. This loss of genetic variation may affect population viability through inbreeding depression and

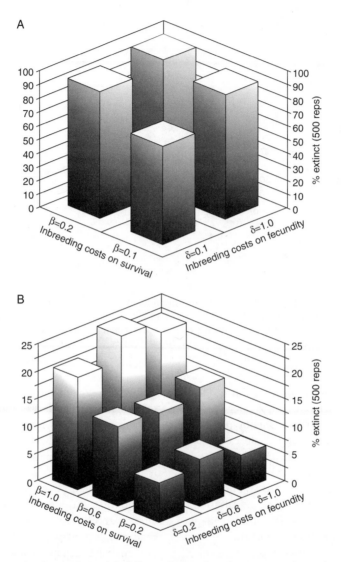

Figure 3.15.   An example of an analysis of the effects of inbreeding depression and population size on the viability of a hypothetical population with an increasing trend ($\lambda = 1.24$), an equal sex ratio, and an initial population set (A) $N_{(O)} = 5$ and (B) $N_{(O)} = 80$. In this analysis, inbreeding was modeled as costs on survival and fecundity and expressed as the percentage of populations in replicate model runs that become extinct. In this model, the combined effects of inbreeding costs on both survival and fecundity resulted in a greater extinction risk than that of each cost factor separately, although initial population size greatly mitigated extinction risk. (From Mills and Smouse 1994, 421; by permission of University of Chicago Press.)

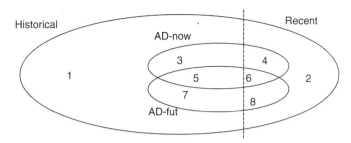

Figure 3.16.   Classification of genetically determined variation for use in conservation planning. *Historical:* Variation that has accumulated over thousands of generations or longer (categories 1, 3, 5, 7). *Recent:* Variation of relatively recent origin (categories 2, 4, 6, 8). *AD-now:* Variation that is of adaptive significance at present (categories 3, 4, 5, 6). *AD-fut:* Variation that will be of adaptive significance in the future (categories 5, 6, 7, 8). (From Sherwin and Moritz 2000, fig. 2.2; reprinted with permission by Cambridge University Press.)

reduced evolutionary potential. Inbreeding depression can lead to population extinction and may be especially important when combined with demographic and environmental stochasticity (Daniels et al. 2000). Evidence of inbreeding depression in wild populations is growing and is especially acute for populations in areas undergoing increasing fragmentation. In red-cockaded woodpeckers, for example, both inbreeding avoidance behaviors and inbreeding depression have been found. Closely related pairs produced 44% fewer yearlings per year than did unrelated pairs (see Daniels et al. 2000 for review). Inbreeding depression has been implicated as the cause of low fertility in Florida panthers (*Felis concolor coryi;* Maehr and Caddick 1995); captive populations of the gray wolf (*Canis lupus;* Laikre and Ryman 1991); and other endangered carnivores. However, there are cases of populations having gone through severe bottlenecks (periods of small population size) and without having suffered any apparent or lasting demographic problems from inbreeding depression. Examples include bird populations as small as 10 pairs persisting for 80 years on the Channel Islands off southern California (Jones and Diamond 1976) and severely reduced populations of northern elephant seals (*Mirounga an-*

gustirostris), which rebounded from about 20 to about 30,000 individuals over 75 years following protection (Bonnell and Selander 1974).

Daniels et al. (2000) recommended that methods to reduce inbreeding could include (1) maintenance of a stable or increasing population through habitat manipulations, (2) enhancing dispersal by retaining multiple populations in a region and by linking disjunct inhabited areas, and (3) translocating individuals to small populations with no immigration.

In comparison to inbreeding depression, *genetic drift* is the random change in frequency of alleles through random mating and allelic assortment over generations. Genetic drift may result in deleterious fixation of alleles in homozygous genotypes, but it can also result in neutral changes and heterozygous conditions as well. Drift may or may not reduce the overall genetic diversity of a population, depending on the size of the population, initial degree of homozygosity, degree of outbreeding, and other factors (Allendorf 1986). The effects of genetic drift on habitat relations and population vitality—and, conversely, the effects of habitat quality and distribution on the likelihood of deleterious effects from genetic drift—are poorly understood at best but are probably important factors over the long run in small, isolated populations or in demes with little or no outbreeding. One example from Hitchings and Beebee (1998) is the empirical verification of the isolation of common toad (*Bufo bufo*) populations because of migration barriers created from urbanization and habitat loss that resulted in genetic drift among the subpopulations and subsequent loss of genetic variation and fitness.

In the 1980s, researchers modeled MVPs by considering only genetic conditions of inbreeding depression and genetic drift. In theory, a minimum viable population is supposedly of a size (number of breeding individuals, assumedly) below which the population is doomed to extinction, and at and above which it is secure. One guideline that was proposed was the "50–500 rule," whereby populations of at least 50 breeding individuals should be maintained for ensuring short-term viability, and at least 500 for long-term viability (Gilpin and Soulé 1986). However, as reviewed above, such guidelines seldom pertain to real-world situations and were devised mostly from genetic considerations alone.

Viable population management goals often specify providing for large, interconnected wildlife populations and maintaining diverse gene pools. Rather, management goals should tier to first understanding the natural conditions of a species in the wild. There are cases in which populations are fully or partially isolated under natural conditions, and artificially inducing outbreeding among isolates, such as through captive breeding programs or manipulation of habitats, may contradict natural conditions. In some cases, naturally balanced polymorphisms, local ecotypes, and local endemics with unique gene pools have derived from, and can be maintained by, partial or full reproductive isolation of populations.

Although usually implemented only when a species or subspecies has reached very low numbers, captive propagation and relocation of individuals among populations are increasingly used management techniques. However, severe genetic problems (e.g., inbreeding, bottlenecks) can occur as a result of such conservation actions. These problems might not be readily apparent in the time scale of management programs because few generations of vertebrates will have elapsed since the implementation of the program (Ramey et al. 2000). An excellent case study of population management, including relocations, exists for the bighorn sheep (Krausman 2000 and articles therein).

Two general types of genetic changes can occur in a captive population that have ramifications for reintroductions or relocations. First, selection can eliminate alleles that are maladaptive in the captive situation yet important for survival in the wild. Second, random genetic drift can cause the cumulative loss of both adaptive and maladaptive alleles. The primary problem in captive propagation is that each successive generation is a sample of the previous generation. Thus the gene pool of the population that will eventually be reintroduced is invariably changed through captive generations. Rare alleles are especially susceptible to loss through genetic drift.

To preserve the genetics of animals that will be reintroduced, captive management must minimize adaptive and nonadaptive genetic changes. The concept of effective population size (Wright 1931) has several related meanings, including the number of individuals at which a genetically ideal population (one with random union of gametes) would drift at the rate of the observed population (also see box 3.4). The rate of genetic drift could be measured as the sampling variance of gene frequencies from parental to offspring generations. The primary goal of captive propagation is, however, to minimize all evolutionary (genetic) change, whether from random drift or selection to the captive environment. Lacy (1994) proposed that at least 90% of the genetic variation in the source (wild) population be maintained in a captive population. Ramey et al. (2000) suggested that it is justifiable to intervene when there has been a severe genetic bottleneck, which they defined as an effective population size of <10 individuals and a lack of gene flow with other outbred populations. Some reintroduction and translocation efforts have been successful with as few as ~10 individuals (Ramey et al. 2000; see also Morrison 2002 for review).

Genetic variation includes many related concepts, such as genetically determined variation in morphology or behavior, variation in chromosomal structure, molecular variation in genes, allelic diversity, and heterozygosity of genes. The phenotype of an organism determines its physical properties and fitness. Thus quantitative genetic variation in phenotypes is of importance in designing genetic management for reintroduction. Much attention has focused, however, on theoretical models and prescriptions related to the management of underlying molecular genetic variation (i.e., the presence of multiple genetic variations within a population and the mean or expected diversity per individual).

Heterozygosity can refer to (1) diversity of alleles within individuals and (2) diversity of alleles among individuals in a population. The proportion of loci for which the average individual is heterozygous is usually termed *observed heterozygosity*. The probability that two homologous genes randomly drawn from a population are distinct alleles is termed the *expected heterozygosity* or *gene diversity*. The guideline that at least 90% of the genetic variation in the source (wild) population should be maintained in a captive population (Lacy 1994) referred to expected heterozygosity, and many management programs have used expected heterozygosity as an index of genetic variability. The possibility of a population to adapt at all to new or changing environmental conditions depends on the presence of sufficient variants, so allelic diversity

may be critical to long-term persistence (Lacy 1994). Lacy (1994) discussed many of the strategies that have been used for maximizing retention of genetic variation in captive breeding programs (see also Morrison 2002 for review). Maximizing retention of heterozygosity usually, but not always, optimizes allelic diversity. Animals that have unique or rare alleles can also have very common alleles at other loci or at the homologous genes. Mating such animals could reduce heterozygosity because they share genes with much of the population, yet not breeding them could result in loss of rare alleles.

Sherwin and Moritz (2000) provided recommendations for averting erosion of genetic variation. This ecological approach to genetic conservation can be coupled with monitoring of genetic indicators to determine if the actions are having the desired effect. Sherwin and Moritz's recommendations included the following:

- Minimizing reduction of effective population size
- Minimizing change of natural levels of gene flow, which might require artificial relocations
- Minimizing loss of separate management units
- Minimizing loss of peripheral populations, especially where clines are present
- Maintaining ecological amplitude of the species through retention of populations in different environments
- Maintaining normal temporal fluctuations

OF SYSTEMATICS AND SUBSPECIES

It should be obvious by now that providing habitat and suitable environments for wildlife should attend to more than just the needs of individual species. Although the idea of species is one of the building blocks of ecological science, it is in some ways an artificial construct. Many taxonomic groups, particularly plants, arthropods, and some herpetofauna, are replete with biological entities that defy easy classification into the standard Linnaean taxonomic system. Reasons are varied and include unusual or complicated reproductive life histories. One example is that of triploid hybrids of the silvery salamander (*Ambystoma platineum,* cf. *jeffersonianum*) and Tremblay's salamander (*A. tremblayi,* cf. *laterale*) in the northeastern United States.

Another example occurs with some clonal populations of the Chihuahua whiptail lizard (*Cnemidophorus exsanguis*) and the checkered whiptail lizard (*C. tesselatus*) that reproduce by parthenogenesis (females give birth to haploid female offspring without contribution of male genes). In some areas, their clones are sympatric (have overlapping ranges) with their parent species and behave as reproductively isolated, separate species, although differences in coloration and scalation are subtle. Checkered whiptail populations tend to be mostly parthenogenic, with males extremely rare.

Why is this important? In the real world of biopolitics, it is vital to clearly and unambiguously recognize species, subspecies, and populations in order to focus activities related to conservation, recovery, recreation, hunting and collection, and restoration. But habitat managers sometimes have to struggle with imperfectly classified organisms and thus encounter problems of correctly identifying habitats and environments deserving conservation activity.

One example is that of the red-legged frog (*Rana aurora*) in the western United States. Two western subspecies have been recognized, *R. a. aurora,* which occurs in the northern part of the range of the species, and *R. a. draytonii,* which occurs in wetlands and streams in coastal drainages of central California. *Draytonii* has been extirpated from 70% of its former range, whereas *aurora* is more common where it occurs farther north. Red-legged frogs found in the intervening area of northern California exhibit intergrade characteristics of both subspecies (Hayes and Krempels 1986), and systematic relationships of the two subspecies are not fully known (Hayes and Miyamoto 1984; Hayes and Krempels 1986). However, significant morphological and behavioral differences between the two subspecies suggest that they may actually be two species in secondary contact (Hayes and Krempels 1986).

The story continues. The U.S. Fish and Wildlife Service designated *R. a. aurora* as a federally threatened taxon in a portion of its range. Should *R. a. aurora* be found to be a separate species, this may narrow future conservation options, such as locating appropriate sites for reintroductions, identifying sources for use in augmentation of native or captive populations, and tracking captive breeding pedigrees, should these be useful conservation measures. It would also raise the question of conservation attention to red-legged frogs in the northern California intergrade zone, should they prove to be hybrids between species rather than hybrids between subspecies. Species hybrids—such as the cases of the triploid *Ambystoma* hybrid cited above, or, potentially, red-legged frogs in northern California—may signal undesirable genetic swamping if the species' sympatry was a result of human activities, but such hybrids may also be a natural outcome of sympatric remixing of closely related but previously separated species. There is no clear, single direction that conservation action should follow in such circumstances. Each case should be evaluated on its own.

The case of the red-legged frog may be one of unknown identity awaiting further taxonomic study. But sufficient other cases, particularly among herpetofauna, question the utility of species and subspecies as the standard—and often the only—measure of entities deserving of special habitat conservation action. In another example, throughout the western United States, several species and a number of described subspecies of garter snakes (*Thamnophis* spp.) defy simple taxonomic description. Intergrades abound, varying in morphometries, habitat selection, and coloration.

One solution to the problem may lie in intensive study of genetic relations among potential species or subspecies using advanced laboratory methods, including comparative nucleic acids, protein sequencing, electrophoresis, immunology, chromosome matching, microsatellite loci, as well as studies of relations between individual gene loci and studies of morphometries and breeding relations (also see Soltis and Gitzendanner 1999). Other genetic techniques in studies of systematics include use of mitochondrial DNA (m-DNA) sequencing, such as used along with microsatellite loci by Delaney and Wayne (2005) to determine that the island scrub-jay (*Aphelocoma insularis*) on California's Santa Cruz Island is a species distinct from the continental western scrub-jay (*A. californica*). Baker et al. (2003) used m-DNA evidence to determine recent separation of two oriole subspecies that should now be considered full species. Many other examples are available.

The next step would be to correlate genetic constitution and variation with patterns of *in situ* habitat use and selection. This correlation would help determine taxonomic entities that relate to specific environmental and geographic conditions, and would help set conservation priorities (e.g., Posadas et al. 2001). However, most wildlife species officially listed as sensitive, threatened, or endangered, or as candidates for these lists, have not been studied to this degree; nor can we realistically expect funding to study the vast majority of these species. As well, removing individuals from at-risk populations for such research may not be feasible or desirable. Still, molecular methods have proven useful (Wayne and Morin 2004). Whether "fuzzy" taxonomic entities can receive conservation attention may be more a matter of biopolitics than of

science. At the least, it raises the question of what to do with dubious taxonomic distinctions and well-defined and potentially viable hybrids between species.

Still other taxonomic designations have focused habitat conservation attention at taxonomic distinctions finer than subspecies. Plant conservationists, and foresters transplanting commercially useful trees, have long focused on *ecotypes*—populations in gene pools adapted to local environmental (usually climatic) conditions. The concept has recently been used in a study of two ecotypes—mountain and northern—of woodland caribou (*Rangifer tarandus caribou*) recently translocated to augment a population of woodland caribou in northern Idaho (Warren et al. 1996). After translocation, the two ecotypes of this one subspecies responded with different patterns of habitat selection, dispersal, and mortality. These findings prompted Warren et al. (1996) to conclude that larger numbers of individuals may need to be translocated to sustain a new or recipient population when donor subpopulations (ecotypes) must be used that do not closely resemble the habitat selection characteristics of extant or extinct resident subpopulations. Another conclusion we can draw is that donor populations should be taken, if possible, from habitats and environments most closely matching those of the recipient population. This requirement was an unexpected complication of an augmentation strategy for an endangered vertebrate, and one not well studied in most other programs of augmentation, translocation, and introduction.

METAPOPULATIONS IN PATCHY HABITATS AND IMPLICATIONS FOR HABITAT MANAGEMENT

The distribution, abundance, and dynamics of a population in a landscape are influenced by species attributes, habitat attributes, and other factors. *Species attributes* include movement and dispersal patterns; habitat specialization; demography, including density-dependence relations; and genetics of the populations. *Habitat attributes* include quality, size, spacing, connectivity, and fragmentation of habitat patches (also see chapter 9) and the resulting availability and distribution of food, water, and cover. Other factors include a host of environmental conditions such as weather, hunting pressure, and influences from other species.

Much work in simulation modeling has focused on the dynamics of habitat occupancy by organisms in metapopulations (for example, Westphal et al. 2003; Moilanen 2004) through use of geographically referenced or spatially explicit population models (see chapter 10). In general, occupancy of habitats by organisms is usually more variable in suboptimal environments, peripheral locations, smaller habitat patches, and patches with greater edge effects adversely affecting required resources, although density-dependent relations may mediate these trends. Also, the effects of competitors, predators, disease, and other nonhabitat factors must not be discounted, although few spatially explicit population models currently include them.

Many models of metapopulations in patchy environments suggest that not all suitable patches will be occupied at any one time. This finding in turn suggests that habitats should be conserved even if they seem unoccupied in any one year, and thus that monitoring wildlife use of habitats in the conditions listed above should proceed for more than one year or season, and perhaps for several. At the same time, it may also be unwise to staunchly preserve every instance of a potentially suitable patch if it is clear that the patch does not serve a function in maintaining the population or in maintaining an interesting genetic variant of the species of interest. The appropriate approach—discussed further below—depends in part on the size and fragility of the population and its habitats, and on conservation objectives.

Metapopulation structure can also be used in designing captive breeding programs. As reviewed by Morrison (2002), dispersing a captive population over a diverse environment can avoid directional selection that depletes genetic variation and can promote selection that enhances genetic diversity. Dispersal also helps protect the population from epidemic disease and other catastrophes. Isolated or partly isolated populations tend to diverge genetically and thus lose different genetic variants. Small subpopulations are, however, subject to extinction, often require intensive management for extended periods of time, and come with potentially serious risks. Animals within each isolated population will become more inbred because of the fewer mate choices and genetic drift. However, if inbreeding depression was not severe, a larger population reconstructed by mixing animals from the isolated populations could be expected to have greater genetic variation and perhaps higher individual fitness than would be the case if the population had never been subdivided. Moving approximately one animal between populations per generation will usually prevent excessive inbreeding within populations, but at a cost of reduced effectiveness of the subdivided population structure in retaining variation overall. If 5 to 10 animals are moved per generation, however, genetic divergence among subpopulations is largely prevented, and the metapopulation is therefore equivalent to a panmictic population. The prudent approach to managing metapopulation structure is to mimic the amount of isolation typical of the wild population (or the best estimate of the characteristics of the wild population) before human-induced decimation and fragmentation.

In the sections that follow, we review concepts and examples of dynamics of populations and habitats in preparation for further discussion of habitat modeling in chapter 10. An appropriate conceptual framework must be established before jumping into any modeling exercise. Thus we develop how animals are distributed over space and time, including how the actual abundance at any place and time can give a misleading indication of survival and productivity of animals.

Dynamics of Populations and Habitats

DISCERNING POPULATION REQUIREMENTS
FROM PATTERNS OF HABITAT SELECTION
AND SPECIALIZATION

The distribution and density of a population are ultimately determined by the collective responses of its individual organisms. It should be remembered that descriptions of population–habitat or species–habitat relations are really just a statistical—even just a categorical—summary of the behavior of individuals. One can describe a central tendency of a statistical population, such as the typical or average habitat conditions in which a particular species is found. But descriptions of variation around that central tendency provide insights into much more interesting biological phenomena. In fact, it is often the spread, rather than the central tendency, that tells us the most about how populations and species occur in and select their environments.

The degree of variation in selection of habitats among individuals of a population describes the degree of specialization or specificity of habitat use—often termed *habitat breadth* (or *habitat use breadth*) or *niche breadth*—and many statistical measures have been devised to measure it (e.g., Feinsinger and Spears 1981; Vázquez and Simberloff 2002). Habitat breadth is really a collective property of a population resulting either from variation in habitat selection among individuals of the population at a particular time, or from variation of individuals' habitat selection patterns over time. Often, in studies lacking individual demarcation, these two components are not differentiated. Thus the resulting estimates of habitat breadth are unpartitioned combinations of individual and temporal variations in habitat selection. If the studies are used to identify habitats for conservation management, then it is not critical that the two components be differentiated, as conservation activities might as well target the entire range of variation in any case, including individual and temporal variations. But the contributions to, and causes of, the observed overall habitat breadth will not be known with certainty.

Habitat specialists—those with narrow habitat breadth. or that are *stenotopic,* as compared to those with wide habitat breadth, or that are *eurytopic*—may be more at risk of local declines or extirpation should the environment change suddenly. Many descriptions of species at risk from regional climate changes use habitat breadth as the measure of risk (e.g., Marcot et al. 1998). For example, in the inland western United States, the white-tailed ptarmigan (*Lagopus leucurus*) is found principally in alpine tundra communities, and the American pipit (*Anthus rubescens*) occurs regularly in alpine tundra and only one other vegetation community, making these two stenotopic species particularly locally vulnerable to upper-elevation warming and loss of alpine tundra habitats (Marcot et al. 1998). Eurytopic species found in alpine tundra may also use other communities as well, and thus are not as subject to local threat.

It is generally not advisable to combine observations of habitat use patterns of individuals from different ecotypes, populations, geographic areas, or ecoregions (Ruggiero et al. 1988). Combining individuals in this way could mask the gamma diversity of habitat selection patterns that vary substantially across geographic areas.

Providing habitat for one ecotype or for organisms in one portion of their range may be senseless for another ecotype or range portion if the organisms there select for different environmental cues (see the example of woodland caribou cited above).

DISTRIBUTION PATTERNS OF POPULATIONS

The overall distribution and local abundance of many wildlife species vary across time and space. Some species have a "bull's-eye" pattern of distribution, with their greatest area of abundance toward the middle of their overall range and only marginal population density in the peripheral portions of the range. In reality, few species show such a unimodal distribution but rather have a multimodal distribution that reflects differences in abundance related to habitat conditions. In essence, many bull's-eyes often occur across species' overall ranges, as reflected in the distribution of the bobolink (*Dolichonyx oryzivorus*) in North America.

Animal distributions typically reflect several aspects of biophysical conditions and species ecology: (1) the geographic range of suitable biophysical conditions; (2) the range of tolerance of biophysical characteristics by the species; and (3) the occurrence of marginally suitable conditions at the peripheries of the geographic range that often act as "sink" habitat to hold nonreproductive individuals or individuals spread from higher-abundance areas during good reproductive years (such as with grassland birds, as studied by Perkins et al. 2003). Such sink habitats can, however, increase the probability that the greater population will persist by holding a large number of individuals at any given time, and may serve as a storehouse of genetic information as long as these individuals can successfully reproduce at some point in time (Donovan and Thompson 2001). Population size thus depends on composition of habitat quality across the landscape, as depicted in figure 3.17. In this figure we see that equilibrium population size was 1000 in landscapes with ≤30% low-quality habitat, suggesting that the presence of low-quality habitats did not negatively impact population size. When 0 to 30% of the population used low-quality habitats, and when landscapes contained 40 to 90% low-quality habitat, equilibrium population size decreased as the percentage of low-quality habitat in the landscape increased. The majority of the individuals in all cases occurred in high-quality habitat because 70% of the individuals used high-quality habitats (see fig. 3.17).

In some cases, species have higher densities in low-quality habitats relative to high-quality habitats. For example, bobolinks have breeding densities up to four times greater in hayfields than in native prairie areas. In most years, however, hayfields are low-quality habitats because nests are often destroyed as fields are hayed (harvested) prior to fledging (Donovan and Thompson 2001). Of course, the "low versus high" quality categorization is artificial because reproductive performance varies along a continuum of time and space. It is quantification of this variation, along with prediction of where and when such variation will occur, that is a major focus of wildlife study and management.

In studying censuses of birds from the North American Breeding Bird Survey, Brown et al. (1995) noted that such highly clumped abundance distributions resemble distributions such as the negative binomial canonical lognormal, used to characterize abundance distributions among species within local ecological communities. They hypothesized that the spatial variation in abundance largely reflects how well local sites meet resource requirements of a species, and concluded that patterns of spatial and temporal variation in abundance should be factors considered when designing nature reserves and in conserving biological diversity. Also important for correctly identifying conservation areas is ac-

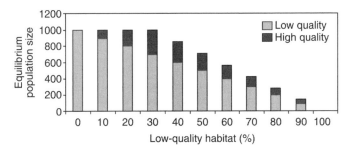

Figure 3.17.   Equilibrium population size and the distribution of individuals in landscapes containing 0–100% low-quality habitat, where ≤30% of the population selected low-quality habitat over high-quality habitat for breeding. (From Donovan and Thompson 2001, fig. 7; by permission of the Ecological Society of America.)

curately delineating the boundaries of species' distributional ranges, such as has been done with North American Breeding Bird Survey data by Fortin et al. (2005).

The abundance distribution of some species is truncated where biophysical conditions come to an abrupt halt, such as along mountain ranges, large rivers, or other major dispersal barriers, or at the edges of continents. An example is the wrentit (*Chamaea fasciata*), a species of chaparral, brush, and thickets, whose distribution truncates at its greatest density along the Pacific coast of the United States (fig. 3.18). Thus one cannot always assume that peripheral distributions of a species mean marginal environmental conditions and lowest population densities.

Identifying areas of high density of organisms or areas of high environmental suitability (remembering that these are not necessarily always synonymous) can be important for management purposes. Wolf et al. (1996) reviewed success of 421 avian and mammalian translocation programs in North America, Australia, and New Zealand and found that release of organisms into the core of their historical range and into habitat of high quality were two key factors contributing to success. Other factors included use of native game species, greater number of re-

leased animals, and an omnivorous diet. Nor should it be forgotten that many factors other than those listed here also contribute to population viability.

Populations of most organisms ebb and flow over time. Changes can be simple trends or complex variations. Causes can be density-independent, such as from population-wide effects of weather, or density-dependent, such as from density-induced variations in natality. Density-dependent relations are often difficult to study but lend well to modeling (Guthrie and Moorhead 2002). Numerous other approaches to detecting and modeling density-dependent population behavior have been proposed (e.g, Wilcox and Elderd 2003).

DYNAMICS OF POPULATIONS AFFECTED BY
MOVEMENTS OF ORGANISMS

Various kinds of movements of wildlife through their habitats and environments impart particular dynamics to their populations. Movements particularly important for habitat management include the following:

- *Dispersal*—one-way movement, typically of young away from natal areas
- *Migration*—seasonal, cyclic movement, typically across latitudes or elevations, to

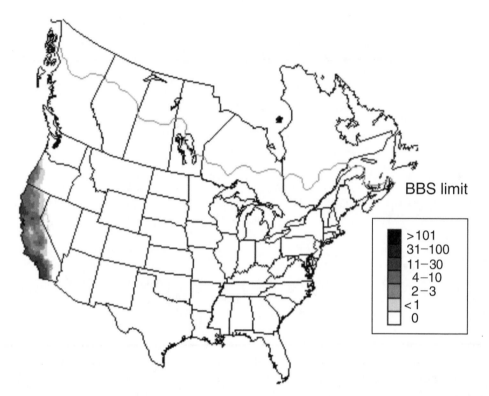

Figure 3.18.   Current summer distribution of wrentits (*Chamaea fasciata*) in North America based on mean bird counts per Breeding Bird Survey route, 1982–1996. Note that highest densities occur at the edge of the overall range, truncated by the continent's margin. (From Sauer et al. 2003a)

track resources or to escape harsh seasonal conditions

- *Home range*—movement throughout a more or less definable and known space, over the course of a day to weeks or months, to locate resources
- *Eruption*—irregular movement into geographic areas normally not occupied, as a response to severe weather or the sudden availability of high-quality resources

Dispersal. The significance of various parameters to the viability of a population is related to the interaction between life history, competitors,

and the landscape. The role of dispersal in the population processes of small, restricted-range species is different from its role in the population processes of relatively wide-ranging species. In addition, in modern, human-impacted landscapes, long movements by an animal are likely to involve both direct encounters with humans and indirect encounters with human-induced factors (e.g., physical barriers, feral animals) (Macdonald and Johnson 2001). Thus dispersal has different consequences and may be selected for different reasons, depending on the distance moved (fig. 3.19). As reviewed by Ronce et al. (2001), relatively short-distance movement—

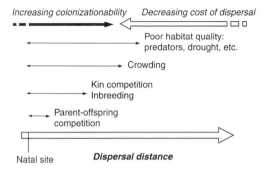

Figure 3.19. Different dispersal distances needed to reduce negative characteristics of natal habitat. (From Ronce et al. 2001, fig. 24.1; reprinted by permission of Oxford University Press.)

such as to an adjacent territory—may allow escape from direct competition with parents without losing the benefit of local adaptations and without much energetic cost. Such short-distance movement does not, however, prevent competition between kin. Alternatively, relatively long-distance movement may be selected for when competition between kin (especially between siblings) is involved. Escaping inbreeding may require movement into a different social group, which would usually involve greater dispersal distances. Although specific dispersal distances are based on the demographics of a population, distances traveled to colonize available habitat are usually longer than those traveled to avoid interactions with kin (Ronce et al. 2001).

Dispersal patterns can also vary by year, gender (Schiegg et al. 2005), time (Morton 1992), and species interactions, including competition (Rodgers and Klenner 1990). A few examples will demonstrate the variations in dispersal patterns among species. Allen and Sargeant (1993) reported that in North Dakota, mean recovery distances of red foxes (*Vulpes vulpes*) tagged as pups increased with age and were greater for males than for females, but that dispersal distance was not related to population density. This

finding was interpreted as meaning that populations can be augmented, but also that disease can be transmitted across long distances regardless of fox population density. Males and females each showed no uniformity in dispersal direction, although littermates did. Beier's (1995) study of dispersal of juvenile cougars (*Felis concolor*) in fragmented habitat in southern California revealed that the animals would use habitat corridors located along natural travel routes with adequate woody cover and low human density, including a roadway underpass. As with the red fox study, littermates dispersed in the same direction.

Among raptors, dispersal distances of fledgling eastern screech-owls (*Otus asio*) was not significantly correlated with either dispersal date or the number of days that juveniles remained on natal territories, and dispersal direction was random (Belthoff and Ritchison 1989). Other studies or reviews of dispersal ecology are available (Clark et al. 2003; Nathan et al. 2003).

Buechner (1987) modeled vertebrate dispersal patterns and, based on comparisons of parameter values and deviations from the geometric pattern between groups within taxa, concluded that males and females may follow different patterns of dispersal. She also indicated that mammals and birds show consistent differences in dispersal distributions. These results are relevant to discussions of natal philopatry, inbreeding avoidance, and proximate mechanisms of dispersal. They can also influence spacing selection of habitat reserves.

One aspect of dispersal important for habitat management is recognition of dispersal barriers or strong filters. Not many studies have empirically identified species-specific dispersal barriers or filters, although they could be critical in determining the occupancy of distributed habitat patches and persistence of low-density and small-sized populations in the wild. For example, Allen and Sargeant (1993) found that a four-lane interstate highway altered dispersal directions of red foxes and apparently caused distant travel from natal areas.

Similarly, until recently, few studies have empirically demonstrated use of habitat corridors, which has served as one of the backbones of many habitat conservation strategies (Lindenmayer and Nix 1993). Haas (1995) found that movements of three migratory bird species among riparian woodlands and shelterbelt woodlots in North Dakota, although rare, occurred more frequently between sites connected by wooded corridors than between unconnected sites. Haas concluded that knowledge of patterns of fledgling, natal, and breeding dispersal of birds in patchy environments would substantially aid decisions on reserve design and on protection or construction of habitat corridors. Hill (1995) found that species of ants, butterflies, and dung beetles—bioindicator groups—responded differentially to linear vegetation corridors in the lowland rainforest of northeastern Australia. Hill concluded that habitat corridors can aid dispersal of only some species and that rainforest-interior obligates would not use the linear corridor strips. Bolger et al. (2001) studied use of corridors by birds and small mammals. Hannon and Schmiegelow (2002) found that corridors may be useful to retain resident birds on harvested landscapes, but that corridors connecting small reserves of forests are unlikely to offset the impacts of fragmentation for most boreal birds. They concluded that assessments of the utility of corridors must, however, be done in the context of the full plant and animal communities that live in the boreal forest. Other recent empirical studies of use of habitat corridors by animals have dealt with butterflies (Haddad and Tewksbury 2005), small mammals (Pardini et al. 2005), and other taxa (Haddad et al. 2003).

Dispersal distances are important parameters for use in some models of genetic diversity and deme size (see box 3.2). Commonly, different age or sex groups have different patterns of habitat selection during different parts of their life cycles or seasons, including dispersal. Understanding age- and sex-specific dispersal patterns and distances helps determine the set of dispersal habitats collectively required by all members of a population. Studies of habitat selection only during breeding tend to miss much of the range of habitats used and required by wildlife. In this spirit, dispersal distances can also be used to help establish spacing patterns of habitat conservation areas.

Four general classes of data for animals that are captured, marked, and released alive can be used to study movements: (1) recovery data, in which animals are recovered dead at a subsequent time; (2) recapture and resighting data, in which animals are recaptured or resighted alive on subsequent sampling occasions; (3) known-status data, in which marked animals are reobserved alive or dead at specified times with probability 1.0; and (4) combined data, in which data are of more than one type (Bennetts et al. 2001).

Conservation has two primary problems: first, reducing deterministic habitat destruction by humans and contamination leading to reduced species populations and, second, minimizing the effects of stochastic processes on these resulting small and isolated populations. Dispersal is critical to determining the ability of a species to cope with both of these processes. Dispersal thus plays a central role in sustaining declining populations above a threshold under which demographic or genetic stochastic events become lethal (Macdonald and Johnson 2001).

MIGRATION

Periodic (typically seasonal and annual) occurrence of organisms among geographic locations and habitats is another type of movement by wildlife. Migration of herding ungulates, large mammalian carnivores, some raptors, many Neotropical migratory songbirds, some amphibians, and other taxa can take place over short or long distances and latitudes. Earlier (see chapter 2, box 2.1) we discussed some of the leading hypotheses concerning the evolution of migration in birds.

The migration status of terrestrial wildlife, particularly birds, can be defined hierarchically, depending on type of movement and distance traveled, as presented in table 3.6. Four general classes of movement are (1) elevational migrants, (2) latitudinal migrants, (3) permanent residents, and (4) nomads (eruptive or irregular movement into new areas). Elevational migrants include species that move downslope during winter so that they move out of, or into, a particular area.

Examples of *elevational migration* are the seasonal movements of the snowcap (*Microchera albocoronata*) and the three-wattled bellbird (*Procnias tricarunculata*) along the Zona Protectora La Selva corridor, which links the eastern lowland Reserva Biológica La Selva with the central highland Reserva Forestal Cordillera Volcánica Central and Parque Nacional Braulio Carrillo along the central cordillera of Costa Rica (Stiles and Clark 1989). The snowcap breeds in the highlands (900 m) during December–June and then moves into the lowlands (100 m) during July–October (thus it would be coded as an E1 or E2 species in table 3.6, depending on the location of interest). The bellbird follows similar patterns but breeds higher up (2000$^+$ m) during March–August and moves to the coastal lowlands (<100 m) during August–January. In northern California, some elevational migrants, such as the western meadowlark (*Sturnella neglecta*) and Brewer's blackbird (*Euphagus cyanocephalus*), move upslope after breeding in valleys or lowlands (code E3, table 3.6).

In the mountains of western North America, some individual *latitudinal migrants,* such as the dark-eyed junco (*Junco hyemalis*), may displace

Table 3.6.   A classification of the migration status of wildlife species

Migration class	Code	Seasons present[a]				
		Sp	B	Su	F	W
Elevational migrants						
Downslope movement during winter						
Breed at studied elevation, move lower in winter	E1		x	x	x	x
Breed at higher elevation, move into study area during winter	E2	x			x	x
Upslope movement during postbreeding season						
Breed in bottomlands, move upslope during postbreeding	E3			x	x	
Latitudinal migrants						
Displacement migrants	L1	x	x	x	x	x
Neotropical migrants and low-latitude Nearctic migrants	L2		x	x	x	
High-latitude Nearctic migrants	L3				x	x
Permanent residents	PR	x	x	x	x	x
Nomads	NOM	———————— any ————————				

Source: Marcot 1985.

[a]Sp = spring; B = breeding; Su = summer (postbreeding dispersal); F = fall; W = winter. Seasons shown here include a breeding season distinguished from spring (postwinter movement and settling period) and summer (postbreeding dispersal period). Seasons pertinent to each migration class may differ depending on geographic location.

other individuals (displacement latitudinal migrants) as they move north or south (code L1). Others, such as the flammulated owl (*Otus flammeolus*), are Neotropical migrants that travel between mid and low latitudes annually (code L2). Still others, such as the yellow-rumped warbler (*Dendroica coronata*), move between high boreal and mid latitudes (code L3).

Permanent residents (code PR), are individuals, such as the northern spotted owl (*Strix occidentalis caurina*), that occupy a location year-round, although some spotted owls in northwest Washington, and the California spotted owl (*S. o. occidentalis*) on the western slopes of the Sierra Nevada, may become elevational migrants, moving onto lower slopes during winter. Note that both permanent residents and displacement migrants may occur in all seasons in a location, but the sources of the individuals differ. Finally, some species, such as red crossbills (*Loxia curvirostra* species complex) and several species of owls in boreal forests (Cheveau et al. 2004), are *nomads* and move irregularly and irruptively in any season.

The use of such a movement classification system can aid habitat management by determining (1) which species are likely to occur in an area in a given season, and thus the resources and habitats required during that season; (2) the number of species expected in an area over seasons, and thus the collective resources and habitats required; and (3) the need for considering habitat conservation in other regions beyond the area of immediate interest. Such a classification system may also aid in identifying habitat corridors used during movement, and thus habitats and geographic areas needing potential conservation focus, as was the case with the designation of Costa Rica's Zona Protectora La Selva corridor, cited above.

The literature on migration biology and physiology is rich and varied. The practitioner is particularly directed to sources reviewing management implications (Alerstam et al. 2003; Skonhoft 2005).

Home range. Although not a parameter of populations per se, how organisms establish and use home ranges nonetheless is important for managing habitat for populations. Many algorithms have been devised for estimating home range area (for example, Ostro et al. 1999; Rosenberg and McKelvey 1999). At the same time, controversy has ensued over appropriate metrics. Different algorithms estimate different aspects of spatial use. Examples are: the overall area circumscribed by movement of an organism in a specified time period (measured or estimated by use of minimum or maximum convex polygon algorithms); the geographic areas of concentrated occurrence of an organism (harmonic mean, adaptive kernel); the general area most used by an organism (95% confidence ellipse); and others.

The controversies over which algorithm to use have focused not so much on technical discourses as on implications for identifying the total area of economically valuable and scarce habitats to protect for viable populations of uncommon species. Examples include preservation or restoration of old longleaf pine (*Pinus palustris*) forests in the U.S. southeast for red-cockaded woodpeckers, and old-growth conifer forests in the U.S. Pacific Northwest for northern spotted owls, pileated woodpeckers (*Dryocopus pileatus*), and mustelids. As it turns out, not surprisingly, different home range algorithms measure different aspects of home range and organisms' use of space and resources, so their results are not identical. Authors often present home-range size using at least two methods of calculation, including the most commonly used minimum convex polygon (MCP) as a standard for comparison among studies. For example, in studying coyotes in Arizona, Grinder and Krausman (2001) presented both the MCP and adaptive kernel methods to depict home range. The adaptive kernel is a popular method because, unlike the MCP, it allows estimating home range size that allows identification of disjunct areas of activity. These activity areas can then be analyzed further (e.g., use versus abundance of habitat, abundance or activity of potential predators, location of roads or buildings) to discover why the activity patterns are occurring. Further, GPS and GIS technologies are providing for more efficient and accurate recording and analysis of data resulting from radio telemetry and other methods of recording animal movements and locations.

It is incumbent on the researcher to better understand what aspect of space and resource usage would best pertain to a particular management question, and then use the algorithm that best fits the organism's biology and the statistical data; and for the habitat manager to better understand the algorithms and better focus questions of habitat use and allocation. Further, the researcher should not just focus on measuring home range sizes and amounts of habitat within home range without equitable treatment of other factors that could influence home range size and habitat selection, such as food supplies, density of conspecifics, effect of body size, competitors, predators, and landforms. In a sense, habitat selection is an optimization process that involves all of these (and other) factors, and home range is an expression of the process; this has important implications for management. Larger-than-average home ranges may or may not mean that insufficient habitat quality or amount has been provided.

Eruptions. Some species undergo periodic eruptions in distributions and dispersal. Formally, the term *eruption* refers to populations that suddenly exceed their normal boundaries or densities outward from a given area, whereas *irruption* refers to invasion into a particular area from outside. (Think of eruptions as being caused by *emigration,* and irruptions as being caused by *immigration.*) Eruptions can be caused by population growth (numerical response; see text below) or by movement of individuals (functional response). In North America during periods of extreme northern winter conditions, some species typical of higher-latitude boreal forests, such as the northern hawk owl (*Surnia ulula*), the snowy owl (*Nyctea scandiaca*), the gyrfalcon (*Falco rusticolus*), and the white-winged crossbill (*Loxia leucoptera*), can occur intermittently farther south in southern Canada and the northern United States. Such eruptive movements push the edges of population ranges into locations not normally occupied.

More locally, seasonal eruptions or wandering movements occur with white-headed woodpeckers (*Picoides albolarvatus*), red crossbills, redpolls (*Carduelis hornemanni* and *C. flammea;* Hochachka et al. 1999), evening grosbeaks (*Coccothraustes vespertinus*), and other species. Irregular movements can help organisms fill vacant environments, thereby colonizing new habitats as founder populations (see chapter 9). For instance, Scott (1994) reported that unusual conditions of prevailing winds aided an irruptive dispersal of black-shouldered kites (*Elanus caeruleus*) over 80 km of open ocean to San Clemente Island in southern California. Such movements may be important for determining potential colonization rates of peripheral habitats and the conservation value of habitat in the periphery of some species ranges. Eruptions of rodents in response to periodic increases in food are well known but sometimes have surprising implications for human populations and health. For example, although numbers of rodents have been shown to cycle with the production of pine cones (e.g., Morrison and Hall 1998), only recently have these rodent cycles been linked to transmissions of hantavirus from rodent to human, which causes the often fatal human disease hantavirus pulmonary syndrome (e.g., Yates et al. 2002). Eruptions of deer are well known, such as the long-term, adverse effects of browsing by introduced deer on native southern beech (*Nothophagus*) forests in New Zealand (Husheer et al. 2003). Eruptions of white-tailed deer (*Odocoileus virginianus*) in the eastern United States are a continuing management problem (Horsley et al. 2003).

STOCHASTIC ENVIRONMENTS AND DESIRED CONDITIONS

Stochastic environments contain biotic and abiotic conditions that vary irregularly over time, usually substantially and unpredictably. All environments change. But not all populations and ecological communities are equally resilient or equally pliable to change. It is a matter of the intensity and duration of the change, coupled with the kinds of populations and ecological communities present, that determines whether a change produces long-lasting or permanent effects or whether populations or communities will return to a prior state. The degree that populations are resilient to change or are pushed to new distributions or densities, or even to extirpation, is largely determined by the initial abundance and the vagility and habitat-specificity of the organisms, although certainly other factors play roles as well.

Populations can respond to a perturbation either functionally or numerically. The *functional response* of a population refers to changes in the behavior of organisms, such as seletion of different prey or use of different substrates for resting or reproduction. Functional responses can also entail a temporary and localized increase in numbers resulting from immigration, or a decrease from emigration. The *numerical response* of a population refers to absolute changes in the abundance of individuals through changes in recruitment.

Disturbance of a habitat might elicit one or both kinds of responses. For example, a crown fire in a subalpine forest of the northern Rocky Mountains might increase suitability of habitat for black-backed woodpeckers (*Picoides arcticus*) in several ways. It might induce a temporary influx of foraging woodpeckers into the forest (a shorter-term, functional response) and also provide snag substrates for increased nesting density of woodpeckers in the area (a longer-term, numerical response).

It is important to distinguish between such responses to understand whether management activities—especially intentional habitat restoration or enhancement activities—are truly serving to increase absolute population size or are simply redistributing organisms. In some cases, simple redistribution of organisms may be the goal, such as warding off foraging waterfowl from grain fields and agricultural lands and into nearby wetland refuges. In other cases, redistributions and local increases—such as displacement from disturbed or fragmented habitats—may belie an overall population decline.

We discuss disturbance dynamics of habitats further in chapter 9.

EVOLUTIONARY ADAPTATIONS AND SELECTION MECHANISMS

The evolution of individual behaviors of adaptive significance directly influences the overall distribution and abundance of populations. The degree of species' specialization in habitat selection has importance for prioritizing habitat conservation activities and identifying at-risk species. Of prime interest to habitat ecologists and managers are evolution of habitat selection behaviors for maximizing energy efficiency, including selection of prey and breeding, resting, and foraging habitats.

Reproductive and hibernation ecology also has important bearing on habitat selection and conservation. For example, the stenotopic endangered Indiana bat (*Myotis sodalis*) selects natural, undisturbed caves as wintering hibernacula. Modifications of cave entrances by humans has degraded the bat's winter habitat and altered its hibernation physiology (Richter et al. 1993). Caves with artificially modified (widened) entrances were found by Richter et al. to have a mean winter temperature 5.0°C higher than that of unmodified caves. Bats in modified caves entered hibernation at a 5% greater body mass and lost 42% more body mass, and small bats (<5.4 g) did not survive the winter. Modification of the entrance to more natural conditions increased the colony from 2000 to 13,000 bats over a decade. On the other hand, some congeners, such as the little brown myotis (*Myotis lucifugus*), have evolved to be more eurytopic. The little brown myotis uses a variety of substrates and locations for roosting, including buildings, snags, exfoliating bark of live trees, caves, rock outcrops and crevices, and mines, and thus probably is not in particular danger from cave modification.

Benkman (1999) studied the significance of crossbill (*Loxia*) adaptations to foraging on conifer seed cones. He found that four taxa or "types" (species or subspecies) of red crossbills in the U.S. Pacific Northwest have diversified morphologically in bill characters in response to alternative adaptive peaks of their foods from four conifer trees. Each adaptive peak corresponded to one conifer tree species whose seeds are produced regularly from year to year, held in cones through late winter when seed is most limiting, and protected from depletion by potential noncrossbill competitors. Benkman concluded that (1) a critical feature of the ecology and evolution of crossbills is reliability of seeds on key conifers during periods of food scarcity; (2) even in populations in highly variable environments, morphological traits have evolved to optimize

food variability and availability; and (3) disruptive selection against intermediate phenotypes is likely serving to maintain, if not reinforce, the distinctiveness of types. Also, (4) the diversity of cone structure and seed size among key conifers is ultimately responsible for the diversification of crossbills. Further studies would determine the effect, if any, that selective foraging by crossbills, or by other spermivores that may take seeds of specific size or viability, would have on the evolution of conifer seed characteristics and conifer tree genetics and morphology.

Wing, bill, and mass measurements of eastern (*Otus asio*) and western (*O. kennicottii*) screech-owls identify them, their subspecies, and sexes. Increasing sexual dimorphism in bill length has been correlated with decline in mass, which might reflect increased subdivision of the food niche related to increased competition in more dense populations and among more coexisting insectivorous owls (Gehlbach 2003). Gehlbach thought that interspecific competition was suggested by such character displacement and habitat segregation.

Herbivory by ungulates and other species can have a major direct impact on vegetation conditions and habitat for other species. However, few studies have determined the specific mechanisms that induce such changes. To determine these mechanisms, Anderson and Briske (1995) conducted a controlled experiment with domestic herbivores in the southern true prairies in Texas. They concluded that selective herbivory of the late-seral-dominant plant species, the perennial grass *Schizachyrium scoparium*, is the chief means by which plant species replacement occurred. This grass has a greater competitive ability and greater tolerance to herbivory pressure than the midseral grass species that made up the community. Thus the herbivores modified the community because of differential evolved tolerance to predation among different grass species. (Coevolution is also discussed in chapter 2.)

The evolution of species relations can have important implications for population persistence and management of habitat for populations. Many examples can be found in the literature. One pertains to obligate pollinator–host relations. Globally, insects—specifically Hymenoptera (bees and wasps), Lepidoptera (moths and butterflies), and Coleoptera (beetles)—are by far the most important pollinators of flowering plants. Recent declines in bees and wasps have sparked critical concern in some ecologists (Buchmann and Nabhan 1996; Cane and Tepedino 2001). Upon loss of pollination vectors, plants may alter reproductive modes, suffer loss of genetic heterogeneity, or become locally extirpated (Washitani 1996). Aside from direct effects on vegetation and habitats as discussed above, herbivory can also have major indirect effects on pollination success and persistence of vegetation (Vázquez and Simberloff 2004).

Some vertebrate wildlife populations figure prominently in pollination ecology. Examples include many hummingbirds of Nearctic and Neotropical regions and their ecological vicariants, the sunbirds of Asia. Bats can play major roles in pollination of a variety of species in the tropics, although such roles are likely reduced or absent in many temperate and boreal ecosystems. Obligate pollinators and seed dispersers of "structural species" plants, such as some trees that serve as habitat or provide resources for a host of other species, were termed *mobile links* by Gilbert (1980). Terbourgh (1986) used the term *keystone mutualist* to refer to species that, if extirpated, would likely result in secondary extinctions of many associated obligate species. Habitat managers might want to identify mobile links and keystone mutualists—particularly important pollination vectors and dispersers of plant disseminules (e.g., plant seeds, fungi spores, lichen thalli)—to better predict ecosystem effects of altered habitat on such species, and to better determine specific habitat requirements to provide for continuation of their ecological services. Below, we further discuss the importance of species' ecological functions to population conservation.

EXOTIC SPECIES

Intrusion into natural environments by exotic species has become a major challenge for habitat conservation and restoration (Allendorf and Lundquist 2003; Lodge and Shrader-Frechette 2003). Exotic species come in all taxonomic groups and all environments. Exotic plants can disturb native ungulate use of rangelands (Trammell and Butler 1995) and severely handicap management of natural conditions in parks (Tyser and Worley 1992). Exotic game-bird and big-game introductions can affect distribution of native plants and animals (OTA 1993). Unintentional invasions of exotic insects, plants, goats, pigs, and many other species in high-elevation parks in the Hawaiian Islands have caused major disruption of native ecological communities and spurred intensive eradication management activities (Cole et al. 1995).

The Office of Technology Assessment (OTA 1993) used the term *nonindigenous species* to refer to alien or exotic species. The OTA concluded that, in a worst-case scenario, major economic losses could ensue with continued spread of harmful nonindigenous species. The office identified several major issues needing attention, including the need for a more stringent national policy, management of nonindigenous species for spread of disease, and control of the spread of weed plants and damage into natural areas.

However, it is not always clear which species are exotic (nonindigenous) and which are native. There are various cases of range expansions that may be natural, that may have been induced or enhanced by human alteration of environments, or that began as a minor introduction by humans. An example of a "natural" invader of North America is the cattle egret (*Bubulcus ibis*), which spread from Africa to South America about 1880, reached Florida and Texas in the 1940s and 1950s, and rapidly expanded north and west in North America (Ehrlich et al. 1988). Exotic escapee species that have spread throughout the continent include the European starling (*Sturna vulgaris*); after two unsuccessful introductions, 60 birds were released into New York's Central Park in 1890, and within 60 years they had spread to the Pacific and have since outcompeted and threatened many other bird species (Ehrlich et al. 1988). Another example is the brown-headed cowbird (*Molothrus ater*), which has greatly expanded its range with the opening of forests and development of agriculture. On the other hand, the spotted dove (*Streptopelia chinensis*) was intentionally introduced in the Los Angeles area, and its spread has been limited to coastal southern California despite its wide natural geographic range throughout Asia.

Horned lizards (*Phrynosoma coronatum*) have disappeared throughout much of their range in coastal southern California. In addition to direct impact of habitat loss due to urbanization, horned lizard populations continue to decline in the remaining fragmented landscapes. Suarez and Case (2002) found that horned lizards are disappearing from these remaining habitat fragments at least in part due to the presence of the exotic Argentine ant (*Linepithema humile*). Hatchling horned lizards on a diet of Argentine ants showed either a negative or a near-zero growth rate, which contributes to the overall population declines being seen.

In recent years, barred owls have spread westward from subboreal forests in the northern United States and southern Canada, and southward down the Cascade Mountains into the Klamath Mountains of northwestern California, coming into secondary contact with the closely related and stenotopic northern spotted owl (Pearson and Livezey 2003). Reasons for the barred owl's recent and quick spread may have to do with its eurytopic habitat use, or perhaps with the degree that clearcutting of conifer and mixed forests in North America has afforded habitats for foraging. Is the barred owl to be considered a desired native species of the West, or an undesirable exotic from the East indirectly introduced through major changes in forest conditions by human activities? Federal land management currently views the barred owl as a native species in the West, but ecological causes of its range expansion have not been determined.

Some exotic species are desired by people for their color or for sport. Such North American species that may be termed "desired nonnatives" include many introduced game birds such as the ring-necked pheasant (*Phasianus colchicus*). Some populations of desired nonnatives —such as the nutria (*Myocaster coypus*), introduced into the United States for its pelt and once sought after—have become harmful to other species or to wetlands or other sensitive ecosystems. Work remains to be done in better defining exotics and desired nonnatives, and in projecting potential harm to native populations and ecosystems before introductions are made.

HUMAN DISTURBANCE ACTIVITIES

Escalating human populations and increased intensity of our consumption of natural resources are major threats to many plant and animal species (Mace and Reynolds 2001). For birds and mammals, exploitation (harvest) is second in importance only to habitat degradation as a cause of threat (fig. 3.20). Below we identify and discuss many of the human-caused factors that are impacting wildlife populations.

Population introductions. Populations of some species have been translocated or introduced to recover species extirpated from an area or to provide opportunities for hunting or fishing of

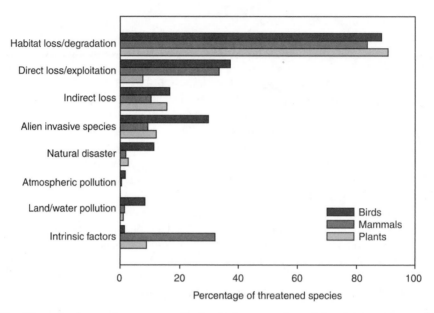

Figure 3.20.   The major threatening processes affecting birds, mammals, and plants. (From Mace and Reynolds 2001, fig. 1.2; by permission of the World Conservation Union.)

exotic species. Introductions, however, do not always succeed, particularly with threatened or endangered species recovery efforts. Introduction success has been related to initial population size, degree of environmental stochasticity, subsequent poaching or exploitation pressures, and other factors.

Berger (1990) reported that, in over 70 years of data with 122 populations of bighorn sheep (*Ovis canadensis*), 100 percent of populations with <50 individuals went extinct within 50 years, whereas populations with >100 individuals persisted for up to 70 years. Berger concluded that causes of extinction were not food shortages, severe weather, predation, or interspecific competition. Likely candidates for causes of extinction were demographic stochasticity caused by small population size and lack of demographic rescue effects (adequate numbers of im-

migrants supplementing the population during critical population lows). However, Berger's study did not quantify food supply relations and focused only on transplant population size. Additional work could examine records of historic vegetation conditions. The vegetation has likely changed in many of the areas Berger evaluated, some due to livestock grazing and some due to fire suppression. Also, the dynamics of individual transplant populations, seasonal occurrence and food quality on local transplant ranges, and whether transplants are from different ecotypes, need additional study. Krausman et al. (1993, 1996) concluded that population size alone may not be an accurate indicator of mountain sheep persistence for at least the desert races they studied in Arizona. They noted that most indigenous populations were founded with <50 individuals, which means that these small populations can, indeed, increase in size given the proper environmental conditions.

Another major disturbance element affecting native wildlife populations is intentional or unintentional introduction of exotic competitors, parasites, predators, or pathogens, as discussed above. In some cases, cultivars or horticultural escapees of plants have wreaked ecological havoc on native habitats and biota. Classic examples include the widespread dispersal of kudzu (*Pueraria lobata*) in southeastern U.S. forests, and the occurrence of dodder (*Cuscuta* spp.) on native flora and agricultural crops. Plants intentionally introduced for one reason may spread beyond their intended beneficial use. What seems amazing in hindsight is that kudzu was intentionally introduced in the South for erosion control and forage in the 1940s because of its rapid growth, ease of propagation, and wide adaptability, but those very characteristics have caused it to stifle native vegetation well beyond its intended range (OTA 1993). Many other examples of introduced species gone madly awry were cited by the OTA (1993).

Feral populations of domesticated goats (*Capra hircus*) and pigs (*Sus scrofa*), as well as introduced plants, have caused horrific ecological damage in isolated and fragile montane or island ecosystems, such as the Channel Islands off southern California or many of the main Hawaiian Islands. Located on the upper slopes of Haleakala Volcano on Maui, Hawaii, is Hosmer Grove, a plantation of exotic trees created in the early part of the 20th century to determine which tree species would do well in Hawaii for reforestation purposes. Curiously, the native koa (*Acacia koa*) and ohia (*Metrosideros collina*) trees were not included in the experiment. Species selected for the test were exotic eucalyptus, pines, acacias, and others. Ironically, none of the species did well or was selected for reforestation use, although at least the pines have escaped from the

grove and continue to encroach into the rare and declining adjacent native plant communities of The Nature Conservancy's Waikamoi Preserve (B. Marcot, pers. obs.). The preserve is one of the last remaining habitats for some of the rare and locally endemic native Hawaiian honeycreepers, particularly the akohekohe, or crested honeycreeper (*Palmeria dolei*) and the Maui parrotbill (*Pseudonestor xanthophrys*). Coupled with incursion into the preserve by feral goats and livestock, ecological pressure from exotic species continues to cause the native vegetation to suffer. Nothing short of ongoing herculean efforts by National Park Service personnel to fence out feral ungulates and manually remove exotic organisms will provide the native plant and animal populations with much of a future.

Controlling populations of ecologically harmful exotic species is typically an expensive and too often a fruitless proposition. For example, Bock and Bock (1992) found that both exotic grass species (lovegrasses, *Eragrostis* spp., native to southern Africa) and local native grass species in an Arizona grassland were equally tolerant of fire, probably because both evolved in fire-prone systems. Thus fire could not be used to eradicate the exotic species and recover the diverse native flora.

Hunting. "Overkill" of animals has been defined as a conspicuous decline in a population of hunted animals without prospect of stabilization or recovery. As reviewed by Murray (2003), overkill has been invoked as an explanation for late Pleistocene declines or extinctions of large mammals in Australia, New Guinea, and the Americas; the extinction of flightless birds in New Zealand in the 14th century; the decline of several species because of the North American fur trade in the 17th and 18th centuries; and the current harvest of wild animals for meat from tropical forests in western and central Africa.

Murray reviewed the major factors leading to overkill, including naïveté of prey regarding humans as predators; vulnerability of animals to easy killing by humans; introduction of new technology (e.g., snares, guns); and trade in animal parts. In contrast, sustainable hunting is encouraged by the availability of alternative food sources, ownership of wildlife resources, and the existence of cultural and spiritual beliefs.

Murray (2003) concluded that two main options are available to reduce overkill. First, we can make animals more difficult to market through restrictions in access, trade, and the use of modern technology in hunting. Second, we can provide resource users with a greater sense of protective ownership toward animals. Stating that we now place emphasis on granting exclusive rights to commercial harvesting, Murray suggested that greater emphasis could be placed on the spiritual and knowledge-based dimensions of ownership, and recommended incorporating traditional beliefs, values, and knowledge of indigenous people, as well as scientific knowledge, into sustainable hunting systems as a means of reducing overkill throughout the world.

From the habitat perspective, Cook and Cable (1990) argued for creation and conservation of windbreaks as an economic value for hunting. In agricultural and pasture fields of England, hedgerows—often the only remaining natural forest cover in an area—serve as key hiding and nesting habitat for many game and nongame birds and small mammals. The use of hedges and hedgerows for wildlife cover is not new. The wildlife literature addressed it as early as 1939 (Edminster 1939), and for centuries hedges and hedgerows have been used to provide cover for huntable game.

Some aspects of habitat conservation, particularly of summer range habitat for big game and wetlands for waterfowl, were originally initiated largely for hunting sport. Hunting activities, including archery, can cause changes in habitat use by animals. Conner et al. (2001) showed, for example, that early-season hunting could cause a change in the timing of elk movements onto private lands (fig. 3.21).

Harvesting techniques are under constant evaluation, in part to discover ways to increase huntable segments of populations. For example, Washington State implemented a "spike-only/ branched-bull-by-permit" (SOBBP) strategy to increase adult bull elk (*Cervus elaphus*) survivorship. Bender et al. (2002) showed that the SOBBP strategy did, indeed, increase total-bull and branched-bull ratios in hunted elk populations. These increased ratios, however, did not influence calf recruitment, likely because of the importance of female condition to production and survival of young.

In some cases, changing public attitudes toward hunting (Heberlein 1991) have led to new legislation. For example, Loker and Decker (1995) reported that controversies over hunting of black bears (*Ursus americanus*) in Colorado resulted in a referendum (Amendment 10) on the statewide ballot in November 1992 to prohibit specific hunting methods and spring hunting of bears. The vote was swayed by concern for the animals' welfare, largely from the nonhunting segment of the voting population. Presenting to the public the results of studies on the effects of hunting on population structure (numbers of individuals by sex and age class) and trend—such as with black bears (Miller 1990), moose (Miller and Ballard 1992), and Dall sheep (*Ovis dalli;* Murphy et al. 1990)—had little to do with the referendum vote, however. Krausman (2002) presented a succinct review of the differing views on how wildlife in general should be managed and used (fig. 3.22).

Should the economics of hunting change the fee system and procedure for funding habitat restoration and conservation activities on public lands (Williams and Mjelde 1994)? What should

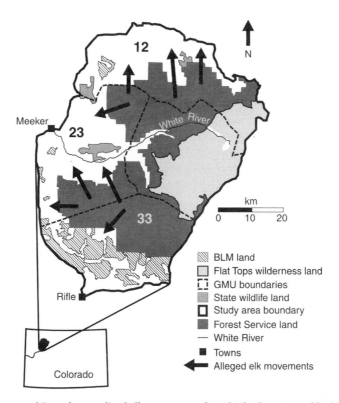

Figure 3.21. Land ownership and generalized elk movements from high-elevation public land to lower-elevation private land, White River area, Colorado (GMU = grazing management units). (From Conner et al. 2001, fig. 1 with permission from the Wildlife Society.)

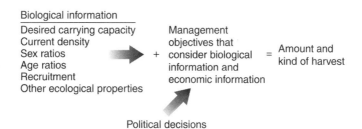

Figure 3.22. Biological information and management objectives are often influenced by political decisions in setting harvest levels. (From Krausman 2002, fig. 10-1.)

be the role of fee hunting in private lands in the United States (e.g., Smith et al. 1992)? The debate has yet to be settled.

Development. The effects of human developments and habitat alteration on wildlife populations are varied. Although much of the literature deals with the adverse effects of development on wildlife, there is also promise and potential for providing habitat conditions in urban environments for selected wildlife species more tolerant of human proximity (McKinney 2002). The kinds of habitat alterations that are acceptable are largely a social issue. Depending on wildlife conservation goals, at least some wildlife in areas of human habitation can be provided for, and the choice is not always one of strict preservation versus total usurpation.

Blair (1996) studied bird community diversity along an urban gradient in southern California and found that bird species richness, Shannon diversity, and bird biomass peaked at moderately disturbed sites, and that species composition varied among all sites. Patterns significantly related to differences in habitat structures, particularly in the percentage of land covered by pavement, buildings, lawn, grasslands, and trees or shrubs. His work suggests manipulable components in areas of more intense human habitation that would better provide for species tolerant to moderate disturbance. Urban park size may also play a role in providing habitat for urban bird populations; Donnelly and Marzluff (2004) found that urban reserves >42 hectares in size were more likely to protect native forest birds in urban settings. However, some native habitats, such as grasslands, and associated native wildlife populations may be significantly more sensitive to urban encroachment such as roads (Forman et al. 2002) and may require special conservation measures.

Bock et al. (2002) found that proximity to suburban edges had a strong negative effect on abundances of native grassland rodents in Colorado. They concluded that distance from suburban edge was a necessary but not a sufficient condition for rodents to achieve high densities, with other unmeasured factors (e.g., habitat quality, predator abundance) as likely contributing factors. The scarcity of avian and mammalian prey may also partially explain why many raptor species tend to avoid suburban boundaries.

In some cases, urbanization can tip the balance between species interactions to the detriment of species more closely associated with undeveloped habitats. For example, Engels and Sexton (1994) verified their hypothesis that urban development in the Austin, Texas, area favored blue jays (*Cyanocitta cristata*) to the detriment of the endangered golden-cheeked warbler (*Dendroica chrysoparia*). These researchers conjectured that, although the two species showed significant and inverse correlations to urban development, it was the foraging activity of jays in canopies of the warbler's juniper–oak woodland habitat that may have been the key factor. Jays may have deterred successful establishment of territories by male golden-cheeked warblers or successful attraction of warbler mates. But not all potential competitors produced adverse effects; the researchers did not find a negative correlation between the warblers and scrub jays (*Aphelocoma coerulescens*), which also occurred in the urban environments, but in lower numbers. These are good examples of why it is important to unravel the causal factors influencing population size and trends rather than focusing merely on habitats or environmental conditions.

Agricultural development is, of course, a major driver of both direct and indirect changes in the environment. Because agriculture encompasses up to 50% of the land area in portions of the United States, its impacts on wildlife are large. As reviewed by Murphy (2003), the term *agricultural practices* refers to decisions about the choice of crops, pesticides, fertilizers, and field maintenance, whereas the term *farmland structure* refers to the relative abundance and distribution of crops, pasture, and uncropped areas. We will not attempt to review the diversity of agricultural practices and farmland structures and the ways in which they can impact wildlife populations. It should be obvious, however, that conversion of natural areas into farmland will impact wildlife in species-specific manners that will vary depending on the intensity of the practices applied (e.g., see discussion below on pesticides).

Roads and other linear openings. Road networks and traffic volumes are steadily expanding in the United States and in most other countries. The total effects of roads and traffic have been estimated to directly affect about one-fifth of the land area of the conterminous United States (see Gibbs and Shriver 2002 for review). The effect of roads on wildlife populations, however, has not been well studied for many native species. Grizzly bear, wolverine (*Gulo gulo*), lynx (*Felis lynx*), Florida panther, and other large carnivores are thought to be particularly sensitive to presence of humans, including even sporadically used roads in otherwise rural or natural environments, although this theory needs empirical testing for some species. Some empirical evidence on grizzly bears (McLellan and Shackleton 1988) supports this theory, and road sensitivity is used in cumulative effects modeling of grizzly bears in the western United States. Mitigation of the adverse effects of roads on Florida panthers by providing highway underpasses has been somewhat successful (Foster and Humphrey 1995), although it is difficult to determine the cumulative effect of development and highways on this endangered subspecies. Vistnes and Nellemann (2001) showed that the combined actions of powerlines, roads, and cabins caused reindeer (*Rangifer t. tarandus*) to change their patterns of habitat use.

The effect of roads on small mammals has not been well studied; the topic needs much

work. Brock and Kelt (2004) found that dirt roads acted as landscape linkages, whereas gravel roads acted as movement barriers to Stephens' kangaroo rat (*Dipodomys stephensi*). In some cases, roads can serve as corridors for weeds and undesired species to spread into native environments (Watkins et al. 2003). In Australia, Seabrook and Dettmann (1996) found that roads in forested and densely vegetated habitats acted as activity corridors for the introduced toxic cane toad (*Bufo marinus*), a species that has the potential to do much ecological harm to native wildlife communities. However, in montane temperate environments, seldom-used rural roads can usefully provide travel lanes for snowshoe hares and other species.

Gibbs and Shriver (2002) implicated road mortality as a component of habitat fragmentation for land turtles and large-bodied pond turtles, and therefore as a potential contributor to their declines in several regions of the United States. Sisk and Battin (2002) reviewed the impacts of fragmentation on avian ecology, including situations in which roads and other artificial openings (e.g., powerline corridors) created edge effects. Although no studies they reviewed found community-scale decreases in bird diversity, four of seven showed increases in density and three of seven showed increased species richness.

Some effects of roads on habitat conditions and ecosystem processes have been studied. Some rural roads can change landscape processes. Amaranthus et al. (1985) reported that logging and forest roads contributed to debris slides in southwestern Oregon, which in turn adversely altered riparian and aquatic stream habitats through increased sedimentation. Off-road vehicles have the potential for altering habitat conditions and directly disturbing some wildlife, although careful management of sites provided for such use can do much to mitigate adverse effects (Lacey et al. 1982).

Another aspect of potential disturbance of wildlife populations from human activities is noise (Fletcher and Busnel 1978). Noise seems to inevitably accompany most human activities, particularly noise associated with roads and recreation. Many wildlife species seem to acclimate to repetitive highway noise; egrets and herons commonly forage or even nest along median strips of busy highways in many coastal areas, and desert ungulates acclimate to jet aircraft (Weisenberger et al. 1996). However, noise sensitivity in wildlife has been poorly studied, and we should not be fooled by a few incidental and salient observations. Studies have shown declines in high-frequency auditory ability by some wildlife in the presence of noise (Viemeister 1983), but few studies have tested the effect of noise on wildlife dependent on high-frequency communication.

Trails (footpaths) have also been shown to be detrimental to some animal species. For example, Swarthout and Steidl (2003) found that high levels of short-duration recreational hiking are detrimental to nesting Mexican spotted owls (*Strix occidentalis lucida*). No-entry buffer zones have been recommended for the spotted owl and other species to reduce disturbance during specific stages of a species' life cycle, such as breeding (e.g., Delaney et al. 1999; Swarthout and Steidl 2001, 2003). Taylor and Knight (2003) showed that human recreational activities such as hiking and biking have a predictable impact on certain wildlife species, including flushing distances (fig. 3.23). See also chapter 7 for a discussion of the impact of human activities on studies of animal behavior and habitat use.

Pesticides. Much has been researched and published on the effects of pesticides on wildlife populations. In fact, it was the use of DDT and organochlorines, and their biological magnification in the higher trophic levels of food webs (e.g., Johnston 1975), that spurred the environ-mental movement of the 1960s. The effects of pesticides on wildlife populations impelled the development of risk analysis procedures in the U.S. Environmental Protection Agency (Tiebout and Brugger 1995) and guidelines for measuring and avoiding toxicity to wildlife (Hudson et al. 1984).

Moulding (1976) monitored the impact of the insecticide SEVIN, used for controlling gypsy moths (*Porthetria dispar*), on a forest bird community in New Jersey. He reported a consistent, gradual decline in bird numbers, species richness, and diversity to 55% below control numbers of birds during 8 weeks following spraying, and declines continued the following summer although spraying had ceased. Canopy-foraging birds were affected more than ground foragers, possibly because of the greater proportion of invertebrates in their diets. In a study of herbicide effects, Sullivan (1990) studied the effects of forest application of glyphosate herbicide on recruitment, growth, and survival in deer mice (*Peromyscus maniculatus*) and Oregon voles (*Microtus oregoni*) and found a temporary suppression of deer mice populations and no effect on vole populations. The glyphosate treatment had little or no direct effect on metabolic or general physiological processes in the young mice or voles, and any individual effect did not manifest itself as changes of population densities and recruitment. Secondary effects of herbicides on habitat alteration were studied by Morrison and Meslow (1983) in western forest plantations, where response of vegetation and bird assemblages to use of the glyphosate 2,4-D was found to be short-lived.

Recently, concern has been raised over pesticides drifting long distances and impacting particularly sensitive species. For example, Davidson et al. (2002) found a strong positive association between declines in amphibian populations and the amount of upwind agricultural land use, suggesting that windborne pesticides may be an im-

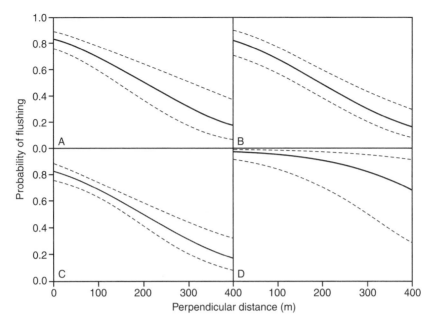

Figure 3.23.   Probability of wildlife flushing with increasing perpendicular distance: (A) bison; (B) pronghorn; (C) deer on-trail; (D) deer off-trail. Dashed lines indicate 95% confidence limits on probability. (From Taylor and Knight 2003, fig. 3; by permission of the Ecological Society of America.)

portant factor in the declines. For other species they studied, declines were strongly associated with local urban and agricultural use, which was potentially caused by habitat destruction. The patterns of association they found were not, however, consistent with declines being caused by ultraviolet radiation and climate change.

Nearly one-fifth of the United States is covered by cropland that for almost two generations has been chemically treated to create uniform monocultures. Pesticide use, along with other farming practices, affects the potential quality of fields as a foraging resource and is coming under increasing study because of a number of unforeseen consequences (see Beecher et al. 2002 for review). As such, various alternative farming practices are being investigated. Organic farming, for example, excludes synthetic fertilizers and pesticides and relies on mechanical and cul-

tural practices, organic on-farm inputs, and natural processes. Beecher et al. (2002) found that organic cropland sustained a greater abundance and species richness of birds than nonorganic croplands. They concluded that these differences were due to a larger resource base on the organic fields.

TRACKING LONG-TERM CHANGES IN
ABUNDANCE AND DISTRIBUTION

In North America, long-term monitoring of population attributes can help to determine the contribution of habitat to species conservation. Standardized surveys of breeding and wintering birds—through U.S. Fish and Wildlife Service Breeding Bird Surveys (BBS; Sauer et al. 2003b) and wintering bird counts (WBCs, or Audubon Christmas Bird Counts; Bock and Root 1981), respectively—provide broad-scale information from which long-term trends may be inferred (Sauer et al. 2003a). Except as occasionally integrated into long-term ecological research (LTER) sites and other lengthy ecological studies, few, if any, established broad-based and long-term monitoring programs have been instituted for other wildlife taxonomic groups, although such studies on amphibians are beginning. Further discussions of using BBS data for assessing and interpreting population trends were presented by Dunn (2002). Analysis methods of BBS data were discussed by Thomas and Martin (1996), and biases inherent in the data-collection methods were reviewed by Keller (1999).

BBS and other similar broad-scale survey and monitoring data by themselves do not explain the causes of population changes. For that, we need to marry geographic-specific evidence of population trends with information on changes of habitat and environmental conditions. One such approach is the well-established national Gap Analysis Program (GAP), which maps habitat conditions interpreted for each species (Scott et al. 1993). There is great potential to closely integrate the BBS and WBC data with GAP and other similar data on habitat distribution, to verify the predictions of species distributions from the GAP models, and to determine the potential contributions of habitat changes to population trends. However, additional studies on factors other than habitat are essential as well. There is a major leap between identifying vegetation and environmental correlates of the distribution of species and explaining the causal mechanisms of population change.

Guidelines for Habitat Management for Metapopulations

Given the dynamics of metapopulations in heterogeneous environments, guidelines for habitat management might include the following four categories:

1. Guidelines for maintaining species richness and overall biodiversity. Guidelines for species richness and overall biodiversity can be applied at three scales and can be derived from understanding how populations vary as a function of habitat complexity at each scale: (a) within habitat patches, or *alpha diversity,* such as with the observed correlations between foliar height diversity and bird species diversity; (b) between habitat patches, or *beta diversity;* and (c) among broader geographic areas, or *gamma diversity.* Beta diversity is often maintained by moderate disturbance regimes that provide for variations across space—typically within subbasins—in vegetation elements, substrates, and abiotic characteristics (including soils) of the environment. Beta diversity appears at different spatial scales for species with different body size, home range area, and vagility. Gamma diversity is often controlled by climate, landform, geographic location, and broad-scale vegetation formation features.

2. Guidelines for maintaining within-patch conditions. It may be desirable to ensure that specific habitat patches remain viable environmental units. In this case, size, topographic location, adjacency of other patches, and susceptibility to disturbances such as floods or fires can influence within-patch conditions (see chapter 9). Within-patch conditions are usually mediated primarily by patch size and within-patch dynamics but can also be mediated by the context of the patch (i.e., vegetation conditions in adjacent patches); the type and intensity of natural disturbances within the patch; and particularly the kind, frequency, and intensity of management activities within the patch.

3. Guidelines for maintaining a desired occupancy rate of habitat patches. A collection of habitat patches may be occupied by organisms over time, depending on the size and quality of the patches, the type and quality of the intervening matrix environment, and the spatial juxtaposition of the patches. In dynamic environments such as subtidal marine substrates or high-elevation fire-adapted forests, the quiltwork of patches may change over time under disturbance regimes and with the presence of species that provide physical surfaces and substrates ("structural" species sensu Huston 1994).

Empirical studies and models of populations in patchy environments can help provide realistic expectations for occupancy rates within and among habitat patches, which has important implications for designing inventory and monitoring studies and interpreting results of such studies. Depending on the species and habitat conditions, not all patches may be occupied all the time. Seemingly unoccupied patches may still be important as "stepping-stones" or key links for the distribution and dispersal of organisms throughout a landscape. Habitat patches can often be mapped and key links identified as part of a set of habitat features important to metapopulation dynamics (table 3.7). Stepping-stones or habitat corridors do not necessarily have to consist of primary habitat (i.e., provide for all life history needs to maximize realized fitness) in order to be valuable in connecting populations across a landscape.

Seemingly unoccupied patches may also serve as occasional sink habitat to house nonreproductive floaters not yet assimilated into the breeding segment of the population. Floaters are often important for maintaining populations and often occupy marginal habitats. Floaters have now been studied in a wide range of species, including gulls (Shugart et al. 1987); wrentits (Geupel et al. 1992); sparrows (Smith 1978); chickadees (Smith 1984, 1989); black kites (*Milvus migrans;* Blanco 1997); red-shouldered hawks (*Buteo lineatus elegans;* Bloom et al. 1993); tree swallows (*Tachycineta bicolor;* Barber and Robertson 1999); and many others.

Other situations that affect desired occupancy rates are listed in table 3.7. Habitat management guidelines can address these situations by ensuring that habitat for naturally disjunct populations and peripheral parts of a species' range are not overlooked (it is in such places that interesting, new "natural experiments" in species lineages take place); by restoring, or at least maintaining, habitat in bottlenecks, corridors, and key links to ensure connectivity of demes and subpopulations; and by accounting for what constitutes primary habitat when engaging in habitat acquisition or restoration programs. In addition, monitoring programs and metapopulation modeling projects should take into account locations of dispersal barriers and filters, areas with unsuitable environments, and population isolates.

4. Guidelines for habitat configuration. Configuration of habitat patches may be directly manipulated as a result of management activities; as an indirect result of other human land-altering activities such as growth of towns and roads; or as a result of stochastic and unpredictable natural disturbance events such as wildfire, mass wasting, vulcanism, and other dynamics. Where it is possible to guide direct management activities and account for indirect effects and natural disturbances, to provide for long-term viability of populations, habitats should be configured so as to afford rescue effects of habitat colonization by organisms over time (see chapter 9 for further discussion of rescue effects). Habitat patches can be managed to remain reachable by dispersing and migrating individuals and by organisms moving within their home ranges. Feeding and nesting habitats can be provided within daily dispersal distances or home range areas, such as has been included in habitat management guidelines for red-cockaded woodpeckers in forests of longleaf pine in the southeastern United States. Secondary or marginal habitats can be provided

113

Table 3.7. Mappable elements of habitat distribution and pattern important to maintaining metapopulations

Mappable element	Definition	Importance for maintaining populations
Habitat isolates	Segments of a species' distribution or habitat distribution that are separated from all other segments by at least one standard deviation of average dispersal distance (to maintain genetic continuity), or twice the average dispersal distance (to maintain demographic continuity), of the species.	Isolated or disjunct populations or reproductive individuals should be modeled separately. Small, isolated populations have higher susceptibility to decline or extirpation from a variety of factors. Subpopulations need to be connected both genetically and demographically to maintain continuity.
Habitat bottlenecks	A narrowing of habitat to accommodate only one connection between demes or populations across an area, rather than redundant connections in several directions.	Severing of habitat bottlenecks may partly or fully isolate demes or populations. Bottlenecks can help identify areas of habitat conservation or restoration priority
Habitat corridors	Areas of more or less stable habitat serving to link population centers. Corridors can facilitate actual movement of organisms between centers or provide for reproductive individuals so that centers are linked by transmission of genes.	Corridors serve to enlarge the effective (reproductive) size of populations, which increases the likelihood of persistence (viability). Some, however, might serve as "predator traps," or avenues of dispersal for undesirable exotic species.
Areas of unsuitable conditions	Areas lacking habitat for a species.	Unsuitable areas can help identify priorities for restoration or habitat creation if needed for population size or connectivity. Through empirical survey and modeling, these areas can help set realistic expectations for presence of organisms.
Habitat barriers	More than just areas of unsuitable conditions, barriers are absolute obstacles to movement and interaction of organisms.	Barriers demarcate the demographic boundary of populations and are important for devising realistic guidelines for population connectivity.
Key link habitat patches	Single patches that serve as critical links ("cut points") in the distribution of habitat in an area.	Key links may be vital conduits for ensuring connectivity within a population.
Degree of habitat suitability	In the simplest terms, two levels of suitability can be described: (1) primary, or source, habitat with sufficient total area, patch size, food, water, and substrates required by the species and that serve to maintain the population; (2) secondary, or sink, habitat that can temporarily hold nonreproductive or floater individuals, in which populations likely are not self-maintaining.	The degree of habitat suitability provides realistic expectations of reproductive performance of organisms within each class and identifies critical core habitats.
Peripheral or marginal habitat	Habitat occurring at the edges of a species' distributional range. (Note that secondary, or sink, habitat, above, might also occur elsewhere in a species' range.)	Peripheral habitats can provide conditions (selection pressures) for the emergence of new subspecies, ecotypes, morphs, or, eventually, species.
Habitat patch density	The number of patches of habitat within a general area. The size of patches and the general area would vary by species body size, life history, and home range size.	Habitat patch density provides a basis for calculating crude and ecological density (Caughley 1977).

close to primary habitats to help serve as sinks for surplus or floater individuals in good reproductive years.

Additional ideas on guidelines for managing and monitoring habitats in landscapes are presented in chapter 9.

Summary: Understanding and Managing Populations in the Context of Habitats and Biodiversity

This chapter has presented a number of topics related to population conservation in the context of habitat ecology. It is important to study and understand population dynamics to be informed of potential responses by species to conditions and changes in habitats and environments, and to understand the sometimes-limited degree to which habitat management alone can provide for populations. Although much conservation work is done under the assumption that wildlife are a function of their habitat, and that providing habitat provides for wildlife, the real world is far more complicated. It is a truism that habitat provides the basis for wildlife conservation, but it does not ensure it.

Considering Relations of Vegetation and Populations in Conservation

Population trends and viability are influenced not just by environmental conditions, but by environmental, demographic, and genetic conditions and variations, and by the presence and roles of other organisms. Few wildlife species exist as ideal genetic populations; most occur as metapopulations with weakly interacting subpopulations, or with isolated or disjunct population centers. These variations complicate the answers provided by simple models of minimum viable population sizes and suggest a need for assessing species individually rather than using a single golden rule for target population size. Small populations can often be at substantially greater risk of decline and extirpation.

Habitat conservation needs to attend to how populations occupy their distributional range, how organisms move through space, and how they respond to stochastic environments. Ultimately, these factors—along with others, including effects of exotic species and human disturbances—affect the evolution and expression of adaptive traits.

Long-term changes in populations can be provided by some ongoing surveys for breeding and wintering birds, but few such surveys have been instituted for other wildlife taxa. In the United States, state natural heritage programs provide sources of data on occurrence of rare plants and selected wildlife species, but not specifically under a trend-monitoring program. Some broad-based programs can help provide coarse-grained information on habitat distribution with which population presence can be projected or integrated, but we cannot ignore the important contributions of nonhabitat factors in affecting population density and trend.

Maintenance of habitats and wildlife populations is but one facet of maintaining overall biological diversity. We have provided a set of simple guidelines for maintaining biodiversity, species, and populations, particularly metapopulations in heterogeneous environments. These guidelines pertain to maintaining species richness and biodiversity, within-patch conditions, desired habitat occupancy rates, and habitat configuration for populations. Additional, species-specific guidelines would be needed on other, nonhabitat factors that influence population density and trend.

Table 3.8. Conditions of organisms, populations, and species warranting particular management attention for conservation of biodiversity

Ecological component	Description	Examples	Importance to biodiversity
Species richness	Number of species of particular taxonomic group	Total number of species by taxonomic class in an area	Total representation of phylogenetic taxa
Functional species redundancy	Number of species having the same key ecological function	Ungulate herbivores in African savannas; primary cavity excavators; primary burrow excavators	Redundancy provides continuation of ecological functions (at least at general level) should some species become extirpated
Isolated or disjunct populations	Demes and populations separated from others of the same species by more than their median dispersal distance	Western jumping mouse (*Zapus princeps*), on isolated mountain ranges of the Great Basin, western North America	Locally isolated populations can adapt to local conditions as first stages of allopatric speciation or development of ecotypes
Edges of distributional ranges of species	Peripheral parts of a species' range, often with suboptimal sink habitat conditions	Coati (*Nasua nasua*), in open forests of southeastern Arizona	Character divergence of organisms or local adaptation to conditions different from those in the core of a species' range can occur at range margins.
Coevolved species complexes	Species, with their environmental requirements, that have evolved in commensal or mutualistic relations	Pollinator–plant relations: for example, bats and calabash tree (*Crescentia*) or saguaro cactus (*Carnegiea*) in southwestern U.S. deserts	Changes in populations of one species can alter those of other species.
Sibling or cryptic species complexes	Closely related species nearly identical in morphology but with different resource use characteristics	Dusky flycatcher (*Empidonax oberholseri*) in early and mid-stage forests, and Hammond's flycatcher (*E. hammondii*) in old-growth forests in western North America	The full range of resource and environmental requirements of all species of a complex is not represented by just one species of a complex
Mobile links (Gilbert 1980).	Organisms significant in the persistence of several plant species that support separate food webs	Bees (Hymenoptera), butterflies and moths (Lepidoptera), flies (Diptera), and some beetles (Coleoptera) as obligate pollinators of many plant species in many parts of the world	Mobile links support the basis for several food webs in a community and numerous dependent species.
Keystone mutualists (Gilbert 1980)	Organisms (typically plants) that provide critical support to large complexes of mobile links	Fig trees (*Ficus* spp.) in many New and Old World tropics	Loss of keystone mutualists would trigger additional loss of mobile links, link-dependent plants, and decline in host-specialist insect diversity
Locally endemic species	A taxonomic species or morphospecies complex found in only one small geographic location	Glacier Bay water shrew (*Sorex alaskanus*), endemic to southeast Alaska	Locally endemic species represent a novel genetic entity.

Table 3.8.   Continued

Ecological component	Description	Examples	Importance to biodiversity
Locally endemic subspecies, forms, and varieties	Taxonomic subspecies or characteristically unique phenotypes of a species found in only one small geographic location	Prince of Wales flying squirrel (*Glaucomys sabrinus griseifrons*), Prince of Wales Island	Locally endemic subspecies, forms, and varieties represent a potential trend toward adaptation to local environmental conditions and the possible emergence of new species lineages.
Ecotypes	Locally adapted genotypes that perform better in their native environments than in other parts of their parent species' range	Mountain and northern ecotypes of woodland caribou (*Rangifer tarandus caribou*) in northern Idaho (see text)	Ecotypes signal evolution of local adaptations and contribute to the overall genetic diversity of the species.
Unique morphs and polymorphic populations	Local occurrences of organisms with unique phenotypic differences in meristics or coloration	Color morphs of northern rough-skinned newt (*Taricha g. granulosa*) in western U.S. ponds	Unique morphs and polymorphic populations could signal adaptive significance, as with differential survival to predation or harsh environmental conditions, or emergence of new genetic entities.

Note: Most of the components listed here would not be adequately represented simply by listing species presence, counting species in an area, or focusing conservation activity at the species level. However, these components may be of great interest in maintaining ecological processes of ecosystems and providing for evolutionary lineages in biogeographic settings fostering uniquely adapted organisms.

Toward a New Focus for Habitat and Population Management in an Ecological and Evolutionary Context

Many ecological entities poorly represented by accepted taxonomies, particularly of species and subspecies, may nonetheless warrant attention for conserving the evolutionary potential of species lineages. Such entities include demes, subpopulations, disjunct populations, ecotypes, ecological functional groups of species, organisms at the periphery of their distributional range, organisms at the limits of their range of physiological tolerance, and organisms with ecological roles central to maintaining the richness of an ecological community (table 3.8).

Groups of organisms in these conditions usually do not fit traditional Linnaean taxonomy and are typically excluded from administrative conservation programs. They nonetheless constitute valid ecological categories deserving of understanding and potential conservation focus, as suggested by lessons and examples presented in this chapter by woodland caribou, garter snakes, red-legged frogs, and many others. Ultimately, public policy makers and managers must decide whether they will merely maintain current conditions for the short term, or attempt to manage evolutionary conditions for the long term.

Literature Cited

Alerstam, T., A. Hedenstrom, and S. Akesson. 2003. Long-distance migration: Evolution and determinants. *Oikos* 103 (2):247–60.

Allen, S. H., and A. B. Sargeant. 1993. Dispersal patterns of red foxes relative to population density. *Journal of Wildlife Management* 57 (3):526–33.

Allendorf, F. W. 1986. Genetic drift and the loss of alleles versus heterozygosity. *Zoo Biology* 5:181–90.

Allendorf, F. W., and L. L. Lundquist. 2003. Introduction: Population biology, evolution, and control of invasive species. *Conservation Biology* 17 (1):24–30.

Amaranthus, M. P., R. M. Rice, N. R. Barr, and R. R. Ziemer. 1985. Logging and forest roads related to increased debris slides in southwestern Oregon. *Journal of Forestry* 83:229–33.

American Ornithologists' Union. 2003. *Check-list of North American birds.* 7th ed. Ithaca, NY: Committee on Classification and Nomenclature of the American Ornithologists' Union.

Anderson, V. J., and D. D. Briske. 1995. Herbivore-induced species replacement in grasslands: Is it driven by herbivory tolerance or avoidance? *Ecological Applications* 5 (4):1014–24.

Arthur, W. 1987. *The niche in competition and evolution.* New York: Wiley.

Askins, R. A. 1995. Hostile landscapes and the decline of migratory songbirds. *Science* 267:1956–57.

Baguette, M., and N. Schtickzelle. 2003. Local population dynamics are important to the conservation of metapopulations in highly fragmented landscapes. *Journal of Applied Ecology* 40 (2):404–12.

Baguette, M., and V. M. Stevens. 2003. Local populations and metapopulations are both natural and operational categories. *Oikos* 101 (3):661–63.

Bahre, C. J. 1991. *A legacy of change: Historic human impact on vegetation of the Arizona borderlands.* Tucson: Univ. of Arizona Press.

Baker, J. M., E. López-Medrano, A. G. Navarro-Sigüenza, O. R. Rojas-Soto, and K. E. Omland. 2003. Recent speciation in the orchard oriole group: Divergence of *Icterus spurius spurius* and *Icterus spurius fuertesi. Auk* 120 (3):848–59.

Barber, C. A., and R. J. Robertson. 1999. Floater males engage in extrapair copulations with resident female tree swallows. *Auk* 116 (1):264–69.

Barbour, M. G. 1996. American ecology and American culture in the 1950s: Who led whom? *Bulletin of the Ecological Society of America* 77:44–51.

Barbraud, C., J. D. Nichols, J. E. Hines, and H. Hafner. 2003. Estimating rates of local extinction and colonization in colonial species and an extension to the metapopulation and community levels. *Oikos* 101 (1):113–26.

Batzli, G. O. 1994. Special feature: Mammal–plant interactions. *Journal of Mammalogy* 75:813–15.

Beecher, N. A., R. J. Johnson, J. R. Brandle, R. M. Case, and L. J. Young. 2002. Agroecology of birds in organic and nonorganic farmland. *Conservation Biology* 16:1620–31.

Beier, P. 1995. Dispersal of juvenile cougars in fragmented habitat. *Journal of Wildlife Management* 59 (2):228–37.

Belthoff, J. R., and G. Ritchison. 1989. Natal dispersal of eastern screech-owls. *Condor* 91 (2):264–65.

Bender, L. C., P. E. Fowler, J. A. Bernatowicz, J. L. Musser, and L. E. Stream. 2002. Effects of open-entry spike-bull, limited-entry branched-bull harvesting on elk composition in Washington. *Wildlife Society Bulletin* 30:1078–84.

Benkman, C. W. 1999. The selection mosaic and diversifying coevolution between crossbills and lodgepole pine. *American Naturalist* 153:S75–91.

Bennetts, R. E., J. D. Nichols, J.-D. Lebreton, R. Pradel, J. E. Hines, and W. M. Kitchens. 2001. Methods for estimating dispersal probabilities and related parameters using marked animals. In *Dispersal,* ed. J. Clobert, E. Danchin, A. A. Dhondt, and J. D. Nichols. Oxford: Oxford Univ. Press.

Berger, J. 1990. Persistence of different-sized populations: An empirical assessment of rapid extinctions in bighorn sheep. *Conservation Biology* 4:91–98.

Blair, R. B. 1996. Land use and avian species diversity along an urban gradient. *Ecological Applications* 6 (2):506–19.

Blanco, G. 1997. Role of refuse as food for migrant, floater and breeding black kites (*Milvus migrans*). *Journal of Raptor Research* 31 (1):71–76.

Block, W. M., M. L. Morrison, J. Verner, and P. H. Manley. 1994. Assessing wildlife–habitat relationships: A case study with California oak woodlands. *Wildlife Society Bulletin* 22:549–61.

Blondel, J. 2003. Guilds or functional groups: does it matter? *Oikos* 100(2):223–231.

Bloom, P. H., M. D. McCrary, and M. J. Gibson. 1993. Red-shouldered hawk home-range and habitat use in southern California. *Journal of Wildlife Management* 57 (2):258–65.

Bock, C. E., and T. L. Root. 1981. The Christmas bird count and avian ecology. *Studies in Avian Biology* 6:17–23.

Bock, C. E., K. T. Vierling, S. L. Haire, J. D. Boone, and W. W. Merkle. 2002. Patterns of rodent abundance on open-space grasslands in relation to suburban edges. *Conservation Biology* 16:1653–58.

Bock, J. H., and C. E. Bock. 1992. Vegetation responses to wildfire in native versus exotic Arizona grassland. *Journal of Vegetation Science* 3 (4):439–46.

Bokdam, J., and M. F. W. de Vries. 1992. Forage quality as a limiting factor for cattle grazing in isolated Dutch nature reserves. *Conservation Biology* 6:399–408.

Bolger, D. T., T. A. Scott, and J. T. Rotenberry. 2001. Use of corridor-like landscape structures by bird and small mammal species. *Biological Conservation* 102:213–24.

Bonnell, M. L., and R. K. Selander. 1974. Elephant seals: Genetic variation and near extinction. *Science* 184:908–9.

Bonser, S. P., and R. J. Reader. 1995. Plant competition and herbivory in relation to vegetation biomass. *Ecology* 76:2176–83.

Breininger, D. R., J. J. Barkaszi, R. B. Smith, D. M. Oddy, and J. A. Provancha. 1998. Prioritizing wildlife taxa for biological diversity conservation at the local scale. *Environmental Management* 22:315–21.

Brennan, L. A., M. L. Morrison, and D. L. Dahlsten. 2000. Comparative foraging dynamics of chestnut-backed and mountain chickadees in the western Sierra Nevada. *Northwestern Naturalist* 81: 129–47.

Brock, R. E., and D. A. Kelt. 2004. Influence of roads on the endangered Stephens' kangaroo rat (*Dipodomys stephensi*): Are dirt and gravel roads different? *Biological Conservation* 118:633–40.

Brown, J. H., D. W. Mehlman, and G. C. Stevens. 1995. Spatial variation in abundance. *Ecology* 76 (7): 2028–43.

Buchmann, S. L., and G. P. Nabhan. 1996. *The forgotten pollinators.* Washington, DC: Island Press.

Buechner, M. 1987. A geometric model of vertebrate dispersal: Tests and implications. *Ecology* 68:310–18.

Burrow, A. L., R. T. Kazmaier, E. C. Hellgren, and D. C. Ruthven III. 2001. Microhabitat selection by Texas horned lizards in southern Texas. *Journal of Wildlife Management* 65:645–52.

Cane, J. H., and V. J. Tepedino. 2001. Causes and extent of declines among native North American invertebrate pollinators: Detection, evidence, and consequences. *Conservation Ecology* 5 (1):1; www.consecol.org/vol5/iss1/art1.

Caswell, H. 2001. *Matrix population models: Construc-*

tion, analysis, and interpretation. Sunderland, MA: Sinauer.

Caughley, G. 1977. *Analysis of vertebrate populations.* New York: Wiley.

Chambers, S. M. 1995. Spatial structure, genetic variation, and the neighborhood adjustment to effective population size. *Conservation Biology* 9 (5):1312–15.

Cheveau, M., P. Drapeau, L. Imbeau, and Y. Bergeron. 2004. Owl winter irruptions as an indicator of small mammal population cycles in the boreal forest of eastern North America. *Oikos* 107 (1):190–98.

Clark, J. S., M. Lewis, J. S. McLachlan, and J. Hille Ris Lambers. 2003. Estimating population spread: What can we forecast and how well? *Ecology* 84 (8):1979–88.

Clements, F. E. 1904. *The development and structure of vegetation.* Botanical Survey of Nebraska 7: Studies in the Vegetation of the State 3. Lincoln: University of Nebraska.

Clements, F. E. 1916. *Plant succession.* Publication no. 242. Washington, DC: Carnegie Institute of Washington.

Clements, F. E. 1920. *Plant indicators.* Publication no. 290. Washington DC: Carnegie Institute of Washington.

Cody, M. L. 2000. Antbird guilds in the lowland Caribbean rainforests of southeast Nicaragua. *Condor* 102:784–94.

Cole, F. R., L. L. Loope, A. C. Medeiros, J. A. Raikes, and C. S. Wood. 1995. Conservation implications of introduced game birds in high-elevation Hawaiian shrubland. *Conservation Biology* 9 (2): 306–13.

Collins, J. T., and Taggart, T. W. 2002. *Standard common and current scientific names for North American amphibians, turtles, reptiles, and crocodilians.* 5th ed. Lawrence, KS: Northern Society for the Study of Amphibians and Reptiles.

Collins, S. L. 1983. Geographic variation in habitat structure of the black-throated green warbler (*Dendroica virens*). *Auk* 100:382–89.

Conner, M. M., G. C. White, and D. J. Freddy. 2001. Elk movement in response to early-season hunting in northwest Colorado. *Journal of Wildlife Management* 65:926–40.

Cook, P. S., and T. T. Cable. 1990. The economic value of windbreaks for hunting. *Wildlife Society Bulletin* 18 (3):337–42.

119

Culotta, E. 1995. Minimum population grows larger. *Science* 270 (5233):31–32.

Daniels, S. J., J. A. Priddy, and J. R. Waters. 2000. Inbreeding in small populations of red-cockaded woodpeckers: Insights from a spatially explicit individual-based model. In *Genetics, demography, and viability of fragmented populations,* ed. A. G. Young and G. M. Clarke, 129–47. Conservation Biology 4. Cambridge: Cambridge Univ. Press.

Dasmann, R. F. 1964. *Wildlife biology.* New York: Wiley.

Davidson, C., H. B. Shaffer, and M. K. Jennings. 2002. Spatial tests of the pesticide-drift, habitat destruction, UV-B, and climate-change hypotheses for California amphibian declines. *Conservation Biology* 16:1588–1601.

DeGraaf, R. M., and N. L. Chadwick. 1987. Forest type, timber size class, and New England breeding birds. *Journal of Wildlife Management* 51:212–16.

DeGraaf, R. M., and D. D. Rudis. 1990. Herpetofaunal species composition and relative abundance among three New England forest types. *Forest Ecology and Management* 32:155–65.

Delaney, D. K., T. G. Grubb, P. Beier, L. L. Pater, and R. M. Reiser. 1999. Effects of helicopter noise on Mexican spotted owls. *Journal of Wildlife Management* 63:60–76.

Delaney, K. S., and R. K. Wayne. 2005. Adaptive units for conservation: Population distinction and historic extinctions in the island scrub-jay. *Conservation Biology* 19 (2):523–33.

Dennis, B. 2002. Allee effects in stochastic populations. *Oikos* 96 (3):389–401.

Diamond, J. M. 1984. "Normal" extinctions of isolated populations. In *Extinctions,* ed. M. H. Nitecki, 191–246. Chicago: Univ. of Chicago Press.

Donnelly, R., and J. M. Marzluff. 2004. Importance of reserve size and landscape context to urban bird conservation. *Conservation Biology* 18:733–45.

Donovan, T. M., and F. R. Thompson III. 2001. Modeling the ecological trap hypothesis: A habitat and demographic analysis for migrant songbirds. *Ecological Applications* 11:871–82.

Dunn, E. H. 2002. Using decline in bird populations to identify needs for conservation action. *Conservation Biology* 16 (6):1632–37.

Edminster, F. C. 1939. Hedge plantings for erosion control and wildlife management. *Transactions of the North America Wildlife and Natural Resources Conference* 4:534–41.

Ehrlich, P. R., D. S. Dobkin, and D. Wheye. 1988. *The birder's handbook.* New York: Simon & Schuster.

Engels, T. M., and C. W. Sexton. 1994. Negative correlation of blue jays and golden-cheeked warblers near an urbanizing area. *Conservation Biology* 8 (1):286–90.

Feinsinger, P., and E. E. Spears. 1981. A simple measure of niche breadth. *Ecology* 62:27–32.

Fleishman, E., R. Mac Nally, and J. P. Fay. 2003. Validation tests of predictive models of butterfly occurrence based on environmental variables. *Conservation Biology* 17 (3):806–17.

Fletcher, J. L., and R. G. Busnel. 1978. *Effects of noise on wildlife.* New York: Academic Press.

Floyd, C. H., D. H. Van Vuren, and B. May. 2005. Marmots on Great Basin mountaintops: Using genetics to test a biogeographic paradigm. *Ecology* 86 (8):2145–53.

Forman, R. T. T., B. Reineking, and A. M. Hersperger. 2002. Road traffic and nearby grassland bird patterns in a suburbanizing landscape. *Environmental Management* 29:782–800.

Fortin, M.-J., T. H. Keitt, B. A. Maurer, M. L. Taper, D. M. Kaufman, and T. M. Blackburn. 2005. Species' geographic ranges and distributional limits: Pattern analysis and statistical issues. *Oikos* 108 (1):7–17.

Foster, M. L., and S. R. Humphrey. 1995. Use of highway underpasses by Florida panthers and other wildlife. *Wildlife Society Bulletin* 23 (1):95–100.

Frankham, R., J. D. Ballou, and D. A. Briscoe. 2004. *A primer of conservation genetics.* Cambridge: Cambridge Univ. Press.

Frankham, R., J. D. Ballou, and D. A. Briscoe. 2002. Introduction to conservation genetics. New York : Cambridge University Press.

Futuyma, D. J. 1986. *Evolutionary biology.* 2nd ed. Sunderland, MA: Sinauer.

Gehlbach, F. R. 2003. Body size variation and evolutionary ecology of eastern and western screech-owls. *Southwestern Naturalist* 48:70–80.

Geupel, G. R., O. E. Williams, and N. Nur. 1992. Factors influencing the variation in abundance of "floaters" in a population of wrentits. Paper presented at the 62nd annual meeting of the Cooper Ornithological Society, Seattle, WA.

Gibbons, J. R. H., and H. B. Lillywhite. 1981. Ecological segregation, color matching, and speciation in lizards of the *Amphibolurus decresii* species complex (Lacertilia: Agamidae). *Ecology* 62 (6):1573–84.

Gibbs, J. P., and W. G. Shriver. 2002. Estimating the effects of road mortality on turtle populations. *Conservation Biology* 16:1647–52.

Gilbert, L. E. 1980. Food web organization and conservation of neotropical diversity. In *Conservation biology: An evolutionary-ecological perspective,* ed. M. E. Soulé and B. A. Wilcox, 11–33. Sunderland, MA: Sinauer.

Gilpin, M. E., and M. E. Soulé. 1986. Minimum viable populations: Processes of species extinction. In *Conservation biology: The science of scarcity and diversity,* ed. M. E. Soulé and B. A. Wilcox, 19–34. Sunderland, MA: Sinauer.

Ginzburg, L. R., L. B. Slobodkin, K. Johnson, and A. G. Bindman. 1982. Quasiextinction probabilities as a measure of impact on population growth. *Risk Analysis* 2:171–81.

Gleason, H. A. 1939. The individualistic concept in plant succession. *American Midland Naturalist* 21:92–110.

Glenn-Lewin, D. C., R. K. Peet, and T. T. Veblen, eds. 1992. *Plant succession: Theory and prediction.* New York: Chapman & Hall.

Gosz, J. R. 1993. Ecotone hierarchies. *Ecological Applications* 3 (3):369–76.

Grinder, M. I., and P. R. Krausman. 2001. Home range, habitat use, and nocturnal activity of coyotes in an urban environment. *Journal of Wildlife Management* 65:887–98.

Grinnell, J. 1917. Field tests of theories concerning distributional control. *American Naturalist* 51:115–28.

Guthery, F. S., and R. L. Bingham. 1992. On Leopold's principle of edge. *Wildlife Society Bulletin* 20:340–44.

Guthrie, C. G., and D. L. Moorhead. 2002. Density-dependent habitat selection: Evaluating isoleg theory with a Lotka-Volterra model. *Oikos* 97 (2):184–94.

Haas, C. A. 1995. Dispersal and use of corridors by birds in wooded patches on an agricultural landscape. *Conservation Biology* 9 (4):845–54.

Haddad, N. M., D. R. Bowne, A. Cunningham, B. J. Danielson, D. J. Levey, S. Sargent, and T. Spira. 2003. Corridor use by diverse taxa. *Ecology* 84 (3):609–15.

Haddad, N. M., and J. J. Tewksbury. 2005. Low-quality habitat corridors as movement conduits for two butterfly species. *Ecological Applications* 15 (1):250–57.

Hannon, S. J., and F. K. A. Schmiegelow. 2002. Corridors may not improve the conservation value of small reserves for most boreal birds. *Ecological Applications* 12:1457–68.

Hayes, M. P., and D. M. Krempels. 1986. Vocal sac variation among frogs of the genus *Rana* from western North America. *Copeia* 1986 (4):927–36.

Hayes, M. P., and M. M. Miyamoto. 1984. Biochemical, behavioral and body size differences between *Rana aurora aurora* and *R. a. draytonii. Copeia* 1984 (4):1018–22.

Heberlein, T. A. 1991. Changing attitudes and funding for wildlife: Preserving the sport hunter. *Wildlife Society Bulletin* 19 (4):528–34.

Heglund, P. J. 2002. Foundations of species–environment relations. In *Predicting species occurrences: Issues of accuracy and scale,* ed. J. M. Scott, P. J. Heglund, M. L. Morrison, J. B. Haufler, M. G. Raphael, W. A. Wall, and F. B. Samson, 35–41. Washington, DC: Island Press.

Hejl, S. J., and J. Verner. 1990. Within-season and yearly variation in avian foraging locations. *Studies in Avian Biology* 13:202–9.

Hilden, O. 1965. Habitat selection in birds. *Annales Zoologici Fennici* 2:53–75.

Hill, C. J. 1995. Linear strips of rain forest vegetation as potential dispersal corridors for rain forest insects. *Conservation Biology* 9 (6):1559–66.

Hitchings, S. P., and T. J. C. Beebee. 1998. Loss of genetic diversity and fitness in common toad (*Bufo bufo*) populations isolated by inimical habitat. *Journal of Environmental Biology* 11 (3):269–83.

Hochachka, W. M., J. V. Wells, K. V. Rosenberg, D. L. Tassaglia-Hymes, and A. A. Dhondt. 1999. Irruptive migration of common redpolls. *Condor* 101 (2):195–204.

Holdridge, L. R. 1947. Determination of world plant formations from simple climatic data. *Science* 105:367–68.

Hopkins, M. S., J. Ash, A. W. Graham, J. Head, and R. K. Hewett. 1993. Charcoal evidence of the spatial extent of the Eucalyptus woodland expansions and raindforest contractions in North Queensland during the late Pleistocene. *Journal of Biogeography* 2:0357–72.

Horsley, S. B., S. L. Stout, and D. S. deCalesta. 2003. White-tailed deer impact on the vegetation dynamics of a northern hardwood forest. *Ecological Applications* 13 (1):98–118.

Hudson, R. H., R. K. Tucker, and M. A. Haegele. 1984. *Handbook of toxicity of pesticides to wildlife.*

Resource Publication 153. Washington, DC: USDI Fish and Wildlife Service.

Husheer, S. W., D. A. Coomes, and A. W. Robertson. 2003. Long-term influences of introduced deer on the composition and structure of New Zealand *Nothofagus* forests. *Forest Ecology and Management* 181 (1–2):99–117.

Huston, M. A. 1994. *Biological diversity: The coexistence of species on changing landscapes.* Cambridge: Cambridge Univ. Press.

Huston, M. A. 2002. Introductory essay: Critical issues for improving predictions. In *Predicting species occurrences: Issues of accuracy and scale,* ed. J. M. Scott, P. J. Heglund, M. L. Morrison, J. B. Haufler, M. G. Raphael, W. A. Wall, and F. B. Samson, 7–21. Washington, DC: Island Press.

Hutto, R. L. 1985. Habitat selection by nonbreeding, migratory land birds. In *Habitat selection in birds,* ed. M. L. Cody, 455–76. San Diego, CA: Academic Press.

Hyde, E. J., and T. R. Simons. 2001. Sampling plethodontid salamanders: Sources of variability. *Journal of Wildlife Management* 65:624–32.

James, F. C. 1971. Ordinations of habitat relationships among breeding birds. *Wilson Bulletin* 83:215–36.

James, F. C., C. A. Hess, B. C. Kicklighter, and R. A. Thum. 2001. Ecosystem management and the niche gestalt of the red-cockaded woodpecker in longleaf pine forests. *Ecological Applications* 11:854–70.

Johnson, D. H. 1980. The comparison of usage and availability measurements for evaluating resource preference. *Ecology* 61:65–71.

Johnson, D. H., and L. D. Igl. 2001. Area requirements of grassland birds: A regional perspective. *Auk* 118:24–34.

Johnson, D. M. 2005. Metapopulation models: An empirical test of model assumptions and evaluation methods. *Ecology* 86 (11):3088–98.

Johnston, D. W. 1975. Organochlorine pesticide residues in small migratory birds, 1964–73. *Pesticides Monitoring Journal* 9:79–88.

Jones, C. G., J. H. Lawton, and M. Shachak. 1996. Organisms as ecosystem engineers. In *Ecosystem management: Selected readings,* ed. F. B. Samson and F. L. Knopf, 130–47. Springer, New York.

Jones, J. L., and J. M. Diamond. 1976. Short-time-base studies of turnover in breeding bird populations on the California Channel Islands. *Condor* 78:526–49.

Jones, K. J., Jr., R. S. Hoffmann, D. W. Rice, C. Jones, R. J. Baker, and M. D. Engstrom. 1992. *Revised checklist of North American mammals north of Mexico, 1991.* Occasional paper 146. Lubbock: Museum of Texas Tech Univ.

Karr, J. R., and R. R. Roth. 1971. Vegetation structure and avian diversity in several New World areas. *American Naturalist* 105:423–35.

Keane, J. J., and M. L. Morrison. 1999. Temporal variation in resource use by black-throated gray warblers. *Condor* 101:67–75.

Keller, C. M. E. 1999. Potential roadside biases due to habitat changes along breeding bird survey routes. *Condor* 101:50–57.

Keller, L. F., and D. M. Waller. 2002. Inbreeding effects in wild populations. *Trends in Ecology and Evolution* 17:230–41.

Kelly, J. P., and C. Woods. 1996. Diurnal, intrasexual, and intersexual variation in foraging behavior of the common yellowthroat. *Condor* 98:491–500.

Kimura, M., and J. F. Crow. 1963. The measurement of effective population number. *Evolution* 17:279–288.

Kindlmann, P., S. Aviron, and F. Burel. 2005. When is landscape matrix important for determining animal fluxes between resource patches? *Ecological Complexity* 2 (2):150–58.

Klopfer, P. H. 1959. An analysis of learning in young Anatidae. *Ecology* 40 (1):90–102.

Koenig, W. D. 2001. Spatial autocorrelation and local disappearances in wintering north American birds. *Ecology* 82 (9):2636–44.

Krausman, P. R. 2000. An introduction to the restoration of bighorn sheep. *Restoration Ecology* 8:3–5.

Krausman, P. R. 2002. *Introduction to wildlife management: The basics.* Upper Saddle River, NJ: Prentice Hall.

Krausman, P. R., R. C. Etchberger, and R. M. Lee. 1993. Persistence of mountain sheep. *Conservation Biology* 7:219.

Krausman, P. R., R. C. Etchberger, and R. M. Lee. 1996. Persistence of mountain sheep populations in Arizona. *Southwestern Naturalist* 41:399–402.

Kuenzi, A. J., and M. L. Morrison. 2003. Temporal patterns of bat activity in southern Arizona. *Journal of Wildlife Management* 67:52–64.

Lacey, R. M., R. S. Baran, H. E. Balbach, R. G. Goettel, and W. D. Severinghaus. 1982. Off-road vehicle site selection. *Journal of Environmental Systems* 12:113–40.

Lacy, R. C. 1994. Managing genetic diversity in captive populations of animals. In Restoration and Recovery of Endangered Plants and Animals, ed. M. L. Bowles and C. J. Whelan, 63–89. Cambridge: Cambridge University Press.

Laikre, L., and N. Ryman. 1991. Inbreeding depression in a captive wolf (*Canis lupus*) population. *Conservation Biology* 5 (1):33–40.

Lande, R. 1987. Extinction thresholds in demographic models of territorial populations. *American Naturalist* 130 (4):624–35.

Lande, R., and G. F. Barrowclough. 1987. Effective population size, genetic variation, and their use in population management. In *Viable populations for conservation,* ed. M. E. Soulé, 87–123. New York: Cambridge Univ. Press.

Legendre, S., J. Clobert, A. P. Meller, and G. Sorci. 1999. Demographic stochasticity and social mating system in the process of extinction of small populations: The case of passerines introduced to New Zealand. *American Naturalist* 153:449–63.

Leopold, A. 1933. *Game management.* New York: Scribner's.

Levine, J. M. 2003. A patch modeling approach to the community-level consequences of directional dispersal. *Ecology* 84 (5):1215–24.

Lindenmayer, D. B., and H. A. Nix. 1993. Ecological principles for the design of wildlife corridors. *Conservation Biology* 7 (3):627–30.

Lodge, D. M., and K. Shrader-Frechette. 2003. Nonindigenous species: Ecological explanation, environmental ethics, and public policy. *Conservation Biology* 17 (1):31–37.

Loker, C. A., and D. J. Decker. 1995. Colorado black bear hunting referendum: What was behind the vote? *Wildlife Society Bulletin* 23 (3):370–76.

Luken, J. O. 1990. *Directing ecological succession.* New York: Chapman & Hall.

MacArthur, R. H., and J. W. MacArthur. 1961. On bird species diversity. *Ecology* 42:594–98.

MacArthur, R. H. 1968. The theory of the niche. In *Population biology and evolution,* ed. R. C. Lewontin, 159–76. New York: Syracuse Univ. Press.

Macdonald, D. W., and D. D. P. Johnson. 2001. Dispersal in theory and practice: Consequences for conservation biology. In *Dispersal,* ed. J. Clobert, E. Danchin, A. A. Dhondt, and J. D. Nichols, 358–72. Oxford: Oxford Univ. Press.

Mace, G. M., and J. D. Reynolds. 2001. Exploitation as a conservation issue. In *Conservation of exploited species,* ed. J. D. Reynolds, G. M. Mace, K. H. Redford, and J. G. Robinson, 3–15. Cambridge: Cambridge Univ. Press.

Maehr, D. S., and G. B. Caddick. 1995. Demographics and genetic introgression in the Florida panther. *Conservation Biology* 9 (5):1295–98.

Marcot, B. G. 1985. *Habitat relationships of birds and young-growth Douglas-fir in northwestern California.* PhD dissertation, Oregon State Univ., Corvallis.

Marcot, B. G. 1996. An ecosystem context for bat management: A case study of the interior Columbia River Basin, USA. In *Bats and forests symposium, October 19–21, 1995, Victoria, BC,* ed. R. M. R. Barclay and R. M. Brigham, 19–36. Working Paper 23. Victoria, BC: Research Branch, British Columbia Ministry of Forests.

Marcot, B. G., and K. B. Aubry. 2003. The functional diversity of mammals in coniferous forests of western North America. In *Mammal community dynamics: Management and conservation in the coniferous forests of western North America,* ed. C. J. Zabel and R. G. Anthony, 631–64. Cambridge: Cambridge Univ. Press.

Marcot, B. G., M. A. Castellano, J. A. Christy, L. K. Croft, J. F. Lehmkuhl, R. H. Naney, K. Nelson, C. G. Niwa, R. E. Rosentreter, R. E. Sandquist, B. C. Wales, and E. Zieroth. 1997. Terrestrial ecology assessment. In *An assessment of ecosystem components in the interior Columbia Basin and portions of the Klamath and Great Basins,* vol. 3, ed. T. M. Quigley and S. J. Arbelbide, 1497–1713. USDA Forest Service General Technical Report PNW-GTR-405. Portland, OR: Pacific Northwest Research Station, USDA Forest Service.

Marcot, B. G., L. K. Croft, J. F. Lehmkuhl, R. H. Naney, C. G. Niwa, W. R. Owen, and R. E. Sandquist. 1998. *Macroecology, paleoecology, and ecological integrity of terrestrial species and communities of the interior Columbia River Basin and portions of the Klamath and Great Basins.* USDA Forest Service General Technical Report PNW-GTR-410. Portland OR: Pacific Northwest Research Station, USDA Forest Service.

Margalef, R. 1963. On certain unifying principles in ecology. *American Naturalist* 97:357–74.

Marshall, K., and G. Edwards-Jones. 1998. Reintroducing capercaillie (*Tetrao urogallus*) into southern Scotland: Identification of minimum viable populations at potential release sites. *Biodiversity and Conservation* 7:275–96.

Martinez-Solano, I., J. Bosch, and M. Garcia-Paris. 2003. Demographic trends and community stability in a montane amphibian assemblage. *Conservation Biology* 17 (1):238–44.

Maruyama, T. 1977. *Stochastic problems in population genetics*. Berlin: Springer-Verlag.

Marzluff, J. M., J. J. Millspaugh, P. Hurvitz, and M. S. Handcock. 2004. Relating resources to a probabilistic measure of space use: Forest fragments and Steller's jays. *Ecology* 85 (5):1411–27.

McCarthy, M. A., S. J. Andelman, and H. P. Possingham. 2003. Reliability of telative predictions in population viability analysis. *Conservation Biology* 17 (4):982–89.

McConnell, W. V. 1999. Red-cockaded woodpecker cavity excavation in seedtree-shelterwood stands in the Wakulla (Apalachicola National Forest, Florida) sub-population. *Wildlife Society Bulletin* 27 (2):509–13.

McKinney, M. L. 2002. Urbanization, biodiversity, and conservation. *BioScience* 52:883–90.

McLellan, B. N., and D. M. Shackleton. 1988. Grizzly bears and resource-extraction industries: Effects of roads on behaviour, habitat use and demography. *Journal of Applied Ecology* 25:451–60.

Menges, E. S. 1991. The application of minimum viable population theory to plants. In *Genetics and conservation of rare plants*, ed. D. A. Falk and K. E. Holsinger, 45–61. New York: Oxford Univ. Press.

Merriam, C. H. 1898. *Life zones and crop zones*. USDA Division of Biological Survey, Bulletin no. 10.

Michalski, F., and C. A. Peres. 2005. Anthropogenic determinants of primate and carnivore local extinctions in a fragmented forest landscape of southern Amazonia. *Biological Conservation* 124 (3):383–96.

Miller, S. D. 1990. Impact of increased bear hunting on survivorship of young bears. *Wildlife Society Bulletin* 18 (4):462–67.

Miller, S. D., and W. B. Ballard. 1992. Analysis of an effort to increase moose calf survivorship by increased hunting of brown bears in south-central Alaska. *Wildlife Society Bulletin* 20 (4):445–54.

Mills, L. S. 1995. Edge effects and isolation: Red-backed voles on forest remnants. *Conservation Biology* 9:395–403.

Mills, L. S., and P. E. Smouse. 1994. Demographic consequences of inbreeding in remnant populations. *American Naturalist* 144 (3):412–31.

Moilanen, A. 2004. SPOMSIM: Software for stochastic patch occupancy models of metapopulation dynamics. *Ecological Modelling* 179:533–50.

Morris, C., ed. 1992. *Academic Press dictionary of science and technology*. San Diego, CA: Academic Press.

Morrison, M. L. 2001. A proposed research emphasis to overcome the limits of wildlife–habitat relationship studies. *Journal of Wildlife Management* 65:613–23.

Morrison, M. L. 2002. *Wildlife restoration: Techniques for habitat analysis and animal monitoring*. Washington, DC: Island Press.

Morrison, M. L., and L. S. Hall. 1998. Responses of mice to fluctuating habitat quality. I. Patterns from a long-term observational study. *Southwestern Naturalist* 43:123–36.

Morrison, M. L., and L. S. Hall. 2002. Standard terminology: Toward a common language to advance ecological understanding and applications. In *Predicting species occurrences: Issues of accuracy and scale*, ed. J. M. Scott, P. J. Heglund, M. L. Morrison, J. B. Haufler, M. G. Raphael, W. A. Wall, and F. B. Samson, 43–52. Washington, DC: Island Press.

Morrison, M. L., and E. C. Meslow. 1983. Impacts of forest herbicides on wildlife: Toxicity and habitat alteration. *Transactions of the North American Wildlife and Natural Resources Conference* 48:175–85.

Morton, M. L. 1992. Effects of sex and birth date on premigration biology, migration schedules, return rates and natal dispersal in the mountain white-crowned sparrow. *Condor* 94:117–33.

Moulding, J. D. 1976. Effects of a low-persistence insecticide on forest bird populations. *Auk* 93:692–708.

Murphy, E. C., F. J. Singer, and L. Nichols. 1990. Effects of hunting on survival and productivity of Dall sheep. *Journal of Wildlife Management* 54 (2):284–90.

Murphy, M. T. 2003. Avian population trends within the evolving agricultural landscape of eastern and central United States. *Auk* 120:20–34.

Murray, M. 2003. Overkill and sustainable use. *Science* 299:1851–53.

Nagorsen, D. W., and R. M. Brigham. 1993. *Bats of British Columbia*. Royal British Columbia Museum handbook. Vancouver, BC: Univ. of British Columbia Press.

Nathan, R., G. Perry, J. T. Cronin, A. E. Strand, and M. L. Cain. 2003. Methods for estimating long-distance dispersal. *Oikos* 103 (2):261–73.

Nathan, R., and S. Volis. 2005. Effects of long-distance dispersal for metapopulation survival and genetic structure at ecological time and spatial scales. *Journal of Ecology* 93 (5):1029–40.

Nunney, L. 1999. The effective size of a hierarchically structured population. *Evolution* 53:1–10.

Nunney, L. 2000. The limits to knowledge in conservation genetics: The value of effective population size. In *Limits to knowledge in evolutionary genetics,* ed. M. T. Clegg, 179–94. New York: Plenum.

Nussbaum, R. A., E. D. Brodie Jr., and R. M. Storm. 1983. *Amphibians and reptiles of the Pacific Northwest.* Moscow: Univ. of Idaho Press.

Oborny, G. M., and B. G. Szabo. 2005. Dynamics of populations on the verge of extinction. *Oikos* 109 (2):291–96.

O'Connor, R. J. 2002. The conceptual basis of species distribution modeling: Time for a paradigm shift? In *Predicting species occurrences: Issues of accuracy and scale,* ed. J. M. Scott, P. J. Heglund, M. L. Morrison, J. B. Haufler, M. G. Raphael, W. A. Wall, and F. B. Samson, 25–33. Washington, DC: Island Press.

Odum, E. P. 1969. The strategy of ecosystem development. *Science* 164:262–70.

Ostro, L. E. T., T. P. Young, S. C. Silver, and F. W. Koontz. 1999. A geographic information system method for estimating home range size. *Journal of Wildlife Management* 63 (2):748–55.

OTA (Office of Technology Assessment). 1993. *Harmful non-indigenous species in the United States.* 2 vols. OTA-F-565. Washington, DC: U.S. Government Printing Office.

Otway, N. M., C. J. A. Bradshaw, and R. G. Harcourt. 2004. Estimating the rate of quasi-extinction of the Australian grey nurse shark (*Carcharias taurus*) population using deterministic age- and stage-classified models. *Biological Conservation* 119:341–50.

Pardini, R., S. M. de Souza, R. Braga-Neto, and J. P. Metzger. 2005. The role of forest structure, fragment size and corridors in maintaining small mammal abundance and diversity in an Atlantic forest landscape. *Biological Conservation* 124 (2):253–66.

Paton, P. W. C. 1994. The effect of edge on avian nest success: How strong is the evidence? *Conservation Biology* 8:17–26.

Patterson, C., and A. B. Smith. 1989. Periodicity in extinction: The role of systematics. *Ecology* 70 (4):802–11.

Pearman, P. B. 2002. The scale of community structure: Habitat variation and avian guilds in tropical forest understory. *Ecological Monographs* 72:19–39.

Pearson, R. G., and K. B. Livezey. 2003. Distribution, numbers, and site characteristics of spotted owls and barred owls in the Cascade Mountains of Washington. *Journal of Raptor Research* 37 (4):265–76.

Perkins, D. W., P. D. Vickery, and W. G. Shriver. 2003. Spatial dynamics of source-sink habitats: Effects on rare grassland birds. *Journal of Wildlife Management* 67 (3):588–99.

Perrin, N., and V. Mazalov. 1999. Dispersal and inbreeding avoidance. *American Naturalist* 154:282–92.

Petit, L. J., D. R. Petit, K. E. Petit, and J. W. Fleming. 1990. Intersexual and temporal variation in foraging ecology of prothonotary warblers during the breeding season. *Auk* 107:133–45.

Posadas, P., D. R. Miranda Esquivel, and J. V. Crisci. 2001. Using phylogenetic diversity measures to set priorities in conservation: An example from southern South America. *Conservation Biology* 15 (5):1325–34.

Ramey, R. R., II, G. Luikart, and F. J. Singer. 2000. Genetic bottlenecks resulting from restoration efforts: The case of the bighorn sheep in Badlands National Park. *Restoration Ecology* 8:85–90.

Ratliff, R. D., and R. D. Pieper. 1982. Approaches to plant community classification for the range manager. *Journal of Range Management,* Monograph series no. 1, 1–10.

Reed, D. H. 2005. Relationship between population size and fitness. *Conservation Biology* 19 (2):563–68.

Reed, D. H., J. J. O'Grady, B. W. Brook, J. D. Ballou, and R. Frankham. 2003. Estimates of minimum viable population sizes for vertebrates and factors influencing those estimates. *Biological Conservation* 113:23–34.

Richter, A. R., S. R. Humphrey, J. B. Cope, and V. Brack Jr. 1993. Modified cave entrances: Thermal effect on body mass and resulting decline of endangered Indiana bats (*Myotis sodalis*). *Conservation Biology* 7 (2):407–15.

Rieman, B. E., D. C. Lee, R. F. Thurow, P. F. Hessburg, and J. R. Sedell. 2000. Toward an integrated classification of ecosystems: Defining opportunities for managing fish and forest health. *Environmental Management* 25:425–44.

Robinson, S. K., F. R. Thompson III, T. M. Donovan, D. R. Whitehead, and J. Faaborg. 1995. Regional forest fragmentation and the nesting success of migratory birds. *Science* 267:1987–90.

Rockwell, R. F., and G. F. Barrowclough. 1995. Effective population size and lifetime reproductive success. *Conservation Biology* 9 (5):1225–33.

Rodgers, A. R., and W. E. Klenner. 1990. Competition and the geometric model of dispersal in vertebrates. *Ecology* 71:818–22.

Ronce, O., I. Olivieri, J. Colbert, and E. Danchin. 2001. Perspectives on the study of dispersal evolution. In *Dispersal*, ed. J. Clobert, E. Danchin, A. A. Dhondt, and J. D. Nichols, 341–57. Oxford: Oxford Univ. Press.

Root, R. B. 1967. The niche exploitation pattern of the blue-gray gnatcatcher. Ecological Monographs 37:317–350.

Rosenberg, D. K., and K. S. McKelvey. 1999. Estimation of habitat selection for central-place foraging animals. *Journal of Wildlife Management* 63 (3):1028–38.

Rotenberry, J. T. 1985. The role of habitation avian community composition: Physiognomy or floristics? *Oecologia* 67:213–17.

Roth, R. R. 1976. Spatial heterogeneity and bird species diversity. *Ecology* 57:773–82.

Ruggiero, L. F., R. S. Holthausen, B. G. Marcot, K. B. Aubry, J. W. Thomas, and E. C. Meslow. 1988. Ecological dependency: The concept and its implications for research and management. *North American Wildlife and Natural Resources Conference* 53:115–26.

Saltz, D., and D. I. Rubenstein. 1995. Population dynamics of a reintroduced Asiatic wild ass (*Equus hemionus*) herd. *Ecological Applications* 5 (2):327–35.

Sauer, J. R., J. E. Hines, and J. Fallon. 2003a. *The North American Breeding Bird Survey, results and analysis 1966–2002*. Version 2003.1. Laurel, MD: USGS Patuxent Wildlife Research Center.

Sauer, J. R., J. E. Fallon, and R. Johnson. 2003b. Use of North American Breeding Bird Survey data to estimate population change for bird conservation regions. *Journal of Wildlife Management* 67 (2):372–89.

Savidge, J. A. 1987. Extinction of an island forest avifauna by an introduced snake. *Ecology* 68:660–68.

Savory, A. 1988. *Holistic resource management*. Washington, DC: Island Press.

Schiegg, K., J. R. Walters, and J. A. Priddy. 2005. Testing a spatially explicit, individual-based model of red-cockaded woodpecker population dynamics. *Ecological Applications* 15 (5):1495–1503.

Scott, J. M., F. Davis, B. Csuti, R. Noss, B. Butterfield, C. Groves, H. Anderson, S. Caicco, F. D'Erchia, T. C. Edwards Jr., J. Ulliman, and R. G. Wright. 1993. Gap analysis: A geographic approach to protection of biological diversity. *Wildlife Monographs* 123:1–41.

Scott, J. M., S. Mountainspring, F. L. Ramsey, and C. B. Kepler. 1986. *Forest bird communities of the Hawaiian Islands: Their dynamics, ecology, and conservation*. Studies in Avian Biology 9. Los Angeles: Cooper Ornithological Society.

Scott, T. A. 1994. Irruptive dispersal of black-shouldered kites to a coastal island. *Condor* 96:197–200.

Seabrook, W. A., and E. B. Dettmann. 1996. Roads as activity corridors for cane toads in Australia. *Journal of Wildlife Management* 60 (2):363–68.

Senner, J. W. 1980. Inbreeding depression and the survival of zoo populations. In *Conservation biology: An evolutionary-ecological perspective*, ed. M. E. Soulé and B. A. Wilcox, 209–24. Sunderland, MA: Sinauer.

Shaffer, M. L. 1983. Determining minimum viable population sizes for the grizzly bear. *International Conference on Bear Research and Management* 5:133–39.

Shaffer, M. L. 1990. Population viability analysis. *Conservation Biology* 4:39–40.

Sherwin, W. B., and C. Moritz. 2000. Managing and monitoring genetic erosion. In *Genetics, demography, and viability of fragmented populations*, ed. A. G. Young and G. M. Clarke, 9–34. Conservation Biology 4. Cambridge: Cambridge Univ. Press.

Shugart, G. W., M. A. Fitch, and G. A. Fox. 1987. Female floaters and nonbreeding secondary females in herring gulls. *Condor* 89 (4):902–6.

Sisk, T. D., and J. Battin. 2002. Habitat edges and avian ecology: Geographic patterns and insights for western landscapes. *Studies in Avian Biology* 25:30–48.

Skonhoft, A. 2005. The costs and benefits of a migratory species under different management schemes. *Journal of Environmental Management* 76 (2):167–75.

Smith, J. L. D., A. H. Berner, F. J. Cuthbert, and J. A. Kitts. 1992. Interest in fee hunting by Minnesota small-game hunters. *Wildlife Society Bulletin* 20 (1):20–26.

Smith, S. M. 1978. The "underworld" in a territorial sparrow: Adaptive strategy for floaters. *American Naturalist* 112:571–82.

Smith, S. M. 1984. Flock switching in chickadees: Why be a winter floater? *American Naturalist* 123:81–98.

Smith, S. M. 1989. Black-capped chickadee summer floaters. *Wilson Bulletin* 101 (2):344–49.

Snyder, M. A. 1993. Interactions between Abert's squirrel and ponderosa pine: The relationship between selective herbivory and host plant fitness. *American Naturalist* 141:866–79.

Soltis, P. S., and M. A. Gitzendanner. 1999. Molecular systematics and the conservation of rare species. *Conservation Biology* 13 (3):471–83.

Soulé, M. E. 1980. Thresholds for survival: Maintaining fitness and evolutionary potential. In *Conservation biology: An evolutionary-ecological perspective*, ed. M. E. Soulé and B. A. Wilcox, 151–70. Sunderland, MA: Sinauer.

Soulé, M. E. 1987. *Viable populations for conservation.* Cambridge: Cambridge Univ. Press.

Stiles, F. G., and D. A. Clark. 1989. Conservation of tropical rain forest birds: A case study from Costa Rica. *American Birds* 43: 420–28.

Suarez, A. V., and T. J. Case. 2002. Bottom-up effects on persistence of a specialist predator: Ant invasions and horned lizards. *Ecological Applications* 12:291–98.

Sullivan, T. P. 1990. Influence of forest herbicide on deer mouse and Oregon vole population dynamics. *Journal of Wildlife Management* 54:566–76.

Swarthout, E. C. H., and R. J. Steidl. 2001. Flush responses of Mexican spotted owls to recreationists. *Journal of Wildlife Management* 65:312–17.

Swarthout, E. C. H., and R. J. Steidl. 2003. Experimental effects of hiking on breeding Mexican spotted owls. *Conservation Biology* 17:307–15.

Sweanor, L. L., K. A. Logan, and M. G. Hornocker. 2000. Cougar dispersal patterns, metapopulation dynamics, and conservation. *Conservation Biology* 14 (3):798–808.

Tansley, A. G. 1935. The use and abuse of vegetational concepts and terms. *Ecology* 16:284–307.

Taylor, A. R., and R. L. Knight. 2003. Wildlife responses to recreation and associated visitor perceptions. *Ecological Applications* 13:951–63.

Terborgh, J. 1986. Keystone plant resources in the tropical forest. In *Conservation biology: The science of scarcity and diversity*, ed. M. E. Soulé, 330–44. Sunderland, MA: Sinauer.

Thomas, C. D. 1990. What do real population dynamics tell us about minimum viable population sizes? *Conservation Biology* 4:324–27.

Thomas, J. W., R. J. Miller, C. Maser, R. G. Anderson, and B. E. Carter. 1979. Plant communities and successional stages. In *Wildlife habitats in managed forests: The Blue Mountains of Oregon and Washington*, ed. J. W. Thomas, 22–39. USDA Forest Service Agricultural Handbook no. 553. Washington, DC: USDA Forest Service.

Thomas, L., and K. Martin. 1996. The importance of analysis method for breeding bird survey population trend estimates. *Conservation Biology* 10 (2):479–90.

Tiebout, H. M., III, and K. E. Brugger. 1995. Ecological risk assessment of pesticides for terrestrial vertebrates: evaluation and application of the U.S. Environmental Protection Agency's quotient model. *Conservation Biology* 9 (6):1605–18.

Trammell, M. A., and J. L. Butler. 1995. Effects of exotic plants on native ungulate use of habitat. *Journal of Wildlife Management* 59 (4):808–16.

Tyre, A. J., H. P. Possingham, and D. B. Lindenmayer. 2001. Inferring process from pattern: Can territory occupancy provide information about life history parameters? *Ecological Applications* 11: 1722–37.

Tyser, R. W., and C. A. Worley. 1992. Alien flora in grasslands adjacent to road and trail corridors in Glacier National Park, Montana (U.S.A.). *Conservation Biology* 6 (2):253–62.

Ucitel, D., D. P. Christian, and J. M. Graham. 2003. Vole use of coarse woody debris and implications for habitat and fuel management. *Journal of Wildlife Management* 67:65–72.

Valle, C. A. 1995. Effective population size and demography of the rare flightless Galápagos cormorant. *Ecological Applications* 5 (3):601–17.

Van Horne, B. 2002. Approaches to habitat modeling: The tensions between pattern and process and between specificity and generality. In *Predicting species occurrences: Issues of accuracy and scale*, ed. J. M. Scott, P. J. Heglund, M. L. Morrison, J. B. Haufler, M. G. Raphael, W. A. Wall, and F. B. Samson, 63–72. Washington, DC: Island Press.

Van Hulst, R. 1992. From population dynamics to community dynamics: Modelling succession as a species replacement process. *Plant succession: Theory and prediction*, ed. D. C. Glenn-Lewin, R. K. Peet, and T. T. Veblen, 188–214. New York: Chapman & Hall.

Vázquez, D. P., and D. Simberloff. 2002. Ecological specialization and susceptibility to disturbance: Conjectures and refutations. *American Naturalist* 159 (6):606–23.

Vázquez, D. P., and D. Simberloff. 2004. Indirect effects of an introduced ungulate on pollination and plant reproduction. *Ecological Monographs* 74 (2):281–308.

Viemeister, N. F. 1983. Auditory intensity discrimination at high frequencies in the presence of noise. *Science* 221:1206–7.

Vistnes, I., and C. Nellemann. 2001. Avoidance of cabins, roads, and power lines by reindeer during calving. *Journal of Wildlife Management* 65:915–25.

Waples, R. S. 2002. Definition and estimation of effective population size in the conservation of endangered species. In *Population viability analysis,* ed. S. R. Beissinger and D. R. McCullough, 147–68. Chicago: Univ. of Chicago Press.

Warren, C. D., J. M. Peek, G. L. Servheen, and P. Zager. 1996. Habitat use and movements of two ecotypes of translocated caribou in Idaho and British Columbia. *Conservation Biology* 10 (2):547–53.

Washitani, I. 1996. Predicted genetic consequences of strong fertility selection due to pollinator loss in an isolated population of *Primula sieboldii. Conservation Biology* 10 (1):59–64.

Watkins, R. Z., J. Chen, J. Pickens, and K. D. Brosofske. 2003. Effects of forest roads on understory plants in a managed hardwood landscape. *Conservation Biology* 17 (2):411–19.

Wayne, R. K., and P. A. Morin. 2004. Conservation genetics in the new molecular age. *Frontiers in Ecology and the Environment* 2 (2):89–97.

Weiermans, J., and R. J. van Aarde. 2003. Roads as ecological edges for rehabilitating coastal dune assemblages in northern KwaZulu-Natal, South Africa. *Restoration Ecology* 11:43–49.

Weisenberger, M. E., P. R. Krausman, M. C. Wallace, D. W. DeYoung, and O. E. Maughan. 1996. Effects of simulated jet aircraft noise on heart rate and behavior of desert ungulates. *Journal of Wildlife Management* 60 (1):52–61.

Westphal, M. I., M. Pickett, W. M. Getz, and H. P. Possingham. 2003. The use of stochastic dynamic programming in optimal landscape reconstruction for metapopulations. *Ecological Applications* 13 (2):543–55.

Whitlock, M. C., and N. H. Barton. 1997. The effective size of a subdivided population. *Genetics* 146:427–41.

Whittaker, R. H. 1975. *Communities and ecosystems.* 2d ed. New York: Macmillan.

Wiens, J. A. 1969. *An approach to the study of ecological relationships among terrestrial birds.* Ornithological Monographs no. 8. Washington, DC: American Ornithologists' Union.

Wilcox, C., and B. Elderd. 2003. The effect of density-dependent catastrophes on population persistence time. *Journal of Applied Ecology* 40 (5):859–71.

Williams, C. F., and J. W. Mjelde. 1994. Conducting a financial analysis of quail hunting within the Conservation Reserve Program. *Wildlife Society Bulletin* 22 (2):233–41.

Willson, M. F. 1974. Avian community organization and habitat structure. *Ecology* 55:1017–29.

Wissinger, S. A., and H. H. Whiteman. 1992. Fluctuation in a Rocky Mountain population of salamanders: Anthropogenic acidification or natural variation? *Journal of Herpetology* 26 (4):377–91.

Wolf, C. M., B. Griffith, C. Reed, and S. A. Temple. 1996. Avian and mammalian translocations: Update and reanalysis of 1987 survey data. *Conservation Biology* 10 (4):1142–54.

Wright, S. 1931. Evolution in Mendelian populations. *Genetics* 16:97–159.

Yates, T. L., J. N. Mills, C. A. Parmenter, T. G. Ksiazek, R. R. Parmenter, J. R. Vande Castle, C. H. Calisher, S. T. Nichol, K. D. Abbott, J. C. Young, M. L. Morrison, B. J. Beaty, J. L. Dunnum, R. J. Baker, J. Salazar-Bravo, and C. J. Peters. 2002. The ecology and evolutionary history of an emergent disease: Hantavirus pulmonary syndrome. *BioScience* 52:989–98.

Young, A. G., and G. M. Clarke, eds. 2000. *Genetics, demography, and viability of fragmented populations.* Conservation Biology 4. Cambridge: Cambridge Univ. Press.

PART II

The Measurement of Wildlife—Habitat Relationships

4 The Experimental Approach in Wildlife Science

One way to approach definition is to consider science as a process of questioning and answering. The questions are, by definition, scientific if they are about relationships among observed phenomena. The proposed answers must, again by definition, be in natural terms and testable in some material way. On that basis, a definition of science as a whole would be: Science is an exploration of the material universe that seeks natural, orderly relationships among observed phenomena and that is self-testing.
—G. G. SIMPSON, *THIS VIEW OF LIFE*

Research is a race between ignorance and irreversible consequences."
RUSSELL SADLER*

Study of the relationships between animals and their habitats is no different from any other scientific endeavor in that the "scientific method" must be used. In the broadest sense, scientific thinking is the logical thought process one might use to discover, for example, why a car will not start. The activities of scientists striving to understand how the universe works, however, are collectively called the scientific method. This method appears relatively easy to apply when described in its simplest form, yet its application, particularly in ecology and related fields, has generated considerable discussion. The discussions have focused on which activities are most useful or important (e.g., Romesburg 1981; Peters 1991) and how to perform them correctly (e.g., Murphy 1990; Nudds and Morrison 1991; Drew 1994). Thus telling someone to follow the scientific method when designing and carrying out a study, although good advice, is inadequate as a set of instructions for what to actually do.

This chapter provides a brief description of the activities associated with the scientific method and their relation to the study of animals and habitats. We also present some issues concerning the scientific method that have generated discussions among ecologists and wildlife biologists. Our purpose is not to resolve the controversies but to identify the kinds of information needed by those who study wildlife–habitat relationships. We also discuss the design and application of experiments in wildlife science and some of the problems inherent in experiments in field and laboratory situations. We focus on the experimental approach because it is a powerful tool that has been underutilized by wildlife biologists. We do not review statistical models, even though the design of an experiment is intimately linked to a statistical model. Such a review is beyond the scope of this chapter, and several texts on statistics and experimental design are available (e.g., Skalski and Robson 1992; Kuehl 1994; Morrison et al. 2001; Scheiner and Gurevitch 2001; Ramsey and Schafer 2002). We end the chapter with a discussion of when to use field and laboratory experiments and what to do when tightly controlled experiments cannot be done because of practical constraints.

The Scientific Method

Carey (1994, 5) defined the scientific method as "a rigorous process whereby new ideas about how some part of the natural world works are put to the test." Descriptions of the scientific method often include three to six "steps" or activities, depending on how they are combined: (1) making observations; (2) searching for patterns among observations; (3) generating potential explanations for patterns detected; (4) deducing predictions from the proposed explanations; (5) testing potential explanations by looking for predicted phenomena, often after some experimental manipulation; and (6) tentatively accepting, rejecting, or modifying proposed explanations, depending on how well predictions were met (fig. 4.1).

Observations are the starting point of science (Simpson 1964, 88). Some philosophers argue about whether what we perceive is real, but most ecologists are willing to accept that our senses, or instruments we develop, provide us with a good approximation of the existing world (Simpson 1964). That is not to say that our observations are free of error, but numerous observations by different observers from different points in space and time generally help separate observational errors from reality (Matter and Mannan 1989).

Observations tell us little about the natural world when made independently; their value becomes apparent, however, when patterns can be recognized among them. If, for example, we were to observe many individuals of a given species in a sample of sites from a particular type of vegetation and recognized the association, some level of prediction might be possible. A simple prediction from this pattern is that we can expect to see individuals of this species when we visit different sites within the same vegetation type. A more complex prediction is that the number of individuals of this species will increase if we increase the extent of this vegetation type through some management activity. Several factors, such as whether the species successfully survives and breeds in the vegetation type, will

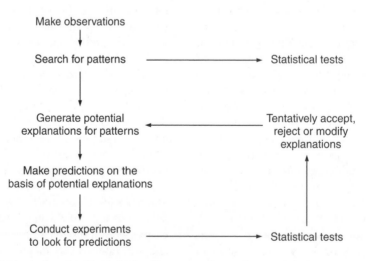

Figure 4.1.   Activities in the scientific method and common places where statistical tests are employed

influence whether the second prediction is accurate. The point, however, is predictions of this kind become possible only after patterns of association are recognized.

The process by which we come to recognize patterns is called *induction* (Hanson 1965), or reasoning from particulars to generals. The "particulars" are the individual observations we make about the natural world, and the "generals" are the associations or patterns we detect among them. Humans have the ability to recognize patterns partly because our brains can catalog and retain information for long periods of time, but recognition of subtle patterns requires keen observation skills and considerable information. Skills associated with observation almost certainly vary among individuals but can be enhanced for most with practice and knowledge.

Scientists usually search for patterns in nature in a more formal manner than casual observers. Scientific observations are often made systematically or come from random samples so that they represent the natural world as accurately as possible. If a relationship between a species and a particular vegetation type is suspected, then a series of surveys for the species might be conducted throughout the year, over several years, in a random sample of areas within the vegetation type to determine if the relationship is real. Furthermore, statistics might be employed to help detect patterns or correlations among variables that would otherwise be difficult to detect (see fig. 4.1).

It is important to remember that a correlation between two variables does not imply a cause-and-effect relationship between them. For example, a positive correlation between the abundances of two species—say, mosquitoes and deer ticks—in an area does not indicate that an increase in the abundance of mosquitoes caused an increase in the abundance of deer ticks, or vice versa. The two species could simply be responding to the effects of an increase in rainfall on their respective resources, and be increasing independently of each other. Conversely, if the abundances of two species are positively correlated and one species (e.g., a flycatcher) depends on another for some needed resource (e.g., a woodpecker, to excavate its nest cavities), then there could be a casual mechanism relating the covariance of the two species. Patterns alone, however, reveal little about these mechanisms.

Once patterns among observations have been identified and confirmed, some people are stimulated to wonder what processes or mechanisms might have created the patterns; that is, they begin to ask why. This speculation can result in the development of potential explanations of the observed patterns. Such "speculative explanations" are called *research hypotheses* (Romesburg 1981, 295), and they articulate a mechanism or process that might be responsible for the pattern of interest.

The term sometimes used to describe the process by which we generate research hypotheses, *induction* (Platt 1964; Popper 1981; Medawar 1984), is the same one used to identify how we recognize patterns. But the mental activities associated with hypothesis generation, although they involve pattern recognition, probably require more information, creativity, and insightfulness than pattern recognition alone. Neither process, however, is well understood or completely under our control (Matter and Mannan 1989). Thus the word *method*, in the term *scientific method*, is somewhat misleading if it is interpreted to mean that all people will be able to perform all steps or activities in science equally if given enough instruction. We emphasize that anyone can improve their chances for participation in the creative aspects of science by being well informed in the area of their interest, and understanding how science works.

Potential explanations of patterns or phenomena in nature are constrained by what is already known about the world. They cannot, for example, contradict the laws of physics or umbrella theories, such as evolution. But even with these constraints, it is possible (and even likely) that more than one research hypothesis will be proposed for a given pattern or phenomenon. Chamberlain (1897) advocated working with multiple research hypotheses because he felt that a single hypothesis is more likely to become accepted without being tested. We do not think that the scientific method requires multiple explanations, but it does require that research hypotheses be viewed as candidate explanations until they are tested. Prior to testing, scientists must remain objective about their hypotheses, no matter how convincing they sound. If more than one potential explanation for a given pattern exists, tests should help determine which is most likely valid. Deciding which of several competing hypotheses to test first depends, in part, on which best fits existing theories and empirical information.

The approach for testing research hypotheses, and either gaining or losing confidence in their validity, is called the *hypothetico-deductive (H-D) method* (Romesburg 1981). In this activity, scientists devise tests that put their hypotheses at risk; that is, they perform experiments or propose novel observations that potentially refute or cast doubt on their candidate explanations. Romesburg (1981) noted that the causal mechanism or process described in a research hypothesis might be difficult to observe directly. Thus "a research hypothesis must be tested indirectly because it *embodies a process*, and experiments can only give facts entailed in the process" (Romesburg 1981, 295; italics added). A general model of an H-D test might be: If my hypothesis is true (i.e., if the mechanistic process I envision is actually ongoing), and if I manipulate a critical element involved in the process while simultane-

ously controlling other critical elements, certain things must happen, or, conversely, certain things cannot happen.

The "things" in the general model of an H-D test are predictions—observable events or patterns—derived from what should happen after a manipulation, given that the tentative explanation being tested is valid. *Deduction* is the word that describes the process by which predictions are generated in an H-D test. Deduction, by definition, is the derivation of a conclusion by reasoning or logic, where the conclusion follows necessarily from the premises. Predictions of an H-D test, then, are logically deduced from the premise of the hypothesis (hence the term *hypothetico-deductive*), and they should be specific to the hypothesis being tested. If the predictions generated in an H-D test could come about by processes other than the one being tested (e.g., by a competing hypothesis), then the test would not be a good one.

The best tests of research hypotheses according to Popper (1981) are *falsification tests*—those in which all elements are controlled except the element critical to the proposed explanation, and those that involve attempts to find results that are logically prohibited from occurring if the hypothesized explanation is true. If experiments produce logically prohibited results, the hypothesis is falsified and should be modified or discarded. Failure to find the prohibited results does not prove the hypothesis to be true but increases confidence in it (Simpson 1964). A high level of confidence in the validity of a given hypothesis should come only after it has been tested repeatedly in a variety of ways.

Tests of the kind Popper described are possible only if one can envision events or patterns that will refute the hypothesis under consideration. If hypotheses are vague or general, they can sometimes be interpreted to explain nearly any related observation in nature or any outcome of a pertinent manipulation. Hypotheses of this kind are considered *unfalsifiable*—that is, not amenable to testing. Unfalsifiable hypotheses can sometimes be improved by describing more explicitly the mechanism or process being tested or by making them more specific. Identifying critical elements to manipulate in an experiment and making what Popper called "risky predictions" are often easier if the details of the proposed process are explicitly described.

Predictions that come from hypotheses about ecological phenomena often involve the responses of individual animals or plants. These responses are not likely to be identical among individuals, no matter how carefully a test is controlled, at least partly because each individual has a unique genome and environmental history. For example, providing the same amount of an important nutrient to individual plants, even under highly controlled conditions, will not result in the same amount of growth in each plant. Statistical tests have traditionally been used to assess whether the predictions of an H-D test are met. In this example, a scientist might have hypothesized that the nutrient in question plays an important function in the growth of plants and articulated a mechanism through which the nutrient aids growth. A prediction of an H-D test might be that plants that receive the nutrient will grow more rapidly than those that do not. A statistical test could be used to compare the average growth of plants in the two groups. The null "statistical hypothesis" in this example would typically be that there is no difference between the mean growth rates of the two groups. Rejecting or failing to reject the null statistical hypothesis determines whether the predictions of the test have been met (see fig. 4.1). Thus assessment of patterns with statistics can play an important role in the validation of research hypotheses, when the statistical test is embedded in the broader framework of an H-D test. Differences between research hypotheses and statistical hypotheses and recent concerns

about the use and misuse of statistical tests in wildlife science are outlined below.

Issues Associated with the Scientific Method

What Are Research Hypotheses?

Many discussions about the scientific method have focused on formulating and testing "hypotheses." What constitutes a research hypothesis is a subject that sometimes causes confusion among students and professional biologists alike because the term is used in a variety of ways. Romesburg (1981) recognized this problem and clearly differentiated between *research hypotheses* (i.e., those that include a mechanistic process and explain a pattern) and *statistical hypotheses*, which he defined as conjectures about classes of facts (i.e., patterns). Unfortunately, the distinction between the two kinds of hypotheses continues to be overlooked (Guthery et al. 2001). Carey (1994, 9) noted, "In the jargon of the scientist, just about any claim that may require testing before it is accepted will be called a hypothesis." Thus a question about whether a species is more abundant in one mountain range than in another might be framed as a hypothesis. But Carey (1994) was clear that statements of this kind are not research hypotheses because *they do not include an explanation*. We suggest that questions that address patterns without explanation are best framed as statistical hypotheses. Tentative explanations for how some part of the world works (i.e., research hypotheses) should be assessed with the H-D method.

Statistical Hypothesis Testing

The traditional approach used in ecology to make inferences about patterns is with statistical tests, in which a statistical null hypothesis either is rejected or not rejected based on the value of a computed test statistic and associated P value. Concerns about the utility of this approach are not new (e.g., Berkson 1938, 1942; Cox 1958b), but recent challenges to it (e.g., Cherry 1998; Johnson 1999; Anderson et al. 2000) suggest that a shift in the way data are analyzed in ecology and wildlife science may be forthcoming (Guthery et al. 2001). At a minimum, students and professional biologists should be aware that concerns exist about the traditional approach and that alternatives are available.

Problems identified with null hypothesis testing include, but are not limited to, the following: (1) it is used when it is not required; (2) the meaning of P values often is misinterpreted; (3) assumptions of statistical tests are ignored or misunderstood; (4) many null hypotheses are known to be false before the test is conducted, or if rejected provide little information; (5) the decision to reject or not reject a null hypothesis is based on a fixed cutoff value (usually alpha = 0.05) that is arbitrarily selected; and (6) statistical significance may or may not be associated with biological significance (Cherry 1998; Johnson 1999; Anderson et al. 2000). These problems are a mixture of misunderstanding or misuse of the traditional procedures, and weaknesses inherent in the procedures. Recommendations for correcting the problems, therefore, include calls for wiser use of the traditional approaches, and for use of alternative approaches.

Cherry (1998) recommended that investigators decide whether they are interested in assessing the presence or absence of effects (i.e., what is assessed in most null hypothesis tests) or in estimating the size of effects. He suggested that the focus of many studies in wildlife science is estimation and recommended that greater emphasis be placed on calculating and presenting point estimates, standard errors, and confidence intervals for characteristics of interest (see box 4.1 for use of intervals). Similarly, Johnson (1999) and Anderson et al. (2001) cautioned against presenting only P values from statistical tests (i.e., "naked P values") without including estimates of effect size and measures of precision. Present-

---

Box 4.1   What is a statistical interval and how should it be used in wildlife studies?

There are three kinds of statistical intervals: confidence intervals, prediction intervals, and tolerance intervals (Hahn and Meeker 1991). *Confidence intervals* (CIs) are used to provide estimates of mean values from additional studies. *Prediction intervals* (PIs) are used to provide estimates of future values of means (or standard deviations [SDs]) based on current variance (or trend) in the sample data. *Tolerance intervals* (TI) are used to provide estimates of the proportion of elemental observations within specific percentages. TIs are best for describing historical or existing data patterns and, in particular, the distribution of values among observations (Mood et al. 1974).

TIs are based on the spread of values for the elemental observations—that is, among individuals of a wildlife population, whereas CIs and PIs are based on values of means or SDs from subpopulations of additional or future samples (studies). So if you are interested in knowing what proportion of individuals in a wildlife population select some resource in a specific way (such as prey size), then use TIs. If you are interested in knowing how tightly you have estimated the *mean* of some parameter, such as some resource selection behavior (such as prey size), across all individuals within the population, then use a CI. If you want to estimate future values of some mean population parameter based on variance in the sample data, use PIs.

It is clear from our reading of the literature that CIs are very often used and interpreted incorrectly—that is, as TIs. Unfortunately, TIs are seldom discussed or presented in standard statistical textbooks, although their basis is well represented in statistical theory and in the published statistical literature, and they are used in ecological practice—for example, for representing the proportions of wildlife populations that use various size and densities of snags and down wood in the Pacific Northwest (Mellen et al. 2003).

ing confidence intervals in addition to or instead of P values provides information about the precision with which a parameter has been estimated, which in turn allows the reader to evaluate whether lack of significance is likely due to lack of an effect or small sample size (Johnson 2002). Confidence intervals also encourage biologists to consider effect size in interpretation of results (Johnson 2002; also see below). In some situations, use of confidence intervals may be more meaningful than statistical tests and could reduce tests of "silly null hypotheses" (i.e., those known to be false before the tests are conducted; Robinson and Wainer 2002). For example, Cherry (1998) noted that the null hypothesis tested in many studies of habitat use and availability is that animals use different environments in proportion to their availability on the landscape. In most cases, this null is known to be false *a priori*. Cherry (1998) recommended that this kind of question might best be answered with estimates and confidence intervals, avoiding null statistical hypothesis tests altogether.

Rejecting or failing to reject a null statistical hypothesis has clear implications about how we view the existence of whatever pattern is under investigation. If the test is part of an H-D experiment, then the outcome of the test either lends support to the research hypothesis under consideration or falsifies it. It is important, therefore, to understand the kinds of errors that can be made when deciding the outcome of a statistical test. A Type I error is made if the null hypothesis is rejected when it is true, and a Type II error is made if the null hypothesis is not rejected when it is false. The probability of a Type I error (alpha) in a statistical test is set by the investigator, and convention has established that alpha is often 0.05. The probability of a Type II error in a test is denoted by beta (β). The reciprocal of beta ($1 - \beta$) is called the *statistical power* of a test and is the probability of correctly rejecting a null hypothesis that is false.

Committing a Type II error would, in the typical design of a statistical test, indicate that no pattern exists when in fact one does. This kind of error could be problematic in wildlife science. For example, in a study about pattern, a Type II error could lead a biologist to conclude that there was no decline in abundance of a species after some human-induced manipulation, when in fact there was. Analysis of statistical power during the design of a study can help avoid Type II errors, as power increases with increasing sample size, n, and effect size. Effect size is broadly defined as "the difference between the null hypothesis and a specific alternative" (Steidl and Thomas 2001, 17), and could be identified as the smallest change that has biological meaning in the context of the study (Steidl et al. 1997). Analyses of statistical power should be done during the design of a study to help determine the number of samples needed to detect a biologically meaningful effect.

Identifying what is a meaningful biological effect is perhaps the most challenging aspect of power analyses (Steidl and Thomas 2001), but it is also among the most beneficial aspects as well (Cherry 1998). Because statistical significance is related to sample size, even very small differences between, for example, the mean growth rates of plants in two groups (see example above) can be statistically different. A more important question is whether the difference found is biologically meaningful. Use of power analyses encourages biologists to think about "meaningful biological effects" before beginning a study.

Anderson et al. (2000) described an alternative to statistical null hypothesis testing called the *information theoretic method*. This method is based on a formal relationship, found by Akaike (1973), between "Kullback-Leibler information (a dominant paradigm in information and coding theory) and maximum likelihood (the dominant paradigm in statistics)" (Anderson et al. 2000, 917). In this approach, what would be statistical hypotheses (both null and alternative) in traditional tests are developed *a priori* as models about relationships within a system of interest. Models are then ranked based on Akaike's information criterion (AIC), which describes how well an empirical set of data fits each model, while minimizing the number of estimated parameters (Anderson et al. 2000). Professional biologists and students who have used Program MARK (White and Burnham 1999) are likely familiar with this approach. In this program, AIC values could be used to evaluate, for example, a set of models that incorporate the influence of sex, age, and time on rates of survival of marked animals. Anderson et al. (2000) argued that the information theoretic approach is superior to null statistical hypothesis testing because no null model is rejected (or not rejected) based on a fixed alpha level, but rather a set of models are evaluated and ranked based on best inference. Guthery et al. (2001) generally favored the use of the information theoretic method but cautioned that because it was by definition a parametric approach, assumptions about probability distributions are required before inferences drawn are useful. Furthermore, they noted that the approach offered no protection against trivial hypotheses. Models developed in this approach could be just as "silly" as many null models in statistical tests.

A second alternative approach to null statistical hypothesis testing is Bayesian statistics. This approach can be used to predict the outcome of environmental changes in terms of likelihood or odds. The approach relies, in part, on the ability of biologists to describe *a priori* likelihood functions for ecological outcomes (i.e., prior probabilities), given specific environmental conditions, such as the relationship between environmental states and sizes of a given population (see box 10.2). The approach has been controversial but recently is being more accepted and used by wildlife ecologists. It may

be of value when standard experimental designs or statistical tests are constrained by real-world problems (e.g., small sample sizes or inadequate replicates or controls). Advantages and disadvantages of Bayesian statistics are outlined in chapter 10 (box 10.2) but focus, not surprisingly, on the use of existing knowledge and expert opinion to estimate prior probabilities, and on the influence of errors in "priors" on predicted outcomes.

Null statistical tests clearly suffer from problems—some inherent, some a product of misuse (but see Robinson and Wainer 2002). Greater use of confidence interval, tolerance intervals (see box 4.1), and identifying meaningful biological effects (i.e., wiser use of traditional methods), and use of alternative methods as outlined above may reduce these problems. But we emphasize, as did Guthery et al. (2001), that statistical tests are primarily a tool to identify or elucidate patterns. They are not an end in themselves and are most powerful when they are part of a broader experimental approach designed to test an explanation.

Challenges to the Scientific Method

The issues described above are subjects that might be considered "in-house," because they primarily involve scientists debating with one another about how they should do their work. The issue described in this section is one that involves public perceptions about the products of science and about scientists themselves. It thus has implications that go beyond improving the details about how science is conducted.

One of the basic tenets of science is that we live in a material universe that can be at least partly understood (Feder 1990). How the universe works is the same for everyone (at least in our view), but our individual understanding and perceptions of these workings may differ. Simpson (1964, vii) described this situation by noting that everyone lives in two worlds, one public and one private: "The public world is the objective, material, outer world that exists around us regardless of what we know or think about it. The private world is just what we do know and think about that public world; it is the world as it seems to us, as we perceive and conceive it." The private world of an individual can be influenced by his or her gender, race, culture, and individual experiences, and can include misconceptions and erroneous ideas. Science is the best way we know of to align and match our private worlds with the public world.

Recent challenges to the content of science (i.e., our current knowledge about how the universe works), and even the scientific process itself, are founded on the idea that individual biases due to, for example, gender or sociopolitical or economic background, severely constrain our ability to objectively perceive and understand reality (e.g., Harding 1986). Individuals who espouse this idea, sometimes called "social relativists" or "postmodernists," claim that what we know about the universe is flawed because of inevitable biases of those who did the work. There can be little doubt that politics and other cultural influences have played a significant role in who has participated in science. For example, until recently, men almost completely dominated the field of wildlife science. Furthermore, even the questions asked in scientific endeavors are likely influenced by societal context. Thus what we know about the universe is, to some extent, culturally driven. However, we disagree that bodies of accepted knowledge are flawed (i.e., do not match reality) because of these influences. If the scientific process is performed properly, the answer derived should be the same no matter who does the work. This is not to say that the process is perfect. Mistakes in the history of modern science are common. But the process tends to be self-correcting so that even if personal biases, or outright fraud (Goodstein 1996),

cause an individual scientist to misinterpret or misrepresent the outcome of an experiment, other scientists doing similar work should soon expose the error. Challenges of social relativists and postmodernists are of concern not because they cause scientists to seriously question their practices or the ideas about which they have developed great confidence, but rather because the general public may interpret such challenges as legitimate, become disillusioned with answers provided by science, and turn to less reliable sources of information to assist them in making decisions (Gross and Levitt 1994; Gross et al. 1996).

Application of the Experimental Approach

In the past, wildlife scientists did not often participate in experimental studies, partly because the questions they asked were generally about patterns of natural phenomena and not the processes that caused them. Species studied were frequently those that were of interest from a management perspective, and questions addressed patterns of relative abundance, habitat use, survival, and productivity. Wildlife science focused on describing these patterns because the information they provided was generally sufficient for developing successful management strategies. Studies that document patterns in nature often require intensive and extensive surveys of animals, or painstaking observations of animal behavior (see chapter 7), but usually not experiments.

Wildlife scientists were not alone in the past in ignoring experiments. Hairston (1989) noted that ecologists used experiments infrequently in the 1950s and 1960s, despite the pleas of advocates of the experimental approach. Of interest is that some of the early field and laboratory experiments performed by ecologists related directly to animal–habitat relationships.

For example, Harris (1952) attempted to test the idea of habitat selection experimentally by presenting individual prairie deer mice (*Peromyscus maniculatus bairdi*) and woodland deer mice (*P. m. gracilis*) with the choice of artificial woods or artificial fields. He found that test animals preferred the artificial habitat that most closely resembled the natural environment of their own subspecies, and concluded that mice were reacting to visual cues provided by the artificial vegetation. Wecker (1963) later attempted to determine whether early experience (i.e., learning) played a role in habitat recognition and selection in deer mice. He found that behaviors associated with habitat selection were primarily controlled by what Mayr (1974) called *closed genetic programs;* that is, they are not greatly affected by early experience. Similar experiments conducted by Klopfer (1963), however, revealed that habitat selection by chipping sparrows (*Spizella passerina*) could be modified to some extent by early experience.

Experiments are a potentially powerful tool for increasing understanding of ecological processes and phenomena, and their use in ecological research, including wildlife science, has increased dramatically in recent years. However, the role of experiments in the scientific method is still sometimes misunderstood. Therefore, students hoping to pursue careers as scientists in wildlife ecology or related fields should familiarize themselves with the experimental approach. Knowledge of experiments should not diminish the value of wildlife studies designed to identify and elucidate patterns, but a more complete understanding of the purpose and design of experiments will facilitate their use when they are needed.

Purposes of Experiments

The primary purpose of an experiment, stated simply, is to test an idea about how the natural world works. In terms presented in the previous section, the purpose of an experiment is to test the validity of a research hypothesis. A secondary purpose of experiments is to corroborate or describe the existence or nature of a pattern (i.e., "fact finding"; Kneller 1978, 116). The design of fact-finding experiments is the same as or similar to those used to test research hypotheses, but there is no explanation or mechanism being tested, and no predictions being made because the purpose is to find out what happens when, through manipulation, some natural event is mimicked.

Considerations of Experimental Design

The model of a hypothetical deductive test, presented in the previous section, forms the conceptual foundation for designing an H-D experiment, but there is no simple formula or methodology that can provide experimental details. Each experiment is an exercise in creativity because the experimenter must be able to identify an element that plays an important role in the research hypothesis being tested, manipulate it without changing other important elements, and logically deduce (predict) what will happen after the manipulation.

The strength of experimental results either to support or to refute a research hypothesis depends on whether the experiment is properly designed and executed. Consideration of several rules and concepts about experimental design should help strengthen experimental results. Below, we briefly review several factors that should be considered during the design of experiments. Our review focuses on field experiments because they are more common in wildlife science than experiments conducted in laboratories.

One important rule associated with experimentation is that experiments must be repeatable—that is, other scientists must be able to duplicate the experiment if they wish. The methods, therefore, must not include any subjective assessments of important variables, and they must be carefully and accurately described, usually in a written report or publication. In fact, field experiments in ecology and wildlife science are rarely repeated. But that is another matter, the implications of which are discussed below.

Another important consideration when designing an H-D experiment is that the experimental units (e.g., individual animals or plants, or plots of land) involved must accurately represent the domain to which the hypothesis is supposed to apply. In other words, inferences from an experiment will apply only to the domain from which samples were taken (Morrison et al. 2001). Thus, the *sampling frame* (Skalski and Robson 1992), or the pool from which samples are drawn, should match the target population if the experiment is to have relevance to the target population. For example, if the hypothesis being tested is about a process that operates within a given vegetation type, the plots used in the experiment should ideally be selected at random from the entire distribution of the vegetation type. In reality, experimental units used in field experiments are often constrained by lack of access to private lands or by lack of travel funds. And, in some situations, experimental units are purposely chosen so that they are similar to each other in order to reduce variation (see below). Thus, for several reasons, experimental units used in an experiment might come from a subset of the "target population," and inferences drawn from the experiment are then limited (see Skalski and Robson 1992).

Manipulation of an element that is central to the hypothesis being tested is a critical component of an H-D experiment. Manipulations in experiments are usually called *treatments*. Ideally, important biotic and abiotic factors not being manipulated are controlled by the experimenter; thus changes in the variable of interest

after the manipulation can be attributed to the treatment. In field experiments, these "other factors" are usually not under control, but when they change during an experiment, their effects must be assessed by collecting information about the variable of interest before and after treatment on areas where the treatment was applied, and on controls. *Controls* are experimental units that are not manipulated but are assessed in the same manner as units that are (Hairston 1989, 24–26). There are no set criteria to determine the period of time over which data on the variable of interest should be collected, but Hairston (1989, 24) noted that lack of adequate baseline (before-treatment) data was a common failing among ecological experiments. Measurements after treatment should extend at least through the period during which the variable of interest is expected to change.

Assessing changes in the variable of interest before and after treatment on areas treated and on controls provides a measure of how much the variable of interest would have changed in the absence of the manipulation. For example, Franzblau and Collins (1980) hypothesized that food availability is a primary element determining the size of territories in birds. They predicted that, if their hypothesis was true, adding food to territories of rufous-sided towhees (*Piplo erythrophthalmus*) should result in a decrease in the size of the birds' territories. The first steps in their experiment were to identify a sample of territories of rufous-sided towhees and measure the size of each territory. As noted above, inferences from an experiment are restricted to the "universe," or sampling frame, from which the samples are drawn. In this experiment, the area from which territories were selected (ideally in a random fashion) represent the area to which the results of the experiment will apply. Food was then added to some territories, the *treatment*, or *experimental, territories,* and other territories were left untreated as controls. The size of each territory was remeasured after the addition of food, and changes in the size of treatment territories were assessed relative to changes in the controls. The value of the controls becomes apparent if, for example, the size of all territories had decreased because of some change in the environment other than the addition of food or because of some effect of the observer. If this had happened, there would have been no way to evaluate changes in the treatment territories without comparing them to changes in the controls.

Application of treatments and selection of experimental units are associated with two concepts, *randomization* and *replication* (Fisher 1947), which are integral to the design of an experiment. Randomization is one way to protect the experiment from unknown sources of experimental bias (Skalski and Robson 1992; Morrison et al 2001). When experimental units are selected, there will inevitably be variation among them, and these differences could influence the effects of the treatment. Therefore, designation of which experimental units receive treatment should be made randomly (or randomly with some constraints) (Cox 1958a) so that "a treatment is not consistently favored or impaired by extraneous or unexpected sources of variation" (Skalski and Robson 1992, 16). For example, Franzblau and Collins (1980) provided food to randomly selected territories within their study area so that differences among territories would not uniformly bias the effects of their treatment. If, instead, they had arbitrarily provided food to territories on the northern half of their study area, some environmental factor in that portion of the study area could have compromised their experimental results by either ameliorating or enhancing the effects of adding food on territory size.

Responses of experimental units to treatments will not be identical no matter how carefully a test is controlled. Therefore the responses of several experimental units must be assessed to increase the precision of the estimate of the variable of interest and thus help determine if the predictions of the experiment were met. This approach is called *replication*, and each set of treatment and control units is called a *replicate*. The number of replicates needed in an experiment is positively related to the amount of variation or error in the experiment that will mask the effects of the treatment. A host of factors can influence experimental error (see Skalski and Robson 1992 for review), but it can be estimated from other similar experiments or preliminary surveys. Error in an experiment can be reduced through appropriate design. For example, if experimental units are to be selected along an environmental gradient that will influence the effects of the treatment, one way to reduce error is to group experimental units along the gradient and assess the effects of the treatment on groups. This technique, called *blocking,* is an attempt to remove experimental error by assigning as much of it as possible to differences between groups or blocks (Skalski and Robson 1992).

Another consideration related to selection of replicates or experimental units is that they should be *independent* of each other. Independent experimental units "are dispersed in time and space so that the response [to a treatment] of any one unit has no influence on the response of any other unit" (Smith 1996, 17). Selecting independent replicates in field experiments is sometimes difficult because of interactions between animals and the movements of animals, water, or other materials between experimental units. For example, if the movements or behavior of towhees in territories receiving food had influenced the size of control territories, or vice versa, the replicates used by Franzblau and Collins (1980) would not have been independent. Another way that independence is sometimes violated in experiments is by sampling repeatedly from one site (e.g., an area treated with prescribed fire) and considering each sample a replicate. The samples in this situation are not independent of each other because they all come from a single application of the treatment. Analyzing samples of this kind as if they were independent replicates is one form of *pseudoreplication* (Hurlbert 1984).

Field Experiments

Field experiments are powerful tools, in part because researchers can bring about specific conditions in study plots that, without manipulation, might never occur or take decades to occur. Ecologists often favor field experiments over laboratory experiments (Hairston 1989) because they are conducted in natural settings, and hence the results are generally considered more believable. The price of increased realism, however, is loss of control over most environmental variables. This loss of control may preclude testing some research hypotheses because the critical elements may be difficult or impossible to manipulate in field situations.

The dependence of field experiments on controls and the spatial scale at which many field experiments must be conducted make them vulnerable to several practical problems (Hilborn and Ludwig 1993; Morrison et al. 2001). Finding sites or populations to serve as controls may be difficult, especially if the experiment concerns endangered species, limited environments, or mechanisms that operate at broad spatial scales. Furthermore, as the spatial scale of an experiment increases, experimental error is also likely to increase because of an increase in heterogeneity within and among experimental units. Increasing sample size to deal with this error may be impractical or impossible because of cost and availability. The small sample sizes often associated with field experiments also make randomization and proper interspersion of treatments difficult to achieve. Thus field experiments are difficult to design and execute, and the difficulty

usually increases as the spatial scale of the experiment increases. Unfortunately, mechanisms that operate at broad spatial scales are sometimes among the most important for understanding the causes of ecological patterns (Lawton 1996b).

Another problem inherent in field experiments (and to some degree all ecological experiments) is that patterns and phenomena in ecology often have multiple and interacting causes (Peters 1991). Experimental designs exist that examine the effects and interactions of more than one treatment at a time, but they require more replicates than simpler experiments, and thus the practical problems outlined above are exacerbated.

Field experiments are rarely repeated (Hairston 1989). This is not surprising, given the costs and difficulties associated with designing and executing them and the fact that many take years to complete. Furthermore, most scientists would probably prefer to test a new idea than repeat someone else's work. However, confidence in a research hypothesis should come only after it has been tested in a variety of ways (Kneller 1978, 117). The lack of repetition of field experiments may mean that some (many?) hypotheses in ecology are accepted prematurely.

The difficulty of designing field experiments and the cost and time needed for their execution may also discourage experiments beyond relatively crude levels of explanation. For example, an experiment could be done to test whether application of prescribed fire causes an increase in the abundance of an endangered species. The experiment might involve comparing the abundance of the species in treatment and control plots before and after burning. The results of such an experiment may support the idea that a causal relationship exists between fire and the abundance of the species, but additional experiments would be needed if biologists were interested in knowing, for example, whether fire increases the abundance of the species by increasing food, or by improving the structure of vegetation for cover relative to the needs of the species, or both. Biologists would also want to know whether the increase in abundance was due to a functional or a numeric response and whether abundance is positively related to survival and productivity. The frequent need to explain patterns in nature at multiple levels is one of the concerns that Peters (1991) expressed about the utility of ecological experiments (also see Gavin 1991).

The level of explanation needed for management purposes depends on the situation. If, in the example above, prescribed fire was consistently effective in increasing the abundance of the endangered species, and if patterns of land ownership and other social constraints permitted burning, then the level of explanation provided by the hypothetical experiment might be sufficient. If, in contrast, burning was not feasible throughout the range of the species, it would be important to know what features of habitat were changed by fire so that alternative manipulations could be devised to produce the same effect.

Laboratory Experiments

The design of ecological experiments in laboratories is conceptually no different from those in field situations. The primary difference between laboratory and field experiments is the degree of control over experimental conditions. In a laboratory, nearly all important variables can be controlled. This level of control conceptually makes designing experiments simpler (although not necessarily easy) because the investigator can include and manipulate the environmental elements needed to test the idea under study and can exclude or control everything else. The questions that can be asked in laboratory experiments are thus more precise than those that can

be asked in field experiments (Hairston 1989). Control of nearly all variables in an experiment also facilitates replication of trials within an experiment, and duplication of the experiment itself, because conditions do not change over time and presumably can be re-created by other scientists.

Significant advances in understanding the natural world have been made in laboratory experiments in many scientific fields, including biology. However, there is reluctance by some, perhaps many, ecologists to accept the findings of laboratory experiments in ecology (Mertz and McCauley 1980), primarily because "artificiality may simply swamp processes of ecological relevance" (Peters 1991, 137).

Lawton (1996a) listed and discussed the primary arguments against experiments in the Ecotron, a set of highly controlled environmental chambers designed to replicate miniature terrestrial ecosystems (sometimes called *bottle communities*). Not surprisingly, the arguments against Lawton's (1996a) "bottle experiments" are the same arguments leveled against many other laboratory experiments in ecology. The arguments focus on the lack of applicability of the experiments to actual wildland environments and include concerns about choice of species composition in experiments and the lack of major perturbations, immigration, emigration, seasons, appropriate scale, and important biological processes (Lawton 1996a). A major concern is that manipulating a single factor in a setting that does not include the full array of ecological interactions may produce results that could either unrealistically emphasize or diminish the importance of that factor (Peters 1991). Because of these concerns, results of laboratory experiments in ecology have been most widely accepted when they have addressed questions about physiological processes, bioenergetics, or relatively simple animal behaviors (Hairston 1989).

Improvement in the design of laboratory settings may help reduce some of the concerns about the applicability of results that come from them. The "big bottle" experiments of Lawton (1996a) represent an effort to make an experimental setting for study of miniature terrestrial ecosystems more realistic, despite the expressed concerns. Lawton (1996a, 668) noted, "Many ecologists seem to understand the need for simple experiments in pots and greenhouses; and yet a minority of colleagues become highly critical and concerned when an attempt is made to make artificial, controlled environments more rather than less realistic!"

One specific concern about the artificiality in laboratory experiments is that the experimental settings (e.g., cages, aquaria) can frustrate or prevent behaviors that animals would perform in natural situations (Peters 1991). Many important ecological interactions, such as competition and predator–prey relations, directly involve animal behavior. Clearly, laboratory settings must be designed to allow animals to respond naturally to experimental conditions if the results are to be more acceptable (see Matter et al. 1989; Glickman and Caldwell 1994).

Animal–habitat relationships obviously involve the behavior of habitat selection. Animals seeking a place to live can respond to a set of environmental conditions in at least two ways—stay and try to establish themselves or leave. Experimental enclosures with no exits are likely adequate settings for assessing the preference of animals for a given environmental condition. For example, if fish are placed in a tank of water with a gradient of temperatures, most generally stay in the part of the tank where the temperature is most acceptable to them (Warren 1971, 186). However, enclosures are not adequate for assessing whether conditions as a whole inside the enclosure represent a suitable environment for an animal, because the animal cannot leave.

Enclosures with exits that allow animals to leave have been used to study a variety of subjects, including social behavior and population size (Butler 1980; Gerlach 1996, 1998), habitat relationships (Wilzbach 1985), and emigration and population regulation (McMahon and Tash 1988; Nelson et al. 2002). Allowing animals to leave an enclosure may represent a significant improvement in the design of ecological experiments in controlled settings because the animals' responses to the conditions inside the enclosure indicate whether they are suitable (Matter et al. 1989; although see Wolff et al. 1996).

How to Proceed

Romesburg (1981) felt that many of the ideas upon which the science of wildlife management has been founded are hypotheses that have not been tested. He argued convincingly that wildlife biologists do a good job of making observations, searching for patterns, and formulating hypotheses, but that attempts to apply the H-D method to verify hypotheses are rare. He encouraged wildlife biologists to participate more actively in the search for causal mechanisms through experimentation. In contrast, Peters (1991) suggested that the search for causal mechanism in ecology was overemphasized and that more effort should be placed in the search for predictive relationships to help us solve, or at least identify, ecological problems.

Given the problems inherent in both laboratory and field experiments and the somewhat contradictory advice in the literature about both the scientific method and its application, a legitimate question for those studying wildlife–habitat relationships is how to proceed with investigations. The answer, in our opinion, depends on the kind of investigation that is undertaken and the resources available to the investigator. In wildlife ecology, the questions at the heart of many investigations are "species driven." That is, studies are done primarily because information about some aspect about the life history of a species is needed for management purposes. Often even basic information about these species, such as the types of vegetation they use, is lacking, so studies that document patterns of this kind are necessary. If an investigation is about identifying or verifying *patterns* in nature, then H-D tests are not necessary or applicable.

Studies of pattern are critical in science and management and should not be viewed as second-class activities (Weiner 1995; Lawton 1996b). In science, patterns stimulate the development of explanations and are therefore a vital part of the scientific method. In management, decisions frequently need to be made quickly and almost always without all the necessary information; predictions derived from patterns alone may be sufficient for management action. For example, if we know that a species is declining in abundance and the habitat on which it depends is also declining, some form of habitat conservation should probably be initiated, even though a causal relationship between the two patterns has not been established definitively.

The important role that patterns play in science and management necessitates that they be accurately identified. Thus studies of patterns should be conducted at several spatial scales over relatively long periods of time to ensure that the true natures of the patterns under investigation are revealed. Also, "attention should be paid to details such as using unbiased sampling techniques, collecting adequate samples, and employing appropriate statistics" (Nudds and Morrison 1991, 759). Following these suggestions will help ensure that the patterns detected are real and accurately described. Inaccurate descriptions of patterns can lead to inappropriate management actions and will hinder development of realistic explanations (Kodric-Brown and Brown 1993).

When the purpose of an investigation is to discover the nature of a process or mechanism underlying a pattern in nature, an H-D test is needed. The species or ecological system to be studied in an H-D test might be chosen on the basis of how amenable the species or system is to control and manipulation. Neither laboratory nor field experiments are perfect for answering questions about ecological processes, yet both have value. Their relative values can perhaps be assessed by thinking of each as a tool designed for a specific function. Like most tools, each will do "some things well, some things badly, and other things not at all" (Lawton 1996a). We suggest that, when possible, investigators make use of both field and laboratory experiments to gain the highest level of confidence in their hypotheses.

Murphy (1990) advocated applying the H-D method to single, large-scale manipulations, such as the creation of refuge boundaries, and treating these manipulations as experiments. He proposed that we could make predictions about what would happen after the manipulation is completed, wait to see what in fact happens, and then make inferences about the processes that cause the outcomes. Sinclair (1991) echoed these ideas, and Nudds and Morrison (1991, 758–59) noted: "Hypothetico-deductive research is not characterized by whether it is experimental, because hypotheses can be tested with data not collected by experiment."

There are almost certainly some situations in which hypotheses can be falsified conclusively by a single observation or a few observations not derived from experiments, but the level of confidence in accepting or rejecting an explanation under investigation is dependent on the number of alternative explanations that could also account for observed outcomes. Ecosystems and the ecological processes that drive them are sufficiently complex that alternative explanations for observed phenomena are usually abundant (Peters 1991). Hence, large-scale manipulations, which are usually not replicated or randomly applied and during which there is often no control or assessment of critical environmental variables, are not likely to be exclusionary tests of underlying processes. Drew (1994, 597) concluded, "The assertion that any question can be made scientifically rigorous by forcing it into [the H-D method] is false. Demographic and environmental indicators can be useful management tools, but when attempts are made to raise management targets to the level of scientific hypotheses—and exclusionary hypotheses at that—we only confuse the issue and give managers a false sense of assuredness."

Manipulating the environment, monitoring what happens, and changing the manipulation if we do not get the results we expect is called *adaptive management* (see chapter 11). This approach is potentially very useful, provided the opportunity for modifying management practices actually exists (chapter 11), and could provide a wealth of information about what happens when we change the environment under specific conditions at a specific point in time. It may also provide hints about processes that cause the results we observe. Caution should be used, however, if adaptive management exercises are treated as H-D tests, because most are not designed as conventional experiments and cannot be analyzed as such.

A legitimate concern, therefore, of many ecologists is that environmental manipulations, particularly those dealing with large areas or occurring over long periods of time, may not ever provide much information about cause–effect relationships. We submit that activities that can cause significant changes in natural environments, but are relatively rare, such as building a telescope on a forested mountaintop, are not likely to afford the opportunity to conduct H-D experiments. Inferences from studies of the effects of these "one-time events" will be limited due to lack of replication and randomization.

But studies of this kind are not without value. Morrison et al. (2001) categorized studies in which either randomization or replication or both were sacrificed as "quasi experiments." The design of one common kind of quasi experiment is called BACI (before-after/control-impact) (Stewart-Oaten 1986). This design relies on measuring an ecological variable of interest (e.g., the abundance of an endangered species) on the site where a disturbance is to take place and on a similar site nearby for enough time before the disturbance to assess how the variable changes over time on both sites. The variable is then measured for a period of time after the disturbance, and the relative difference in the variable between the disturbed site and the nearby site before and after disturbance is evaluated (Stewart-Oaten 1986; Morrison et al 2001). Inferences beyond the disturbed site usually are not an objective of this kind of study, so random sampling from a larger target population is not as critical as in an experiment (Stewart-Oaten 1986). Complications that might be caused by differences between the disturbed site and nearby site (Morrison et al. [2001] called these *reference sites*) are potentially reduced by tracking the variable of interest long enough to understand how the variable behaves on the two sites. Bayesian statistics (see above, chapter 10, and box 10.2) may also provide an effective alternative means to analyze the results of manipulations that lack classical experimental designs. Bayesian statistics and quasi experiments are not substitutes for careful experimentation, but may be useful when experimentation is not possible (Hilborn and Ludwig 1993).

Approaches used to understand how the world works vary according to circumstances and subject matter. Some subjects and circumstances may not lend themselves to experimentation. Understanding in these situations may need to be based on relatively weak inferences derived from patterns or creative analyses of manipulations without replication or randomization. However, properly designed experiments embedded in H-D tests are perhaps the best tool available for understanding ecological processes and, when feasible, should be favored.

Conclusion

It is critical that individuals participating in the study of wildlife–habitat relationships have a thorough understanding of all activities involved in the scientific method. This understanding does not mean that each individual will perform all the activities in every study he or she undertakes. Such an understanding does, however, help ensure that the approach taken in a study will be appropriate for the kind of question being asked. The benefit of the scientific method is its power of helping humans understand the world in which they live. Thus all scientific activities—observing, recognizing patterns, and formulating and testing hypotheses— should be carried out so as to move toward understanding.

Literature Cited

*Epigraph taken from a paper presented by Russell Sadler at the New Perspectives Conference, USDA Forest Service, Roanoke VA, 3 December 1991.

Akaike, H. 1973. Information theory as an extension of the maximum likelihood principle. In *Second International Symposium on Information Theory*, ed. B. N. Petrov and F. Csaki, 267–81. Budapest: Akademiai Kiado.

Anderson, D. R., K. P. Burnham, and W. L. Thompson. 2000. Null hypothesis testing: Problems, prevalence, and an alternative. *Journal of Wildlife Management* 64:912–23.

Anderson, D. R., W. Link, D. J. Johnson, and K. Burnham. 2001. Suggestions for presenting the results of data analysis. *Journal of Wildlife Management* 65:373–78.

Berkson, J. 1938. Some difficulties of interpretation encountered in the application of the chi-squared test. *Journal of the American Statistical Association* 33:526–36.

Berkson, J. 1942. Tests of significance considered as evidence. *Journal of the American Statistical Association* 37:325–35.

Butler, R. G. 1980. Population size, social behavior, and dispersal in house mice: A quantitative investigation. *Animal Behavior* 28:78–85.

Carey, S. S. 1994. *A beginner's guide to scientific method.* Belmont, CA: Wadsworth.

Chamberlin, T. C. 1897. Studies for students: The method of multiple working hypotheses. *Journal of Geology* 5:837–48.

Cherry, S. 1998. Statistical tests in publications of the *Wildlife Society Bulletin. Wildlife Society Bulletin* 26:947–53.

Cox, D. R. 1958a. *Planning experiments.* New York: Wiley.

Cox, D. R. 1958b. Some problems connected with statistical inference. *Annals of Mathematical Statistics* 29:357–72.

Drew, G. S. 1994. The scientific method revisited. *Conservation Biology* 8:596–97.

Feder, K. L. 1990. *Frauds, myths, and mysteries: Science and pseudoscience in archaeology.* Mountain View, CA: Mayfield.

Fisher, R. A. 1947. *The design of experiments.* 8th ed. London: Oliver and Boyd.

Franzblau, M. A., and J. P. Collins. 1980. Test of a hypothesis of territory regulation in an insectivorous bird by experimentally increasing prey abundance. *Oecologia* 46:164–70.

Gavin, T. A. 1991. Why ask "why": The importance of evolutionary biology in wildlife science. *Journal of Wildlife Management* 55:760–66.

Gerlach, G. 1996. Emigration mechanisms in feral house mice: A laboratory investigation of the influence of social structure, population density, and aggression. *Behavioral Ecology and Sociobiology* 39:159–70.

Gerlach, G. 1998. Impact of social ties on dispersal, reproduction and dominance in feral house mice (*Mus musculus domesticus*). *Ethology* 68:684–94.

Glickman, S. E., and G. S. Caldwell. 1994. Studying behavior in artificial environments: The problem of "salient elements." In *Naturalistic environments in captivity for animal behavior research*, ed. E. F. Gibbons Jr., E. J. Wyers, E. Walters, and E. W. Menzel Jr., 197–216. Albany: State University of New York Press.

Goodstein, D. 1996. Conduct and misconduct in science. In *The flight from science and reason,* ed. P. R. Gross, N. Levitt, and M. W. Lewis. New York: New York Academy of Sciences.

Gross, P. R., and N. Levitt. 1994. *Higher superstition.* Baltimore: Johns Hopkins University Press.

Gross, P. R., N. Levitt, and M. W. Lewis, eds. 1996. *The flight from science and reason.* New York: New York Academy of Sciences.

Guthery, F. S., J. J. Lusk, and M. J. Peterson. 2001. The fall of the null hypothesis: Liabilities and opportunities. *Journal of Wildlife Management* 65:379–84.

Hahn, G. J., and W. Q. Meeker. 1991. *Statistical intervals: A guide for practitioners.* New York: Wiley.

Hairston, N. G. Sr. 1989. *Ecological experiments: Purpose, design, and execution.* Cambridge: Cambridge University Press.

Hanson, N. R. 1965. *Patterns of discovery.* Cambridge: Cambridge University Press.

Harding, S. 1986. *The science question in feminism.* Ithaca: Cornell University Press.

Harris, V. T. 1952. *An experimental study of habitat selection by prairie and forest races of the deer mouse,* Peromyscus maniculatus. Contributions of the Laboratory of Vertebrate Biology, no. 56. Ann Arbor: University of Michigan.

Hilborn, R., and D. Ludwig. 1993. The limits of applied ecological research. *Ecological Applications* 3:550–52.

Hurlbert, S. H. 1984. Pseudoreplication and the design of ecological field experiments. *Ecological Monographs* 54:187–211.

Johnson, D. H. 1999. The insignificance of statistical significance testing. *Journal of Wildlife Management* 66:763–72.

Johnson, D. H. 2002. The role of statistical hypothesis testing in wildlife science. *Journal of Wildlife Management* 66:272–76.

Klopfer, P. H. 1963. Behavioral aspects of habitat selection: The role of early experience *Wilson Bulletin* 75:15–22.

Kneller, G. F. 1978. *Science as a human endeavor.* New York: Columbia University Press.

Kodric-Brown, A., and J. H. Brown. 1993. Incomplete data sets in community ecology and biogeography: A cautionary tale. *Ecological Applications* 3:736–42.

Kuehl, R. O. 1994. *Statistical principles of research design and analysis.* Belmont, CA: Duxbury Press.

Lawton, J. H. 1996a. The Ecotron facility at Silwood Park: The value of "big bottle" experiments. *Ecology* 77:665–69.

Lawton, J. H. 1996b. Patterns in ecology. *Oikos* 75:145–47.

Matter, W. J., and R. W. Mannan. 1989. More on gaining reliable knowledge: A comment. *Journal of Wildlife Management* 53:1172–76.

Matter, W. J., R. W. Mannan, E. W. Bianchi, T. E. McMahon, J. H. Menke, and J. C. Tash. 1989. A laboratory approach for studying emigration. *Ecology* 70:1543–46.

Mayr, E. 1974. Behavior programs and evolutionary strategies. *American Scientist* 62:650–59.

McMahon, T. E., and J. C. Tash. 1988. Experimental analysis of the role of emigration in population regulation of desert pupfish. *Ecology* 69:1871–83.

Medawar, P. 1984. *Pluto's republic.* New York: Oxford University Press.

Mellen, K., B. G. Marcot, J. L. Ohmann, K. Waddell, S. A. Livingston, E. A. Willhite, B. B. Hostetler, C. Ogden, and T. Dreisbach. 2003. *DecAID: The decayed wood advisor for managing snags, partially dead trees, and down wood for biodiversity in forests of Washington and Oregon.* Version 1.10. Portland, OR: USDA Forest Service, Pacific Northwest Region, and Pacific Northwest Research Station; USDI Fish and Wildlife Service, Oregon State Office.

Mertz, D. B., and D. E. McCauley. 1980. The domain of laboratory ecology. In *Conceptual issues in ecology,* ed. E. Saarinen, 229–44. Dordrecht, the Netherlands: D. Reidl.

Mood, A. M., F. A. Graybill, and D. C. Boes. 1974. *Introduction to the theory of statistics.* New York: McGraw-Hill.

Morrison, M. L., W. W. Block, M. D. Strickland, and W. L. Kendall. 2001. *Wildlife study design.* New York: Springer.

Murphy, D. D. 1990. Conservation biology and scientific method. *Conservation Biology* 4:203–4.

Nelson, A. R., C. L. Johnson, W. J. Matter, and R. W. Mannan. 2002. *Canadian Journal of Zoology* 80:2056–60.

Nudds, T. D., and M. L. Morrison. 1991. Ten years after "reliable knowledge": Are we gaining? *Journal of Wildlife Management* 55:757–60.

Peters, R. H. 1991. *A critique for ecology.* Cambridge: Cambridge University Press.

Platt, J. R. 1964. Strong inference. *Science* 146:347–53.

Popper, K. 1981. The myth of inductive hypothesis generation. In *On scientific thinking,* ed. R. D.

Tweney, M. E. Doherty, and C. R. Mynatt, 92–99. New York: Columbia University Press.

Ramsey, F. L., and D. W. Schafer. 2002. *The statistical sleuth.* Pacific Grove, CA: Duxbury.

Robinson, D. H., and H. Wainer. 2002. On the past and future of null hypothesis significance testing. *Journal of Wildlife Management* 66:263–71.

Romesburg, H. C. 1981. Wildlife science: Gaining reliable knowledge. *Journal of Wildlife Management* 45:293–313.

Scheiner, S. M., and J. Gurevitch, eds. 2001. *Design and analysis of ecological experiments.* 2nd ed. New York: Oxford University Press.

Simpson, G. G. 1964. *This view of life.* New York: Harcourt, Brace, and World.

Sinclair, A. R. E. 1991. Science and the practice of wildlife management. *Journal of Wildlife Management* 55:767–73.

Skalski, J. R., and D. S. Robson. 1992. *Techniques for wildlife investigations: Design and analysis of capture data.* New York: Academic Press.

Smith, R. L. 1996. *Ecology and field biology.* 5th ed. New York: HarperCollins.

Steidl, R. J., J. P. Hayes, and E. Schauber. 1997. Statistical power analysis in wildlife research. *Journal of Wildlife Management* 61:270–79.

Steidl, R. J., and L. Thomas. 2001. Power analysis and experimental design. In *Design and analysis of eco-logical experiments,* 2nd ed., ed. S. M. Scheinerand J. Gurevitch, 14–36. New York: Oxford University Press.

Stewart-Oaten, A., W. W. Murdoch, and K. R. Parker. 1986. Environmental impact assessment: "Pseudoreplication" in time? *Ecology* 67:929–940.

Warren, C. E. 1971. Biology and water pollution control. Philadelphia: Saunders.

Wecker, S. C. 1963. The role of early experience in habitat selection by the prairie deer mouse, *Peromyscus maniculatus bairdi. Ecological Monographs* 33:307–25.

Weiner, J. 1995. On the practice of ecology. *Journal of Ecology* 83:153–58.

White, G. C., and K. P. Burnham. 1999. Program MARK: Survival estimation from populations of marked animals. *Bird Study* 46 (suppl.): 120–138.

Wilzbach, M. A. 1985. Relative roles of food abundance and cover in determining the habitat distribution of stream-dwelling cutthroat trout (*Salmo clarki*). *Canadian Journal of Fisheries and Aquatic Science* 42:1668–72.

Wolff, J. O., E. M. Schauber, and W. D. Edge. 1996. Can dispersal barriers really be used to depict emigrating animals? *Canadian Journal of Zoology* 74:1826–30.

5 Measuring Wildlife Habitat: What to Measure and How to Measure It

[I]f you are not interested you will see nothing but the road you walk on, and if you have no desire to acquire knowledge and assume you can learn in a fortnight what cannot be learnt in a lifetime, you will remain ignorant to the end.

JIM CORBETT, *JUNGLE LORE*

In this chapter we first review two of the major aspects of measuring wildlife habitat: what to measure and how to measure it. Implicit in our discussion is a careful evaluation of the niche gestalt of the animal. As discussed elsewhere, we must find ways to identify what the animal perceives as important to its survival: the *what* question. That is, we must try to see through the eyes of the animal and understand its sensory perceptions. We review literature on niche descriptions, and on resource partitioning and allocation to identify factors of importance in a study.

We then discuss common methods used in measuring wildlife habitat: the *how* question. We identify major techniques, discuss their relationship to analysis of community structure, and outline their pros and cons. Techniques include the use of organism-centered plots versus various randomization procedures, how vegetation and other environmental features have been measured, and measurements of habitat diversity and heterogeneity.

"When" to measure, sample-size requirements, observer biases, and related topics are covered in chapter 6. "Whom" to measure—an important part of wildlife–habitat analysis—is covered briefly in chapter 4. A severe weakness in any one of these aspects of designing a study, or a series of weaknesses across several of them, will place severe limitations on the scientific credibility of results and the applicability of study results to management situations. Wiens (1989, 307–12) previously developed a categorization of measuring habitat similar to ours.

This and chapter 6 can perhaps best be described as an exercise in "pattern seeking." That is, in most cases we are seeking to describe patterns of habitat use, rather than testing specific hypotheses about the causes of the relationships we observe in nature. Although one can almost always develop a hypothesis, there is no reason to do so just for the sake of developing one. A hypothesis such as "testing the null hypothesis of no difference in used versus available habitat" is trivial on several fronts. Most notably, if the hypothesis was not rejected, we would have to conclude that the animal was randomly distributed, a situation we know simply does not occur. In addition, we can always find a difference between used and available habitat, depending on how finely we divide our measurement variables. Pattern seeking is a necessary early step in analysis of an animal's habitat and falls into the natural history tradition that has marked the history of animal ecology. Pattern seeking can most readily be contrasted with various experimental methodologies, as developed in chapter 4.

What to Measure

Green (1979, 10) asked, "What criteria should be applied to choice of a system and variables in it for use in applied environmental studies?" He answered his question with respect to four criteria:

1. Spatial and temporal variability in biotic and environmental variables that would be used to describe or predict impact effects
2. Feasibility of sampling with precision at a reasonable cost
3. Relevance to the impact effects and a sensitivity of response to them
4. Some economic or aesthetic value, if possible

These criteria apply either in descriptive studies or in analyses of impact (e.g., chemical spill, forest harvesting) effects. Understanding the variability inherent in the system of interest is critical in designing a study; the variability must be generally understood and the sampling designed to capture it. This variability includes natural, stochastic, or systematic change, as well as measurement and sampling error. There is always a cost–benefit analysis that researchers conduct, either formally or informally, when choosing variables for measurement; one must explicitly identify the precision necessary to

reach project goals and then match all sampling to this needed precision. "Mismatching" of precisions is a common mistake (i.e., waste of time) in wildlife studies. For example, there is no reason to determine the density of small mammals to two or three significant figures (e.g., 3.25 animals/ha) if these data are to be correlated with ocular measurements of vegetative cover that can only be honestly estimated into broad categories (e.g., grass cover = 10–20%).

A classic "waste of time" inherent in habitat studies is to measure everything possible for the duration of the study, and then run correlations to try to identify variables important in describing the distribution of the animal of interest. As noted above by Green, the variables measured should have relevance and sensitivity to the question at hand. Researchers often fail to follow the lessons learned in other studies and insist, instead, on reinventing the proverbial wheel at the start of each new study. The fear of missing something important certainly drives such decisions and is, as such, somewhat understandable. However, given that most studies are operating on limited time and scarce budgets, more thought should go into variable selection before field studies are started. If questions remain regarding the necessity of including certain variables, then preliminary analyses of data should be used to further reduce the variables being collected in the field. Such preliminary analyses provide more time to increase the precision of the remaining variables as necessary and, especially, to increase the sample size.

Finally, Green notes that the variables selected could have some economic or aesthetic value. This criterion may or may not be a factor, depending on the purpose of the study. Naturally, variables of economic interest might be included in a study if the intended audience for the results has specific interests. For example, foresters would usually be more interested in the relationship between birds and an economically valuable tree species than in a relationship between birds and shrub cover.

Conceptual Framework

Efforts to understand and predict variation in the abundance and distribution of species have been hampered by inadequate theories, mismatches between processes and sampling scales, and inappropriate statistical methods. We have lacked a strong theoretical framework that addresses multiple interacting processes and limiting factors; such a framework would help resolve the patterns underlying the complexity of nature and also help to identify appropriate sampling scales and to develop statistical methods more appropriate for quantifying causal relationships than those traditionally used (e.g., Huston 2002). Lack of such a framework has been a major impediment to sound resource management and conservation planning. For example, there are literally thousands of models describing the habitat use of terrestrial vertebrates that use regression, correlation, various multivariate methods, as well as habitat suitability index models, habitat evaluation procedures, and gap analysis program techniques. These efforts are usually only relevant to the specific location, and often only to the specific time, of the study. What they strictly predict is an envelope of environmental and vegetation requirements within which the species *may* occur. These models have little utility for tracking changes in distribution (O'Connor 2002). Additionally, we do not have strategies for building upon a model as new information is gathered. Van Horne (2002) noted that animal ecologists work within or among discrete vegetation types, or within some management or other type of artificial "study" area. Thus resource managers are left searching for a means of synthesizing and then applying data from the myriad independent studies by ecologists.

O'Connor (2002) concluded that the key focus should be on factors as limiting agents in species abundance. His argument had two key assumptions: first, that most species can be limited by any of a variety of factors; second, that the influence of any ecologically relevant factor is not additive to the influence of any other factor. Rather, typically only one factor is limiting in any particular situation. Clearly, a paradigm shift is needed in the way we conceptualize, and thus study, the factors driving the distribution and abundance of animals if we are to advance understanding and implement meaningful management and conservation actions (e.g., Huston 2002; O'Connor 2002).

Many hypotheses have been proposed to explain the persistence of animals, including vegetation structure or floristics, predator activity, food availability, and abiotic conditions. Most of these hypotheses are not mutually exclusive (e.g., habitat and predator activity), and most have been shown to be operative only in certain locations and at certain times. These findings further strengthen the conclusion that multiple interacting processes and limiting factors are responsible for the patterns underlying the complexity of nature and also contribute to identification of appropriate sampling scales. Thus trying to isolate a specific "driving" hypothesis appears to be misdirected in most applications. Although we are not suggesting that experimental and observational study of specific factors is unwarranted, what we are proposing is that a different overall framework is needed to improve our ability to predict animal persistence across space and time.

Most ecological processes are influenced by multiple factors, only some of which an organism can control. Although any resource or regulator can potentially occur at levels that limit the rate of a process, it is unlikely that the same single factor will always be limiting because spatial and temporal variation in resources and regulators results in a continual shifting of resources. Such variation results in a continual shifting among potentially key limiting factors. This shifting between limiting factors complicates our understanding and lessens our predictive power. When processes are regulated according to the law of minimum, variation in a nonlimiting factor has little or no effect on the rate of the process of interest. Process rates will be correlated only with variation in the limiting resource, and only under these conditions will regression analysis reflect the potential effect of the limiting resource on the process. Thus maximum potential response to any specific resource or regulator of a process can only be quantified when all other resources or regulators occur as nonlimiting factors (Huston 2002). Most ecological research, however, is conducted using a mixture of measurements made under both limiting and nonlimiting conditions and thus produce data sets with high variance (and low correlation coefficients) that prevent identification of the actual response of a process to a specific factor. Thus the high variance that is often found in correlations between ecological processes and presumed causal factors may not be sampling error or random "noise," but rather mechanistic consequences of shifts between limiting resources or the effects of other limiting factors, such as mortality or dispersal.

MODELING APPROACH

Models that make specific predictions are likely to be limited in their application because of variability in both time and space. Even the most complete statistical model cannot include all biologically important causal relationships for a given location. A model that works well in one location may not be useful in another (Van Horne 2002). Usually, general models are simpler than models that aim to make specific predictions. Using fewer terms enhances the clarity of such models and increases their applicability

to a broad range of systems. General models may either describe broad patterns or be more theoretical than specific models. General models lose the ability to make specific predictions because they are less likely to be directly related to processes than are either specific models or theoretical models. Researchers construct theoretical models to understand the important relationships that govern population numbers. A theoretical model would describe when and how each type of regulator variable is primary across a wide range of populations of a given species or across a broad range of species. A specific model would document the nature of population regulation for an intensively studied population but would have little explanatory power across the species range or among species. Thus most current modeling approaches cannot be universally validated, because they are too general to make precise and testable predictions, apply only within tight and often unknown boundary conditions, or have failed to incorporate process and therefore rely on correlations that may be spurious (Van Horne 2002).

A major problem in using correlation is the shape of the variable–response curve. Classic niche concepts envisioned a normal Gaussian response by the organism to a gradient in the variable or resource. Where such a curve exists, we may record an increasing, nonexistent, or decreasing response, depending on the range of the variable measured (see fig. 3.7). Yet most correlative approaches assume that if no linear response is detected, a given variable is not an important determinant of species occupancy. There are, of course, many possible nonlinear responses in ecology. Hypothesizing the best model fit a priori is often difficult, and trial-and-error exercises of model fitting weaken final inferences.

Understanding how a species responds under varying environmental conditions across its range is critical for restoration, conservation, and management of species. In addition, any framework developed must be readily accessible to the resource managers who ultimately make on-the-ground management decisions. Habitat variables, especially when sampled over short time frames, provide little information about key processes such as birth and death rates, that are needed to make sound management decisions (e.g., Tyre et al. 2001). Below we introduce and review many of the methods that have been used to assess the use of resources by wildlife, including relatively traditional as well as relatively recent methodologies. Each method has fundamental strengths and weaknesses, and each method has merit, depending on the questions being asked in a particular study.

SPATIAL SCALE

In chapter 1 we defined the key ecological differences between ecological scale and ecological level. Here we will be concerned with spatial scale as it relates to the measurement of environmental features that influence the distribution, abundance, and viability of animals. Correct determination of the scale of analysis is the cornerstone of habitat analysis and model development. That is, we should match the scale of analysis with the scale at which we wish to apply our results for management purposes. For example, there is likely little reason to spend time collecting "microhabitat" variables if one is only interested in describing the distribution of an animal across general vegetation types (i.e., relatively "macrohabitat"). Green (1979) asserted that it is not the degree to which a model meets perfection that renders it valid, but rather its adequacy for fulfilling a prescribed purpose. There is no reason for high precision if the biological change to be detected and managed for is large.

The scales at which wildlife–habitat relationships can be examined fall along a continuum that is not unlike how an animal selects "proper" habitat (chapter 2). The way an animal perceives its environment and the way we relate these perceptions to an organized method of study have the utmost importance in model development. We can examine habitat use at its broadest, or biogeographic, scale, passing through successively finer-scale evaluations until we reach the level of the individual. As we move from relatively broad-scale, or macrohabitat, variables to relatively fine-scale, or microhabitat, variables (fig. 5.1), both the scale of our measurements and the methodologies we must employ change. As noted below, mixing scales (scale mismatch) into a single analysis has been a common problem in habitat analysis and ecology, a problem that is beginning to be resolved.

It is important to note—as discussed above— that these fine-scale models almost always vary between locations and time periods, and certainly between populations. The magnitude of these variations determines the generality of the model (*generality* refers to the ability of a model developed at one time and place to be applicable at other times and places). Much of the wildlife–habitat relationships literature has been criticized because of its time- and place-specificity (e.g., Irwin and Cook 1985). This criticism reflects a misunderstanding of the relationship between the precision of the variables measured and the scale of application possible. We should add that researchers conducting most wildlife–habitat relationships studies seldom acknowledge the generality of their data and models, and thus the locations and conditions under which managers should use them. Thus the decision to develop relatively "extensive" broad-scale models or more "intensive" fine-scale models should be based on the objectives of the particular study. The more extensive approach typically cannot tell us such things as how an animal reacts to changes in litter depth, local density of trees by species, or the occurrence of a predator in a specific patch of vegetation; such approaches are likely necessary for management of localized populations of animals. Once the

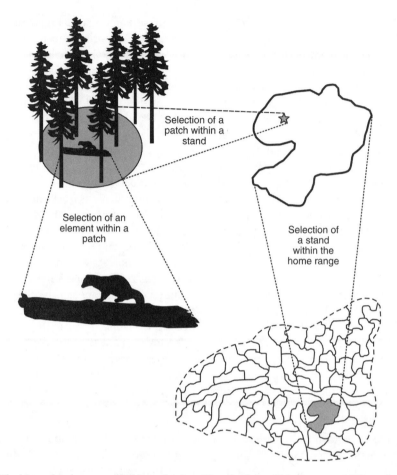

Figure 5.1   The hierarchical nature of habitat selection. Here the fisher (*Martes pennanti*) is used as an example. (Reproduced from Weir and Harestad 2003, fig. 1; reprinted by permission of the Wildlife Society.)

specificity is determined, then the researcher can determine the type, resolution, and geographic extent of data collection required.

Van Horne (1983) outlined a hierarchical approach to viewing wildlife–habitat relationships (fig. 5.2). Van Horne's level 1 (*level* here refers to a categorization of studies, not to biological levels) applies to intensive, site-specific studies of individual species. Her level 2 uses more generalizable variables and likely allows application (or

relatively easy adaptation) of a model to other locations. Her level 3 is the most extensive approach and develops relationships for a host of species. Van Horne's scheme could easily be divided into many more categories but serves to indicate how habitat relationships can be studied, depicted, modeled, and applied along a continuum of resolution. Figure 5.3 depicts the major spatial scales at which ecological field studies are typically conducted.

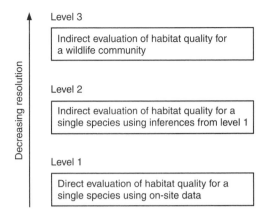

Figure 5.2. Hierarchical description of habitat quality assessments. (Reproduced from Van Horne 1983, fig. 1; reprinted by permission of the Wildlife Society.)

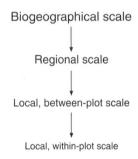

Figure 5.3. An outline of the spatial scales at which ecological field studies are conducted.

That wildlife managers are often frustrated at the failure of most models to "work" in their specific location is understandable. Yet we have a dilemma: models based on broad measurements of vegetation can seldom be applied to local situations (e.g., a small management unit or refuge), whereas models developed at a fine scale can seldom be adequately applied (generalized) to other locations (Block et al. 1994). The patterns of habitat use that emerge from our studies are sensitive to the scale of the comparison, because different relationships may exist in different subsets of the samples being compared—the habitat and niche discussion we raised above. As shown by Wiens and Rotenberry (1981; see also Wiens 1989, 56–57), a species will exhibit different patterns of habitat use, depending upon which portion of its distribution across the landscape is sampled by the observer (fig. 5.4). In figure 5.4, sampling across the entire habitat gradient depicted would result in yet another ("averaged") pattern of habitat use. None of the depictions would be wrong per se; the "correctness" would be based on the scale of the question being asked.

How can this dilemma be solved? It is possible in many situations to adapt a model developed for use in one location to the conditions that exist at another location. Again, we need not constantly reinvent the wheel. Models that include fundamental, mechanistic explanations of the activities and responses of animals to environmental conditions are more likely to be applicable across time and space relative to those based solely on habitat factors (with the caveat that we usually should not mix scales in a typical statistical analysis; examples of analyses are given below).

As summarized by Wiens (1989, 239), therefore, studies across broad geographic areas are likely to overlook important details that account for the dynamics of local populations. Far from being idiosyncratic "noise," the variations within or among local populations may contain important mechanistic information about the factors causing the organisms' response. These variations tend to disappear at broader scales because of the consequences of averaging unless the study is designed to determine such variation, as by stratified subsample or blocking designs.

We cannot specify here the detail required to develop proper habitat relationships for every situation. What we are trying to develop is the

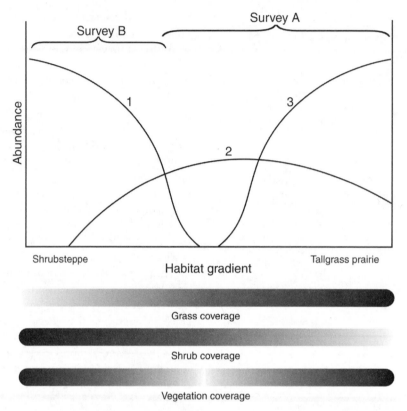

Figure 5.4.   Effects of scale on study of patterns of habitat association. In this schematic diagram, species 1, 2, and 3 have characteristic distributions on a gradient from shrubsteppe through short- and mixed-grass prairies to tallgrass prairie. Grass coverage, shrub coverage, and overall vegetation height change on the gradient, as shown below the graph. In survey *A*, a large portion of the gradient is sampled, but extreme shrubsteppe sites are omitted, whereas in survey *B*, only shrubsteppe and a few grass–shrub sites are studied. The species will exhibit different patterns of habitat association in the two surveys. Species 1, for example, will exhibit a strong negative association with grass coverage and a positive association with shrub coverage in survey *A* but may fail to show either association in survey *B*. (Reproduced from Wiens and Rotenberry 1981, fig. 3. reprinted with permission by the Ecological Society of America.)

thought process that researchers must use on a study-specific basis. We will present and discuss, however, several examples of the types of variables collected by researchers attempting to develop predictive relationships. The list of potential studies is long, and our examples should not be taken as indicating only the best, or even the most common, techniques. Rather, we want re-

searchers to see the range of variables being collected by workers in the field.

Measurements of the Animal

We have discussed the question of how refined our estimates of the numbers of an animal population need be for model development. For example, must the estimate be at the level of presence or absence of a species, index of abundance (e.g., numbers/count), or absolute density (numbers/unit area)? It is beyond the scope of this book to review the methods available for estimating animal abundance. In fact, many excellent works have been published on this subject (e.g., Cooperrider et al. 1986; Ralph et al. 1993; Bookhout 1994; Heyer et al. 1994; Buckland et al. 2001). Here we wish to briefly touch on a method of evaluating wildlife–habitat relationships that centers on counting animals and then relating these numbers to environmental features.

Although we are not covering this topic in detail, it is crucial that our estimates of the number of animals present be reliable. That is, are our estimates of absolute or even relative abundance a fair reflection of the number of animals actually present? If we use multiple observers, are their estimates comparable? Can estimates derived by one observer or group of observers in one study area be used to validate a model developed by other people at a different location? Throughout this book we discuss the varying effects of observer error, errors encountered because of hidden biases, and other factors that work against the development of habitat models. Here we are interested in errors associated with estimating numbers of animals. We remind the reader that there are many problems associated with counting animals and reiterate that poor estimates of animal numbers will negate conclusions drawn on habitat relationships based on even the most carefully collected environmental variables.

As developed above, an entire class of habitat models is based on correlating animal numbers to some features of the animal's biotic and abiotic environment. The purpose of these models is to develop adequate predictions of the presence, abundance, or density of the species based on environmental features (see later in this chapter; also chapter 10). The methods vary according to the available data; techniques for handling presence–absence data are not necessarily the same as those suitable for density data. Researchers and managers can use these models to predict the changes that will occur in animal abundance given changes in variables in the model.

A tremendous amount of research has been devoted to developing density estimates for vertebrates. Much of the theory in ecology, as well as applications in resource management and conservation, depends on reliable expression and comparison of numerical abundance (Smallwood and Schonewald 1996; Buckland et al. 2001). Additionally, the development of many wildlife–habitat relationships models is based, in part, on estimations of abundance (e.g., regression analysis; see below).

However, the expression of density (or an index thereof) can vary widely depending upon the spatial scale of the study. That is, simply converting number of animals to some standard area—for example, extrapolating birds in a 1.5-ha study site to birds per 40 ha (100 acres)—does not standardizes estimates for comparisons with other studies. Density for any one population is not likely to remain constant across spatial scales. For example, Smallwood and Schonewald (1996) found that $\log_{10}$ population density estimates consistently decreased linearly with $\log_{10}$ spatial extent of study area for species of terrestrial Carnivora. They showed that size of the study area accounted for most of the variation in population estimates. Similarly, Smallwood and Smith (2001) found that shrew (*Sorex*) abundances were often estimated at study areas too small to encompass "populations."

Therefore, if density is to be related to ecological variables, then it should be estimated from a representative spatial scale or scales; otherwise, the pattern will be masked by high variability in density among scales. The choice of spatial scale must be based on the species' relationship with the landscape, not on the spatial resolution of the technology used for observation, or some artificial or convenient area (Addicott et al. 1987; Wiens 1989; Smallwood and Schonewald 1996). Variations in abundance are influenced by study area size, year of study, site selection, sampling method, trap type, and various other factors (Smallwood and Smith 2001). Determination of the effect of study area size on abundance values should be incorporated into the preliminary sampling phase of all studies; unfortunately, virtually no researcher takes the time to do so (see chapter 12).

To ensure accumulation of an adequate sample size, most studies of habitat use occur where the researcher has previous knowledge that the species of interest is in adequate abundance. The home range or a finer-scale activity location (e.g., foraging or nest site) within the home range is usually the focus of study. Results of these studies are then often extrapolated (often not by the original researcher) to a much larger area than was used to collect the data. However, such an extrapolation can only be made reliably if animals and their habitat are uniformly distributed across the landscape. The distribution of animals is, of course, aggregated in some manner.

Working with Swainson's hawks (*Buteo swainsoni*), for example, Smallwood (1995) showed that most of the variation in density among study sites was due to the size of the study area chosen by the researcher. Further, studies conducted at larger spatial scales yielded different habitat associations than those at more conventional, home range scales. This relationship holds for other species (as reviewed above; Smallwood and Schonewald 1996). For example, Tarvin and Garvin (2002) showed that the scale of analysis influenced the probability of nest success, including conclusions drawn on the influence of specific habitat features as well as nest predation. These studies clearly indicate that researchers must give close attention to the size of

the study area used; most such decisions are made either from convenience or justified based on the size of the species' home range. Additionally, managers must use caution when trying to extrapolate abundances derived from small, relatively high-density areas.

Measurements of Environmental Features

Next we describe some common techniques used to measure habitat. As described throughout this book, habitat can be viewed along a continuum of spatial scale—the key word here being *continuum*. That is, some modifiers of habitat that we have invented or will invent, be they microhabitat, macrohabitat, or mesohabitat, are meaningful only in the context of specific spatial and temporal extents. Thus it is important that researchers clearly elucidate their conceptual basis for making the measurements that they report.

Two basic and obvious aspects of vegetation can be distinguished: the structure or physiognomy, and the taxa of the plants, or floristics. Many authors had initially concluded that vegetation structure and "habitat configuration" (size, shape, and distribution of vegetation in an area), rather than particular plant taxonomic composition, most determined patterns of habitat occupancy by animals, especially birds (see Hilden 1965; Wiens 1969; James 1971; Anderson and Shugart 1974; Willson 1974; James and Wamer 1982; Rotenberry 1985; Rotenberry and Wiens 1998). As Rotenberry observed, however, more recent studies have shown that plant species composition plays a much greater role in determining patterns of habitat occupancy than previously thought. As detailed below, the relative usefulness of structural versus floristic measures is foremost a function of the spatial scale of analysis.

Many earlier researchers failed to adequately place their studies into a specific spatial scale, thereby obscuring the relative (and proper) roles that structure and floristics can play in predicting habitat relationships at different scales (Levin 1992). As further noted by Rotenberry, the same species that appears to respond to the physical configuration of the environment at the continental scale may show little correlation with physiognomy at the regional or local scale. Thus many animals may be differentiating between gross vegetative types on the basis of physiognomy (that is, they occupy a general area that is "proper" in its structural configuration), with further refinement of the distribution (and thus abundance) within a local area based on plant taxonomic considerations. Note that this scenario of Rotenberry's relates closely to Hutto's (1985) hypothesized mode for the process of habitat "selection" in animals (see chapter 2). Rotenberry quantified his ideas using data from the shrubsteppe studies of Wiens and his coworkers. Rotenberry's analysis of these data indicated that when the correlation between physiognomy and flora was statistically sepa-

rated, a significant relationship remains between bird abundance and flora, but not between birds and physiognomy (fig. 5.5).

Thus we return to the important theme that the variables measured and the level of model refinement required should be based on the scale of interest. Simple presence–absence studies of animals at regional or broader scales likely do not require floristics analysis of vegetation. Broad categorization by physiognomy— probably including differentiation no lower than life form, such as deciduous and evergreen, or by general ecological classes or vegetative types—is probably adequate. Plant taxonomy becomes increasingly important, however, as our studies becoming increasingly site-specific. As a general rule, then, one would always be safe in collecting data on as fine a scale as time and budget allow, using common sense not to unduly overdo it; lumping a posteriori is always possible. A better approach is to begin a study with a preliminary evaluation of the variables and sampling methods necessary to achieve the desired level of

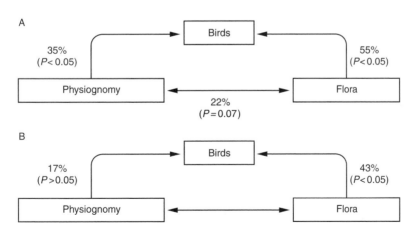

Figure 5.5.   (A) Coefficients of determination ($r^2 \times 100$) between similarity and distance matrices based on avian, floristic, and physiognomic composition of eight grassland study sites. Significance levels of association are given in parentheses. (B) Partial coefficients of determination, as above. Correlation between physiognomy and flora has been partialed out. (Reproduced from Rotenberry 1985, fig. 1; reprinted with permission by Springer Science and Business Media.)

refinement; necessary sample sizes can also be determined from such preliminary work (see chapter 4; also Morrison et al. 2002).

MACROHABITAT

With the rise in interest of studying animal diversity, researchers developed various measures that sought to relate the numbers and kinds of animals to some measure of the gross structure of the vegetation. Most famous is the foliage height diversity–bird species diversity (FHD-BSD) constructs of MacArthur and MacArthur (1961). In figure 3.2 we saw that the diversity of birds rose as vegetation became increasingly stratified. A plethora of studies followed the MacArthurs' early work, with most results showing similar results. Wiens (1989, 129–34) provided a review, with cautions, of FHD-BSD and related issues.

In vertically simple vegetation, such as brushlands and grasslands, FHD would not be expected to provide a good indicator of animal diversity (at least for most vertebrates). Recognizing this problem, Roth (1976) developed a method by which the dispersion of clumps of vegetation—here, shrubs—forms the basis for a measure of habitat "heterogeneity." Roth was able to relate BSD to this measure of habitat heterogeneity, or patchiness. Other workers also built relationships between measures of horizontal vegetation development and animal communities (see Wiens 1969; Morrison and Meslow 1983; Rotenberry and Wiens 1998; see also chapter 8).

Returning to figure 3.2, note that there is considerable scatter around the regression line. Thus the usefulness of the general FHD-BSD principle as a predictor decreases as the scale of application becomes increasingly fine. Karr and Roth (1971) suggested that the scatter in a FHD-BSD relationship likely resulted from important, but unmeasured, variables influencing the avian community. Measures of diversity sacrifice complexity for simplicity and thus are useful primarily at larger spatial scales. These indices collapse detailed information on plants—species composition, foliage condition (vigor), and arthropod abundance—into a single number.

The FHD-BSD relationships do indicate, however, that species respond to complexity in their environment. But to what specifically do they respond? The observer can increase FHD artificially simply by changing the number of categories used in the calculations. Thus while FHD gives some indication of foliage strata and the response of animals to such vegetation development, it is much too arbitrary and much too general to result in more than a gross and relative examination of wildlife–habitat relationships among different areas.

As discussed above, indirect measures of foliage structure (e.g., FHD) are extremely gross and provide only limited predictive power. However, measures of canopy closure are being used as centerpieces of management guidelines for many species of wildlife, including the federally endangered northern spotted owl (*Strix occidentalis caurina*) and the northern goshawk (*Accipter gentilis*). Thus if management guidelines are to be interpreted and applied adequately, we must know that our methodologies are truly reflecting something of ecological causation.

A more useful measure of foliage structure for many species may be foliage volume, the actual surface area of foliage available for consumption or as substrates for insects and other prey. It is extremely tedious, however, to measure actual foliage volume. Plant ecologists have done so by cutting down trees and then measuring and counting their leaves, needles, and other parts. Such data are necessary for accurate predictions of photosynthetic rates and measurements of tree growth (see Carbon et al. 1979). After ecologists have completed such work, they can develop statistical models that relate some more easily measured aspect of the plant to foliage volume (VanDeusen and Biging 1984). Plant ecologists also use dry volume of plant material as a measure of volume.

Researchers have developed even more indirect measures of foliage volume. In a comparisons of habitats of the chestnut-back chickadee (*Parus rufescens*) and the black-capped chickadee (*P. atricapillus*), Sturman (1968) used equations for shapes that approximated the structure of conifers and hardwoods. Methods such as Sturman's (see Mawson et al. 1976) can be viewed as compromises between labor-intensive techniques that involve sampling whole trees and the qualitative estimates of FHD.

Many other methods of estimating foliage development have been devised (see reviews by Campbell and Norman 1989; Ganey and Block 1994). A popular technique involves the use of a spherical densiometer. Most easily described as a round, concave mirror onto which a grid has been etched, this device is used to estimate canopy closure directly over the device. Ganey and Block (1994) tested the relative results between the spherical densiometer and the sighting tube (a piece of PVC pipe or similar material through which the observer observes canopy closure). Ganey and Block found that the densiometer resulted in higher estimates of canopy closure than the sighting tube: 57% of all estimates showed >50% cover for the densiometer, whereas only 38% did so for the sighting tube. They could not, of course, determine which method was the most accurate. Their results do show, however, that the method used can have a significant impact on the conclusions regarding canopy cover, which in turn, could have substantial ecological ramifications.

Many of the currently used habitat models operate at the macrohabitat scale, including most statewide wildlife–habitat relationships (WHR) constructs (Block et al. 1994), gap analysis program (GAP) models (Scott et al. 1993), and habitat suitability index (HSI) models (USDI Fish and Wildlife Service 1981; see Rumble et al. 1999; Mitchell et al. 2002 for examples). Most of these models use broad-scale categorizations of vegetation types (often mislabeled as "habitat types") as a predictor of animal presence or abundance. However, many developers of these models substantially mismatch scales in the variables used to develop their models. Such mismatching of scales (e.g., entering micro- and macrohabitat variables into the same analysis) is especially evident in HSI and WHR models. It ignores current theories concerning the hierarchical nature of habitat selection and makes it difficult to interpret model output. Models developed at the macrohabitat scale are very useful in our understanding of broad habitat relationships, but they should be limited to application at the broad scale.

Quite obviously, how we categorize vegetation is not a trivial matter if we expect our macrohabitat models to perform adequately. Unfortunately, many workers categorize vegetation types based on observer-based, qualitative judgments that are thus difficult to replicate in other studies. The lack of any testing before most of these models are used to make management decisions further confounds our ability to evaluate their appropriateness (Block et al. 1994).

MICROHABITAT

To determine the relevant variables to collect and analyze, we must determine what aspects of the environment that organisms recognize as relevant. As noted by Krebs (1978, 40), "We must be careful here to define the perceptual world of the animal in question before we begin to postulate the mechanisms of habitat selection." Krebs (1978, 42) identified two basic factors that must be kept separate when discussing habitat selection: *evolutionary factors,* which ultimately confer reproductive fitness and survival value on habitat selection; and *behavioral factors,* which provide the proximate mechanisms by which animals select habitat. The proper behavioral factors—that is, those that have direct ties to the survival of the animal—must be identified. Thus we are interested in behaviors that result from stimuli from the landscape; breeding, display, and feeding sites; food; and other animals (Hilden 1965; Krebs 1978, 42). As developed above, as we approach relatively fine, or micro-habitat, scale of measurement, we move closer to the need to quantify niche-related factors.

Dueser and Shugart (1978) listed four criteria that can help guide selection of microhabitat variables for measurement:

1. Each variable should provide a measure of the structure of the environment that is either known or reasonably suspected to influence the distribution and local abundance of the species.
2. Each variable should be quickly and precisely measurable with nondestructive sampling procedures.
3. Each variable should have intraseasonal variation that is small relative to interseasonal variation.
4. Each variable should describe the environment in the immediate vicinity of the animal.

Dueser and Shugart noted that the final criterion reflected their concern for describing the environment in sufficient detail to detect subtle differences among microhabitats that appeared to be grossly similar. Their first criterion tells us that previous natural history information, plus a good deal of biological common sense, will help narrow the choice of variables. We must be careful, however, not to follow past mistakes or misinformation regarding the natural history of the animal and, further, not to let our preconceived notions and biases eliminate potentially important variables. It is often worthwhile to have your list of variables reviewed by a biologist who is familiar with the area but not with the species; for instance, a plant ecologist can probably offer valuable advice to a wildlife biologist in planning a study. Dueser and Shugart were apparently concerned with nondestructive sampling because of their desire to make repeated observations at a site. Also, it would be difficult to repeat a study in the same area if the site had been severely disturbed (assuming that the disturbance was not part of the planned study). Their concern regarding sampling variation is both a biological and a statistical issue. They indicated that measurements should be sufficiently precise that variations within relatively short periods of time (within seasons or intraseasonal) are not obscured by the interseasonal variations, which are likely to be much larger. In studies of microhabitat, we usually wish to determine how these often-subtle measures of habitat vary; thus a high degree of precision is required. Dueser and Shugart's final point is intuitive: microhabitat should be measured in close proximity to the animal. It is not a mistake to measure variables from many scales; it may be a mistake to include them in the same analysis (unless a clear biological and statistical justification is provided).

Whitmore (1981) also outlined general criteria for selecting variables, listing three categories that involved "practical aspects" of variable selection:

1. Variables should be measurable at the desired level of precision.
2. Variables should be biologically meaningful.
3. Variables should be relevant to the species in question.

In reference to his second criterion, Whitmore asked what possible direct meaning measurements of tree roots would have for a canopy-feeding warbler. Regarding his third criterion, he asked how directly important the percentage of grass cover in the ratio of grasses to forbs could be to a bark-foraging bird. It can be argued that these variables may have some indirect relationship to a variable of more obvious importance to an animal; for example, root condition could relate to canopy volume or condition, which in turn might relate to insect density and thus to the behavior of the warbler. However, such indirect measures increase the variance (weaken the relationship) between the warbler and the truly relevant variable. Many indirect measurements in the same analysis thus greatly compound the error in the results, making for weak conclusions.

A Major Weakness:
Indirect Measurements of Habitat

We should note here that most of our measurements are only indirect reflections of what the animal is likely responding to in making decisions on habitat selection—a problem with the correlative techniques we developed above. Few would argue that insectivorous birds must have an adequate supply of the proper prey. Yet few studies of such birds quantify prey because of the extreme difficulty in doing so (Morrison et al. 1990). We therefore often make indirect measurements of prey by assuming that more foliage likely results in more prey. Yet indirect measures do not end here. Measuring foliage volume is also extremely difficult. Thus, as described above, we usually use some indirect measure of foliage volume. What we have, then, is an indirect measure of foliage volume being used to indirectly index insect abundance; errors are certainly grossly compounded, and it is difficult to know the biological meaning of our results in terms of specific causes. Such problems are not restricted to bird studies: big-game biologists often make indirect estimates of food requirements, including micronutrient needs, through

various indirect estimates of forage abundance (e.g., ocular estimates or line intercepts of shrub cover); herpetologists use numerous measures of ground, shrub, and canopy cover to indirectly describe the microsite conditions (e.g., soil moisture) available to salamanders.

The prevalence of indirect measures—and indirect measures to indirectly measure other variables—is one of the primary weaknesses in the area of habitat relationships research and modeling. In fact, it will be difficult to further increase the accuracy of our models, and thus our management decisions, until we begin to make more direct measurements of the factors that influence the distribution, abundance, and behaviors of wildlife. We have done a lot of the relatively easy work; substantial advances await implementation of the much more difficult measurements. We are not calling for abandonment of indirect measures. Rather, we are saying that such measures have limited predictive power in many situations, and that we should openly acknowledge the limits such measures place on our results.

Focal-Animal Approach

Most studies of microhabitat selection are variations of the "focal-animal" approach (see chapter 7 for focal-animal sampling in behavioral research). These methods use the presence of an animal as an indication of the habitat being used by the species. No correlation between abundance and the environment is involved. Rather, the location of individual animals is used to demark an area from which environmental variables are measured. As detailed in the following section, an animal's specific location might serve as the center of a sampling plot. Or, a series of observations of an individual might be used to delineate an area from which samples are then made (e.g., see Wenny et al. 1993). In either case, the major assumption of this approach is that measurements indicate habitat preferences of the animal.

Animals do not, however, spend equal amounts of time at each activity and location during a day. Feeding, drinking, calling, resting, grooming, and other activities each consume different amounts of time and energy, and often take place in different locations within the home range (e.g., calling from exposed locations or resting in the shade). The amount of time spent in any one activity may not indicate the importance of the activity to the animal; for example, drinking may take only a few minutes each day, but without water the animal is unlikely to survive. Again, the researcher must be aware of these behaviors when designing a study based on the focal-animal approach; chapter 7 provides a more detailed description of behavioral sampling. For example, many studies have used the location of a singing male bird or a foraging individual as the center of plots describing the habitat of the species (e.g., James 1971; Holmes 1981; Morrison 1984; VanderWerf 1993). But how well do such plots indicate the species' habitat preference, or even the individual's territory preference? Collins (1981) examined habitat data collected at perch and nest sites for several species of warblers. He found that 29% of the nest sites had vegetation structures that significantly differed from the corresponding perch sites within the territories. Not surprisingly, he found that basing habitat on only perch sites overestimated the tree component of habitat. A study based only on perches or only on nests is not necessarily flawed. Rather, it means that such studies only describe "perch habitat" or "nest site habitat," and, as such, describe only part of what one might call the "breeding habitat" of the species. Similar examples can be found in the literature for all major groups of wildlife: for example, bedding sites of deer (Ockenfels and Brooks 1994); trapping (Morrison and Anthony 1989; Kelt et al. 1994; Morzillo et al. 2003) and telemetry (Hall and Morrison 1997; Ucitel et al. 2003) locations of small mammals; and movements of amphibians (Griffin and Case 2001).

Use versus Availability: Basic Designs

As discussed throughout this book, the comparison of the use of an item or habitat element (food, habitat characteristic) relative to its abundance or availability is a cornerstone of wildlife–habitat relationships analyses. Thomas and Taylor (1990) reviewed 54 papers published in the *Journal of Wildlife Management* that analyzed use and availability of food items or habitat characteristics (for this discussion we ignore the conceptual differences between measurable "abundance" and what an organism perceives as "availability"). Thomas and Taylor concluded that use-availability studies could be categorized into three basic designs (see also Manly et al. 2002, 4–9, for discussion and examples):

Design 1. The availability and use for all items are estimated for all animals (population), but organisms are not individually identified, and only the item used is identified. Availability is assumed to be equal for all individuals. Here, Thomas and Taylor found that habitat studies often compare the relative number of animals or their sign of presence in each vegetation type to the proportion of that type in the study area. About 28% of the studies reviewed fell into this category.

Design 2. Individual animals are identified, and the use of each item is estimated for each animal. As for design 1, availabilities are assumed equal for all individuals and are measured or estimated for the entire study area. Studies that compare the relative number of relocations of marked animals in each vegetation type to the proportion of that type in the area fall into this category; 46% of the studies reviewed were placed into this category.

Design 3. This design is the same as design 2, except that the availability of the items is also estimated for each individual animal. Studies in this category often estimate the home range or territory for an individual and compare use and availabilities of items within that area; 26% of the studies were in this category.

Thomas and Taylor (1990) and Manly et al. (2002) provide a good review of studies that fit each of these categories, as well as guidelines for sample sizes necessary to conduct such analyses. Virtually all classes of statistical techniques have been used to analyze use-availability (or use-nonuse) data, depending on the objectives of the researcher, the structure of the data, and adherence to statistical assumptions (i.e., univariate parametric or nonparametric univariate comparisons, multivariate analyses, Bayesian statistics, and various indices). These techniques have been well reviewed (e.g., Johnson 1980; Alldredge and Ratti 1986, 1992; Thomas and Taylor 1990; Aebischer et al. 1993; Manly et al. 2002). Compositional analysis, only recently applied to habitat analysis, should be considered for use in these studies (Aebischer et al. 1993). *Compositional analysis* is an application of multivariate analysis of variance (MANOVA) that is applicable to selection among categories of a single categorical variable. Individual animals are used as replicates, and the technique falls under designs 2 and 3. Resource use is defined in terms of the proportion of different types of resources within the estimated use area (e.g., home range) (Manly et al. 2002, 192–94).

In all of these designs, however, composition of available habitat is measured only once, and then compared to observations of habitat used by the animals; the latter is usually collected over a period of time (e.g., a "season") to gather an adequate sample size. Habitat used by animals is, in essence, an average use over a defined period. Of course, habitat availability is not constant, and changes in various components of it could substantially influence habitat use, the ramifications of which depending largely on the goals of the investigation.

To rectify this potential problem in use-availability studies, Arthur et al. (1996) developed a method of estimating habitat use involving multiple observations of habitat availability. Their method, which is too detailed to develop here, allows for quantitative comparisons among habitat categories and is not affected by arbitrary decisions about which categories to include in a study (see also Manly et al. 2002).

What to Measure

As we showed earlier in this book, the distribution of plants is tied closely to regimes of temperature, soil, and moisture. Animals, in turn, are often linked in some fashion to plants for shelter, food (directly or indirectly), or both. But to what specific aspect of vegetation are animals responding? What are the stimuli causing the behaviors that we call resource use? Notwithstanding our comments above on indirect measurements, the fact remains that most studies have used, and will continue to use, indirect measures of habitat selection.

To answer these questions, we now turn to specific examples of variables collected by researchers seeking to describe the habitat use patterns of animals. The examples we use come from widely read and cited papers. We offer these few examples as good starting points for students planning similar evaluations of habitat use by a particular species. We strongly recommend, however, that researchers concentrating on a particular taxon not restrict themselves to literature relating to that group only. For example, a researcher designing a study to examine habitats of ground-foraging birds would likely gain valuable information on types of variables and sampling design by reviewing papers on small mammals and reptiles. We will also give brief summaries of the objectives of each paper; remember that the variables should follow closely from the objectives.

Birds initially received the most attention with regard to the analysis of habitat use patterns. This research attention is likely a reflection of the conspicuousness of birds: most are active during the day, most give at least some vocalizations during all parts of the year, and they are inexpensive to observe (one only needs binoculars and a notebook). However, since the late 1980s, other vertebrate groups have also begun to receive increasing attention.

James (1971) conducted one of the first and most-cited studies quantifying bird–habitat relationships. As discussed earlier (see chapter 3), James based her study on the conceptual framework of the niche and a bird's niche gestalt. She used 15 measures of vegetation structure to describe the multidimensional "habitat space" of a bird community in Arkansas and followed closely the methods she developed with Herman Shugart Jr. (James and Shugart 1970); these methods are described in the next section ("How to Measure").

The conceptual framework on which James based her work (niche gestalt) and the general analytical techniques she used (multivariate analysis of focal-bird observations) have led to a plethora of studies that expanded on her basic ideas. Her strategy and methods are still in wide use.

Dueser and Shugart (1978) were among the first workers to quantify microhabitat use patterns of small mammals in a multivariate sense. The variables they selected, however, would apply regardless of the analytical techniques used. They had as their goal the description of microhabitat differences among the small-mammal species of an upland forest in eastern Tennessee.

Their specific objectives were to characterize and compare microhabitats of species within the forest and to examine the relationships of species abundances and distributions to the relative availability of selected microhabitats. This study is a good example of the pattern-seeking nature of most habitat descriptions. Although one could devise a null hypothesis here, it would be trivial: for example, test the null hypothesis of no difference in habitat use between species.

Dueser and Shugart gathered information for vertical strata at each capture site of a small mammal: overstory, understory, shrub level, forest floor, and litter-soil level. Table 5.1 lists the variables they collected. Note that they did not collect species-specific information on plants beyond designations of "woodiness," "evergreenness," and the like, which can be considered an unfortunate omission for a microhabitat analysis, the ramifications of which for the final results are unknown. They did, however, record the number of woody and herbaceous species. They paid special attention to features of the forest floor, such as litter-soil compactability, fallen-log density, and short-herbaceous-stem density. They found that certain of these soil variables played a significant role in describing the differences in microhabitats of the species studied. Except for the lack of detailed information on plant taxonomy, we consider Dueser and Shugart's study a good example of a very detailed set of variables used to differentiate among species of co-occurring animals.

Reinert (1984) sought to differentiate microhabitats of timber rattlesnakes (*Crotalus horridus*) and northern copperheads (*Agkistrodon contortrix*), which occur sympatrically in temperate deciduous forests of eastern North America. His approach was a multivariate habitat description consistent with the Hutchinsonian definition of the niche (Green 1971) and based on the concept of the niche gestalt (James 1971).

Table 5.1. Variables and sampling methods used by Dueser and Shugart in measuring forest habitat structure

Variable	Methods
1. Percentage of canopy closure	Percentage of points with overstory vegetation, from 21 vertical ocular tube sightings along the center lines of two perpendicular 20-m² transects centered on trap
2. Thickness of woody vegetation	Average number of shoulder-height contacts (trees and shrubs), from two perpendicular 20-m² transects centered on trap
3. Shrub cover	Same as (1), for presence of shrub-level vegetation
4. Overstory tree size	Average diameter (in cm) of nearest overstory tree, in quarters around trap
5. Overstory tree dispersion	Average distance (m) from trap to nearest overstory tree, in quarters
6. Understory tree size	Average diameter (cm) of understory tree, in quarters around trap
7. Understory tree dispersion	Average distance (m) from trap to nearest understory tree, in quarters
8. Woody stem density	Live woody stem count at ground level within a 1.00-m² ring centered on trap
9. Short woody stem density	Live woody stem count within a 1.00-m² ring centered on trap (stems ≤0.40 m in height)
10. Woody foliage profile density	Average numbers of live woody stem contacts with a 0.80-cm-diameter metal rod rotated 360°, describing a 1.00-m² ring centered on the trap and parallel to ground at heights of 0.05, 0.10, 0.20, 0.40, 0.60, . . . , 2.00 m above ground level
11. Number of woody species	Woody species count within a 1.00-m² ring centered on trap
12. Herbaceous stem density	Live herbaceous stem count at ground level within a 1.00-m² ring centered on trap
13. Short herbaceous stem density	Live herbaceous stem count within a 1.00-m² ring centered on trap (stems <0.40 m in height)
14. Herbaceous foliage profile	Same as (10), for live herbaceous stem contacts
15. Number of herbaceous species	Herbaceous species count within a 1.00-m² ring centered on trap
16. Evergreenness of overstory	Same as (1), for presence of evergreen canopy vegetation
17. Evergreenness of shrubs	Same as (1), for presence of evergreen shrub-level vegetation
18. Evergreenness of herb stratum	Percentage of points with evergreen herbaceous vegetation, from 21 step-point samples along the center lines of two perpendicular 20-m² transects centered on trap
19. Tree stump density	Average number of tree stumps ≥7.50 cm in diameter, per quarter
20. Tree stump size	Average diameter (cm) of nearest tree stump ≥7.50 cm in diameter, in quarters around trap
21. Tree stump dispersion	Average distance (m) to nearest tree stump ≥7.50 cm in diameter, in quarters around trap
22. Fallen-log density	Average number of fallen logs ≥7.50 cm in diameter, per quarter
23. Fallen-log size	Average diameter (cm) of nearest fallen log ≥7.50 cm in diameter, in quarters around trap
24. Fallen-log dispersion	Average distance (m) from trap to nearest fallen log ≥7.50 cm in diameter, in quarters around trap
25. Fallen-log abundance	Average total length (+0.50 m) of fallen logs ≥7.50 cm in diameter, per quarter
26. Litter-soil depth	Depth of penetration (<10.00 cm) into litter-soil material of a hand-held core sampler with 2.00-cm-diameter barrel
27. Litter-soil compactability	Percentage compaction of litter-soil core sample (26)
28. Litter-soil density	Dry weight density (g/cm²) of litter-soil core sample (26), after oven-drying at 45°C for 48 h
29. Soil surface exposure	Same as (18), for percentage of points with bare soil or rock

Source: Dueser and Shugart 1978, appendix. Reprinted with permission of the Ecological Society of America.

Table 5.2.   Structural and climatic variables used by Reinert in differentiating microhabitats of timber rattlesnakes (*Crotalus horridus*) and northern copperheads (*Agkistrodon contortrix*)

Mnemonic	Variable	Sampling method
ROCK	Rock cover	Coverage (%) within 1-m^2 quadrant centered on snake location
LEAF	Leaf litter cover	Same as ROCK
VEG	Vegetation cover	Same as ROCK
LOG	Fallen-log cover	Same as ROCK
WSD	Woody stem density	Total number of woody stems within 1-m^2 quadrant
WSH	Woody stem height	Height (cm) of tallest woody stem within 1-m^2 quadrant
MDR	Distance to rocks	Mean distance (m) to nearest rocks (>10 cm max. length) in each quarter
MLR	Length of rocks	Mean max. length (cm) of rocks used to calculate MDR
DNL	Distance to log	Distance (m) to nearest log (≥7.5 cm max. diameter)
DINL	Diameter of log	Max. diameter (cm) of nearest log
DNOV	Distance to overstory	Distance (m) to nearest tree (≥7.5 cm dbh [diameter at breast height])
DBHOV	Dbh of overstory tree	Mean dbh (cm) of nearest overstory tree within each quarter
DNUN	Distance to understory	Same as DNOV (trees <7.5 tree cm dbh >2.0 m height)
CAN	Canopy closure	Canopy closure (%) within 45° cone with ocular tube
SOILT	Soil temperature	Temperature (°C) at 5-cm depth within 10 cm of snake
SURFT	Surface temperature	Temperature (°C) of substrate within 10 cm of snake
IMT	Ambient temperature	Temperature (°C) of air at 1 m above snake
SURFRH	Surface relative humidity	Relative humidity (%) at substrate within 10 cm of snake
IMRH	Ambient relative humidity	Relative humidity (%) 1 m above snake

Source: Reinert 1984, table 1. Reprinted with permission of the Ecological Society of America.

Reinert was interested not only in features of the ground but also in the weather conditions immediately surrounding the snake. He measured temperature and humidity at several locations near an animal, as well as the structure of the surrounding vegetation. Here again, however, no information on plant taxa was included (table 5.2).

Morrison et al. (1995) used time-constrained surveys to describe the microhabitats of amphibians and reptiles in southeastern Arizona mountains. Observers walked slowly, searching the ground and tree trunks and turning over movable rocks, logs, and litter to examine protected locations while a stopwatch ran. When an animal was found, the survey time was stopped, and a 5-m-diameter plot was centered on the animal's location. Microhabitat conditions, such as substrate temperature and various descriptors of the vegetation and other habitat characteristics, were then measured for the plot.

Ockenfels and Brooks (1994) radio-tracked 22 Coues white-tailed deer to describe diurnal bedding sites. The proper diurnal sites are critical to this deer in the very hot summer temperatures in the southern Arizona study area. Using 40-m^2-circular plots, Ockenfels and Brooks measured topography, temperature, vegetative characteristics, and percentage of shade at bedding sites and at similar, nonused sites.

Welsh and Lind (1995) analyzed the habitat affinities of the Del Norte salamander (*Plethodon elongatus*) in relation to landscape, macrohabitat, and microhabitat scales. They presented a detailed rationale for the selection of methods, including choice of analytical techniques, data screening, and interpretation of output. The variables they measured, separated by spatial scale, are shown in table 5.3. A similar example for multiple species of amphibians was given by Welsh and Lind (2002).

Table 5.3.   Hierarchic arrangement of ecological components represented by 43 measurements of the forest environment taken in conjunction with sampling for the Del Norte salamander (*Plethodon elongatus*)

Hierarchic scale[a]
Variable category
Variables[b]

II. Landscape scale[c]
  A. Geographic relationships
    Latitude (degrees)
    Longitude (degrees)
    Elevation (m)
    Slope (%)
    Aspect (degrees)
III. Macrohabitat or stand scale
  A. Trees: density by size[d]
    Small conifers (C)
    Small hardwoods (C)
    Large conifers (C)
    Large hardwoods (C)
    Forest age (in years)
  B. Dead and down wood: surface area and counts
    Stumps (B)
    All logs—decayed (C)
    Small logs—sound (C)
    Sound-log area (L)
    Conifer log-decay area
    Hardwood log-decay area (L)
  C. Shrub and understory composition (>0.5 m)
    Understory conifer (L)
    Understory hardwoods (L)
    Large shrub (L)
    Small shrub (L)
    Bole (L)
    Height II–ground vegetation (B) (0.5–2 m)
  D. Ground-level vegetation (<0.5 m)
    Fern (L)
    Herb (L)
    Grass (B)
    Height I–ground vegetation (B) (0–0.5 m)
  E. Ground cover
    Moss (L)
    Lichen (B)
    Leaf (B)
    Exposed soil (B)
    Litter depth (cm)
    Dominant rock (B)
    Codominant rock (B)

Table 5.3.   Continued

Hierarchic scale[a]
Variable category
Variables[b]

  F. Forest climate
    Air temperature (°C)
    Soil temperature (°C)
    Solar index
    % canopy closed
    Soil pH
    Soil relative humidity (%)
    Relative humidity (%)
IV. Microhabitat scale
  A. Substrate composition
    Pebble (P) (% of 32–64 mm-diameter rock)
    Cobble (P) (% of 64–256-mm-diameter rock)
    Cemented (P) (% of rock cover embedded in soil/litter matrix)

Source: Welsh and Lind 1995, table 1.
[a]Spatial scales are arranged in descending order from coarse to fine resolution (see Wiens 1989).
[b]The abbreviations used for the variables are as follows:
C = count variables (numbers per hectare)
B = Braun-Blanquet variables (percentage of cover in 1/10-ha circle)
L = line transect variables (percentage of 50-m line transect)
P = percentage within 49-m^2 salamander search area
[c]Level I relationships (the biogeographic scale) were not analyzed because all sampling occurred within the range.
[d]Small trees = 12–53 cm dbh (diameter at breast height); large trees = >53 cm dbh.

How to Measure

In this section we review some of the common methods used to measure wildlife habitat. We do not present a survey of all literature available for all taxa; Cooperrider et al. (1986) provided a thorough review of basic sampling techniques for all major taxa of wildlife (see also Bookhout 1994).

Sampling Principles

Recall from chapter 3 our discussion of how vegetation traditionally forms the template for how we view wildlife–habitat selection (see also discussion of key environmental correlates in chapter 11). It is not surprising that we turn to plant ecologists for advice on many of our fundamental sampling methods. Indeed, even a cursory review of the methods sections in wildlife publications shows a reliance on standard, classical methods of quantifying the structure and floristics of vegetation: point quarter, circular plots and nested circular plots, sampling squares, line intercepts, and so on. These methods are used for good reason: they have been developed and extensively tested by plant ecologists in a multitude of environmental situations for decades. This is not to say we should not be innovative. Standard methods do, however, provide an established starting point from which wildlife biologists can adapt specific methods as needed and easy comparability between studies. There are many fine books available that review sampling methods in vegetation ecology (e.g., Daubenmire 1968; Mueller-Dombois and Ellenberg 1974; Greig-Smith 1983; Cook and Stubbendieck 1986; Bonham 1989; Schreuder et al. 1993).

We must be careful, of course, not to simply accept as fact all statements concerning habitat-use patterns given in the literature, for two reasons. First—and this factor is more likely to occur in books than in the primary literature (i.e., original publications in scientific journals)—authors tend to repeat what previous authors have written; such citations are known as secondary, tertiary, and so on, depending upon how far removed they are from the original paper. To avoid inserting old biases into new studies (see following section on preliminary sampling), we must carefully question how vegetation structure was categorized in earlier studies.

Second, we should realize that most studies directly pertain only to the time and place in which the study was conducted. Such studies do serve as a fine starting point for development of a new study; and, as aforementioned, we should avoid repeating the mistakes of others and constantly reinventing the wheel. However, although repeatability and corroboration (or, alternatively, falsification) are cornerstones of scientific understanding, there is no reason to keep repeating what others have already done simply because "it hasn't been done *here* before." In such cases, a rather brief study designed to see if the earlier work can be corroborated might be indicated. Unfortunately, though, in real-world management, the reverse is more often the case: studies are often applied to environments and locations far different than intended. In such cases, local corroborative studies are useful for determining how pertinent or applicable findings are to local situations.

Preliminary Sampling

We must remember that the variables we measure—and the means by which we measure them—will themselves play a substantial role in determining the results of our analyses. For example, if we do not record vegetation data by taxonomic classification, then floristics can play no role in our analysis or in subsequent management applications. Likewise, if we categorize vegetation structure into 2-m height intervals, we are making the implicit assumption that vegetation so profiled has meaning to the animal(s) under study. As caution above, when designing a study, researchers must walk the fine line between simply repeating what others have done and inventing new methods, and between economically measuring only the most pertinent variables and more finely dividing variables to test for new associations.

Unfortunately, the methods sections in most papers provide little if any information as to *why* the authors chose the methods they used. For example, it is usually stated simply that "vegetation was placed into 2-m height intervals," or that "trees were categorized into the following diameter classes." A common method is to record vegetation in 11.3-m-radius plots. The relatively young researcher must wonder how such an odd measure was chosen! This measure is a holdover from the old English measurement system: an 11.3-m radius results in a 0.04-ha plot, which corresponds to the 0.1-acre plot that has been the standard for vegetation measurements. Modern researchers use this radius because it is "standard" in the literature. However, we know of no studies that have shown that this radius holds any more ecological validity than any other radius; yet its use continues. It probably does afford some degree of sampling efficiency. We guess that it is as good a radius as any other radius, given the lack of studies showing the contrary; either we will all be correct, or we will all be wrong!

Clearly, there is the need for preliminary sampling to establish the most predictive measurements of an animal's habitat. Although we certainly understand the limits that time and money place on the intensity and duration of a study, it makes little sense to spend one's effort in the field based primarily on established dogma and untested techniques. As a measure of the need for such studies, we can safely state that our most popular publications (as measured by the number of reprint requests and comments from colleagues) are consistently those that report on evaluations of methods and analytical techniques, rather than those that center on descriptors of an animal's use of habitat in a particular area.

Preliminary sampling allows one both to test the predictability of the field methods, as well as to ensure sure that adequate sample sizes are being accumulated (sample size analysis is discussed in chapter 6). Referring back to our previous discussion of variable selection, few of the variables listed in table 5.1, for example, played any significant statistical role in describing habitat of the species under study. Yet all of these variables were collected for the duration of the study. Recording such data is time consuming. Would not a more efficient procedure involve collecting initial samples and then conducting preliminary analyses to determine which variables were duplicative (highly intercorrelated) and which had little or no predictive power?

Sampling Methods

The most popular methods of measuring microhabitat originated with a protocol developed by Frances C. James and Henry Herman Shugart Jr. (1970). They developed a quantitative method of obtaining vegetation data "in a simple and regular manner." Their original intent was to provide a method that could augment the data on bird populations being gathered in the National Audubon Society's breeding bird censuses and winter bird population studies throughout the United States. But as noted earlier, their strategy has found extremely wide applicability throughout the ecological community. As we detail later in this chapter, James and Shugart started by evaluating various sampling methods as to their relative efficiencies. Their method gathered data on the density, basal area, and frequency of trees, canopy height, shrub density, percentage of ground cover, and percentage of canopy cover. They established 0.1-acre (0.04-ha) plots to estimate tree density and frequency. To estimate shrub density they made two transects at right angles to each other across the 0.1-acre plots, counting the number of woody stems intercepted by their outstretched arms. An ocular tube was used to estimate vegetation cover. They also provided details on how the sampling equipment could be constructed and examples of data sheets.

We have previously discussed the importance of James's (1971) paper to our conceptualization of how animals perceive their environment—the niche gestalt. James felt that the size of plot used would give an adequate description of the vegetation within an individual bird's territory; we will see below that she actually tested this assumption by comparing several methods (preliminary sampling!). She also acknowledged a potential bias in concentrating on song perches, and assumed that song-perch habitat reflects the fuller array of habitat elements for each bird species. As we have seen (Collins 1981, discussed previously), this assumption may not hold. But regardless of the problems inherent in using only one behavior as the basis for habitat evaluation, the methods used by James have had a positive and pronounced influence on most analyses of wildlife habitat that followed.

James and Shugart (1970) compared four of the standard methods recommended by plant ecologists for making quantitative estimates of vegetation. Two were plotless methods—namely, the quarter method (Cottam and Curtis 1956; Phillips 1959) and the wandering quarter method (Catana 1963); the other two were areal methods—namely, arm-length transects (Rice and Penfound 1955; Penfound and Rice 1957) and circular plots (Lindsey et al. 1958). James and Shugart compared the average work accomplished using these four sampling methods in 30 minutes of field effort by one observer, assuming that the observer was familiar with the method and species of plants in the study area (see table 5.4). Of course, the amount of work accomplished will vary with the terrain, the density of the vegetation, and the actual number of observers involved. James and Shugart found that results from the two plotless methods (quarter and wandering quarter methods) tended to overestimate the total tree density and to underestimate tree density by species. Results from the two areal methods they used—1/100-acre rectangles and 1/10-acre circles—gave fairly accurate estimates of total density and density by species.

Table 5.4. Average work accomplished in 30 minutes of field effort recording the species and diameters of trees in an upland Ozark forest in Arkansas

Sampling method	Number of units	Number of trees identified and measured
Quarter method	12 quarters	48
Wandering quarter method	40 trees	40
1/10-acre circles	2 circles	57
1/100-acre rectangles	6 rectangles	19

Source: James and Shugart 1970, table 1.

Circular plots are easy to establish, mark, measure, and relocate, and estimates of animal numbers within such plots can be statistically related to vegetation data in a straightforward manner. Plots provide for the sampling of vegetation and animals at specific locations in space and time. Thus it is also easy to pinpoint plots using global positioning systems (GPS) and subsequently to input their data into geographic information systems (GIS). If plots can be considered independent data points (a function of the sampling design and behavior of the animals), then one's sample size is equal to the number of plots sampled. Or if the plots are used to sample from a single study area, then plots sampled can be averaged, and associated measures of variance can be calculated. Noon (1981) presented a useful description and example of both the transect and the areal plot sampling systems. The problem with transects is that they cover relatively large areas and thus make it difficult to relate specific animal observations (or abundances) to specific sections of the transect. Transects are, however, widely used to provide an overall description of the vegetation of entire study areas.

In summary to this point, fixed-area plots and transects can be used to provide site-specific, detailed analyses of wildlife–habitat relationships. The majority of sampling methods used since the 1970s to develop wildlife–habitat relationships—for subsequent multivariate analyses—have used fixed-area plots (usually circular) as the basis for development of a sampling scheme that may then incorporate subplots, sampling squares, and transects. We next describe examples of some of the more widely used methods.

Dueser and Shugart (1978) developed a detailed sampling scheme that combined plots of various sizes and shapes, as well as short transects (see fig. 5.6). Although designed for analysis of small-mammal habitat, the techniques can easily be adapted for most terrestrial vertebrates. Dueser and Shugart established three independent sampling units, centered on each trap: a 1.0-m² ring, two perpendicular 20-m² arm-length transects, and a 10-m-radius circular plot. The 1.0-m² circular plot provided a measure of vertical foliage profile from the ground through 2 m height, for both herbaceous and woody vegetation. Also, four replicate core-sample estimates of litter-soil depth, compactability, and dry weight density were made on the perimeter of this central ring. The two arm-length transects provided measures of cover type, surface characteristics, and density and evergreenness of the four strata of vegetation. Data recorded for each quarter of the 10-m-radius plot included the species, dbh (diameter at breast height), distance from the trap to the nearest understory and overstory trees, numbers

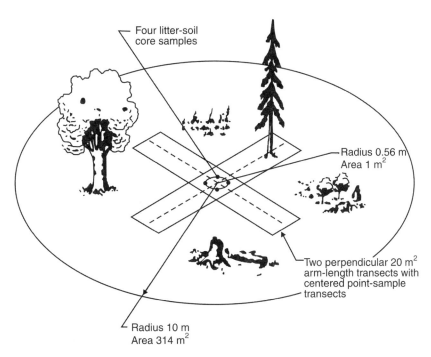

Four litter-soil
core samples

Radius 0.56 m
Area 1 m²

Two perpendicular 20 m²
arm-length transects with
centered point-sample
transects

Radius 10 m
Area 314 m²

Figure 5.6.   Habitat variable sampling configuration used by Dueser and Shugart in their study of small-mammal habitat use. (Reproduced from Dueser and Shugart 1978, fig. 1 reprinted by permission of the Ecological Society of America.)

of stumps and fallen logs, basal diameter and distance of nearest stump and fallen log, and total length of fallen logs.

We consider Dueser and Shugart an excellent starting point for any study being established in forested areas; the basic sampling strategy could also be adapted for other vegetation types. Although the variables they measured (see table 5.1) all relate to "microhabitat," some relate to the specific trap location (e.g., litter samples, soil compactability), whereas others relate more to the conditions surrounding the trap (e.g., data collected along the arm-length transects, distance to trees). Thus to a minor degree, the authors were mixing scales of measurement within the general construct of measuring microhabitat. As discussed previously, *microhabitat* is a general term that should be clearly defined for all research applications. Authors should explain how each variable fits within the overall concept of spatial scale, and how it thus meets study objectives.

In his analysis of snake populations, Reinert (1984) adopted techniques similar to those used in the bird study by James (1971) and the small-mammal study by Dueser and Shugart (1978). That is, Reinert applied the basic conceptual framework used by the earlier authors—the niche gestalt and multivariate representation of the niche—in developing the rationale for his methods. Here again we see the commonality in methods running across studies of wildlife–habitat relationships.

Reinert made several modifications to the sampling methods used by James and by Dueser and Shugart. Notably, he used a 35-mm camera equipped with a 28-mm wide-angle lens to photograph 1-m^2 plots from directly above the location of a snake. He then determined the various surface cover percentages by superimposing each slide onto a 10 × 10 square grid. Reinert, then, more rigorously quantified his measure of cover values than most workers, who usually use ocular estimates. As discussed in chapter 6, much error can be entered into a data set when ocular estimates are used to measure plant cover (Block et al. 1987). The specific sampling scheme used by Reinert is summarized in figure 5.7, and his variable list was previously presented in table 5.2. Note the similarity between Reinert's design and that of Dueser and Shugart, including the minor mixing of spatial scales. Reinert added several environmental variables that measured air, surface, and soil temperature and humidity. The values these variables take are obviously dependent on the time of day and the general weather conditions at the time of measurement; such constraints do not influence (are not correlated with) the other variables measured. In addition, temperature data take on a much different statistical distribution than most vegetation variables. Such mixing of variable types in the same analysis should be undertaken with caution and, in fact, should probably be avoided (unless partitioned out in a stepwise analysis).

Radiotelemetric Methods

Attaching radio transmitters to animals is a method often used to quantify movements. The use of radiotelemetry usually results in a more complete and thus more accurate depiction of the activities and areas used by animals, and thus provides a more reliable basis for developing habitat relationships than is available with visual observations and traps (e.g., Ribble et al. 2002; Briner et al. 2003). With the availability of increasingly small transmitters and batteries, researchers have been studying increasingly small animals. Radio transmitters are extremely useful in quantifying habitat use because they allow the researcher to locate animals that might otherwise be unobservable (e.g., unable to be located visually), thus reducing one of the biases associated with habitat assessment. However, careful study designs are needed to avoid introducing

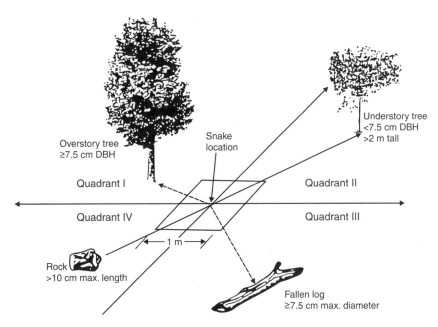

Figure 5.7.   Sampling scheme used by Reinert for snake locations. (Reproduced from Reinert 1984, fig. 1; reprinted by permission of the Ecological Society of America.)

other biases into the research, such as the methods used to select animals for attachment of the transmitter in the first place, the timing and frequency of obtaining relocations, the number of animals used, and other issues. Thus the old practice of simply tagging animals and following them (i.e., "collar 'em and foller 'em") is giving way to much more sophisticated sampling strategies built around a specific project goal. Many fine books and articles are available on radio tagging, including White and Garrott (1990); Kenward (2000); and Millspaugh and Marzluff (2001); see also Turchin (1998) for analyses of animal movements. Additionally, because the method of transmitter attachment can have negative impacts on the behavior and health of an individual, care must be taken when implementing this technique (e.g., Fleskes 2003).

We will not go into further detail here on sampling methods used in radio tagging. Rather, we present a few examples showing the diversity of radiotelemetric methods as applied to habitat or niche analysis. Matthews and Pope (1999) used radio transmitters to quantify the seasonal movements and habitat use of mountain yellow-legged frogs (*Rana muscosa*). They usually relocated frogs multiple times per day and recorded their general habitat association (defined as rocks, bedrock, undercut bank, willow, silt); whether they were in water or on land; their exposure (covered or exposed); and the air temperature. These data were then compared with the abundance of these variables to infer habitat selection. Fellers and Pierson (2002) attached radio transmitters to Townsend's big-eared bats (*Corynorhinus townsendi*) with surgical adhesive; the radios weighed 0.44 g, ~4% of bat body weight, and transmitted for about 10 days. Telemetry is also useful in reducing potential observer-induced biases by allowing observers to remain out-of-site and at relatively longer distances from the animals than usually possible using visual observations. Remaining sufficiently away from animals is especially useful when investigating interactions between species. For example, Kamler et al. (2003) studied the impacts of coyotes on swift foxes (*Vulpes velox*) using radio-collared animals, which allowed them to examine the proximity of the species while remaining unobserved themselves.

Although suitable only for relatively large animals at this time, radio collars are available that can transmit animal locations to satellites; other collars can act as GPS receivers and record animal movements at set intervals. Thus the traditional process of observers locating (e.g., triangulating) radio-equipped animals using hand-held, vehicle, or aircraft-mounted antennae/receivers is giving way to more automated systems. For example, Anderson and Lindzey (2003) fitted cougars with GPS receivers programmed to record animal locations at 16:00, 19:00, 22:00, 02:00, 05:00, and 08:00 each day, hours corresponding to the nocturnal activity period of the cougars.

Literature Cited

Addicott, J. F., J. M. Aho, M. F. Antolin, D. K. Padilla, J. S. Richardson, and D. A. Soluk. 1987. Ecological neighborhoods: Scaling environmental patterns. *Oikos* 49:340–46.

Aebischer, N. J., P. A. Robertson, and R. E. Kenward. 1993. Compositional analysis of habitat use from animal radio-tracking data. *Ecology* 74:1313–25.

Alldredge, R. J., and J. T. Ratti. 1986. Comparison of some statistical techniques for analysis of resource selection. *Journal of Wildlife Management* 50:157–65.

Alldredge, R. J., and J. T. Ratti. 1992. Further comparison of some statistical techniques for analysis of

resource selection. *Journal of Wildlife Management* 56:1–9.

Anderson, C. R., Jr., and F. G. Lindzey. 2003. Estimating cougar predation rates from GPS location clusters. *Journal of Wildlife Management* 67:307–16.

Anderson, S. H., and H. H. Shugart Jr. 1974. Habitat selection of breeding birds in an east Tennessee deciduous forest. *Ecology* 55:828–37.

Arthur, S. M., B. F. J. Manly, L. L. McDonald, and G. W. Garner. 1996. Assessing habitat selection when availability changes. *Ecology* 77:215–27.

Block, W. M., M. L. Morrison, J. Verner, and P. N. Manley. 1994. Assessing wildlife-habitat-relationships models: A case study with California oak woodlands. *Wildlife Society Bulletin* 22:549–61.

Block, W. M., K. A. With, and M. L. Morrison. 1987. On measuring bird habitat: Influence of observer variability and sample size. *Condor* 72:182–89.

Bonham, C. D. 1989. *Measurements for terrestrial vegetation.* New York: Wiley.

Bookhout, T. A., ed. 1994. *Research and management techniques for wildlife and habitats.* 5th ed. Bethesda, MD: Wildlife Society.

Briner, T., J.-P. Airoldi, F. Dellsperger, S. Eggimann, and W. Nentwig. 2003. A new system for automatic radiotracking of small mammals. *Journal of Mammalogy* 84:571–78.

Buckland, S. T., D. R. Anderson, K. P. Burnham, J. Laake, D. L. Borchers, and L. Thomas. 2001. *Introduction to distance sampling: Estimating abundance of biological populations.* Oxford: Oxford Univ. Press.

Campbell, G. S., and J. M. Norman. 1989. The description and measurement of plant canopy structure. In *Plant canopies: Their growth, form and function,* ed. G. Russell, B. Marshall, and P. G. Jarvis, 1–19. Cambridge: Cambridge Univ. Press.

Carbon, B. A., G. A. Bartle, and A. M. Murray. 1979. A method for visual estimation of leaf area. *Forest Science* 25:53–58.

Catana, A. J., Jr. 1963. The wandering quarter method of estimating population density. *Ecology* 44:349–60.

Collins, S. L. 1981. A comparison of nest-site and perch-site vegetation structure for seven species of warblers. *Wilson Bulletin* 93:542–47.

Cook, C. W., and J. Stubbendieck, eds. 1986. *Range research: Basic problems and techniques.* Denver: Society for Range Management.

Cooperrider, A. Y., R. J. Boyd, and H. R. Stuart, eds. 1986. *Inventory and monitoring of wildlife habitat.*

Denver: USDI Bureau of Land Management Service Center.

Cottam, G., and J. T. Curtis. 1956. The use of distance measures in phytosociological sampling. *Ecology* 37:451–60.

Daubenmire, R. 1968. *Plant communities: A textbook of plant synecology.* New York: Harper & Row.

Dueser, R. D., and H. H. Shugart Jr. 1978. Microhabitats in a forest-floor small mammal fauna. *Ecology* 59:89–98.

Fellers, G. M., and E. D. Pierson. 2002. Habitat use and foraging behavior of Townsend's big-eared bat (*Corynorhinus townsendi*) in coastal California. *Journal of Mammalogy* 83:167–77.

Fleskes, J. P. 2003. Effects of backpack radiotags on female northern pintails wintering in California. *Wildlife Society Bulletin* 31:212–19.

Ganey, J. L., and W. M. Block. 1994. A comparison of two techniques for measuring canopy closure. *Western Journal of Applied Forestry* 9:21–23.

Green, R. H. 1971. A multivariate statistical approach to the Hutchinsonian niche: Bivalve mollusks of central Canada. *Ecology* 52:543–56.

Green, R. H. 1979. *Sampling design and statistical methods for environmental biologists.* New York: Wiley.

Greig-Smith, P. 1983. *Quantitative plant ecology.* 3rd ed. Berkeley and Los Angeles: Univ. of California Press.

Griffin, P.C., and T. J. Case. 2001. Terrestrial habitat preferences of adult arroyo southwestern toads. *Journal of Wildlife Management* 65:633–44.

Hall, L. S., and M. L. Morrison. 1997. Den and relocation site characteristics and home ranges of *Peromyscus truei* in the White Mountains of California. *Great Basin Naturalist* 57:124–30.

Heyer, W. R., M. A. Donnelly, R. W. McDiarmid, L. C. Hayek, and M. S. Foster. 1994. *Measuring and monitoring biological diversity: Standard methods for amphibians.* Washington, DC: Smithsonian Institution Press.

Hilden, O. 1965. Habitat selection in birds. *Annales Zoologici Fennici* 2:53–75.

Holmes, R. T. 1981. Theoretical aspects of habitat use by birds. In *The use of multivariate statistics in studies of wildlife habitat,* ed. D. E. Capen, 33–37. USDA Forest Service General Technical Report RM-87. Fort Collins, CO: USDA Forest Service, Rocky Mountain Research Station.

Huston, M. A. 2002. Introductory essay: Critical issues for improving predictions. In *Predicting species occurrences: Issues of accuracy and scale,* ed. J. M.

Scott, P. J. Heglund, M. L. Morrison, J. B. Haufler, M. G. Raphael, W. A. Wall, and F. B. Samson, 7–21. Washington, DC: Island Press.

Hutto, R. L. 1985. Habitat selection by nonbreeding, migratory land birds. In *Habitat selection in birds,* ed. M. L. Cody, 455–76. New York: Academic Press.

Irwin, L. L., and J. G. Cook. 1985. Determining appropriate variables for a habitat suitability model for pronghorns. *Wildlife Society Bulletin* 13:434–40.

James, F. C. 1971. Ordinations of habitat relationships among breeding birds. *Wilson Bulletin* 83:215–36.

James, F. C., and H. H. Shugart Jr. 1970. A quantitative method of habitat description. *Audubon Field Notes* 24:727–36.

James, F. C., and N. D. Wamer. 1982. Relationships between temperate forest bird communities and vegetation structure. *Ecology* 63:159–71.

Johnson, D. H. 1980. The comparison of usage and availability measurements for evaluating resource preference. *Ecology* 61:65–71.

Kamler, J. F., W. B. Ballard, R. L. Gilliland, P. R. Lemons II, and K. Mote. 2003. Impacts of coyotes on swift foxes in northwestern Texas. *Journal of Wildlife Management* 67:317–23.

Karr, J. R., and R. R. Roth. 1971. Vegetation structure and avian diversity in several New World areas. *American Naturalist* 105:423–35.

Kelt, D. A., P. L. Meserve, and B. K. Lang. 1994. Quantitative habitat associations of small mammals in a temperate rainforest in southern Chile: Empirical patterns and the importance of ecological scale. *Journal of Mammalogy* 75:890–904.

Kenward, R. E. 2000. *A manual for wildlife radio tagging.* 2nd ed. San Diego: Academic Press.

Krebs, C. J. 1978. *Ecology: The experimental analysis of distribution and abundance.* New York: Harper & Row.

Levin, S. A. 1992. The problem of pattern and scale in ecology. *Ecology* 73:1943–67.

Lindsey, A. A., J. D. Barton, and S. R. Miles. 1958. Field efficiencies of forest sampling methods. *Ecology* 39:428–44.

MacArthur, R. H., and J. W. MacArthur. 1961. On bird species diversity. *Ecology* 42:594–98.

Manly, B. F. J., L. L. McDonald, D. L. Thomas, T. L. McDonald, and W. P. Erickson. 2002. *Resource selection by animals: Statistical design and analysis for field studies.* 2nd ed. Dordrecht, Netherlands: Kluwer Academic.

Matthews, K. R., and K. L. Pope. 1999. A telemetric study of the movement patterns and habitat use of

Rana muscosa, the mountain yellow-legged frog, in a high-elevation basin in Kings Canyon National Park, California. *Journal of Herpetology* 33:615–24.

Mawson, J. C., J. W. Thomas, and R. M. DeGraaf. 1976. *Program HTVOL: The determination of tree crown volume by layers.* USDA Forest Service Research Paper NE-354. Newton Square, PA: USDA Forest Service, Northeastern Research Station.

Millspaugh, J. J., and J. M. Marzluff, eds. 2001. *Radio tracking and animal populations.* San Diego: Academic Press.

Mitchell, M. S., J. W. Zimmerman, and R. A. Powell. 2002. Test of a habitat suitability index for black bears in the southern Appalachians. *Wildlife Society Bulletin* 30:794–808.

Morrison, M. L. 1984. Influence of sample size on discriminant function analysis of habitat use by birds. *Journal of Field Ornithology* 55:330–35.

Morrison, M. L., and R. G. Anthony. 1989. Habitat use by small mammals on early-growth clear-cuttings in western Oregon. *Canadian Journal of Zoology* 67:805–11.

Morrison, M. L., W. M. Block, L. S. Hall, and H. S. Stone. 1995. Habitat characteristics and monitoring of amphibians and reptiles in the Huachuca Mountains, Arizona. *Southwestern Naturalist* 40: 185–92.

Morrison, M. L., W. M. Block, M. D. Strickland, and W. L. Kendall. 2002. *Wildlife study design.* New York: Springer-Verlag.

Morrison, M. L., and E. C. Meslow. 1983. Bird community structure on early-growth clearcuts in western Oregon. *American Midland Naturalist* 110:129–37.

Morrison, M. L., C. J. Ralph, J. Verner, and J. R. Jehl Jr., eds. 1990. Avian foraging: Theory, methodology, and applications. *Studies in Avian Biology* 13:1–515.

Morzillo, A. T., G. A. Feldhamer, and M. C. Nicholson. 2003. Home range and nest use of the golden mouse (*Ochrotomys nuttalli*) in southern Illinois. *Journal of Mammalogy* 84:553–60.

Mueller-Dombois, D., and H. Ellenberg. 1974. *Aims and methods of vegetation ecology.* New York: Wiley.

Noon, B. R. 1981. Techniques for sampling avian habitats. In *The use of multivariate statistics in studies of wildlife habitat,* ed. D. E. Capen, 42–52. USDA Forest Service General Technical Report RM-87. Fort Collins, CO: USDA Forest Service, Rocky Mountain Research Station.

Ockenfels R. A., and D. E. Brooks. 1994. Summer diurnal bed sites of Coues white-tailed deer. *Journal of Wildlife Management* 58:70–75.

O'Connor, R. J. 2002. The conceptual basis of species distribution modeling: Time for a paradigm shift? In *Predicting species occurrences: Issues of accuracy and scale,* ed. J. M. Scott, P. J. Heglund, M. L. Morrison, J. B. Haufler, M. G. Raphael, W. A. Wall, and F. B. Samson, 25–33. Washington, DC: Island Press.

Penfound, W. T., and E. L. Rice. 1957. An evaluation of the arms-length rectangle method in forest sampling. *Ecology* 38:660–61.

Phillips, E. A. 1959. *Methods of vegetation study.* New York: Holt, Rinehart, & Winston.

Ralph, C. J., G. R. Geupel, P. Pyle, T. E. Martin, and D. F. DeSante. 1993. *Handbook for field methods for monitoring landbirds.* USDA Forest Service General Technical Report PSW-144. Berkeley, CA: USDA Forest Service, Pacific Southwest Research Station.

Reinert, H. K. 1984. Habitat separation between sympatric snake populations. *Ecology* 65:478–86.

Ribble, D. O., A. E. Wurtz, E. K. McConnell, J. J. Bueggie, and K. C. Welch Jr. 2002. A comparison of home ranges of two species of *Peromyscus* using trapping and radiotelemetry data. *Journal of Mammalogy* 83:260–66.

Rice, E. L., and W. T. Penfound. 1955. An evaluation of the variable-radius and paired-tree methods in the blackjack-post oak forest. *Ecology* 36:315–20.

Rotenberry, J. T. 1985. The role of habitat in avian community composition: Physiognomy or floristics? *Oecologia* 67:213–17.

Rotenberry, J. T., and J. A. Wiens. 1998. Foraging patch selection by shrubsteppe sparrows. *Ecology* 79: 1160–73.

Roth, R. R. 1976. Spatial heterogeneity and bird species diversity. *Ecology* 57:773–82.

Rumble, M. A., T. R. Mills, and L. D. Flake. 1999. *Habitat capability model for birds wintering in the Black Hills, South Dakota.* Research Paper RMRS-RP-19. Fort Collins, CO: USDA Forest Service, Rocky Mountain Research Station.

Schreuder, H. T., T. G. Gregoire, and G. B. Wood. 1993. *Sampling methods for multiresource forest inventory.* New York: Wiley.

Scott, J. M., F. Davis, B. Csuti, R. Noss, B. Butterfield, C. Groves, H. Anderson, S. Caicco, F. D'Erchia, T. C. Edwards Jr., J. Ulliman, and R. G. Wright. 1993. *Gap analysis: A geographic approach to pro-*

tection of biological diversity. Wildlife Monograph 123.

Smallwood, K. S. 1995. Scaling Swainson's hawk population density for assessing habitat use across an agricultural landscape. *Journal of Raptor Research* 29:172–78.

Smallwood, K. S., and C. Schonewald. 1996. Scaling population density and spatial pattern for terrestrial, mammalian carnivores. *Oecologia* 105:329–35.

Smallwood, K. S., and T. R. Smith. 2001. Study design and interpretation of shrew (*Sorex*) density estimates. *Annals Zoologica Fennici* 38:149–61.

Sturman, W. A. 1968. Description and analysis of breeding habitats of the chickadees, *Parus atricapillus* and *P. rufescens. Ecology* 49:418–31.

Tarvin, K. A., and M. C. Garvin. 2002. Habitat and nesting success of blue jays (*Cyanocitta cristata*): Importance of scale. *Auk* 119:971–83.

Thomas, D. L., and E. Y. Taylor. 1990. Study designs and tests for comparing resource use and availability. *Journal of Wildlife Management* 54:322–30.

Turchin, P. 1998. *Quantitative analysis of movement: Measuring and modeling population redistribution in animals and plants.* Sunderland, MA: Sinauer Associates.

Tyre, A. J., H. P. Possingham, and D. B. Lindenmayer. 2001. Inferring process from pattern: Can territory occupancy provide information about life history parameters? *Ecological Applications* 11: 1722–37.

Ucitel, D., D. P. Christian, and J. M. Graham. 2003. Vole use of coarse woody debris and implications for habitat and fuel management. *Journal of Wildlife Management* 67:65–72.

USDI1 Fish and Wildlife Service. 1981. *Standards for the development of suitability index models.* USDI Fish and Wildlife Service Ecological Services Manual 103. Washington, DC: Government Printing Office.

VanderWerf, E. A. 1993. Scales of habitat selection by foraging 'elepaio in undisturbed and human-altered forests in Hawaii. *Condor* 95:980–89.

VanDeusen, P. C., and G. S. Biging. 1984. *Crown volume and dimension models for mixed conifers of the Sierra Nevada.* Northern California Forest Yield Cooperative Research Note no. 9. Berkeley: University of California, Department of Forestry and Resource Management.

Van Horne, B. 1983. Density as a misleading indicator of habit quality. *Journal of Wildlife Management* 47:893–901.

Van Horne, B. 2002. Approaches to habitat modeling: The tensions between pattern and process and between specificity and generality. In *Predicting species occurrences: Issues of accuracy and scale,* ed. J. M. Scott, P. J. Heglund, M. L. Morrison, J. B. Haufler, M. G. Raphael, W. A. Wall, and F. B. Samson, 63–72. Washington, DC: Island Press.

Weir, R. D., and A. S. Harestad. 2003. Scale-dependent habitat selectivity by fishers in south-central British Columbia. *Journal of Wildlife Management* 67:73–82.

Welsh, H. H., Jr., and A. J. Lind. 1995. Habitat correlates of Del Norte salamander, *Plethodon elongatus* (Caudata: Plethodontidae), in northwestern California. *Journal of Herpetology* 29:198–210.

Welsh, H. H., Jr., and A. J. Lind. 2002. Multiscale habitat relationships of stream amphibians in the Klamath-Siskiyou region of California and Oregon. *Journal of Wildlife Management* 66:581–602.

Wenny, D. G., R. L. Clawson, J. Faaborg, and S. L. Sheriff. 1993. Population density, habitat selection and minimum area requirements of three forest-interior warblers in central Missouri. *Condor* 95: 968–79.

White, G. C., and R. A. Garrott. 1990. *Analysis of wildlife radio-tracking data.* San Diego: Academic Press.

Whitmore, R. C. 1981. Applied aspects of choosing variables in studies of bird habitats. In *The use of multivariate statistics in studies of wildlife habitat,* ed. D.E. Capen, 38–41. USDA Forest Service General Technical Report RM-87. Fort Collins, CO: USDA Forest Service, Rocky Mountain Research Station.

Wiens, J. A. 1969. *An approach to the study of ecological relationships among grassland birds.* Ornithological Monographs no. 8. Lawrence, KS: Allen Press.

Wiens, J. A. 1989. *The ecology of bird communities.* Vol. 1, *Foundations and patterns.* Cambridge: Cambridge Univ. Press.

Wiens, J. A., and J. T. Rotenberry. 1981. Habitat associations and community structure of birds in shrubsteppe environments. *Ecological Monographs* 51:21–41.

Willson, M. F. 1974. Avian community organization and habitat structure. *Ecology* 55:1017–29.

6 Measuring Wildlife Habitat: When to Measure and How to Analyze

Perfect as the wing of a bird may be, it will never enable the bird to fly if unsupported by the air. Facts are the air of science. Without them a man of science can never rise.

IVAN PAVLOV

In this chapter we first develop the concept of "when" to measure wildlife habitat. *When* to measure involves both between-season and within-season analysis. It has been clearly shown that resource requirements of animals vary considerably between seasons; the use of different winter and summer ranges by big game is one example. However, species that are permanent residents in an area frequently switch to different foraging substrates and food sources as seasons change. Further, within a given time period, resource use and activity patterns often vary with time of day, stage of the breeding cycle, and other divisions of an animal's life cycle. Thus it is critical to understand the temporal aspects of resource use prior to designing a study of habitat.

The distribution and abundance of animals vary with variations in food supply, weather conditions, predator activity, and many other biotic and abiotic factors. As we show throughout this book, these factors must drive how we design and then interpret our studies of wildlife habitat. Few studies, however, have explicitly rec-

ognized these variations in species distributions and abundances; instead they have based their conclusions on time- and site-species analyses.

In the final sections of this chapter we discuss the use of multivariate statistics in analysis of wildlife habitat. We first examine the rationale for using multivariate statistics, relating these techniques to our conceptualization of wildlife-habitat studies. We then review the all-important assumptions associated with these statistical techniques. We follow with a classification and discussion of several of the more commonly used methods and their applications to wildlife studies. Sample-size requirements and a review of statistical computer packages are also included. A formal course in multivariate statistics is not required for understanding this chapter. Many fine multivariate texts are available and should be consulted for a general overview of this class of analyses (e.g., Morrison 1967; Cooley and Lohnes 1971; Pimentel 1979; Afifi and Clark 1984; Dillon and Goldstein 1984; Digby and Kempton 1987; Krzanowski 1990).

When to Measure

As is commonly known, the behavior, location, and needs of animals change, often substantially, throughout the year. Too many researchers, however, ignore such *temporal variations* in habitat use, in at least two major ways. First, while often acknowledging that temporal variations do occur, they sample from such a narrow time—usually only during spring and/or summer—that the results are so time- and location-specific that they apply only minimally to other situations. Second, they may sample from across some broad time period (such as the "summer season") and then "average out" the relationships over the period. Given adequate samples, we are not faulting the first strategy with regard to the detail of the data collected; applicability is the question. Without knowledge of an animal's total requirements, management recommendations have limited and perhaps faulty implications.

Few would argue, of course, that the preferred study design would be repetition of a study for every appropriate biological period (e.g., prebreeding, breeding). It is easy to conceptualize what "averaging" may mean to resulting descriptions of habitat relationships: in figure 6.1 we see that the average use over time by two hypothetical species is not, in fact, a close approximation of their actual behavior. Rather, the average indicates that the animals use this tree species basically identically. Here we see, however, that the finer we stratify our sampling, the greater the number of levels of resolution we have available for subsequent analysis. The question of how finely to measure is a study-specific problem, and further, within a particular study, the level of resolution will probably vary with time (e.g., between seasons). Hypothetical scenarios such as that depicted in figure 6.1 should concern researchers. Such concerns can be evaluated early in a study by making sure that sampling is sufficiently intensive that such relationships can be identified. Subsequent sampling can be adjusted after such preliminary evaluations of data are completed.

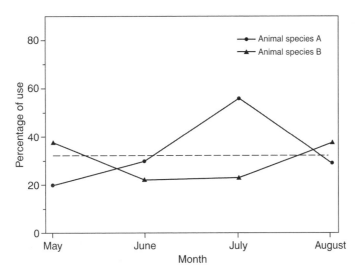

Figure 6.1   Use of a species of tree by two hypothetical animal species during "summer." The dashed horizontal line represents approximate average of use values for both species (calculated separately) across time.

Schooley (1994) reviewed 43 papers published in the *Journal of Wildlife Management* between 1988 and 1991 that examined habitat use of terrestrial vertebrates. He found that most studies pooled data on habitat use among years, evidently without testing for annual variation. Using the black bear (*Ursus americanus*) as an example, he illustrated the misleading inferences that can result from pooling data across years. Patterns of annual variation in habitat use were lost when data were pooled (fig. 6.2).

No distinct separation actually exists between sampling wildlife populations within or between seasons (intra- or interseaonally, respectively). Further, even within relatively small areas, the seasons can differ widely for species both within

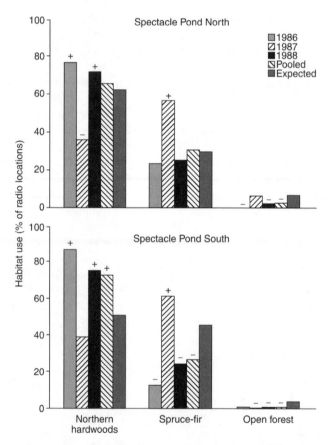

Figure 6.2.   Habitat selection by female black bears at two study areas in northern Maine during fall (den entry on 1 September), 1986–1988. Habitat use is presented for individual years and for all years pooled. Expected use is estimated from availability of habitats. Habitat availability was estimated separately for each year, but only 1987 data are presented because availability differed little between years. A + indicates use was greater than availability; a − indicates use was less than availability; no symbol indicates use was equal to availability. Selection was based on Bonferroni-adjusted 90% CIs. (Reproduced from Schooley 1994, fig. 1; by permission of The Wildlife Society)

and, especially, between vertebrate groups. In most regions, for example, the breeding season for many rodents can start as early as February, months before many breeding bird species even arrive in the area. Further, within the birds, many resident species begin breeding several months before nonresidents. Few studies acknowledge these sampling problems; or if they do, they sample from the "middle" of the season (which, for logistical reasons, usually conforms to the summer recess of most universities).

Because of limited funding, availability of personnel, cold weather, and other problems related to management of studies, most researchers have concentrated their studies during the summer. But it is now well known that the survival of the animal may depend on the nonbreeding period, particularly the wintering period. Intuitively, we would expect that the fall and winter periods—when populations are at their greatest numbers (because of offspring), resources are declining (trees and arthropods are dying or going dormant), animals are physiologically stressed by dispersal or migration movement, and the weather is becoming more harsh—would be the most difficult times for an animal. Further, because of these changes in environmental conditions, the "habitat" itself changes. Large-scale migrations by numerous species of animals provide a good indication of the lengths to which animals must go to find favorable living conditions. The importance of nonbreeding periods to animals was popularized by Fretwell (1972), although he was certainly not the first worker to recognize the significance of such periods to animals. If an animal changes its pattern of habitat use or a change occurs in the environment itself, then a summer-based study cannot be used with any confidence to predict the subsequent responses of the animal.

At a minimum, researchers should acknowledge these sampling limitations and the ramifications of such weaknesses on management rec-

ommendations. Ideally, researchers will avoid many of these problems by designing studies that actually determine if such potential temporal variations are influencing their results in major ways.

Long-Term Temporal Changes

We have just discussed the need to evaluate intra- and interseasonal variations in habitat use. The magnitude of the temporal variation, however, is a question of scale, and the scale can run across as well as within years. Thus relationships developed within a year or over a period of a few years may not hold for a different period of years. We all know how environmental conditions can change drastically across years: "normal" weather conditions seldom occur but are in reality a crude average of extremes in rain, temperature, and wind. An animal's reaction to such conditions over time will provide the most important data with which to evaluate habitat relationships.

As reviewed by others (e.g., Likens 1983; Wiens 1984; Strayer et al. 1986; Leigh and Johnston 1994), a tradition has developed over the past several decades, especially among North American scientists, of the pursuit of short-term studies. This situation arose from constraints imposed by short funding duration, the need to finish graduate programs within short periods of time (actually a noble goal!), and the pressure placed on researchers to publish. But by restricting the duration of our investigations, we adopt a snapshot approach to studying nature. Our wildlife studies usually run from 1 to 3 years and, at best, give us only a partial view of most ecological situations; at worst, they give us a false interpretation.

Long-term studies are especially suited to exploring four major classes of ecological phenomena: *slow processes, rare events, subtle processes,* and *complex phenomena* (Strayer et al.

1986; also discussed in chapter 9 in the context of disturbance dynamics). Forest succession, invasion of exotic species, and vertebrate population cycles are prominent examples of slow processes that have obvious importance in formulating management decisions. The substantial impact that the El Niño oscillations, catastrophic natural events (e.g., floods, disease outbreaks, fires), and population eruptions are examples of rare events that are certainly missed by short-term studies. Processes that change over time in a regular fashion, but in which the year-to-year variance is large relative to the magnitude of the longer-term trend, are examples of the subtle processes that short-term studies cannot evaluate, Finally, although scientists certainly realize that nature is complex, seldom do we provide the time necessary for the phenomena to reveal enough of their characteristics to allow meaningful interpretation.

But how long is long enough? Strayer et al. (1986) provided two rather different definitions of the concept of long-term. Their first definition considers the length of the study in terms of natural processes. Here a study is considered long-term if it continues for as long as the generation time of the dominant organism, or long enough to include examples of the important processes that structure the ecosystem under study; the length of study is measured against the dynamic speed of the system being studied. Obviously, such a criterion demands that the researcher has a good understanding of the system of interest. Strayer et al.'s other approach was to view the length of study in a relative fashion, with long-term studies being those that continue for a longer time than most other such studies. Here they are accepting the constraints applied by human institutions, not the rate of natural processes.

Of course, not all studies need be long-term to provide useful results. Descriptive studies of essentially static patterns (e.g., morphology, ge-

netic characteristics of species); of processes at the individual level (e.g., growth, behavior, physiology); or of patterns of adaptation do not necessarily require long time periods. Ecological studies conducted at broad spatial scales (e.g., relating presence–absence to vegetation types) are also usually relatively static. The principal disadvantages of long-term studies are, of course, practical rather than ecological.

Four classes of short-term studies can potentially provide insight into long-term relationships: (1) *retrospective studies;* (2) *substitution of space for time;* (3) *use of systems with fast dynamics as analogs for systems with slow dynamics;* and (4) *modeling* (Strayer et al. 1986). Further, a series of short-term studies can be incorporated into a longer-term plan for research; such a plan works especially well in a professor–graduate student relationship. Substitution of space for time and modeling are common substitutes for longer-term work. In substitution studies, sites with different characteristics are used instead of following a few sites for an extended period (a chronosequence). An example is examining the relationship between birds and plant succession over 1 year using sites of different ages (e.g., 1, 5, 15, 30, or more years postharvest or postburn). To provide valid results, such a design requires that all the sites have similar histories and characteristics. Naturally, a large number of replicates enhances the reliability of such a study. Of course, these studies cannot capture the historical events that shaped each site; they can only swamp the effect through a large sample size; ideally, this approach will yield adequate results. Needed is an increase in experimental manipulations of vegetation, food, competitors, predators, and other parameters, in association with demographic studies.

Can short-term studies, then, tell us much about animals? Wiens (1984, 202–3) thought not, answering that "a short-term approach is likely to produce incomplete or incorrect perceptions of a complex reality . . . perhaps obtaining results that are superficial and quite possibly incorrect." Short-term studies can provide useful information if they have a specific focus that is not likely to be obscured by the background variation in habitat relationships. However, available evidence clearly indicates the necessity of the implementation of longer-term studies if the development of wildlife–habitat relationships is to advance. We should not be surprised that the relatively "easy" studies have been conducted, or that the task presented to new students will ease with time. Graduate students cannot be expected, of course, to conduct the long-term work in their thesis projects. It is the duty of graduate advisors to help direct students toward projects whose results are not likely to be swamped by unknown long-term variance. Long-term studies done in parallel with carefully matched, short-term substitution studies are needed. Wiens (1989, 174–96) provided additional review material on the usefulness of long-term ecological studies.

Sample Size Requirements

Regardless of the care taken in designing a study, all is for naught if an insufficient number of observations are made. To most, this statement must seem obvious. However, very little attention has been paid in the scientific literature to this fundamental question of study design and statistical analysis. We are not sure why this is so, and can only advance the suggestion that researchers have generally tried to collect the largest sample size their budgets and time would allow. If researchers must limit their sampling because of time or monetary constraints, they would be better advised to limit the scope of study and produce one good result rather than many weak ones.

Most general statistics texts discuss determination of sample size necessary to perform certain analyses. The specific methods vary according to the type of data being collected. In general, the sample size needed depends on four factors: (1) *the variance of the populations;* (2) *the size of the difference between sampling units to be detected;* (3) *the statistical probability* (Type I error); and (4) *the assurance one desires in detecting the difference* (Type II error). The researcher can establish the final three factors; obviously, any two of these factors will determine the third (Zar 1984). Note that an initial estimate of the population variance is required. These variances can be estimated either from the literature or from data collected during a preliminary analysis of the population of interest. See box 6.1 for details on calculating sample size.

Another method of estimating necessary sample size involves continually evaluating the data as increasing sample sizes are obtained. Such *sequential sampling* is especially useful when one has no idea of the true population variance, and especially when one is gathering data on numerous variables (as is often the case in habitat studies). Further, continual evaluation of data gives the added benefit of providing in-depth familiarity with the data set, familiarity that may identify unexpected or interesting facts about the species in time to modify the sampling protocol. Such a procedure fits as part of the preliminary sampling strategy that we previously discussed. Mueller-Dombois and Ellenberg (1974) suggested plotting the standard error as a function of the number of samples and denoting an adequate sample size when the curve decreased to less than 5 to 10% gain in precision for a 10% increase in sample size.

A major problem encountered by many wildlife researchers is the adequate sampling of rare populations (see Queheillalt et al. 2002 for review). Two major issues are involved here. First is the issue of being forced to repeatedly sample the same individual animals. Such repeated sampling usually violates statistical independence,

Box 6.1   How many samples are needed?

In wildlife research—directly manipulative experiments as well as observational studies—it is always better to plan ahead for sampling design than to struggle with incomplete data after the fact. Correctly designing a study typically entails asking the following kinds of questions:

- What and how many conditions (control and treatment groups) am I trying to test?
- How variable are the groups?
- How confident do I need to be in detecting differences?
- How precise should the test be?
- How many samples do I need?

Much has been written on sampling designs in response to these and related questions (see text for discussion and references). Here, we focus on the last question, that of sample size.

Determining the sample size needed to detect differences between means in a multigroup study is a relatively simple procedure, but one that is seldom conducted in wildlife monitoring or research studies. As a result, many investigators (including managers interested in data on monitoring of wildlife populations or habitats) struggle with samples inadequate for statistically answering questions about differences and must resort to pooling samples, which is often undesirable.

To illustrate the calculation, let's work with an example. Let us assume we want to detect differences in the mean number of large snags (dead standing trees) in four successional stages of forest development. The question is then, How many (what sample size of) vegetation sampling plots (or forest stands) are needed per successional stage? Our successional stages make up sample groups. Comparing means between more than two groups, in this case four groups, entails use of analysis of variance (ANOVA) techniques and use of Student's t statistics to determine mean differences. We will assume a balanced ANOVA design with equal sample sizes in each group. Also, making the calculation entails using an initial sampling from which preliminary estimates of sample standard deviations and coefficients of variation are derived.

First, we need to specify the level of certainty of detecting a difference and the percentage of difference between means to detect. In our hypothetical study, we can specify, for example, an 80% certainty of detecting a 5% difference between two of the four group means at a confidence level of 0.01. This is a reasonable expectation, although a confidence level of 0.05 is also typically used in such studies.

Since $a = 4$, the ANOVA error MS (mean square) has $v = a(n-1) = 4\ (n-1)$ degrees of freedom. The sample size n is thus:

$$n \geq 2(\sigma/\delta)^2(t_{\alpha(v)} + t_{2(1-P)(v)})^2$$

where

σ = population standard deviation
δ = smallest true difference that is desired to be detected
t = values of Student's t (found in statistical tables)
v = degrees of freedom (df) of the sample standard deviation with a groups each with n replications in a balanced ANOVA
α = significance level (here, 0.01)
P = desired probability (here, 80%)

Note that only the ratio σ/δ is needed, not the individual factors; this ratio is specified in our example here at 5%.

Next, assume that a sample resulted in a coefficient of variation (CV $= \sigma/y$, where $y =$ sample mean) among the groups at 6%. We proceed by making an initial guess at n, say $n = 20$ (always a safe starting point unless samples are extremely variable, in which case begin with $n = 40$). Then $v = 4(20 - 1) = 76$ df. Since CV = 6%, then $s = 6y/100$, where $s =$ the sample standard deviation. We wish δ to be 5% of the mean; that is, $\delta = 5y/100$. Using s as an estimate of σ, which is obtained from initial exploratory sampling, we obtain $\sigma/\delta = (6y/100)(5y/100) = 6/5$, so that

(box continued on next page)

Box 6.1   Continued

$$n \geq 2(6/5)^2(t_{0.01(76)} + t_{2(1-0.80)(76)})^2$$

$$= 2(6/5)^2(2.64 + 0.847)^2$$

$$= 2(1.44)(12.16)$$

$$= 35.0$$

Next, we try another value for *n*, say, *n* = 35. So, *v* = 4(35 − 1) = 136; and $n \geq 2(1.44)(2.61 + 0.845)^2 = 34.4$, or $n \approx 35$ again. Thus, it seems that 35 samples per population (or forest plots per successional stage) are necessary.

Our example depicts sampling forest vegetation plots. The approach can be used to determine the number of samples for monitoring programs, such as those needed to detect differences between treatments and controls in adaptive management studies (see chapter 12).

---

which prevents each observation from being treated as a "sample." Although beyond the scope of this book, statistical methods such as time-series analysis and repeated-measures analysis of variance are available to handle such situations (see Raphael 1990; Morrison et al. 2001). In addition, decent biological rationales can often be used to minimize repeated sampling (e.g., sampling an individual animal only once per sampling session). True independence is never achieved, but compromises are often necessary (and should be fully explained and defended in any publication).

Small-mammal trapping presents a special problem with repeated sampling: the same individuals will certainly be captured within and also between trapping sessions. Researchers have handled this dilemma in various ways. Kelt et al. (1994) concluded that the repeated captures of the same individuals at trap stations could introduce a bias, so they used only first-time captures in their analyses. In contrast, Morrison and Anthony (1989) considered each capture of a new individual or recapture at a new location a single sample. For example, if two *Microtus* individuals were captured at the same trap, the habitat characteristics associated with the trap were entered into the analysis twice. Recaptures of the same individual at the same trap were not ana-

lyzed. Morrison and Anthony followed this procedure to weight the analysis toward favored trap sites and their associated habitat characteristics. They rationalized that if that procedure was not used, trap sites with one individual of a species and those with multiple individuals would likely show false similarity in habitat use. Their procedure was thus an attempt to weight habitat use by animal abundance. We recommend, however, that researchers avoid the problem of independence and design studies to fit within repeated-measures designs. Or if use of repeated-measures designs is not possible (e.g., an adequate sample of recaptures cannot be obtained), analysis of trapping grid habitat should be based on the area used by each individual within defined sampling seasons. Here, a crude "used area" delineated by trap locations serves as the sampling area for vegetation analysis. After all, the use of a trap—into which an animal has been attracted through the use of baits—is unlikely to represent an unbiased sample of "habitat."

Second, detecting the presence of rare animals cannot typically be accomplished through the use of standard sampling methods. The abundance of rare animals usually approximates a Poisson distribution. Such distributions are characterized by many zeros and only an occasional location of the animal of interest. The sample size necessary for detecting species in such situations has been developed by Green and Young (1993; see also Morrison et al. 2001); a generalized view of sampling efforts needed is provided in figure 6.3. As noted by Green and Young, "rarity" is a relative concept; they define anything that has a density of <0.1 per sampling unit size (e.g., quadrant size) as rare. They present formulae that are applicable to specific sampling situations.

Block et al. (1987) studied the sample size necessary in making estimates of vegetation characteristics in the mixed-conifer zone of the western Sierra Nevada. Locations of foraging birds (although the results of their study are applicable to any forest animal) served as the center of 0.04-ha circular plots in which structural and floristic variables were quantified. Measurements were taken using both visual (ocular) estimates and more objective techniques involving standard measuring devices (diameter tape for tree diameter at breast height [dbh], clinometer for tree heights, measuring tape for distances).

For each variable they randomly selected, with replacement, 10 subsamples of 5, 10, 15, 20, 30, 40, 50, and 60 plots. They then calculated the mean for each sample, and then the mean of the means for each subset. Stability of the estimates (i.e., an adequate sample size) was defined as the point at which the estimate of the mean remained within one standard deviation of the subsequent estimates with little variation in the magnitude of the confidence interval (CI).

Block and his colleagues found that the type of variable collected, whether visually estimated or measured, affected the sample size required. Minimum sample size requirements for measured variables ranged from 20 (for average tree dbh) to 50 (for average tree height), and those for estimated variables ranged from 20 (for average tree dbh) to more than 75 (for average height to the first live tree branch). More importantly, the final point estimates of variables often differed between estimates and measurements (fig. 6.4).

The Block et al. (1987) study shows that researchers must be concerned not only with the size of the sample gathered but also with the way

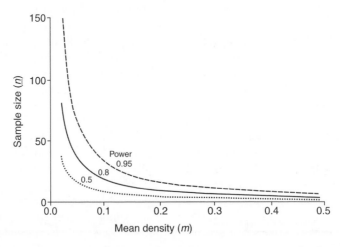

Figure 6.3.   The necessary sample size, n, as a function of mean density, m, for various degrees of power, $1 - \beta$, when sampling the Poisson distribution. (Reproduced from Green and Young 1993, fig. 3; by permission of the Ecological Society of America.)

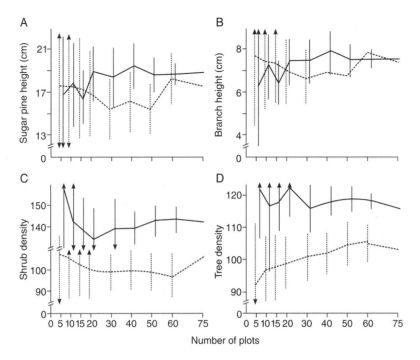

Figure 6.4.   Influence of sample size on the stability of estimates (dashed horizontal lines) and measurements (solid horizontal lines) of bird-habitat characteristics. Dashed or solid vertical lines represent 1 standard deviation from point estimates for estimates and measurements, respectively. Variables shown are *A*, average height of sugar pine; *B*, average height to the first live branch of sugar pine; *C*, average number of shrubs within sample plots; *D*, average number of trees within sample plots. (Reproduced from Block et al. 1987, fig. 2; by permission of the Cooper Orthinological Society.)

in which variables were estimated. Note again (see fig. 6.4) that the two techniques of making observations often resulted in different values. Many such variables entered into a multivariate analysis can alter the resulting conclusions regarding habitat use by an animal, especially if accompanied by small sample sizes (Morrison 1984b). These researchers used "standard" methods of recording data, which raises questions regarding the conclusions drawn by other workers using similar techniques. In fact, a review of studies of wildlife–habitat relationships indicates that extremely small sample sizes were used in many instances (Morrison 1984a, b; Morrison et al.

1990). Students are encouraged to examine carefully the sample sizes used, and the techniques used to gather data, in the publications they read in support of their studies.

Observer Bias

The scientific method is designed, in part, to help us avoid allowing our preconceived notions to determine the outcome of our studies. We all carry numerous biases into the design of any study: our training predisposes us to prefer certain sampling methods; we might believe that animals respond in specific ways to environmental conditions, which tends to narrow our sampling focus; time and monetary constraints limit our ability to sample in the way we would prefer; and so forth. Not only do biases influence the way we design a study, but additional biases are inserted into field sampling when multiple observers—likely all with different training and experience—are used to gather data. As noted by Gotfryd and Hansell (1985, 224), "Ignoring observer-based variability may lead to conclusions being precariously balanced on artifacts, spurious relations, or irreproducible trends." This is a strong statement that is, unfortunately, ignored in the design and implementation of many studies of wildlife and their habitats. Here we will concentrate on the evaluation and subsequent reduction of bias in field sampling.

Gotfryd and Hansell (1985) used four observers to independently sample eight plots located within an oak–maple forest near Toronto, Canada. They followed methods detailed by James and Shugart (1970) and measured the variables given in table 6.1. They found that observers differed significantly in their measurements on 18 of the 20 vegetation variables measured. As these authors noted, their study addressed only the precision of estimates between observers; no measure of the accuracy of their results was conducted.

Block et al. (1987) used several univariate and multivariate analyses to test for differences among three observers in estimating plant structure and floristics. They found that ocular estimates by the three observers differed significantly for 31 of the 49 variables they measured. Perhaps the most confounding aspect of using multiple observers they found was the unpredictable nature of variation among observers. Multiple comparisons of estimates for the 31 significant variables resulted in all possible combinations of observers. Thus, when samples from different observers are pooled, sampling bias can

Table 6.1. Vegetational habitat variables and their mnemonics, used by Gotfryd and Hansell

Mnemonic	Variable
TRSP	Number of tree species
SHRSP	Number of shrub species
*SDEN	Density of woody stems <7.6 cm diameter at breast height (dbh)
*CC	Canopy cover
*GC	Ground cover
BAA	Basal area (BA) of trees 7.6 to 15.2 cm dbh
BAB	BA of trees 15.2 to 23 cm dbh
*BA1	BAA + BAB
*BA2	BA of trees 23 to 53 cm dbh
*BA3	BA of trees >53 cm dbh
CH1	Maximum canopy height
CH4	Maximum canopy height in quadrant having lowest canopy
*CHAV	Average of canopy height maxima by quadrant
*CHRNG	CH1 to CH4
*CHCV	CHRNG/CHAV
DTR1	Distance to nearest tree >15.2 cm dbh
DTR4	Maximum of by-quadrant nearest tree distances
*DAV	Average of by-quadrant nearest tree distances
*DRNG	DTR1 to DTR4
*DCV	DRNG/DAV

Source: Gotfryd and Hansell 1985, table 1.
*Variables used in multivariate analyses.

increase. As we show later in this chapter, this variability has, especially when combined with low sample sizes, an especially profound influence on multivariate analyses.

Ganey and Block (1994) used three observers to sample plots for canopy closure, employing two different estimation techniques (spherical densiometer and sighting tube; results described above). They found significant variation among observers in estimates of canopy cover for both methods. Results were, however, relatively more consistent for the sighting tube.

The biases associated with estimations of animal abundances should also be considered carefully in habitat studies, because many of our analytical procedures correlate animal numbers with features of the environment. Obviously, a study that has low bias among habitat characteristics can be ruined by biased count data, and vice versa. For example, Dodd and Murphy (1995) evaluated the accuracy and precision of nine techniques used to count heron nests. Although they found rather high error rates among the techniques, observer bias was low for most methods. Interestingly, the highest observer bias was found for their point counting technique, a result apparently due to the varying choices made by observers of the optimum vantage point for which to count nests at colonies.

Researchers can reduce interobserver variability by following closely a set of well-defined criteria for selecting and training observers. Although designed for bird censusing, the steps outlined by Kepler and Scott (1981) for bird counting methods can be applied generally to most types of sampling. Carefully screen applicants initially to eliminate the more obvious visual, aural, and psychological factors that increase observer variability. In addition, researchers should organize a rigorous observer training program, which will further reduce inherent variation but will not eliminate it. In a field experiment, Scott et al. (1981) found that observers, after training, could estimate the distance of a singing bird with an accuracy of 10 to 15% of true distance. Also, periodic training sessions with observers should be conducted to counter observer "drift" and thus recalibrate their recording to standard and known values (see Block et al. 1987). It is the case, though, that observers can be trained to develop more precise ocular estimates of vegetation conditions, although such training does not necessarily reduce bias.

Plant ecologists have long recognized differences among data collection techniques (see Cooper 1957; Lindsey et al. 1958; Schultz et al. 1961; Cook and Stubbendieck 1986; Hatton et al. 1986; Ludwig and Reynolds 1988; Kent and Coker 1994). Although the cost of measuring plant structure and floristics in an adequate number of plots is usually great in both time and money, the ramifications of not following a rigorous sampling design are severe. Again, it is better to limit the scope of a study so that data are collected properly; preliminary sampling and analysis help to identify potential problems.

Multivariate Assessment of Wildlife Habitat

In this section we first examine the rationale for using multivariate techniques, relating these methods of analysis to our conceptualization of wildlife–habitat studies. We then review the all-important assumptions associated with these methods. We follow with a classification and brief discussion of some of the current techniques and their applications to wildlife studies, concentrating on the frequently used multiple regression and discriminant analysis procedures. Our intent here is to briefly introduce the application of multivariate techniques to the analysis of wildlife habitat data, concentrating on problems encountered in their use. As mentioned at the beginning of this chapter, a formal course in multivariate statistics is not a prerequisite for understanding this chapter. We do, however, strongly recommend such a course in the study plan of all graduate students. Such a course assists your ability to evaluate the literature, even if you never conduct a multivariate analysis of your own data. Many good multivariate texts are available and should be consulted for details not included herein (e.g., Cooley and Lohnes 1971; Pimentel 1979; Dillon and Goldstein 1984; Johnson 1992; Manly 1994; Timm 2002; Shaw 2003; Dillon and Goldstein's text is an especially readable work.

Conceptual Framework

Multivariate analysis is a branch of statistics used to analyze multiple measurements that have been made on one or more samples of individuals. Multivariate analysis is distinguished from other forms of statistical procedures in that multiple variables are considered in combination as a system of measurement. Because the variables are typically dependent among themselves, we cannot separate them and examine each individually (Cooley and Lohnes 1971, 3).

Multivariate statistical techniques were not originally designed for analysis of wildlife habitat and behavioral data. Indeed, multivariate techniques have been used since the late 1880s, with a progression of new, more sophisticated methods following throughout the 1900s (Cooley and Lohnes 1971, 4). Only since the 1960s have researchers placed emphasis on the quantitative analyses of wildlife habitat.

The application of multivariate analyses of wildlife data is the product of a synthesis beginning in the early 1970s that united several lines of ecological research. As outlined by Shugart (1981), this synthesis involved linking two analytical tools with three ecological concepts—niche theory, microhabitat, and individual response (fig. 6.5). The availability of high-speed computers along with prewritten multivariate computer programs (i.e., "canned" statistical packages) allowed easy access to this general field of analysis. Hutchinson's reformulation of the niche concept (1957) in terms of an "*n*-dimensional hypervolume" caused ecologists to alter their view of wildlife habitat and the ways in which they analyzed data (Stauffer 2002).

Green (1971) was one of the first ecologists to formally apply multivariate analyses to Hutchinson's concept. The *n*-dimensional concept of the niche and the *n*-dimensional sample space of multivariate analysis are analogous in many ways, and this similarity led to an obvious application of multivariate methods to ecological data (Shugart 1981).

Assumptions

Four common assumptions are associated with parametric multivariate analyses: *multivariate normality; equality of the variance-covariance matrices (group dispersions); linearity;* and *independence of the error terms (residuals)*. Violation of any of these assumptions can bias or taint analysis results and conclusions derived from them. Unfortunately, many published papers in the wildlife literature have failed to discuss these assumptions and their ramifications on results. As we discuss below, low sample size is often a major factor in such violations.

NORMALITY

The assumption of multivariate normality is more than simply an assumption that each variable is itself normally distributed, as in the univariate sense. Unfortunately, tests of this assumption are cumbersome and yield only approximations. Variable-by-variable examination of normality, however, will certainly help identify variables that depart greatly from normality. Standard univariate transformations (e.g., log, square-root transformations) can be applied as appropriate.

VARIANCE

In ecology, however, it is well known that the distribution and behavior of animals varies along gradients of environmental variables (e.g., soil moisture, canopy cover, air temperature). As noted by Pimentel (1979, 177), however, biologists seem too concerned with the mean responses of animals to environmental gradients, rather than with the distribution of animals along such gradients. Concentrating on mean responses certainly lowers the ability of research results from one location (and thus one point on

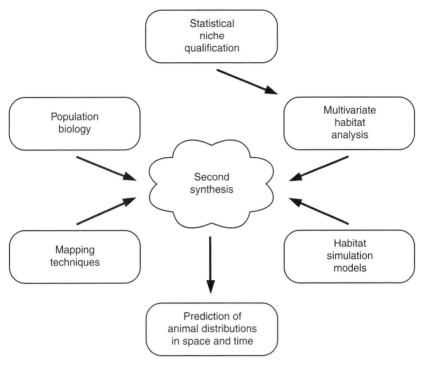

Figure 6.5. Schematic diagram of the scientific research elements that combined in a synthesis to produce multivariate habitat analysis. (Reproduced from Shugart 1981, fig. 1.)

the gradient) to be applied to another geographic location (and thus likely from another point on the gradient). If populations have unequal dispersions, then the populations are different, even if their central tendencies (means) are the same. Thus trying to force normality to meet the formal assumption of equality of dispersions is biologically unsound. In fact, the magnitude of this dispersion can be used as a multivariate measure of niche breadth (see Carnes and Slade 1982). Nothing prevents one from comparing populations having unequal dispersions. Fortunately, research indicates that tests of equality of group centroids are rather insensitive to moderate departures of multivariate normality and homoscedasticity. If sample sizes are large and equal

between groups, the inequality of dispersions has no real effect on interpretation of results. Application of this rationale requires careful planning to ensure that large and equal samples are collected for all groups under study—a caveat that has been ignored in most studies of wildlife–habitat relationships. As noted by Wiens (1989, 66), a considerable portion of the variance that is so easily discarded in statistical analyses may contain important insights into the dynamics of the system under study.

LINEARITY

Linearity is important in multivariate analyses in two main ways. First, most models are based on a linear relationship. Second, the correlation coefficient—which forms the basis of most multivariate calculations—is sensitive only to the linear component of the relationships between two variables. Also, assuming that linear relations are a more parsimonious approach than the use of nonlinear multivariate models, nonlinear models should be used only if linear ones prove to be poor fits. Fortunately, many relationships can be approximated using linear models, even though nonlinear components may exist. The data transformations noted above may help to linearize a nonlinear relationship. However, nonlinearity changes the probabilities in tests of significance; for example, one is likely to fail to reject null hypotheses of equality of group centroids but to reject null hypotheses of equality of group dispersions (Pimentel 1979, 178–79). Researchers would be well served by first examining the linearity of their data, variable by variable, before plunging them into the black box of a canned multivariate statistical package. Highly nonlinear data should not be combined into a single analysis with linear data.

Thus the selection of linear models is usually more a matter of statistical convenience than a decision based on ecological reality. As noted above, we should expect biological entities often to have nonlinear and nonnormal distributions. Linear analyses are, however, relatively straightforward; their statistical properties are well understood. Further, a model has little utility if other researchers and, especially, managers find it difficult to understand. Thus it can be argued that linear models should be used unless the data are *highly* nonlinear.

There are two basic methods for applying nonlinear methods to statistical models. The first is to transform variables by various orders of polynomial transformations. Linear models are, specifically, first-order polynomial models. Second-order, or quadratic, models are obtained by squaring a variable (X^2), resulting in a U-shaped relationship. Third-order, or cubic,

models are obtained by cubing a variable (X^3), resulting in a curving relationship. Higher-order models are possible, but they become increasingly difficult to visualize and are unlikely to be applicable. An interesting example is provided by Burger et al. (1994), who regressed arcsine-transformed percentage of predation on nests onto natural a log-transformed area (ln [size]) and the square of ln-transformed area (ln [size]^2). This polynomial regression explained 77% of the variation in predation rates among study tracts of varying sizes (fig. 6.6). However, it is exceedingly difficult to interpret these transformations in terms of real-world natural history.

Because it is relatively difficult to interpret high-order models, you should first try to linearize data through appropriate transformations. It is, after all, difficult to explain that a population's abundance is a function of cubed foliage volume, squared snag density, and shrub cover. If a linear model can meet your objectives, then use it. However, if a linear model does not provide an adequate prediction after appropriate testing (see below), then nonlinear alternatives should be used before declaring that the linear model did not reveal significant environmental correlations (see also Nadeau et al. 1995).

Meents et al. (1983) evaluated the use of first-, second-, and third-order independent variables on predictions of bird abundance using multiple regression. They found that the polynomial variables resulted in significant regressions in many cases in which linear variables did not do so. They also showed that while linear relationships dominated during certain parts of the year, nonlinear relationships were important during other periods of time. Nonlinear relationships explain, in part, why models must often be season-specific. In addition, they showed why nonlinear relationships can have important management implications. For example, woodpecker abundance might be related to snag density in a curvilinear

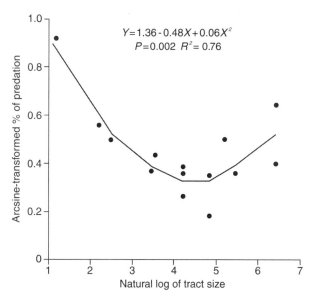

Figure 6.6.   Arcsine-transformed predation on artificial nests ($n = 540$) in 15 prairie fragments regressed onto a natural log-transformed tract area (ln [size] and ln[size]^2). (Reproduced from Burger et al. 1994, fig. 1 by Permission of The Wildlife Society.)

fashion where relatively low or high densities of snags result in a lower bird abundance than found at moderate snag densities. Thus woodpeckers may disappear when snags decline below some critical nonzero threshold. In this simple example, linear models could mask important biological relationships that could result in faulty management decisions. Other examples are apparent (e.g., soil-moisture tolerance in salamanders, micronutrient content of herbivore forage, grass density for rodents). Other examples of nonlinear analysis of wildlife–habitat data are included below with specific analytical techniques (e.g., logistic regression).

Purely nonlinear models exist in a form other than the standard linear least squares model (with polynomial transformations). Draper and Smith (1981, ch. 10) gave examples of such nonlinear models, and Seber (1989) and Ratkowsky (1990) provide thorough development of non-

linear modeling as applied to regression analysis. On computers, nonlinear models are solved as approximations through relatively complicated iterative numerical calculations. The form of the equations and the iterative methods are beyond the scope of this book. Nonlinear models are, however, being incorporated into most of the larger statistical software packages.

INDEPENDENCE

Independence and random sampling are cornerstones of all biological investigations, whether they are later analyzed by uni- or multivariate methods. In studies of biological populations in which the population of interest is hard to define and sampling procedures may affect the probability of an individual being included in the sample, truly random sampling is difficult to achieve. Behavioral responses may vary among sexes, age classes, locations, and the like, rendering the gathering of a truly random sample highly unlikely. Thus we should tightly restrict the definition of the population sampled. For example, populations designated by sex, age, time, and location should provide reliable interpretations. When these factors are ignored, unequal representation of the factors between samples might imply differences that do not exist between populations (Pimentel 1979,176). Unfortunately, few published studies specifically or adequately define their statistical, sampled population. Failure to define the population has clear and adverse management implications: users will be unable to appropriately apply results to the proper time and place, thus overextending results beyond the proper target population (Tacha et al. 1982).

Exploratory versus Corroborative Analyses

Scientists often collect wildlife data in observational studies not specifically designed to test a statistical hypothesis. Here, the researcher does not have sufficient information to state and test a priori a null hypothesis. Analysis of such data is often called *exploratory analysis.* In contrast, if a researcher has some prior information regarding the theoretical structure of the data—such as that gathered from an earlier exploratory study—then *confirmatory studies and analyses* are performed. If data fail to meet the assumptions associated with the particular method(s) used, however, then results of these studies are also stated in terms of "exploratory" or "descriptive" analyses. That is, because formal tests of null hypotheses require that all assumptions be met (or at least not grossly violated), one cannot technically test these hypotheses under violations of assumptions. Unfortunately, many researchers use this rationale—calling their results descriptive—as an excuse for not gathering an adequate number of samples in the first place. Authors must clearly state the severity of the vio-

lations of assumptions, provide details on the specific form of these violations, and interpret how these violations could bias results, interpretations, and conclusions.

Classification of Multivariate Techniques

We can divide multivariate techniques into two general categories:(1) one-group, data reduction or ordination procedures, also called *dependence models;* and (2) two-or-more group, classification procedures, also called *interdependence models.* Both categories, while containing primarily parametric procedures, also include nonparametric methods. Multivariate techniques can also be subdivided into linear and nonlinear techniques.

A chart (table 6.2) of many of the most common statistical methods reveals the interrelated nature of multivariate methods. Methods are broadly classified by the number of dependent and independent variables of interest and the goal of the researcher analyzing the data. That is, does the researcher wish to examine the structure, or interdependence, of variables (PCA); to examine the relationship (correlation) among variables (multiple R); to separate groups (MANOVA); or to develop predictive equations (DA)? Dillon and Goldstein (1984, fig. 1.5–1) and Harris (1985, table 1.1) gave similar classifications of multivariate techniques. Below we briefly outline some of the specific methods found within these categories and provide examples of the applications to wildlife–habitat data. Because most studies of wildlife–habitat relationships have concentrated on principal components analysis (PCA), multiple regression (MR), multivariate analysis of variance (MANOVA), discriminant analysis (DA), and logistic regression (LR), we will concentrate our discussion on how these methods are used and interpreted. There are many additional analyses, such as detrended correspondence analysis (DCA), non-

Table 6.2. Classification of common statistical techniques

Major research question	Number of dependent variables	Number of independent variables	Analytical method[a]
Degree of relationship between variables	One	One	Bivariate r
		Multiple	Multivariate R
	Multiple	Multiple	Canonical R
Significance of group differences	One	One	One-way ANOVA; t-test
		Multiple	Factorial ANOVA
	Multiple	One	One-way MANOVA; T^2
		Multiple	Factorial MANOVA
Prediction of group membership	One	Multiple	One-way DA; logistic regression
	Multiple	Multiple	Factorial DA
Structure	Multiple	Multiple	PCA; FA; metric and nonmetric multidimensional scaling; correspondence analysis

Source: Tabachnick and Fidell 1996. Reprinted with permission of Pearson Education Copyright 1983.
[a]Analysis of variance (ANOVA); multivariate analysis of variance (MANOVA); Hotelling's T^2; discriminant analysis (DA); principal components analysis (PCA); factor analysis (FA).

metric multidimensional scaling (NMDS), and reciprocal averaging (RA), that could be applied to habitat analysis. Capen (1981), Verner et al. (1986), and Scott et al. (2002) should be consulted for many more specific examples.

Data Structure: Ordination and Clustering

Methods within this broad category of analyses seek to reduce a complex data set (i.e., many variables) to a few number of dimensions (axes) that are internally correlated but unique (not correlated) with regard to other derived dimensions. Ordination and clustering are similar in that no groups are assumed to exist prior to analysis. Because wildlife biologists usually collect data on numerous, usually intercorrelated variables, we need a method of reducing the number of variables to a manageable level. A common technique involves a two-step procedure. First, variables that are highly intercorrelated (e.g., an r of >0.7) are identified. Second, for each correlated pair identified, the one with the least power of separating groups is removed

from the analysis (as identified through a t- or F-test). Unfortunately, unless the excluded variable had a perfect correlation with the included variable, this method results in some degree of loss of information from the data set. The lower the correlation (r), the greater the loss. Thus "data reduction" techniques that retain all the information in the variables are desirable.

Principal components analysis is a method that identifies new sets of orthogonal (mutually perpendicular and thus not correlated) axes in the direction of greatest variance among observations. The first axis is the line in a direction through the observations that is oriented such that the projections of the observations onto the axis have maximum variance. The second axis is in the direction of the greatest variance perpendicular ("orthogonal") to the first axis—that is, that does not duplicate the variance "explained" in the first axis. Additional axes are derived until all of the explainable variance is accounted for.

Kelt et al. (1994) used multivariate analyses of distributions of small mammals to identify im-

199

portant habitat associations in a temperate rain-forest in Chile. They centered their vegetation sampling at each trap station and collected during winter and summer detailed data on canopy, shrub, and ground cover (table 6.3). They then used principal components analysis (PCA) to identify vegetation variables that helped distinguish the plots. In summary, variables for ground cover were important descriptors for both seasons, and shrub and tree variables were relatively less important in describing the vegetation. The relative importance of characteristics of ground cover appeared greater in winter than in summer, when other variables were of generally equal importance (see table 6.3). Eigenvalues greater than 1 were retained for ecological evaluation. Kelt and his colleagues then plotted captures of rodents onto the PCA-reduced habitat dimensions and showed that the species tended to overlap considerably in their use of habitat characteristics (fig. 6.7). This study serves as a good example of how PCA can be used to assist in data reduction (i.e., to simplify interpretation of the many correlated vegetation variables), then to help understand how the community of small mammals separate along the PCA axes, and then to interpret the principal causes of such separation in terms of the original variables.

Ihl and Klein (2001) used snow depth, snow hardness, above-snow graminoids, above-snow hummocks, and vegetation cover variables in a PCA to assess the relative importance of variables in the separation of muskoxen (*Ovibos moschatus*) and reindeer (*Rangifer tarandus*) ranges, feeding sites, and cratering (feeding) microsites. When plotted, the relative position of muskox and reindeer in "habitat space" can be depicted and conclusions drawn over the similarity and differences in habitat use. For example, figure 6.8 shows that broad overlap exists between the species in feeding sites, while relatively less overlap exists in general habitat use and

crater sites. Additionally, this figure allows interpretation of the variables responsible for the separation both between general categories of activity (e.g., feeding site versus crater site), as well as more subtle differences within each category (e.g., muskox feeding site versus reindeer feeding site). Note, for example, that feeding sites are separated from crater sites along PC3 but not along PC1, whereas feeding sites and general habitat are not separated along PC3 but are separated along PC1. For this latter comparison, habitat is typified by relatively deeper snow and more graminoids than crater sites.

General Interpretations

It is often difficult to determine which variables are involved with what principal component and to what degree. By comparing component loadings, we can determine those variables most related to a component. Dillon and Goldstein (1984, 69) observed that it is easiest to adopt heuristics for the purpose of interpreting the pattern. The procedure given by Dillon and Goldstein can simplify interpretation considerably. The results in tables 6.4 and 6.5 serve as examples.

1. Starting with the first variable and first component and moving horizontally from left to right across the components, circle the variable loading with the largest absolute value (e.g., in table 6.5, the "litter" variable has the highest loading with PC1). Then find the second highest variable on any component and circle it; continue this procedure for each variable.

2. After evaluating all the variables, examine each loading for "significance." This assessment can be made on the basis of the statistical significance (see table 6.5) of the correlation coefficient (loading) or on the basis of the practical significance (table

Table 6.3. Eigenvectors for the significant components resulting from a principal components analysis conducted on summer and winter vegetative parameters.

Season parameter	PC1	PC2	PC3	PC4	PC5	PC6	PC7
				Summer			
Eigenvalues	3.570	2.790	2.270	1.640	1.330	1.160	1.020
Canopy cover	−0.166	0.180	0.003	−0.189	0.018	0.408*	−0.331*
Number of logs	0.206	−0.144	−0.212	0.274*	−0.285*	0.250*	0.007
Distance to nearest shrub	−0.015	0.197	−0.023	0.491*	0.163	−0.294*	−0.084
Width of nearest shrub	0.016	0.243	0.457*	0.208	−0.004	0.306*	0.146
Height of nearest shrub	0.128	0.212	0.415*	0.213	0.095	0.373*	−0.110
Number of species of shrubs	−0.092	0.063	0.226	−0.004	−0.199	0.086	0.754*
Shrub cover	0.273*	−0.116	0.251*	−0.192	−0.009	0.112	−0.293*
Number of trees	−0.283*	0.154	−0.204	−0.253*	0.292*	0.132	0.049
Number of species of trees	−0.174	0.177	−0.228	−0.179	0.459*	0.177	0.233
Bare ground cover at 2 m	0.204	0.367*	0.012	−0.254	−0.204	−0.167	0.074
Herbaceous ground cover at 2 m	−0.395*	−0.217	0.203	0.147	0.015	−0.068	−0.168
Tree-plus-trunk ground cover at 2 m	0.276*	−0.117	−0.300*	0.117	0.169	0.254*	0.143
Bare-ground cover at 4 m	0.260*	0.366*	0.076	−0.283*	0.023	−0.118	−0.086
Herbaceous ground cover at 4 m	−0.421*	−0.215	0.135	0.030	−0.121	−0.045	0.018
Tree-plus-trunk ground cover at 4 m	0.256*	−0.135	−0.215	0.278*	0.227	0.207	0.109
Soil hardness	−0.175	0.154	0.136	0.172	0.409*	−0.035	−0.010
Foliage density at 15 cm	−0.024	−0.360*	0.096	−0.228	0.043	0.321*	−0.074
Foliage density at 30 cm	0.120	−0.328*	0.185	−0.217	0.153	0.042	0.175
Foliage density at 50 cm	0.187	−0.242	0.245	−0.189	0.241	−0.262*	0.159
Foliage density at 100 cm	0.235	−0.139	0.206	0.068	0.400*	−0.230	−0.064
				Winter			
Eigenvalues	3.380	2.360	1.530	1.040			
Canopy cover	−0.060	−0.117	0.299*	−0.162			
Logs	0.196	0.303	0.020	0.254*			
Distance to nearest shrub	0.122	−0.177	−0.132	0.746*			
Shrub cover	−0.020	0.392*	−0.121	−0.493*			
Number of species of shrubs	0.045	0.071	0.097	−0.035			
Number of trees	−0.102	−0.390*	0.472*	−0.073			
Number of species of trees	0.010	−0.390*	0.446*	−0.085			
Bare-ground cover at 2 m	0.407*	−0.299*	−0.202*	−0.180			
Herbaceous ground cover at 2 m	−0.494*	0.059	−0.025	0.112			
Tree-plus-trunk ground cover at 2 m	0.185	0.394*	0.385*	0.089			
Bare-ground cover at 4 m	0.448*	−0.163	−0.169	−0.140			
Herbaceous ground cover at 4 m	−0.502*	−0.020	−0.110	0.038			
Tree-plus-ground cover at 4 m	0.184	0.352*	0.465*	0.151			

Source: Kelt et al. 1994, table 3. Reprinted with permission of Alliance Communication Group a division of Allen Press Inc.
*Denotes loadings ≥0.25.

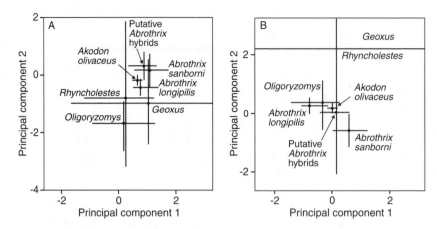

Figure 6.7. Results of a principal components analysis of habitat variables associated with capture sites of small mammals at La Picada, Chile. (A) Mean principal component scores during summer 1983–1984. (B) Mean principal component scores during winter 1984. Projections are 95% CI. (Reproduced from Kelt et al. 1994, fig. 2; reprinted with permission of the Alliance Communication Group a division of Allen Press Inc.)

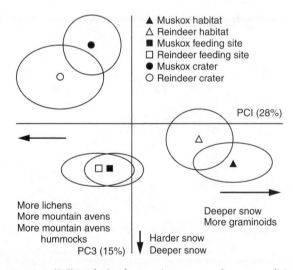

Figure 6.8. Principal component (PC) analysis of vegetation cover and snow conditions on muskoxen and reindeer habitats, feeding sites, and craters on the northern Seward Peninsula, Alaska, during late winter. Means and standard error ellipses for PC1 versus PC3. (Reproduced from Ihl and Klein 2001, fig. 4; reprinted with permission of The Wildlife Society.)

Table 6.4.   Summary of the first four principal components

Variable	Mnemonic	PC1	PC2	PC3	PC4
Eigenvalue		3.72	1.79	1.56	1.29
Percentage of variance		28.6	13.7	12.0	9.9
Percentage ground cover	GC	0.111	0.577	0.261	−0.051
Percentage shrub cover	SC	−0.532	−0.314	0.160	0.551
Percentage conifer	CO	0.238	0.709	−0.423	0.258
Canopy height	CH	−0.739	0.466	−0.281	0.111
Number of tree species	SPT	0.056	−0.510	−0.016	0.508
Trees 7.5–15 cm dbh	T1	0.677	0.019	0.264	0.299
Trees 15.1–23 cm dbh	T2	0.876	0.205	−0.154	0.160
Trees 23.1–30 cm dbh	T3	0.818	0.059	−0.368	−0.032
Trees 30.1–38 cm dbh	T4	0.267	−0.196	−0.549	−0.601
Trees 38.1–53 cm dbh	T5	−0.397	−0.076	−0.711	0.002
Trees 53.1–68 cm dbh	T6	−0.657	0.208	−0.167	0.270
Trees >68 cm dbh	T7	−0.511	0.169	0.166	0.175

Source: Collins 1983, table 2. Reprinted with permission of the American Orthinologists' Union.

Table 6.5.   Results of principal components analysis using weighted averages of eight habitat variables for thirty-four bird species

Component	PC1	PC2	PC3
Variation explained	67.46%	23.47%	5.50%
Cumulative variation	67.46%	90.93%	96.43%

Variable	Correlations with original variables		
Litter	0.85**	0.23	0.42**
Slash	0.38*	0.86	0.32
Herbaceous vegetation	−0.93**	−0.29	0.07
X height of herbaceous vegetation	−0.96**	−0.01	−0.04
Canopy layers	0.98**	−0.14	0.01
Maximum canopy height	0.90**	−0.41*	0.08
Canopy cover	0.95**	0.11	−0.20
Trees <12.7 cm	0.27	0.90**	−0.33*

Source: Maurer et al. 1981, table 3.
$*P < 0.05.$
$**P < 0.01.$

6.4). For statistical significance, in most instances with sample sizes less than 100, the smallest loading would have to be greater than 0.3 in order to be considered significant. Practical significance involves some reasonable or practical rule for the minimum amount of a variable's variance that must be accounted for by a component. Practical significance is especially important in ecological studies in which biological significance may require a much higher loading than indicated by the associated statistical significance (the latter of which is largely based on sample size). Significant loadings should be underlined.

3. Examine this pattern matrix to identify the variables that have not been circled and therefore do not "load" on any component. The researcher then must decide whether to rest the analysis only on those variables with significant loadings or to evaluate critically each variable with regard to the research objective and biological knowledge.

4. On the basis of the results of the previous steps, attempt to assign a biologically meaningful interpretation to the pattern of component loadings. Variables with higher loadings have greater influence; variables with negative loadings have inverse influence. Assign a name that reflects, to the extent possible, the combined meaning of the variables that load on each such component. In table 6.4, for example, PC1 could be interpreted as a forest-height component separating areas with large trees (T6, T7), tall canopies (CH), and a low amount of shrub cover (SC) from areas characterized by smaller trees (T1–T3; see Collins 1983). Note that both the magnitude and direction (+ or –) of the loading were used to infer this relationship.

In practice, having many variables with moderate-sized loadings complicates this step.

Various other parametric and nonparametric methods fall within this category of multivariate techniques, including factor, principal coordinates, and correspondence analyses; nonmetric multidimensional scaling; cluster analysis; and their relatives. Miles (1990), for example, compared results using several of these methods. The nonparametric techniques are, of course, designed for situations in which data are highly nonlinear, or when sample sizes are too low for normality-linearity to be adequately determined. No method is without drawbacks, and further, few people have much practical experience with analysis of data using nonparametric techniques (i.e., even many of the more popular statistical packages include only a few of these methods). Thus we recommend that students proceed cautiously when selecting an analytical technique. Unless your data depart substantially from parametric assumptions, it may be better to stick with the more common and well-understood techniques. Because of the qualitative nature of the interpretation of the output from all these multivariate methods, one must fully understand how a technique operates in order to assign meaningful biological interpretation to the results. The greatest problem involves not so much the specific method used, but rather the absence of any follow-up confirmatory or validation studies (e.g., Marcot et al. 1983; Raphael and Marcot 1986; Fielding and Haworth 1995). Even if multiple techniques are used to "confirm" study results, their use does not justify inadequate sample sizes, biased sampling methods, or gross violations of statistical assumptions. It is best to remember that the components derived from these techniques cannot be properly considered to be "niche dimensions" or "habitat dimensions" except by means of an arbitrary, operational decision by the researcher (Wiens 1989, 65).

Assessing Relationships: Multiple Regression Analysis

Regression analysis is the most widely used method of data analysis. Regression provides three general types of results. First, regression can predict or estimate one response variable from one or more predictor variables. The *estimated variable* is called a predicted, criterion, or (most commonly) *dependent variable.* The one or more variables that estimate a dependent variable are termed predictors, covariates, or (most commonly) *independent variables.* Second, regression analysis can determine the best formula (based on available data) for predicting some relationship. Third, the success (precision) of a regression analysis can be ascertained, usually through use of correlation coefficients (Pimental 1979, 33).

Although popular, regression analysis has many problems associated with its use that can substantially bias biological interpretations of data. Pimentel (1979) provides an especially sobering review of regression analysis. As in all multivariate techniques, the multivariate extension of simple linear regression magnifies these problems.

Draper and Smith (1981, ch. 8) provided a flow diagram of the steps necessary to ensure proper development of predictive models, a frequent goal of biologists using multiple regression (MR) (fig. 6.9). Their diagram identifies three primary stages in such a plan: planning the analysis, developing the models, and verifying (validating or testing) the initial model outputs.

Regression analysis allows us to identify how much of the observed variation in the dependent variable is explained by the independent variables and how much is not (the error term, e). A measure of the relative importance of each of these sources of variation (independent variables) is termed the *coefficient of multiple determination, R^2. R^2* ranges from 0 for no linear relationship to 1 for a perfectly linear relationship. The value of R^2 is thus a measure of the explanatory value of the linear relationship (Wesolowsky 1976, 43).

However, when sample sizes are small in relation to the number of parameters fitted (number of independent variables), it is possible to get a large R^2 even when no linear relationship exists. Conversely, a low value does not necessarily indicate a "bad" relationship. Here, one or more individual coefficients may be significant, and the corresponding parameters rather than the overall regression model may be of primary interest.

In addition, a low R^2 may simply show insufficient variation in mean values. Here again we see the overriding role that spatial scale plays in determining the results of our studies. In figure 6.10A, we see that fitting a line through all of the points results in a good model for predicting the abundance of a bird in relation to tree density. However, when our interest becomes more site-specific (fig. 6.10B), we see that tree density alone is a poor predictor of bird abundance. As we move increasingly into the realm of microhabitat analysis, finer, site-specific measures of an animal's habitat must be measured to explain its abundance. Our simple, univariate example illustrates the relationship between sampling scale and results of regression analysis. You can almost always find a significant regression result by sampling across a wide enough range of conditions—for example, by including young-growth forest in an analysis of the relationship between numbers of canopy-dwelling birds and forest structure.

There are many specific ways in which your data can be processed and statistically significant variables identified within the broad category of

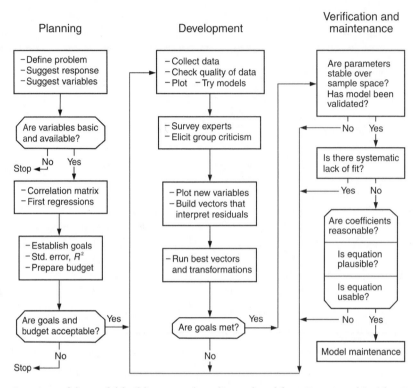

Figure 6.9.   Summary of the model-building procedure. (Reproduced from Draper and Smith 1981; reprinted with permission of John Wiley and Sons Inc.)

multiple regression. The value of these different variable selection procedures—such as forward or backward inclusion, stepwise, and all-possible-subsets—is beyond the scope of this chapter. Additionally, to us the way in which variables are entered into a multiple regression is far less important from the standpoint of ecology than is the way in which the data were collected in the first place. There are numerous examples of the application of different variable selection procedures as applied to wildlife data in the literature. For example, Welsh and Lind (2002) used all-possible-subsets regression to examine differences in the densities of stream amphibians among occupied sites relative to

various independent variables describing stream condition.

Separation: Multivariate Analysis of Variance

Univariate analysis of variance, or ANOVA, examines the effects of one or more factors on only one variable. There are, however, numerous applications in ecology in which we wish to examine the effects of one or more factors on more than one variable—for example, when we are studying the effects of various measurements of the environment (i.e., habitat variables). Thus where we have more than one measurement on

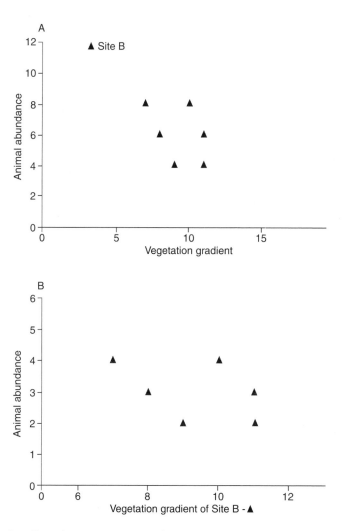

Figure 6.10. Examples of how the extent, or range, of an environmental gradient from which samples are taken can influence conclusions. (A) Sampling from an arbitrary, relatively wide vegetation gradient shows a positive relationship between animal abundance and vegetation. (B) Sampling from only a subset of the gradient ("Site B") results in a negative abundance–vegetation relationship.

each individual, we have a multivariate analysis of variance, or MANOVA.

The most commonly encountered MANOVA statistic is termed Wilks' lambda. Unlike the typical univariate test statistic (e.g., t, F), the null hypothesis is rejected for small values of Wilks'

lambda. Lambda ranges from 0 to 1 and measures the amount of variability among the data that is not explained by the effect of the levels of the factor. Thus a measure of the variability among the data that is explained by the experimental factor is 1 – lambda. Computer programs usually present lambda that has been transformed into a value for F, or chi-square with an associated P, and in such cases a large F of chi-square is associated with a small P. Wilks' lambda is the most commonly used MANOVA test statistic, although other measures are also available (Zar 1999, ch. 16).

MANOVA is used in wildlife ecology when a specific experimental design has been developed to test a stated hypothesis. For example, Schmid et al. (2003) used MANOVA to test whether differences in use and availability proportions of Kemp's ridley turtle (*Lepidochelys kempii*) habitat (i.e., benthic substrates) were significantly different from zero. Bond et al. (2002) used MANOVA on land-cover-type data matrices (used or available) as dependent variables and on group (used or available land cover type) and individual cottontails as independent variables.

Classification: Discriminant Analysis

Discriminant analysis (DA) is widely applied throughout the scientific disciplines, including wildlife science. DA refers to a general group of methods, each of which has slightly different objectives. The overall goal of DA, however, is the classification of individuals into specific groups (such as species, vegetation types). Researchers can use methods of DA to evaluate similarities and differences among sites or individual samples; it thus resembles PCA in its ordination capabilities. Unlike PCA, however, DA starts with sets of groups (two or more) and a sample from each group. Thus, while the goals of PCA and some applications of DA are similar, the experimental designs for collecting data differ markedly.

Dillon and Goldstein (1984, ch. 10) provided an excellent description of how DA works that we will summarize here, along with material from Pimentel (1979, ch. 10) and Neff and Marcus (1980). Under DA, researchers evaluate one categorical dependent variable and a set of independent variables. Although there is no requirement that these independent variables be continuous in nature, DA often performs poorly when independent variables are categorical. The categorical dependent variable is a grouping factor that places each observation into one and only one predefined group.

For example, we might be interested in examining differences among species or study sites on the basis of a series of environmental characteristics. After all individuals are assigned to these groups, we might further wish to "discriminate" among the groups based on the values of the independent (predictor) variables (the habitat characteristics). DA is thus a method of separating groups based on measured characteristics, and determining the degree of dissimilarity of observations and groups and the specific contribution of each independent variable to this dissimilarity (as in the variable loadings described for PCA).

After developing the linear discriminant functions, the researcher can then use these functions to identify or classify "unknowns" into the group predicted by the discriminant analysis. Such a classification analysis is often used to determine how well the DA can identify members of the groups used; table 6.6 presents a simple example. In table 6.6, species not well separated from each other by the DA will show high classifications (misclassifications) for a different species; here, many orange-crowned warblers were misclassified as MacGillivray's warblers, indicating high overlap between the habitats used by two of the species in this study.

INTERPRETATION OF
DISCRIMINANT ANALYSES

The use of discriminant loadings will give the most straightforward and useful interpretation of a discriminant analysis. Similar to principal component loadings, *discriminant loadings* give

Table 6.6.   Classification matrix derived from a discriminant function program showing actual and predicted species (group) membership for singing male warblers based on habitat use on the deciduous and nondeciduous tree sites.

Actual group	Predicted group membership (in percentages)*		
	Orange-crowned	MacGillivray's	Wilson's
	Deciduous tree sites		
Orange-crowned warbler	*13*	48	39
MacGillivray's warbler	9	*88*	3
Wilson's warbler	26	15	*59*
	Nondeciduous tree sites		
Orange-crowned warbler	*16*	58	26
MacGillivray's warbler	16	74	10
Wilson's warbler	26	13	*61*

Source: Morrison 1981, Table 5. Reprinted with permission of the American Ornithologists' Union.
*Italicized numbers denote percentages correctly classified.

the simple correlation of an independent variable with a discriminant function. Most statistical packages will produce discriminant loadings.

A stepwise selection procedure is commonly used to reduce the set of independent variables that best separate the groups being considered. Stepwise procedures function similarly for all multivariate techniques, and they are independent of the actual multivariate analysis. That is, they "screen" variables by a set of statistical criteria to determine if they should be allowed to enter into the actual multivariate calculations. In DA, an *F*-test is usually used to order each variable by its ability to separate the groups. Depending on the *F*-value (or corresponding *P*-value) used, variables are entered into the analysis until some user-defined cutoff is reached (e.g., $P < 0.05$ or 0.1). There are many modifications of the stepwise procedure and several other less used but potentially more appropriate (e.g., all-possible-subsets) methods for variable entry. Researchers should review Dil-

lon and Goldstein (1984, 234–42) regarding specific steps to take when using variable selection procedures.

Table 6.7 shows a typical presentation of results of a discriminant analysis. For the "all species" analysis, note that two significant functions were derived. Box's *M* is the test of the equality of the variance-covariance matrices among species—the multivariate equivalent of the univariate test of equality of variances. Here, Box's *M* was significant for all comparisons, meaning that the species did have different variances (in multivariate habitat use). A significant Box's *M* means that the formal test of the null hypothesis of equality of group centroids (i.e., are the species different in multivariate habitat use?) is technically invalid. As noted earlier, we should expect that species differ in the way they use habitat characteristics, both in terms of their mean use and the distribution of values about the mean. Thus transforming data to meet normality results in loss of ecological information

Table 6.7.   Discriminant analysis of small mammal habitat use, western Oregon

Group discriminant function	Relative variance (%)	Cumulative variance (%)	Wilks'	X	df	P	Box's $M(P)$
All species							
DF I	75	75	0.70	93.5	12	<0.001	
DF II	25	100	0.91	24.8	5	<0.001	<0.001
Rodents only[a]							
DF I	98	98	0.72	68.8	6	<0.001	<0.001
Sorex species only							
DF I	100	100	0.79	9.0	2	0.011	<0.001

Source: Morrison and Anthony 1989, table 4. Reprinted with permission of Alliance Communication Group a division of Allen Press Inc.
[a]DF II accounted for 2% of the variance and was not significant at $P = 0.45$.

about the animals. We are thus left with a dilemma: Should we try to meet formal statistical assumptions, or should we let the data speak for themselves? In the face of such a situation, most biologists present their results as "descriptions" of the habitat use of the species, rather than as formal tests of null hypotheses. The important points are to make sure that you have an adequate sample size (so you know that the distribution of samples is accurate) and that you fully discuss violations of assumptions.

As noted by Wiens (1989, 65), just because a discriminant analysis derives a primary multivariate dimension that is statistically significant and thus provides separation in habitat use among the species measured, it does not necessarily follow that this dimension represents the primary means of habitat separation in the community. Wiens (1989, ch. 9) provides additional examples and cautions concerning the use of discriminant analysis and other multivariate procedures as applied to wildlife studies.

Findley and Black (1983; see also Findley 1993) summarized their conception of how a Zambian insectivorous bat community would appear in multivariate space (fig. 6.11). Their drawing (see fig. 6.11) provides a good example of how projections of species can be interpreted;

this interpretation applies to most methods. Each sphere represents the morpho- or ecospace occupied by one species. The volume of the sphere equals the total niche volume and the diameter the amount of intraspecific variation, while the center is the species' centroid (i.e., the mean value of all its morphological or ecological variables). Overlap is shown if the spheres intersect each other; overlap is a function of both interspecific distance and intraspecific variability. In figure 6.11, the region of the community centroid is occupied by a number of closely packed species with low intraspecific variability (i.e., narrow niches).

Rexstad et al. (1988) conducted a simulation study to examine, in detail, the ramifications of violating the assumptions of discriminant analysis. Although some of their conclusions should be tempered (see Taylor 1990; also Rexstad et al. 1990), the message contained in their paper remains: there is little published evidence to suggest the widely held belief that discriminant analysis is robust to violations of the variance–covariance matrix, or that using the analysis for "descriptive" purposes results only in ecologically meaningful analyses. Because of these shortcomings, many people are now using logistic regression analysis as a replacement for dis-

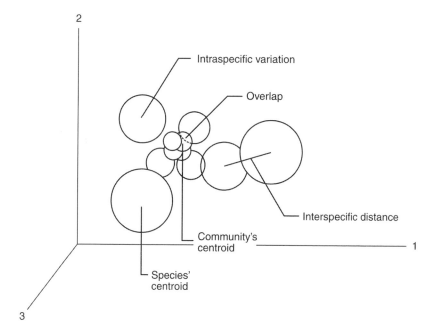

Figure 6.11.   Tentative model of a community in attribute space. The attributes could be morphology or diet, or perhaps other physiological, behavioral, or ecological parameters. The community is composed of some closely packed, rather than invariable, species, and some more distant, more variable kinds. (Reproduced from Findley and Black 1983, fig. 1; reprinted with permission of the Ecological Society.)

criminant analysis; logistic regression is the topic of our next section.

Brown and Krausman (2003), for example, used DA to separate habitat use of three leporid species in southeastern Arizona. Creosote, mesquite, bursage, and thermal cover >1.5 m tall distinguished among species along the first discriminant axis for form (above-ground resting sites) and activity of the three species (fig. 6.12). Palo verde, saguaro, thermal cover plants <0.5 m and >1.5 m tall, and escape cover plants >1.5 m tall best distinguished among species along the second discriminant axis. Note, however, that two of the species (antelope [*Lepus alleni*] and black-tailed jackrabbit [*L. californicus*]) broadly overlapped, whereas the desert cottontail (*Sylvilagus audubonii*) overlapped minimally with the other species.

Logistic Regression Analysis

Logistic regression analysis (LR) has been gaining increased use in recent years as a nonparametric, nonlinear alternative to two-group discriminant analysis. LR can be used to analyze independent variables that are true categorical data (such as coat color) and those that have been summarized into categories (such as height intervals); it can also analyze continuous data. LR can also be used to develop predictive models.

Nadeau et al. (1995) used logistic regression to create a habitat model based on muskrat (*Ondatra zibethicus*) presence along wetlands of James Bay, Quebec. They developed models based on the presence of burrows only and another using the presence of muskrat feeding sign and droppings as dependent variables. Independent

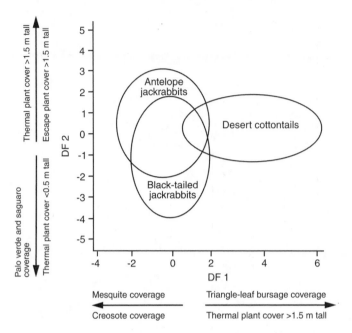

Figure 6.12.   Distribution of form scores showing separation of three leporid species, southeastern Arizona, along two discriminant axes (DF 1 and DF 2) based on 22 continuous habitat characteristics used in discriminant analysis. Ellipsoids show the standard deviation of form scores along each axis. (Reproduced from Brown and Krausman 2003, fig. 1 reprinted with permission of The Wildlife Society.)

variables were a mix of continuous (e.g., water depth, floating plant cover) and categorical (e.g., bank slope, dominance of peat soil) variables. Chandler et al. (1995) used LR to predict the probability of bald eagle use of a shoreline segment on the basis of a series of habitat variables. Brennan et al. (1986) showed that LR could develop more rigorous habitat models than other linear techniques under certain conditions.

LR is potentially more robust than similar parametric procedures to deviations from multivariate normality and equal covariation. Further, various studies have found that LR provides better between-group separations and classification success than does discriminant analysis. Further, the nonlinear tendencies of ecological data are usually distorted by linear analyses.

Thus the nonparametric procedures should provide a more meaningful interpretation of ecological phenomena when linear models fail to adequately explain variation (see Efron 1975; Press and Wilson 1978; Brennan et al. 1986).

LR does have several limitations. First, software with LR routines has not been as widely available as that with parametric procedures, including discriminant analysis and multiple regression. Although LR gained in popularity during the 1990s, relatively few people have direct experience with running and interpreting LR, including resource managers. Second, all commercially available software packages limit LR to the two-group situation. Thus, even within a single study, the researcher often needs to run both two-group (e.g., use vs. nonuse) and greater than two-group (e.g., comparison of community separation) analyses. Clearly, mixing analyses makes comparison of results difficult. The choices of method should be carefully evaluated and then clearly elucidated in all research papers.

For example, Martin and Morrison (1999) used LR to compare used areas to available areas (habitat) and to nest sites for buff-breasted flycatchers (*Empidonax fulvifrons*) in southeastern Arizona. Among other results, they found that successful nest sites differed from unsuccessful nest sites in having more live vegetation 0–1 m high; less silverleaf oak (*Quercus hypoleucoides*) cover 1–2 m high; and more species of shrubs. The LR model correctly classified 52% of successful nest sites, 85% of unsuccessful nest sites, and 72% of all the sampling plots. Johnson et al. (2002) used LR to develop classification models for amphibian species in the Great Lakes region. Their best-fit LR models for the individual species tested ranged from 56% to 85%.

Cautions and Solutions

As discussed in our opening comments in this chapter, and in more detail in chapters 5 and 8, there are important limitations to the pattern-seeking, descriptive methods of habitat use typically found in multivariate analyses to which researchers need attend. Kaufman and Kaufman (1989) aptly summarized the cautions that should be applied to results of descriptive studies:

1. Statistical separation of species on a habitat axis does not prove that the animals recognize and respond to the characteristics measured; best, it provides testable hypotheses to evaluate with experimentation.
2. Even though average positions along habitat axes occupied by a series of species

are different, interspecific overlap may be considerable and must be explained (this caution would also apply to intraspecific, sex–age comparisons).
3. Most studies are not replicated in time or space, so that the generality of most studies is unknown.
4. Differences in resource use do not necessarily elucidate the mechanisms that are ultimately responsible for the patterns observed.
5. Understanding cause and effect of observed patterns of habitat use will require experimental manipulations and not just additional studies using refined descriptions of habitat use (chapters 4 and 5).

It is critical that students carefully examine the magnitude of differences for each independent variable between groups in a multivariate analysis. Remember that the variable(s) entered and the order of entry are primarily a statistical decision based on some selection criterion (e.g., F-value). A variable with a small difference in means and small variance could have a larger F-value and be entered into an analysis before a variable with a large difference in means but also a relatively large variance. As discussed earlier, biologists tend to concentrate on statistical significance rather than on an ecological interpretation of means *and* variances of variables prior to entry into an analysis. Thus the ecological interpretation made after an analysis in such situations can be misleading. Multivariate methods worsen this tendency because of the "black box" that field data disappear into and the myriad of statistical parameters that return on the printout. The high natural variance usually seen in biological populations makes application of models to other places and times difficult (see Fielding and Haworth 1995 for a detailed analysis and discussion).

Another concern in wildlife analyses in general, and in habitat studies in particular, is *spatial autocorrelation*. Most habitat features do not occur independently of surrounding features—that is, where an animal occurs may be based on the features at that point in space, as well as on nearby features. Recall our earlier discussions about how habitat is usually measured in observer-defined areas around an animal. Marzluff et al. (2004) reviewed this issue and presented a new method for analyzing resource selection that incorporates a probabilistic measure of use, known as the *utilization distribution*. Rather than assuming that resource use is uniform within a home range boundary, differential use can be quantified using the utilization distribution.

Multicollinearity is another issue that confounds habitat analyses. *Multicollinearity* occurs when explanatory variables are correlated among each other. Multicollinear explanatory variables are difficult to analyze because their effects on the response variable can be due either to true synergistic relationships among the variables or to spurious correlations. Graham (2003) reviews this subject in detail and offers several relatively simple solutions.

Model Selection

Much of the work conducted by ecologists is exploratory in nature and results in extremely large data sets composed of numerous variables. Historically, ecologists have subjected these data sets to numerous statistical procedures, often beginning with various data reduction techniques (as we have reviewed above). Further, the multivariate methods outlined in this chapter are often used to create multiple habitat models and descriptions. Thus we need unbiased methods for selecting the "best" of what is often a host of potential results. A summary of the various model selection techniques is presented in Burnham and Anderson (2002, 35–37). Available methods include stepwise procedures, cross-validation, goodness-of-fit tests, and the adjusted R^2. All of these approaches have strengths and weaknesses and are applicable according to the form of the data.

Another method that has gained wide use recently is Akaike's information criterion (AIC). Briefly, AIC selects the best-fit model without overfitting the data with too many parameters. The model with the smallest AIC is the best approximating model for the data. Note that AIC does not determine if the model is "good" ecologically; it only helps select among models that were developed by the scientist (Burnham and Anderson 2002, 62); this statement applies to all model selection procedures. Anderson and Burnham (2002) discuss many of the misuses of AIC. A few recent papers using AIC and related terms include Woodward et al. (2001); Kuehl and Clark (2002); and Otis (2002). Burnham and Anderson (2002) also provide many detailed examples.

As summarized by Burnham and Anderson (2002, 96), the best scenario is to develop a set of multiple working hypotheses based on a thorough understanding of the available literature and experience with the system (i.e., natural history). Then one uses scientific knowledge, experience, and expertise to define a set of a priori candidate models that represent each of the working hypotheses. These steps allow clear development of the research problem, followed by careful planning of the study design, including gathering of adequate sample sizes. Lastly, one can select appropriate statistical techniques, including model selection procedures, to compare the various hypotheses. We recommend that scientists concentrate more on estimations of effect size (magnitude of difference) and associated confidence intervals rather than overly relying on *P*-values.

Sample Size Analyses

Inadequate sample size is a chronic problem found in many papers using multivariate methods. Regardless of the refinement of study design and the care taken in recording data, proper ecological interpretation of multivariate analyses is difficult at best; interpretations based on inadequate samples are a waste of time and potentially misleading. Earlier in this chapter we identified methods of determining proper sample sizes in the univariate case; those methods apply here on a variable-by-variable basis. In multivariate analyses, however, our problems are increased by orders of magnitude, given our desire to simultaneously interpret ecological phenomena in many dimensions for many species.

Johnson (1981) outlined general guidelines for determining adequate sample sizes in multivariate studies, noting that more observations were needed when the number of independent variables was large. Many published studies, however, have only slightly more observations than variables; some have even fewer. Johnson thought that an appropriate minimum sample size for multivariate analyses was 20 observations, plus 3 to 5 additional observations for each variable in the analysis. We suggest that an additional 5 to 10 observations for each variable would provide a more conservative target for the sample size. Larger sample sizes do not, however, provide an answer for poorly designed studies or biased data. As Johnson (1981, 56) noted, "Calls for larger samples are the 'knee-jerk' reaction when variability is excessive."

In a study of habitat use by birds in Oregon, Morrison (1984b) found that a minimum of 35 plots was necessary to obtain stable results; stability was determined when means and variances did not change with an increase in sample size. Morrison's review of the wildlife literature showed, however, that very few studies met the minimum criteria established by Johnson (1981). In addition, few researchers even discussed the issue of sample size. Collecting "all the data that I could" is no excuse for publishing results based on inadequate sample sizes. A similar study by Block et al. (1987) found that even larger numbers of plots—up to 75 or more—were needed in their bird habitat analysis.

The minimum size of the samples needed is a study-specific question; the papers reviewed above give approximations of the range of samples you can expect to need. Remember, however, that these minimum sample sizes apply to each biological period being considered (e.g., winter, fall), and the appropriate n might vary among periods. All studies should include a justification of the amount of data used in the analyses.

Computer Statistical Packages

Previously only available on mainframe computers, numerous, powerful statistical packages are now also available for microcomputers. Unfortunately, the availability and ease of use of these packages has likely led to an increasing number of misuses of statistical methods. Novice users of these packages should be aware that the default settings for the specific analytical methods must often be adjusted for each application. Further, each package has a set of options and statistics that must be specifically requested. Many university computer centers offer short courses in the use of statistical packages; an increasing number of statistics departments are offering more in-depth courses on the use and interpretation of such software. Stauffer (2002) noted that a problem with multivariate statistics is that anyone capable of entering data in a form that can be analyzed by a computer package could conduct multivariate analyses, whether or not the data were appropriate for such an analysis. These problems have lessened in the literature over time as referees and editors have become more sophisticated in their knowledge of appropriate multivariate analyses (e.g., addressing assumptions, evaluating sample sizes).

The user's manuals for some of the packages offer descriptions of the analytical methods being used and step-by-step instructions on how to interpret results. These features are especially strong in the widely used SPSS software, which offers an excellent user's manual and both basic and advanced manuals that explain most statistical procedures. White and Clark (1994), in their review of microcomputing for wildlife, preferred the SAS software, primarily because of its useful programming capabilities, which are lacking in SPSS. Although the ability to program within SAS certainly offers long-term advantages to many users, it is an option that most wildlife biologists will seldom, if ever, use. We find that SPSS offers what most researchers require, and that the user's manuals make effective teaching tools. We have heard researchers express the view that SPSS is actually too easy to use, in the sense that "ease of use" equates with "ease of misuse." There is truth in this view, and we caution that the ability to run your data through a computer does not mean that you have necessarily done the analysis correctly. Another package within the SAS family of analytical tools is JMP, which has recently gained popularity as an alternative to the much larger number of analytical tools offered in SAS and is often used in university classes.

Several statistical and ecological textbooks also include examples using output from one or more of the statistical packages; some even provide a disk containing specialized programs for the analysis of ecological data (Berenson et al. 1983; Tabachnick and Fidell 1983; Afifi and Clark 1984; Harris 1985; Ludwig and Reynolds 1988; Kent and Coker 1994; Littell 2002; Dytham 2003).

Literature Cited

Afifi, A. A., and V. Clark. 1984. *Computer-aided multivariate analysis.* Belmont, CA: Lifetime Learning.

Anderson, D. R., and K. P. Burnham. 2002. Avoiding pitfalls when using information-theoretic methods. *Journal of Wildlife Management* 66:912–18.

Berenson, M. L., D. M. Levine, and M. Goldstein. 1983. *Intermediate statistical methods and applications: A computer package approach.* Englewood Cliffs, NJ: Prentice Hall.

Block, W. M., K. A. With, and M. L. Morrison. 1987. On measuring bird habitat: Influence of observer variability and sample size. *Condor* 72:182–89.

Bond, B. T., L. W. Burget Jr., B. D. Leopold, J. C. Jones, and K. D. Godwin. 2002. Habitat use by cottontail rabbits across multiple spatial scales in Mississippi. *Journal of Wildlife Management* 66:1171–78.

Brennan, L. A., W. M. Block, and R. J. Gutierrez. 1986. The use of multivariate statistics for developing habitat suitability index models. In *Wildlife 2000: Modeling habitat relationships of terrestrial vertebrates,* ed. J. Verner, M. L. Morrison, and C. J. Ralph, 177–82. Madison: Univ. of Wisconsin Press.

Brown, C. F., and P. R. Krausman. 2003. Habitat characteristics of 3 leporid species in southeastern Arizona. *Journal of Wildlife Management* 67:83–89.

Burger, L. D., L. W. Burger Jr., and J. Faaborg. 1994. Effects of prairie fragmentation on predation on artificial nests. *Journal of Wildlife Management* 58:249–54.

Burnham, K. P., and D. R. Anderson. 2002. *Model selection and multimodel inference: A practical information-theoretic approach.* 2nd ed. New York: Springer-Verlag.

Capen, D. E., ed. 1981. *The use of multivariate statistics in studies of wildlife habitat.* General Technical Report RM-87. Ft. Collins, CO: USDA Forest Service, Rocky Mountain Research Station.

Carnes, B. A., and N. A. Slade. 1982. Some comments on niche analysis in canonical space. *Ecology* 63:888–93.

Chandler, S. K., J. D. Fraser, D. A. Buehler, and J. K. D. Seegar. 1995. Perch trees and shoreline development as predictors of bald eagle distribution on Chesapeake Bay. *Journal of Wildlife Management* 59:325–32.

Collins, S. L. 1983. Geographic variation in habitat structure of the black-throated green warbler. (*Dendroica virens*). *Auk* 100:382–89.

Cook, C. W., and J. Stubbendieck, eds. 1986. *Range research: Basic problems and techniques.* Denver: Society for Range Management.

Cooley, W. W., and P. R. Lohnes. 1971. *Multivariate data analysis.* New York: Wiley.

Cooper, C. F. 1957. The variable plot method for estimating shrub density. *Journal of Range Management* 10:111–15.

Digby, P. G. N., and R. A. Kempton. 1987. *Multivariate analysis of ecological communities.* London: Chapman & Hall.

Dillon, W. R., and M. Goldstein. 1984. *Multivariate analysis: Methods and applications.* New York: Wiley.

Dodd, M. G., and T. M. Murphy. 1995. Accuracy and precision of techniques for counting great blue heron nests. *Journal of Wildlife Management* 59:667–73.

Draper, N. R., and H. Smith. 1981. *Applied regression analysis.* 2nd ed. New York: Wiley.

Dytham, C. 2003. *Choosing and using statistics: A biologist's guide.* 2nd ed. Malden, MA: Blackwell.

Efron, B. 1975. The efficiency of logistic regression compared to normal discriminant analysis. *Journal of the American Statistical Association* 70:892–98.

Fielding, A. H., and P. F. Haworth. 1995. Testing the generality of bird-habitat models. *Conservation Biology* 9:1466–81.

Findley, J. S. 1993. *Bats: A community perspective.* Cambridge: Cambridge Univ. Press.

Findley, J. S., and H. Black. 1983. Morphological and dietary structuring of a Zambian insectivorous bat community. *Ecology* 64:625–30.

Fretwell, S. D. 1972. *Populations in a seasonal environment.* Princeton, NJ: Princeton Univ. Press.

Ganey, J. L., and W. M. Block. 1994. A comparison of two techniques for measuring canopy closure. *Western Journal of Applied Forestry* 9:21–23.

Gotfryd, A., and R. I. C. Hansell. 1985. The impact of observer bias on multivariate analyses of vegetation structure. *Oikos* 45:223–34.

Graham, M. H. 2003. Confronting multicollinearity in ecological multiple regression. *Ecology* 84:2809–15.

Green, R. H. 1971. A multivariate statistical approach to the Hutchinsonian niche: Bivalve mollusks of central Canada. *Ecology* 52:543–56.

Green, R. H., and R. C. Young. 1993. Sampling to detect rare species. *Ecological Applications* 3:351–56.

Harris, R. J. 1985. *A primer on multivariate statistics.* 2nd ed. Orlando, FL: Academic Press.

Hatton, T. J., N. E. West, and P. S. Johnson. 1986. Relationships of error associated with ocular estimation and actual cover. *Journal of Range Management* 39:91–92.

Hutchinson, G. E. 1957. Concluding remarks. *Cold Spring Harbor Symposium on Quantitative Biology* 22:415–27.

Ihl, C., and D. R. Klein. 2001. Habitat and diet selection by muskoxen and reindeer in western Alaska. *Journal of Wildlife Management* 65:964–72.

James, F. C., and H. H. Shugart Jr. 1970. A quantitative method of habitat description. *Audubon Field Notes* 24:727–36.

Johnson, C. M., L. B. Johnson, C. Richards, and V. Beasley. 2002. Predicting the occurrence of amphibians: An assessment of multiple-scale models. In *Predicting species occurrences: Issues of accuracy and scale,* ed. J. M. Scott, P. J. Heglund, M. L. Morrison, J. B. Haufler, M. G. Raphael, W. A. Wall, and F. B. Samson, 157–70. Washington, DC: Island Press.

Johnson, D. H. 1981. The use and misuse of statistics in wildlife habitat studies. In *The use of multivariate statistics in studies of wildlife habitat,* ed. D. E. Capen, 11–19. General Technical Report RM-87. Ft. Collins, CO: USDA Forest Service, Rocky Mountain Research Station.

Johnson, R. A. 1992. *Applied multivariate statistical analysis.* 3rd ed. Englewood Cliffs, NJ: Prentice Hall.

Kaufman, D. W., and G. A. Kaufman. 1989. Population biology. In *Advances in the study of* Peromyscus *(Rodentia),* ed. G. L. Kirkland Jr. and J. N. Layneeds, 233–70. Lubbock: Texas Tech Univ. Press.

Kelt, D. A., P. L. Meserve, and B. K. Lang. 1994. Quantitative habitat associations of small mammals in a temperate rainforest in southern Chile: Empirical patterns and the importance of ecological scale. *Journal of Mammalogy* 75:890–904.

Kent, M., and P. Coker. 1994. *Vegetation description and analysis: A practical approach.* New York: Wiley.

Kepler, C. B., and J. M. Scott. 1981. Reducing count variability by training observers. *Studies in Avian Biology* 6:366–71.

Krzanowski, W. J. 1990. *Principles of multivariate analysis: A user's perspective.* Oxford: Oxford Univ. Press.

Kuehl, A. K., and W. R. Clark. 2002. Predator activity related to landscape features in northern Iowa. *Journal of Wildlife Management* 66:1224–34.

Leigh, R. A., and A. E. Johnston, eds. 1994. *Long-term experiments in agricultural and ecological sciences.* Oxford, UK: CAB International.

Likens, G. E. 1983. A priority for ecological research. *Bulletin of the Ecological Society of America* 64:234–43.

Lindsey, A. A., J. D. Barton, and S. R. Miles. 1958. Field efficiencies of forest sampling methods. *Ecology* 39:428–44.

Littell, R. C. 2002. *SAS for linear models.* 4th ed. Cary, NC: SAS Institute.

Ludwig, J. A., and J. F. Reynolds. 1988. *Statistical ecology: A primer on methods and computing.* New York: Wiley.

Manly, B. F. J. 1994. *Multivariate statistical methods: A primer.* 2nd ed. New York: Chapman & Hall.

Marcot, B. G., M. G. Raphael, and K. H. Berry. 1983. Monitoring wildlife habitat and validation of wildlife–habitat relationships models. *Transactions of the North American Wildlife and Natural Resources Conference* 48:315–29.

Martin, J. A., and M. L. Morrison. 1999. Distribution, abundance, and habitat characteristics of the buff-breasted flycatcher in Arizona. *Condor* 101:272–81.

Marzluff, J. M., J. J. Millspaugh, P. Hurvitz, and M. S. Handcock. 2004. Relating resources to a probabilistic measure of space use: Forest fragments and Steller's jays. *Ecological Monographs* 85:1411–27.

Maurer, B. A., L. B. MacArthur, and R. C. Whitmore. 1981. Habitat associations of breeding birds in clearcut deciduous forests in West Virginia. In *The use of multivariate statistics in studies of wildlife habitat,* ed. D. E. Capen, 167–72. General Technical Report RM-87. Ft. Collins, CO: USDA Forest Service, Rocky Mountain Research Station.

Meents, J. K., J. Rice, B. W. Anderson, and R. D. Ohmart. 1983. Nonlinear relationships between birds and vegetation. *Ecology* 64:1022–27.

Miles, D. B. 1990. A comparison of three multivariate statistical techniques for the analysis of avian foraging data. *Studies in Avian Biology* 13:295–308.

Morrison, D. F. 1967. *Multivariate statistical methods.* 2nd ed. New York: McGraw-Hill.

Morrison, M. L. 1981. The structure of western warbler assemblages: Analysis of foraging behavior and habitat selection in Oregon. *Auk* 98:578–88.

Morrison, M. L. 1984a. Influence of sample size and sampling design on analysis of avian foraging behavior. *Condor* 86:146–50.

Morrison, M. L. 1984b. Influence of sample size on discriminant function analysis of habitat use by birds. *Journal of Field Ornithology* 55:330–35.

Morrison, M. L., and R. G. Anthony. 1989. Habitat use by small mammals on early-growth clear-cuttings in western Oregon. *Canadian Journal of Zoology* 67:805–11.

Morrison, M. L., C. J. Ralph, J. Verner, and J. R. Jehl Jr. 1990. *Avian foraging: Theory, methodology, and applications.* Studies in Avian Biology no. 13. Bend, OR: Cooper Ornithological Society.

Morrison, M. L., W. M. Block, M. D. Strickland, and W. L. Kendall. 2001. *Wildlife study design.* New York: Springer-Verlag.

Mueller-Dombois, D., and H. Ellenberg. 1974. *Aims and methods of vegetation ecology.* New York: Wiley.

Nadeau, S., R. Decarie, D. Lambert, and M. St-Georges. 1995. Nonlinear modeling of muskrat use of habitat. *Journal of Wildlife Management* 59:110–17.

Neff, N. A., and L. F. Marcus. 1980. *A survey of multivariate methods for systematics.* New York: American Museum of Natural History.

Otis, D. L. 2002. Survival models for harvest management of mourning dove populations. *Journal of Wildlife Management* 66:1052–63.

Pimentel, R. A. 1979. Morphometrics. Dubuque, IA: Kendall/Hunt.

Press, S. J., and S. Wilson. 1978. Choosing between logistic regression and discriminant analysis. *Journal of the American Statistical Association* 73:699–705.

Queheillalt, D. M., J. W. Cain III, D. E. Taylor, M. L. Morrison, S. L. Hoover, N. Tuatoo-Bartley, L. Rugge, K. Christopherson, M. D. Hulst, M. R. Harris, and H. L. Keough. 2002. The exclusion of rare species from community-level analyses. *Wildlife Society Bulletin* 30:756–59.

Raphael, M. G. 1990. Use of Markov chains in analyses of foraging behavior. *Studies in Avian Biology* 13:288–94.

Raphael, M. G., and B. G. Marcot. 1986. Validation of a wildlife–habitat-relationships model: vertebrates in a Douglas-fir sere. In *Wildlife 2000: Modeling habitat relationships of terrestrial vertebrates,* ed. J. Verner, M. L. Morrison, and C. J. Ralph, 129–38. Madison: Univ. of Wisconsin Press.

Ratkowsky, D. A. 1990. *Handbook of nonlinear regression models.* New York: Dekker.

Rexstad, E. A., D. D. Miller, C. H. Flather, E. M. Anderson, J. W. Hupp, and D. R. Anderson. 1988. Questionable multivariate statistical inferences in wildlife habitat and community studies. *Journal of Wildlife Management* 52:794–98.

Rexstad, E. A., D. D. Miller, C. H. Flather, E. M. Anderson, J. W. Hupp, and D. R. Anderson. 1990. Questionable multivariate statistical inferences in wildlife habitat and community studies: a reply. *Journal of Wildlife Management* 54:189–93.

Schmid, J. R., A. B. Bolten, K. A. Bjorndal, W. J. Lindberg, H. F. Percival, and P. D. Zwick. 2003. Home range and habitat use by Kemp's ridley turtles in west-central Florida. *Journal of Wildlife Management* 67: 196–206.

Schooley, R. L. 1994. Annual variation in habitat selection: patterns concealed by pooled data. *Journal of Wildlife Management* 58:367–74.

Schultz, A. M., R. P. Gibbens, and L. DeBano. 1961. Artificial populations for teaching and testing range techniques. *Journal of Range Management* 14:236–42.

Scott, J. M., P. J. Heglund, M. L. Morrison, J. B. Haufler, M. G. Raphael, W. A. Wall, and F. B. Samson, eds. 2002. *Predicting species occurrences: Issues of accuracy and scale.* Washington, DC: Island Press.

Scott, J. M., F. L. Ramsey, and C. P. Kepler. 1981. Distance estimation as a variable in estimating bird numbers from vocalizations. *Studies in Avian Biology* 6:334–40.

Seber, G. A. F. 1989. *Nonlinear regression.* New York: Wiley.

Shaw, P. J. A. 2003. *Multivariate statistics for the environmental sciences.* New York: Oxford Univ. Press.

Shugart, H. H., Jr. 1981. An overview of multivariate methods and their application to studies of wildlife habitat. In *The use of multivariate statistics in studies of wildlife habitat,* ed. D. E. Capen, 4–10. General Technical Report RM-87. Ft. Collins, CO: USDA Forest Service, Rocky Mountain Research Station.

Stauffer, D. F. 2002. Linking populations and habitats: Where have we been? Where are we going? In *Predicting species occurrences: Issues of accuracy and scale,* ed. J. M. Scott, P. J. Heglund, M. L. Morrison, J. B. Haufler, M. G. Raphael, W. A. Wall, and F. B. Samson, 53–61. Washington, DC: Island Press.

Strayer, D., J. S. Glitzenstein, C. G. Jones, J. Kolasa, G. E. Likens, M. J. McDonnell, G. G. Parker, and S. T. A. Pickett. 1986. *Long-term ecological studies: An illustrated account of their design, operation, and importance to ecology.* Occasional Paper no. 2. Millbrook, New York: Institute of Ecosystem Studies.

Stubbendieck, J. L. 1986. *Range research: Basic problems and techniques.* Denver: Society for Range Management.

Tabachnick, B. G., and L. S. Fidell. 1983. *Using multivariate statistics.* New York: Harper & Row.

Tabachnick, B. G., and L. S. Fidell. 1996. *Using multivariate statistics.* New York: Harper Collins.

Tacha, T. C., W. D. Warde, and K. P. Burnham. 1982. Use and interpretation of statistics in wildlife journals. *Wildlife Society Bulletin* 10:355–62.

Taylor, J. 1990. Questionable multivariate statistical inferences in wildlife habitat and community studies: A comment. *Journal of Wildlife Management* 54:186–89.

Timm, N. H. 2002. *Applied multivariate analysis.* New York: Springer-Verlag.

Verner, J., M. L. Morrison, and C. J. Ralph, eds. 1986. *Wildlife 2000: Modeling habitat relationships of terrestrial vertebrates.* Madison: Univ. of Wisconsin Press.

Welsh, H. H., and A. J. Lind. 2002. Multiscale habitat relationships of stream amphibians in the Klamath-Siskiyou region of California and Oregon. *Journal of Wildlife Management* 66:581–602.

Wesolowsky, G. O. 1976. *Multiple regression and analysis of variance.* New York: Wiley.

White, G. C., and W. R. Clark. 1994. Microcomputer applications in wildlife management and research. In *Research and management techniques for wildlife and habitats,* 5th ed., ed. T. A. Bookhout, 75–95. Bethesda, MD: Wildlife Society.

Wiens, J. A. 1984. The place of long-term studies in ornithology. *Auk* 101:202–3.

Wiens, J. A. 1989. *The ecology of bird communities.* Vol. 1, *Foundations and patterns.* Cambridge: Cambridge Univ. Press.

Woodward, A. A., A. D. Fink, and F. R. Thompson III. 2001. Edge effects and ecological traps: effects on shrubland birds in Missouri. *Journal of Wildlife Management* 65:668–75.

Zar, J. H. 1999. *Biostatistical analysis.* 4th ed. Upper Saddle River, NJ: Prentice Hall.

7 Measuring Behavior

The animal shall not be measured by man. In a world older and more complete than ours they move finished and complete, gifted with extensions of the senses we have lost or never attained, living by voices we shall never hear.

HENRY BESTON, THE OUTERMOST HOUSE

Many studies of habitat use have focused on behavior because of its obvious importance in understanding the distribution, abundance, and needs of animals. Analysis of behavior shows how animals *actively* use their environment. In contrast, presence–absence or relative abundance studies rely on indirect correlations between the animal and features of the area it is found in to infer preference for habitat components, substrates, or foods.

Although they are usually more informative than purely correlational studies, behavioral observations by themselves are fraught with uncertainties. In particular, how are the behaviors we observe related to the survival and, ultimately, to the reproductive success of the individual? It should be evident that the study of behavior must begin with the development of a sound theoretical framework that elucidates how the animal perceives and then uses its environment. Determination of an animal's perception of its environment is difficult to achieve through behavioral observations. Thus three general areas of study have contributed to our understanding

of how animals perceive their environment: development of theoretical models, laboratory experiments, and field studies. Each of these areas of study is incorporated into this chapter.

Regardless of the theoretical basis on which we construct our behavioral studies, we must employ proper methods of data collection and analysis. Such considerations include specific methods by which animals are located and visually observed, biases associated with observation methods, the application of statistical procedures, and adequate sample size. Unfortunately, scant attention has historically been given to all of these concerns in any given behavioral study.

In this chapter we will review the general theoretical framework on which studies of wildlife behavior have been conducted. The principal methods used to observe animal behavior and to assess resource abundance and use are then discussed. Because of the obvious importance that the gathering of food plays in the life of an animal, methods of studying foraging behavior and diet are also covered. Finally, we discuss how studies of energetics can be used to advance our knowledge of the habitat-use patterns of wildlife.

Theoretical Framework

We should remember that the study of wildlife habitat is essentially the study of animal behavior in the broadest sense. How organisms select habitat reflects their evolutionary history as well as the current ecological conditions of the environment and the influence of other animals. In the end, all behavior—whether for foraging, dispersal, migration, reproduction, or predator avoidance—is modified and guided by the selective advantage of increasing fitness of individuals and vitality of populations.

The theoretical framework we developed in chapter 3 on the perception of habitat by an animal—its niche gestalt, stimulus summation, and so forth—sets the background by which we can observe and then interpret animal behavior. How an animal views its environment—namely as a series of different spatial scales, from the general vegetation type down to specific features of the microsite—is reflected in an individual's behavior. By observing this behavior in a systematic manner, we can learn much about why animals succeed, or fail to succeed, in a particular environment.

Ploger and Yasukawa (2003) summarized four components of the study of animal behavior: (1) *causation*—what causes an animal to perform a certain behavior? (2) *development*—how does the behavior change as the animal develops? (3) *evolution*—what is the evolutionary history of the behavior? and (4) *function*—how does the behavior help the animal to survive and reproduce successfully? We briefly discuss each of these four essential components of animal behavior as we set the stage for describing fundamental principles of recording behavioral data and then using the results to manage wildlife.

1. *Causation of behavior.* Behaviors we observe have multiple sources, including environmental stimuli, mental processes, neural pathways, and hormones. Thus the study of causation involves ecology, psychology, neurobiology, endocrinology, ethology, and other disciplines. Individually and together these disciplines examine the way in which a behavior is triggered—the *proximate mechanisms.* Why a behavior is elicited usually has direct application to wildlife study, such as why an individual abandons a nest, flees at certain noise, chooses to settle in a particular location, and so on. Behavioral responses are seldom "all or nothing," but rather occur along a continuum of stimuli.

2. *Development of behavior.* Behaviors emerge as an individual develops. Initial behaviors are primarily innate and focus primarily on development and survival (e.g., the begging response of a young animal). Throughout life, depending on the species, animals to varying degrees adopt or learn additional behaviors on the basis of events they experience and survive. In reality, most behaviors have an innate basis that can be modified by experience. Thus there is usually not a clear dichotomy between innate and learned behaviors. The specific behavior an individual learns, when it learns it, and the intensity of stimuli under which the behavior was learned all provide valuable information to the biologist when developing management guidelines.

3. *Evolution of behavior.* The behaviors we observe are based on an evolutionary history. As such, studies of the evolutionary basis for behavior are concerned with the *ultimate mechanism,* or reason for the activity (versus the proximate mechanism, as noted above). Understanding the evolutionary history or setting in which a species evolved is useful in wildlife studies because it places our observations in a broad context, and also helps us understand why a species (or subspecies) is changing in abundance and distribution (e.g., knowing the

environmental conditions under which a species evolved).

4. *Function of behavior.* Understanding the function of the behavioral repertoire of animals tells us how those behaviors promote survival and reproduction (plus a host of related activities). A key component of the study of behavior is determining function, as function is a core component of the ultimate reason for the behavior. Once we know the function of different behaviors, we can focus our attention on those behaviors that promote survival and reproductive success.

Key Terms for Application to Wildlife Study

Three terms often used in behavioral studies are *use, selection,* and *requirement.* In this chapter we define *use* as a demonstrated presence of a particular item in an animal's behavioral repertoire—for example, a den, perch, or foraging site, or a prey item in the diet. *Selection,* in contrast, has typically been defined as the use coupled with evidence that the frequency of occurrence of the behavior is significantly greater (statistically and biologically) than the frequency in the animal's environment. More recently, Hall et al. (1997) suggested that *selection* should be used in reference to the process by which innate and learned behavioral decisions are made. They further defined *preference* as the consequence of this selection process, which is measured as the disproportional use of some resources over others. There is much confusion in the literature regarding these terms, and we suggest that the standardizations suggested by Morrison and Hall (2002) be adopted. We have tried to avoid confusion in the use of terms without altering the original intent of the researchers.

Finally, *requirement* is the presence of, or a particular minimum amount of, a resource that the animal must obtain in order to live and reproduce. Unfortunately, few studies have identified

requirements. Observational studies in the wild cannot usually be designed so that the researcher can conclude whether an animal selects a resource because of a behavioral or a physiological requirement. Most studies of wildlife behavior therefore report the use of various resources, although they may inappropriately label their data as showing preference (Hall et al. 1997). However, if many studies conducted across different time periods and locations consistently show preference of a particular resource or behavior, then one can likely infer that the species is exhibiting a behavior of adaptive significance, thereby implicating a requirement (Ruggiero et al. 1988). Additional terms are defined below as we introduce new behavioral concepts.

Measuring Behavior

Martin and Bateson (1993, 19–23) and Ploger (2003) outlined the general steps to follow in designing and implementing a behavioral study. Here we are specifically interested in the choice of variables, choice of recording methods, accumulation of adequate sample sizes, and data analysis. Additional reviews of research methods in behavioral studies have been given in many publications (see especially Altmann 1974; Hazlett 1977; Colgan 1978; Kamil and Sargent 1981; Kamil et al. 1987; Gottman and Roy 1990; Morrison et al. 1990; Sommer and Sommer 1991; Lehner 1996; Ploger and Yasukawa 2003). Every issue of the behavioral journals (e.g., the *Journal of Animal Behaviour, Behaviour, Behavioral Ecology*), including those devoted to the study of specific taxa, such as primates (e.g., the *International Journal of Primatology, Primates*), presents many papers containing examples of behavioral methods and analyses.

Most behavioral observations can be broken down into three general categories: *structure, consequence,* and *spatial relation* (Martin and

Bateson 1993, 57–58). Researchers must recognize these categories before defining specific variables to record; otherwise, problems will likely arise in the analysis and interpretation of data. *Structure* describes the appearance, physical form, or temporal pattern of the animal's behavior, described in terms of posture and movements. For example, saying that a deer "reaches down and removes a leaf from a bush" describes the structure of the behavior. *Consequence,* in contrast to structure, describes the effects of the animal's behavior. Here, behavior is described without reference to how the effects are achieved (see also Dewsbury 1992). Saying that the deer is "browsing" describes the consequences of the behavior but says nothing about how the browsing took place (i.e., the structure). *Spatial relation* describes behavior in terms of the animal's spatial proximity to features of the environment, including other animals. For example, "approaching a bush" describes the relation of the animal to a potential foraging substrate. Recording the relation or orientation of the animal adds substantial information to the behavioral description.

A common mistake is for behavior to be described in terms of presumed consequences, rather than in neutral terms that do not apply a meaning likely tainted by observer biases. For example, labeling a behavior as "stalking" applies a consequence that may be unwarranted with further, more objective study (see Martin and Bateson 1993, 57–58). Such *anthropomorphism* is a common mistake in behavioral studies (Lehner 1996, 85; Ploger 2003).

Several commonly used terms in behavioral studies must first be defined before observations can be recorded (from Martin and Bateson 1993, 62–66). *Latency* is the time from some specified event (as the beginning of the recording session) to the onset of the first occurrence of the behavior of interest. *Frequency* is the number of occurrences of the behavior pattern per unit time (the total number of occurrences should not be used as frequency unless accompanied by a time unit). *Duration* is the length of time for which a single occurrence of the behavior pattern lasts. *Intensity* is widely used but has no specific definition. According to Martin and Bateson (p. 65), intensity is best viewed as a measure of the amplitude or magnitude of a behavior—for example, how loud was a call, or how high was a jump. They use the term *local rate* as an index of intensity, where local rate is the number of component acts per unit time spent performing the activity (e.g., how many bites were taken within a specific period of time).

Selection of Variables

In chapter 5 we discussed selection of variables in habitat analysis. Much of that discussion applies, in general, to analysis of animal behavior. Whereas habitat analysis focuses on the description of the environment surrounding the animal, behavioral descriptions concentrate on the specific actions of the animal within its habitat.

Operational definitions (or *operationalization*) allow researchers to translate concepts into specific terms so that they can be measured reliably over time and by different observers. Without clear definitions, comparisons between studies are difficult, and entire data sets can be rendered useless. Once a concept is operationalized, it must be used consistently by all observers (Lehner 1996, 115; Glover 2003).

Animals perform a myriad of activities during a 24-hour period. They sleep, groom, engage in intra- and interspecific interactions, feed, and so on. To quantify these behaviors in any practical way requires that we devise some form of record keeping. One could first compile a catalog of behaviors that describes the behavioral repertoire of the species. Known as *ethograms* (*etho-* as in the biological study of behavior, or *ethology;* Martin and Bateson 1993, 6), such descriptions are useful starting points for the design of a be-havioral study (Ploger 2003). However, etho-grams have been published for only a few species. Developing an ethogram is especially appropri-ate during preliminary sampling. Although this subject is too detailed to describe here, we urge readers to review Schleidt et al. (1984) on the de-velopment of ethograms.

Slater (1978) provided five basic points for classifying behaviors. (1) Behavioral categories must be discrete—that is, acts within a category must have clear points of similarity that they do not share with acts outside the category. (2) Be-havioral acts must be homogeneous. All the acts in a category should be similar in form so that there is little danger of massing two different be-haviors in the same category. (3) It is better to split than to lump. Within reason, two similar be-haviors with possibly different consequences should be placed into separate categories; they can always be lumped later. (4) Names of cate-gories should avoid applying a causal or func-tional implication, as mentioned earlier in our initial discussion of consequence. Rather, names should clinically describe the behavior in terms of actions, not outcomes or motives. (5) The number of categories must be manageable. Too many categories will reduce an observer's reac-tion time and lower accuracy. Automated record-ing devices, including computer-aided data entry programs, can help overcome this problem to a degree (see Raphael 1988; Martin and Bateson 1993, ch. 7; Paterson et al. 1994; Samuel and Fuller 1994).

Lehner (1996) recommended that variable selection begin by listing all the variables that are known to affect the behavior under study or are thought to have some influence—for example:

1. Environmental variables
    a. Biotic
        i. Members of social group
        ii. Predator–prey relationships
        iii. Vegetative characteristics of the habitat
    b. Abiotic
        i. Temperature
        ii. Wind speed
        iii. Cloud cover
2. Organismal variables
    a. Genotype
        i. Sex
        ii. Parental stock
    b. Phenotype
        i. Behavioral characteristics
        ii. Description of behavior
        iii. Frequency
        iv. Rate
        v. Duration

By listing all of the potentially important vari-ables, you are working toward determining the potential causation of a behavior as well as helping to identify potential sources of variation (that can be eliminated or measured). Because the number of variables that could affect a be-havior can be large, you should expect to spend a good deal of time developing this list. Addition-ally, it is always best to have several colleagues re-view your list prior to continuing the study.

Recording Methods

Once researchers have developed behavioral cat-egories, they must rigorously define methods for recording these behaviors. Martin and Bateson (1993, ch. 6) and Ploger (2003) provided excel-lent frameworks for recording behavioral obser-vations by defining sets of sampling rules and recording rules (fig. 7.1). Below we summarize these rules, and provide brief examples of their applications to wildlife-habitat studies.

Sampling Rules

Before collecting data, you must objectively de-cide which individuals to observe (sample) and

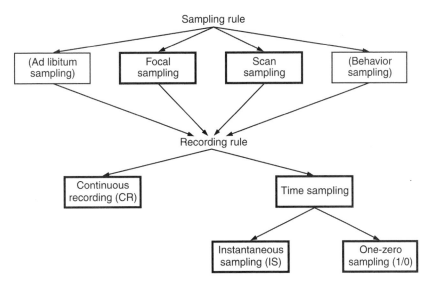

Figure 7.1.   The hierarchy of *sampling rules* (determining who is watched and when) and *recording rules* (determining how their behavior is recorded). (Reproduced from Martin and Bateson 1993, 88, fig. 6.1; reprinted with permission of Cambridge University Press.)

when to observe them. In chapter 4 we described describe reasons for selecting species for study and the potential use of model species or systems for experimentation; and in chapter 6 (and elsewhere) we discussed the importance of temporal and spatial stratification of data collection. Here we concentrate on observations of individual animals within the chosen species and within the context of a proper spatial and temporal study design.

AD LIBITUM SAMPLING

When rare events are the activity of interest a generalized behavior sampling, or *ad libitum* methodology, is preferred. Such a methodology involves watching the entire group for long periods of time and recording each instance of the behavior of interest (Ploger 2003); such sampling can also be incorporated as a subpart of either focal or scan sampling (see below) (Martin and Bateson 1993, 87). Masataka and Thierry

(1993) used ad libitum sampling in their study of macaques. Ad libitum sampling is seldom the primary method of a behavior study, however, because other methods—as described below—provide a more structured means of sampling from the study population.

FOCAL-ANIMAL SAMPLING

Focal-animal sampling, or just focal sampling, involves observing one individual, even if it is located within a group of animals, for a specified period of time. This method has the clear advantage of allowing adequate records to be collected on various classes of a species—for example, data by sex and age. Further, if animals are individually tagged or show unique markings (e.g., coat colors and patterns of wild horses, size and shape of antlers or horns), then one can accumulate information on individual variation rather than simply lumping across individuals in a group. For example, Carranza (1995) used the shape and branching pattern to individually identify red deer (*Cervus elaphus*). Stamps and Krishnan (1994) chose one individual as a focal individual from a dyad of lizards (*Anolis aeneus*) to determine the outcome of social interactions.

Focal-animal approaches are widely used to quantify behavior, competitive interactions, habitat and microhabitat use, and so on. The focal target for observation need not be an individual animal but can include any designated group of individuals. For example, Hale et al. (2003) used groups of brown jays (*Cyanocorax morio*) for behavioral observations, Jones and Bock (2003) used mixed-species foraging flocks, and Courchamp et al. (2002) used packs of painted hunting dogs (*Lycaon pictus*); other examples abound (e.g., Trainer et al. 2002; Salewski et al. 2003).

A problem with focal sampling involves the conspicuousness of individual animals. By focusing on individuals, researchers may produce records of behavior that vary considerably in length. For example, foraging birds are obvious when they are near branch tips and lower in a tree, but they usually disappear when they move upward and inward in the foraging substrate. Likewise, the response of ungulates to disturbance is easy to observe and quantify (e.g., number of steps or bounds, head movements) when they are at the edge of an opening, but observations of their behavior become difficult when they move into the trees. Further, our attempts to follow animals and observe their activities undoubtedly influence their behavior in ways largely unknown to us (more on this subject below). In addition, visibility bias can be introduced into the data when researchers try to follow animals in dense vegetation.

Thus sampling periods must be short, and are therefore biased toward those individuals who are most conspicuous. Unfortunately, no sampling method employing only the observer's eyes

can overcome this problem, although many researchers have made attempts to do so. For example, students of bird foraging behavior often observe a bird for a short period of time (usually a few seconds) before beginning data recording. The rationale here is that they are allowing the bird to move into a less conspicuous position than the one in which it was initially observed. In addition, singing birds are usually ignored because such individuals are usually in a high, conspicuous position (Hejl et al. 1990).

The advent of small radio transmitters promises to solve this problem at least partially. Although one still cannot see specifically what the out-of-sight animal is doing, at least the general location and movement of the animal can be approximated without causing undo disturbance. Many radios can also be fitted with mercury switches or vital-rate sensors to detect body position, heart rate, body temperature, and other activities. Event recorders can automate much of the recording of these data (e.g., Berdoy and Evans 1990). Further, miniaturized video cameras are being developed that will allow observation of behavior from the animal's point of view. Clearly, technology is rapidly advancing our ability to observe behavior in increasingly unbiased ways. Williams (1990) provided a review of the use of telemetry in studies of bird foraging behavior, and many publications and books are available on the use of radio-tagging wildlife (e.g., Mech 1983; Kenward 1987; White and Garrott 1990; Samuel and Fuller 1994).

Losito et al. (1989) developed a modification of focal sampling to address the problem of individuals disappearing from view, a problem especially evident in areas with limited visibility (e.g., dense brush). Termed *focal-switch sampling*, their method allows switching to a new focal individual (nearest neighbor) if the original focal animal goes out of view. They found that focal switching was more efficient than standard focal sampling because it saved 24% of samples from premature termination while also reducing recording biases in several analyses.

Scan Sampling

Although used relatively less than focal-animal sampling, scan sampling is appropriate when observations are taken of a large group of animals. Using scan sampling, observers "view" or "scan" across a large group of animals and collect data at regular intervals, recording the behavior of each individual or a subset of the individuals; individual identification is usually not made, however. The behavior of each individual scanned is intended to be an instantaneous sample. A scan sample usually restricts the observer to recording only a few, simple categories. Scan sampling takes only a few seconds to several minutes, depending on the amount of information being recorded, the group size, and the activity level of the group (e.g., a grazing herd versus a stampeding one). For example, Davoren et al. (2003) used scan sampling to study the foraging strategies of common murres (*Uria aalge*) from boats. They established survey transects and scanned for birds from fixed points for a set period of time (fig. 7.2). Using this method, they were able to relate the foraging behavior of murres to capelin, their primary prey species.

Bias in scan sampling is influenced largely by the period of time that elapses between scans, or samples. Rare or inconspicuous behaviors will likely be observed at a lower frequency than that at which they actually occur. For example, Harcourt and Stewart (1984) showed that studies using scan sampling of foraging gorillas resulted in underestimates of actual foraging time. Focal-animal sampling resulted in a more accurate estimate of the actual feeding time, because the researchers could follow individuals continuously. Biases that can occur using scan sampling include (1) differential visibility between activity categories and seasons, (2) poor correlation between group composition and the representation of age–sex classes in scans, (3) diurnal variation in sample sizes, and (4) increased habituation to the observers with duration of the study (Newton 1992).

Scan sampling can also be combined with focal sampling during the same recording session. Here, researchers record the behavior of focal individuals in detail, but at specific intervals they also scan sample the group for simpler behavioral categories. Workers have shown, for flocking birds and herding mammals, that the time an individual spends foraging is related to the number and proximity of group members; for example, an individual must watch more vigilantly for predators as group size decreases (Hamilton 1971; Pulliam 1973; Caraco et al. 1980; Kildaw 1995). Using both methods in this manner thus allows the researcher to place focal-animal observations in an overall context of group activities.

For example, L'Heureux et al. (1995) used scan and focal sampling to study mother–yearling associations in bighorn sheep (*Ovis canadensis*) in Alberta, Canada. Scan sampling, at 15-minute intervals, was used to record the distance between a mother and yearling. Focal sampling was also used by observing sheep for at least 40 minutes, until they were out of sight, or for a maximum of 2 hours. Carranza (1995) also used both scan and focal sampling to study male-female interactions in red deer—the focal method to record interactions among territorial males, and scan samples at 1-minute intervals to record the number of females in a territory.

Recording Rules

The means of actually recording behavior after the sampling rule has been chosen are called

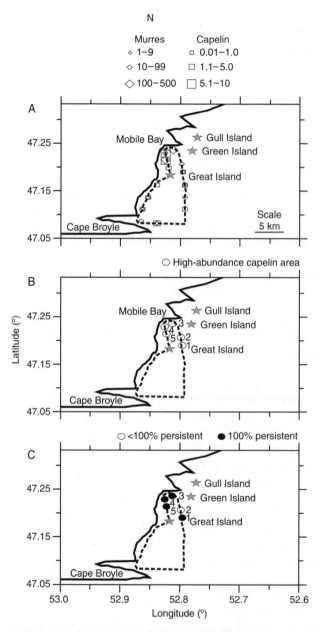

Figure 7.2. An example of scan sampling as used to study the foraging strategies of common murres (*Uria aalge*). (A) Distribution and abundance of common murres and capelin (*Mallotus villosus*) around Great Island, Witless Bay. (B) Location of capelin "hot spots" (capelin abundance higher than the mean). (C) Persistence of capelin presence at each "hot spot" on all visits. Dashed lines indicate transect routes. (Reproduced from Davoren et al. 2003, fig. 3 reprinted with permission of the Ecological Society of America.)

recording rules. Figure 7.1 depicts the relationship between sampling rules and recording rules. We can divide recording rules into two basic categories: *continuous recording* and *time sampling*. Although we would prefer to be consistent and use the term *time "recording"* to clearly identify its classification as a recording rule, we will follow the terminology used by Martin and Bateson (1993); changing terms would only further confusion. Note that focal sampling (a sampling rule) is not synonymous with continuous recording (a recording rule); and scan sampling (a sampling rule) is not synonymous with instantaneous sampling (a recording rule).

CONTINUOUS RECORDING

In continuous recording, researchers record each occurrence of a behavioral act, as well as the time of the activity and pertinent environmental information. Continuous recording gives true frequencies, latencies, and durations of behavior *if* an animal can, indeed, be watched continuously for a sufficient period of time ("sufficient" must be determined through sample size analyses). Termination of a recording session will usually result in an unreliable estimation of the durations of certain behaviors. Continuous recording is necessary when the sequence of behaviors is of interest. The method is, however, tedious, and in practice only a few categories can be measured reliably.

In studies of foraging behavior, observers often record the behavior of an animal continuously for a specific period of time. After some time has elapsed, the worker observes the same or a different individual for the same period of time. Within continuous recording periods, the duration of each behavior is timed. For example, Block (1990) watched foraging white-breasted nuthatches (*Sitta canadensis*) for 10 to 15 seconds at a time, noting the duration of time spent searching for and procuring prey. Because a foraging bird can seldom be continuously observed

for more than a few seconds to a few minutes, accurate estimates of latency and especially duration are difficult to obtain. Loughry (1993) used 5-minute continuous focal-animal sampling to study the behavior of black-tailed prairie dogs (*Cynomys ludovicianus*). In their study of foraging study in the wapiti (*Cervus elaphus*), Wilmshurst et al. (1995) defined a cropping sequence as beginning when the animal put its head down to graze and ending when it lifted it head. Cropping rate was calculated as the number of bouts taken during the cropping sequence divided by time. Weckerly (1994) watched foraging black-tailed deer for periods of 7 to 10 minutes.

TIME SAMPLING

Time sampling is a general category that involves recording behavior periodically; each observation session is divided into successive short periods of time called *sample intervals*. However, time sampling has been criticized because there is no standardized time period used among scientists. Thus the data on frequencies (of behavioral acts) lacks any universal interpretation (Quera 1990). Time sampling is therefore reported as relative durations or prevalence of the acts. There are two basic types of time sampling: *instantaneous sampling* and *one-zero sampling*.

Instantaneous sampling, sometimes referred to as *point sampling*, records the instant (point) at which the activity of an individual is recorded. The structure of the data obtained depends largely on the length of the sample interval. If the interval is long relative to the average duration of the behavior, then one can obtain a measure of the proportion (not frequency) of all sample points at which the behavior occurred. If the sample interval is short relative to the average duration of the behavior, however, then instantaneous sampling can approximate the results of continuous recording. In a detailed analysis of instantaneous sampling, Poysa (1991) concluded that averages of a great number of individuals give reliable estimates of true time budgets even for behavioral sequences of short duration. However, Poysa showed that if duration of a particular act is very short compared with other acts under study, then rate measures should be used instead of instantaneous sampling. It is also critical that samples from each individual being sampled are long and roughly equal.

Instantaneous sampling is a common method used to record wildlife behavioral data. Typically, the observer follows an animal while it remains in view (or a limit is placed on the observation period) and records specific behavioral acts at set intervals. For example, Stamps and Krishnan (1994) recorded the location of every marked lizard every 20 minutes in a study of territoriality. Students of bird foraging commonly use instantaneous sampling (Hejl et al. 1990). Paterson et al. (1994) recommended focal-animal sampling, instantaneous scan sampling, and focal-time sampling (see Baulu and Redmond 1978) as the most flexible and most statistically meaningful techniques.

The second type of time sampling, *one-zero sampling,* is similar to instantaneous sampling in that the recording session is subdivided into short sample intervals. Here, however, the observer merely records whether or not a particular behavior occurred during the sample period, recording no information on the frequency or duration of the act. Unlike instantaneous sampling, one-zero sampling consistently overestimates duration, because the behavior is recorded as though it occurred throughout the sample interval, whereas it need have occurred only once. In fact, Altmann (1974) argued against the use of one-zero sampling in all applications. In contrast, Martin and Bateson (1993, 97–98) provided a strong rationale for using one-zero sampling in certain situations and included examples of how this method should be applied. Bernstein (1991) concluded that one-zero sampling may be preferred when acts are clustered and the goal is to predict the probability of at least one such act occurring in a given time interval. Clearly, caution should be employed when using any method, but especially one that involves such controversy. One-zero sampling is seldom used in behavioral studies.

Sampling Concerns

Included in our discussions above on sampling and recording rules were various cautionary notes on biases associated with specific methods. In the following section we discuss sampling issues that are common to all behavioral observations, including independence, observer bias, and sample size requirements.

INDEPENDENCE

Most statistical analyses carry the assumption that the data represent random samples from populations and that each datum is statistically independent (from other data points; Zar 1999). Thus the number of *individuals* for which data are taken characterizes or defines one's total sample. Machlis et al. (1985) noted that the objective of research is to obtain measurements from an adequate number of individuals, not to obtain large samples of measurements; which they termed the *pooling fallacy.* Although pooling may not be a fatal error in certain situations (see below), studies of behavior are replete with examples of such pooling errors (see also Hurlbert 1984).

In behavioral sampling, researchers commonly collect a series of instantaneous records on a single individual and then consider each point sample or each interval as an independent sample. That is, recording six instantaneous samples every so often (be it 1 sec, 1 min, or 1 hr) on the same individual results in a sample size of one. One procedure, termed *aggregation,* would be to average these six samples, resulting in $n = 1$

individual, where n is the sample size. However, this procedure can lead to false conclusions if the distribution of the behavior is bimodal (or, more generally, nonnormal), in which case the mean score for an individual will be the one that rarely occurs (Leger and Didrichsons 1994). Another sample interval taken a short time later likewise would not represent another sample. Another procedure would be to randomly select, a posteriori, one sample from the repeated samples originally taken on an individual.

The problem, then, becomes one of determining when samples become independent. Martin and Bateson (1993, 86) noted that samples "must be adequately spaced out over time." Defining *adequate* is difficult, however, and is related to the goal of the study. That is, if one's objective is to describe behavior at the population level (i.e., individual variation within a population), then sampling new individuals across the population is indicated. Resampling an individual should be avoided so results are not biased toward a certain segment of the population; independence is thus assured. This statement must be tempered, of course, in cases in which the "population" is localized and rare. Likewise, if the goal is to examine a small segment of a larger population, independence *and* an adequate sample size will be difficult to achieve. Time-series analyses and repeated-measures designs are clearly indicated in such situations (described below).

Leger and Didrichsons (1994) noted that pooling is especially common in field studies of endangered species or other small populations, because multiple samples are taken from each individual to accumulate large sample sizes. They noted that the central question about pooling is whether the population can be represented in an unbiased manner by repeated sampling of the same individuals. If, for instance, all individuals in a theoretical population had the same mean and variance on the behavior of interest, then it would not matter if one obtained 100 data points by sampling 100 individuals once each, 50 individuals twice, or even one individual 100 times.

Leger and Didrichsons (1994) hypothesized that pooling was a reliable procedure *if* intrasubject variance exceeded between-subject variance. They evaluated several data sets, and concluded that pooling could provide estimates of population means and variances that were at least as reliable as those provided by single sampling (of individuals) and aggregation, provided that sample sizes were about equal among individuals, or that intrasubject variance was higher than between-subject variance. When intrasubject variance exceeds between-subject variance, then unequal sample sizes among individuals become problematic.

McNay et al. (1994) examined independence of movements using telemetry data with Columbian black-tailed deer. They found that, even with 6-week intervals between samples (eight samples/year), observations were still dependent for over 50% of the deer tested. Because most animal location data sets are likely to have a skewed distribution of data points, McNay and colleagues recommended placing emphasis on sampling animal locations systematically through time rather than trying to determine a time interval that will provide independent location samples. In other words, do not try to achieve independence of samples at the expense of gathering an adequate understanding of animal behavior. As noted by McNay and colleagues, time intervals between samples should be chosen with the understanding that potential gains in behavioral information are decreased with increasing time intervals between samples. Swihart and Slade (1985) and White and Garrott (1990, 148) provide additional discussion of this topic.

Within-individual variation is an important, although seldom analyzed, aspect of behavioral research. Researchers have tended to treat populations as homogeneous units, thus largely ignoring individual differences in behavior. Implicit in such an attitude is that variance about a mean (behavior) is irrelevant and not of biological interest. In a monograph on this subject, Lomnicki (1988) contended that further advances in population ecology will require consideration of individual differences, such as unequal access to resources. Thus repeatedly collecting samples on known individuals provides us with important ecological data. For example, changes in foraging strategies are common during ontogenetic size development, a phenomenon especially evident in reptiles (see Webb and Shine [1993] on blindsnakes [Typhlopidae] and Wikelski and Trillmich [1994] on iguanas [*Amblyrhynchus*]). Likewise, changes in energy stores can influence the trade-off between foraging time and vigilance in Belding's ground squirrels (*Citellus beldingi;* Bachman 1993).

Thus it should be clear that some advanced planning will avoid most problems involving data independence. We can define some recording rules that seek to maximize independence of observations, while realizing that complete independence will not be achieved in all cases. A few such "rules" are as follows: (1) Only one individual within a group (flock, herd) will be recorded per sampling session. (2) Observers will systematically cover the study area (which is as large as possible, especially if site-specific data are not needed), seeking out new individuals or groups; that is, avoid resampling the same group. (3) Sampling sessions will be stratified by time period, both within and between days. (4) Sessions will be distributed throughout the identified biological period (e.g., breeding) to avoid grouping samples within short periods of time. (5) In small populations, an attempt will be made to individually identify animals.

OBSERVER BIAS

Avoiding bias is virtually impossible. Observer bias takes various forms: the influence of ob-

servers on an animal, intra- and interobserver consistency in recording data, preconceived notions regarding how an animal "should" behave, differences in observer abilities, and so forth. Each bias is serious; when combined, they can make interpretation of results difficult.

Researchers conducting behavioral studies should be aware that their presence and activities likely influence an animal's activities. Researchers often state that their presence "did not markedly alter an individual's behavior," or that "we waited until the animal returned to normal activities before recording data." Such statements seem an exercise in self-delusion! Wild animals are constantly vigilant for predators and competitors; human presence likely heightens that awareness. Such high awareness or responsiveness is termed *sensitization*. Further, it is likely that the animal knew you were there long before you ever saw it. The waning of responsiveness is termed *habituation* and is considered by most to be a form of learning (Immelmann and Beer 1989). Animal species vary widely in their ability to learn. However, research has shown that birds and mammals have the ability to perform both temporal and numerical operations in parallel (Roberts and Mitchell 1994). For example, many corvids can remember the location of food caches for months, and also remember which caches they visited previously. Shettleworth (1993) reviewed the topic of learning in animals. Rosenthal (1976) presented a detailed analysis of the effects of the researcher in behavioral studies.

Animals that appear to become habituated to the presence of observers have thus adopted a modified pattern of behavior that allows them to keep the observer under surveillance. Animals adjust their behavior according to the costs and benefits associated with different courses of action available to them (e.g., hiding, fleeing). Further, detection of a potential predator (or human observer) may precede the observable response by a significant period of time. For example,

Roberts and Evans (1993) showed that sanderlings (*Calidris alba*) were acting to minimize both the number of flights they made and the distance of each flight when approached by a human, subject to not tolerating any close approaches. We reviewed some of the responses of wildlife to human activities in chapter 3. Numerous studies have been published documenting how animals respond to human activities, including human recreational activities (e.g., Knight and Gutzwiller 1995). These factors should be considered when designing behavioral studies, as the response of animals to the presence of humans may impact conclusions of a study that is not concerned with such impacts per se.

Intraobserver reliability is a measure of the ability of a specific observer to obtain the same data when measuring the same behavior on different occasions (Martin and Bateson 1993, 32–34)—in other words, the ability of an observer to be *precise* in his or her measurements. *Precision* describes the repeatability of a measurement and is not synonymous with accuracy. Because we seldom know what the actual behavioral pattern is—that is, the "true" pattern—we cannot directly measure an observer's accuracy. Assessing intraobserver reliability in field-based studies is difficult; animals seldom repeat their behavior in exactly the same fashion. One test is to videotape animals and then repeatedly to present (in some random fashion) individual sequences to the observer. Researchers can use the results of such trials to estimate the degree of observer reliability.

Interobserver reliability measures the ability of two or more observers to obtain the same results on the same occasion (Martin and Bateson 1993, 117; Glover 2003). To what extent is interobserver reliability a problem in field studies? Ford et al. (1990), for example, found that comparisons of foraging behaviors of individuals of the same species in different areas or years, recorded by different individuals, needed to be treated cautiously. Problems were particularly evident when observers had not previously agreed on standard methods of observation or classification of terms. Differences in experience among observers apparently accounted for much of the interobserver variability noted.

Intra- and interobserver reliabilities have been substantially improved by careful, rigorous, and repeated training. Each new observer is taught how data should be recorded, initially working with others in the field, comparing data and discussing reasons for decisions; videotape, as noted above, or captive animals can also be used. Observer reliability increases, in our experience, when observers become informed about and comfortable with why a particular behavior is categorized in a certain manner. Training should continue throughout the study, with frequent sessions to "recalibrate" the observers. Each behavior should be carefully defined in writing. It usually helps to define a behavior by its structure (e.g., *probe* means "insert bill beneath surface of substrate"). Commonly, in protracted studies, definitions and criteria tend to "drift" with the passage of time, as observers become more familiar with behaviors and possibly lazier in their evaluations. Careful and repeated training will help solve this problem. Further, efforts have been made to standardize terminology in various disciplines. For example, Remsen and Robinson (1990) developed a standardized terminology and definitions for avian foraging studies. The methodology of Schleidt et al. (1984) for a standard ethogram is a useful starting point for behavioral studies.

SAMPLE SIZE REQUIREMENTS

Researchers *must* incorporate evaluation of sample size requirements into the design phase of each study. Such planning guides the collection of data, avoiding both under- and overkill of sampling efforts. Further, it immediately informs you if you are trying to accomplish too much for the time and resources available. In our experience with graduate students, we have found that our first job when reviewing research proposals is to prevent the well-meaning student from attempting too much! It is, indeed, far better to do one thing very well, than a bunch of things poorly. Many, many workers have been forced to combine data across seasons, years, ages, and sexes because they did not plan properly.

Throughout this book we have discussed how temporal and sexual differences affect differential habitat use by animals. Studies of behavior show that, even within what we consider a "season," combining data over even a few months can obscure important patterns of resource use. In designing studies, researchers must carefully determine the number and types of variables for which they will have time to collect data.

For example, Brennan and Morrison (1990) found that significant variation in the use of tree species by foraging chestnut-backed chickadees (*Parus rufescens*) occurred throughout the year (fig. 7.3). Using some of the same data, Morrison (1988) showed how lumping of data can result

in inappropriate interpretations: lumping tends to "average out" many possibly important ecological relationships (see fig. 6.1). Likewise, Sakai and Noon (1990) showed that Pacific-slope flycatchers (*Empidonax difficilis*) significantly altered their foraging behavior within the breeding period.

These results are not surprising. Animals must respond to changes in resource availability and abundance and the demands placed on them by both abiotic and biotic factors. Many of the equivocal results obtained in behavioral studies likely result from lumping of data. Thus not only must researchers evaluate sample sizes needed for reliable interpretations, they must also carefully evaluate these samples over relevant periods of time—periods that are usually shorter than those presented in most papers.

Activity Budgets and Energetics

Natural historians have long been interested in quantifying how animals allocate their time

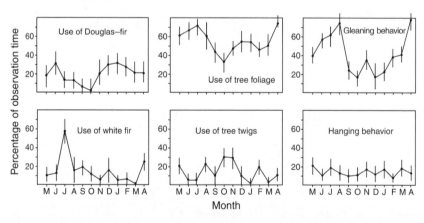

Figure 7.3.   Seasonal variation in the use of tree species, substrates, and foraging modes by chestnut-backed chickadees (*Parus rufescens*). Data graphed at 1-month intervals. Dots represent mean values; vertical bars represent 1 standard deviation. (Reproduced from Brennan and Morrison 1990, fig. 4; reprinted with permission of the Cooper Ornithological Society.)

among the requirements for foraging, sleeping, moving, breeding, and so on. Developing these *activity budgets* (or *time budgets*) is the first step in developing an understanding of the relationship between time allocation and survivorship, reproductive success, body condition, and other aspects of natural history. A simple time budget for the loggerhead shrike (*Lanius ludovicianus*) is given in table 7.1. Defler (1995) presented a good example of time budgets in monkeys (see also Newton 1992). As noted by Weathers et al. (1984), however, quantifying time allocation is easy relative to assessing the energy expended while conducting the activities. In this section we present a brief review of time-energy budgets as they relate to advancing our understanding of animal behavior and, ultimately, survival and fitness.

Assessments of animal energetics commonly assume that the goal of an animal is to maximize its net energy balance. Many theoretical and empirical studies have examined how the animal achieves this balance. Thus the neurological and physiological capabilities of the animal itself link it to its environment. One should think of the environment in its functional relation to the animal, rather than merely as the geographic area

and physical structure of the habitat in which the animal lives, in order to understand the dynamic relationships between an animal and its environment (Moen 1973, 21).

Thermal energy is exchanged between the animal and its environment by radiation, conduction, convection, and evaporation. Each of the methods of thermal energy exchange change relative to one another as the animal's environment alters; rain, wind, and other abiotic factors affect energy exchange over the short term. Changes in such factors as plant cover, for example, influence energy exchange over longer periods of time.

A wild animal has an ecological metabolic rate that is an expression of the energy "cost of living" for the purpose of daily activities and other physiological processes. Ecological metabolism varies from one activity to another and from one species to another. There are various direct and indirect methods for determining metabolic rates of animals under varying environmental conditions, both in the laboratory and in the field.

Total basic metabolic rates (BMRs) generally increase with body size. Thus total energy and food requirements should be higher in large

Table 7.1. Time budgets of loggerhead shrikes, in hours spent per activity

Bird Number	Run Number	Perching		Eating	Preening	Flying	Hopping	Other	Total
		Night	Day						
218	1	11.96	10.79	0.41	0.22	0.10	0.03	0.66	24.17
220	2	12.54	9.44	0.49	0.11	0.07	0.06	0.28	22.99
232	3	11.47	10.70	0.05	0.12	0.08	0.02	0.38	22.82
221	4	11.67	10.40	0.22	0.04	0.20	0.04	0.09	22.66
230	5	11.43	9.42	0.70	0.10	0.10	0.03	1.88	23.66
225	6	12.28	6.73	0.01	0.01	0.19	0.05	2.74	22.01
218	7	11.97	10.30	0.30	0.35	0.08	0.11	0.39	23.50
232	8	12.17	10.49	0.00	0.71	0.05	0.00	0.58	24.00
Mean		11.94	9.78	0.27	0.21	0.11	0.04	0.88	23.23
(SD)		(0.39)	(1.34)	(0.25)	(0.23)	(0.06)	(0.03)	(0.93)	(0.73)

Source: Weathers et al. 1984, table 1. Reprinted with permission of the American Ornithologists' Union.

animals relative to small animals (all else considered equal). Mass specific metabolic rates, however, decrease with increasing body size. Thus energy requirements and food consumption *per unit of body mass* should be higher in small animals relative to large animals. Such relationships influence the ways in which animals behave, including the amount of time spent foraging, their degree of territoriality, patterns of movement (including seasonal movements and migrations), and a host of related factors. Hence a knowledge of energy requirements goes a long way in helping explain why animals are using specific patches of habitat, and their activities within those patches. For example, Rychlik and Jancewicz (2002) related body size, food habits, and activity patterns in shrews of differing body size. Likewise, animals adjust energy reserves according to food availability, which in turn is often based on environmental conditions such as weather (Kelly and Weathers 2002). Here again, we see that an understanding of energy needs relates to activity patterns, which goes a long way in explaining the reasons for observed patterns of habitat use.

Studies of physiology set in an ecological context—or *physiological ecology*—are relatively rare in the field of wildlife management. As developed throughout this book, most wildlife studies are based on indirect measures of an animal's requirements—for example, recording foraging location rather than foods eaten, or recording rates of movement rather than energy expended. It seems intuitive, however, that tying energy expenditures to observed behaviors will go a long way toward furthering our understanding of why animals behave—and ultimately survive and reproduce—in the manner that we observe.

The time-energy budget (TEB) method is a commonly used technique for estimating total daily energy expenditure in animals. TEB has two parts. First, one develops an activity budget as described above. Second, one converts these activity data to energetic equivalents from estimates of energy costs for each activity as determined from controlled studies or the literature (Haufler and Servello 1994). Numerous workers have conducted laboratory studies to determine the relationship between activity and energy expenditure; Thus approximations can usually be made for the species one is studying on the basis of general allometric equations. Naturally, these values are only approximations, but they do provide a general understanding of activity–energy relationships (described below). Weathers et al. (1984) showed up to a 40% error rate for time-budget estimates that assign to behaviors energy equivalents that have been derived from the literature rather than empirically.

Finch (1984) determined the daily activity budget of Abert's towhees (*Pipilo aberti*) by quantifying the duration of four activities: perching, ground foraging, flying, and nest attendance. These data were transformed into percentages of the observation periods and the activity day. Daily energy expenditure was then determined using published estimates of basal and thermostatic requirements and estimates of the energy requirements for the four activities. Hobbs (1989) developed a model of energy balance in the mule deer (*Odocoileus hemionus*) that predicted changes in body condition of does and fawns and predicted the relationship of those changes to rates of mortality. He used literature values to develop a detailed model of energy balance based on animal activity. Another good example of the development of an activity budget and subsequent estimation of energy expenditure was provided by Dasilva (1992) for a colobine monkey (*Colobus polykomos*).

A more direct method of determining energy expenditures is the doubly labeled *water technique*. This technique involves injection (labeling) of oxygen (oxygen-18) and hydrogen (tritium or deuterium) isotopes into an animal prior to its release and calculating the rate of

CO_2 production, which can be equated to metabolic rate from the relative turnover rates of isotopes measured upon resampling (recapture) of the animal; this value is termed the *field metabolic rate* (FMR) (Haufler and Servello 1994).

A large and growing body of literature reports on FMR as determined using doubly labeled water (Nagy 1987; Speakman and Racey 1987; Nagy and Obst 1989). In birds, for example, FMR estimates (in kJ/day) include 118 in the 22-g tree swallow (*Tachycineta bicolor*); 343 in the 220-g Eurasian kestrel (*Falco tinnunculus*); 997 in the 1089-g little penguin (*Eudyptula minor*); and 2401 in the 3706-g grey-headed albatross (*Diomedea chrysostoma*) (Nagy and Obst 1989).

Table 7.2 presents mean values for the rate of energy expenditure associated with the activities given for the shrike in table 7.1. From such data one can determine how energetic costs vary with observed variations in behavior in the field and, further, relate these energy costs to survival, reproductive output, and body condition. For example, Mock (1991) used daily allocation of time and energy of western bluebirds (*Sialia mexicana*) to examine the trade-off between parental survival and survival of their young. She concluded that the species regulates overall daily energy expenditures through differential use of thermal environments and activity budgets. Speakman and Racey (1987; see also Entwistle et al. 1995) used doubly labeled water to study the energetics of the brown long-eared bat (*Plecotus auritus*). Other methods of determining energy costs include estimating mass loss after activity periods (e.g., long flights), oxygen consumption, and heart rate telemetry (see Goldstein 1990 for review).

Foraging and Diet

In the following sections we first discuss foraging theory and then introduce the core research areas of diet, sampling techniques, identification of prey, and statistical indices.

Optimal Foraging Theory

Foraging behavior research has advanced our understanding of animals and animal management in many important ways: in assessing habitat and designing reserves; in livestock husbandry and grazing management; in better understanding the limitations of resources to herbivores and matching their dietary preferences with resource availability; in understanding and predicting foragers' impact on their resources and environment; and in control of pests (Ash et al. 1996). This knowledge has provided the basis for advances in many aspects of wildlife management. For example, we now know that herbivore diet is closely linked to habitat and patch use; habitats and patches are objects that can be manipulated.

Table 7.2. Cost of activity in loggerhead shrikes

	Number of shrikes	Number of observations	kJ/hr	Multiple of H_b
Basal metabolism (H_b)	9	27	1.79 ± 0.20	1.00
Alert perching	6	68	3.51 ± 0.60	1.98
Preening	3	4	3.87 ± 0.71	2.18
Eating	5	21	3.87 ± 0.63	2.19
Hopping	2	5	4.05 ± 0.67	2.28
Flying	—	—	23.7	13.20

Source: Weathers et al. 1984, table 3. Reprinted with permission of the American Ornithologists' Union.

Animals are generally adapted to most efficiently exploit specific types of foraging substrates. A substrate is the specific location and surface at which the animal directs foraging. The animal must compare the risks associated with foraging in a particular manner with risks associated with other methods and other locations. The choice of foraging method and location is based not only on the number of prey present but also on the quality of that prey. That is, do numerous low-quality items that are easy to obtain have a net energy benefit over scarce but high-quality items that are difficult to obtain? Schoener (1969) hypothesized that animals can achieve this net balance by two extreme strategies: energy maximization or time minimization. Energy maximizers try to obtain the greatest amount of energy possible within a given period of time. Time minimizers seek to minimize the time required to obtain a given amount of food. Time minimizers thus have more time available for other activities, such as grooming and parental care (see also Morse 1980, 53–54). As Stephens and Krebs (1986) recognized, however, maximization of net energy is not necessarily a desirable goal. The confusion, they observed, concerns equating the net rate of energy consumption with the ratio of benefit to cost (or foraging efficiency). They noted that maximizing efficiency ignores the time required to harvest resources, and it fails to distinguish tiny gains made at a small cost and larger gains made at a larger cost. For example, a gain of 0.01 calorie at a cost of 0.001 calorie gives the same benefit:cost ratio as a gain of 10 calories costing 1 calorie. The 10-calorie alternative, however, yields 1000 times the net profit of the 0.01 alternative (Stephens and Krebs 1986, 8–9).

These and related considerations of costs and benefits form the basis of a large area of scientific investigation collectively known as *optimal foraging theory*. The term *optimal foraging* is a poor choice of words, however, because it implies there is some optimal strategy to follow; foraging theory per se makes no such claim (Stephens 1990). Although some have questioned the usefulness of optimal foraging theory (e.g., Pierce and Ollason 1987), it is our contention that all studies must be firmly based in theory, and that optimal forging theory provides a guide for development of a research strategy. In addition, comparing actual field data against an "optimal" model allows determination of how divergent optimal and realized strategies are.

Stephens and Krebs's (1986) book provides an excellent coverage of foraging theory. Here we will examine these theories and their associated models only as they relate to our descriptions of how animals might perceive their environment. Ideas about perception lead, in turn, to how we should observe and record animals as they exploit food resources. Thus we will briefly explore how models of foraging behavior can help us design our studies.

In considering how animals determine whether or not to forage in an area, we have two rather distinct alternative views: Is the animal selecting from among various prey distributed throughout a generally suitable area? or is the animal distinguishing between various patches of prey? Foraging theorists thus distinguish between prey-choice models and patch-choice models. An animal can forage "optimally" within either of these two basic constructs. These models have clear implications for the design of foraging studies. That is, if prey is distributed in a patchy manner, then our assessment of prey abundance or availability must be of proper spatial and temporal scale to recognize these patches. An overall average of prey abundance over a large area would fail to identify their patchy nature. Likewise, studies conducted at too small a scale could fail to identify any patches, or could identify patches that the animal might not recognize.

Both prey-choice and patch-choice models assume that a foraging animal sequentially searches for prey, encounters the prey, and then decides whether or not try and consume the prey. However, the form of the decision taken by the animal upon encountering a prey item differs between these two models. In the prey-choice model, an animal must decide whether it should take the item or continue searching. One can then state various rules upon which the animal should base its decision (e.g., based on prey size). By contrast, in the patch-choice model, the animal decides how long it should forage in the patch encountered. Both models consider how an animal can best make these decisions with the goal of maximizing the long-term average rate of energy intake.

These models thus develop a general theoretical constraint upon which we can base our foraging studies. Clearly, the rate and method of searching, the frequency of encounter of a prey item, and the type of attack used all tell us a great deal about the ecology of a species. Further, variations in search, encounter, and prey consumption may lend insight into the current physiological condition of the animal. Thus measuring aspects of foraging rate, the frequency and types of encounter, and related aspects of foraging is important in understanding animal distribution, and survival and reproductive success in occupied locations.

Foraging theory predicts, when the animal considers prey items, that the decision to take a specific item will depend not only on the item's abundance but also on the abundance of other food items. Food items can be ranked by their ratio of food value to handling time (Morse 1980, 54). Handling time is the amount of time necessary to pursue, capture, and consume the prey (Stephens and Krebs 1986, 14). Thus foraging theory suggests that we should be concerned not only with how an animal goes about foraging, but also with the types and relative ranking

of abundance and quality of prey encountered by the animal. Sampling the abundance of only items consumed by the animal tells us little about the reasons for that decision.

The foraging patterns of animals has also been evaluated using the concept of "giving-up density" (GUD), which is the level at which an individual ceases to use a particular resource. Most of the resource used is some type of food that is mixed within a matrix of materials such as sand, resulting in a diminishing rate of return (to the individual) as the food is consumed (Yunger et al. 2002).

Foraging theory has been applied as a framework for the design of studies of herbivore ecology. As outlined by Wilmshurst et al. (1995), optimal foraging theory assumes that foraging decisions by herbivores should be strongly influenced by physiological and environmental constraints on the rates of nutrient uptake. Two such constraints commonly invoked for vertebrate grazers are the effect of plant density on the short-term rate of food uptake (the availability constraint) and the effect of digestive capacity on the long-term rate of energy assimilation (the processing constraint). The short-term rate of food intake should be positively related to plant size, bite size, and plant density. Using this model as a guide, we would thus design our studies to measure plant size and density, and relate these environmental variables to bite size and actual food intake.

Working with elk, Wilmshurst et al. (1995) hypothesized that at low biomass, the processing rate of forage should be high, but the short-term rate of intake low; whereas at high forage biomass, the processing rate should be low, but the short-term rate of intake high. The "foraging maturation hypothesis" states that the net rate of energy intake should be maximized accordingly on patches of intermediate plant biomass. Wilmshurst and colleagues concluded that the preference for grass patches of intermediate biomass and fiber content could help explain patterns of animal aggregations and seasonal migration.

Schmitz (1992) tested whether white-tailed deer selected their diet in accordance with a foraging model that predicts that to maximize fitness, deer have to balance a trade-off between maximizing growth and offspring production with minimizing risk of starvation. He found that deer appeared to show plasticity in their diet in response to temporal and spatial changes in perceived risks and gains. Deer appeared to balance gains in fitness due to reproduction with losses in fitness due to energetic shortfalls. When starvation risk was eliminated, deer tended to select diet breadths that simply maximized their mean rate of energy intake.

Foraging theory directly relates to wildlife management. As developed by Nudds (1980) for ungulates, the type of foraging model most closely followed by an animal (energy maximizing, equal food value, nutrient optimizing, unequal food value) can guide land management decisions concerning the amounts and types of food to emphasize. Nudds concluded that deer, as well as other temperate-latitude ungulates, are primarily habitat specialists but become food generalists in winter. The foraging behavior of deer in winter adhered most closely to the predictions of the energy maximizing models; it seemed energetically less costly to remain in sheltered areas and fast than to forage in exposed areas. Translating these conclusions to a management scenario, Nudds suggested that manipulating winter habitats of deer by increasing only the abundance of "preferred" food would not be warranted. Management would more beneficially be directed toward the physical structure of the winter habitat. Although some of Nudds's suggestions have been criticized (Jenkins 1982; but see Nudds 1982), his 1980 paper is important in that it directly links theory with management. Likewise, Kotler et al. (1994) showed how

study of patch use in Nubian ibex (*Capra ibex*) can be used by managers to modify habitat to reduce predation risk, thus allowing the animals to use available food more efficiently. As noted by Jenkins (1982), foraging theory, combined with good empirical work on food preference, may lead to valuable new insights about problems in wildlife management.

Foraging models used within a management context should address the spatial and temporal scales at which management can be controlled. These management scales are usually larger than the scales of the models; management scales are often determined by socioeconomic factors not under the control of the researcher. Although management may not specifically address the details of grazing behavior, for example, an understanding of the details of the foraging process is essential to provide a contextual framework for decision making (Ash et al. 1996).

Diet

A dichotomy exists in the literature regarding the emphasis placed on quantification of animal diets. Wildlife biologists and economic entomologists have expended much effort to determine the actual food items consumed by animals. Korschgen (1980) observed that, in the late 1800s and early 1900s, studies of food habits examined the economic importance of bird feeding habits, concentrating on the plunder of agricultural crops, poultry, and livestock. The greatest activity in food-habit studies took place in the 1930s and 1940s, emphasizing waterfowl and upland game birds. Regarding ungulate food habits, papers dealing with diets dominated the literature prior to 1950. The proportion of research reporting on food availability, food digestibility, and food requirements has grown steadily since that time.

In contrast, scientists studying the ecological relationships of animals seldom attempt to quantify the occurrence of specific prey items in the diet. Rather, ecological studies have concentrated on indirect measures of food use, such as foraging location (Hutto 1990; Rosenberg and Cooper 1990). Although morphological differences between species undoubtedly reflect some degree of evolutionary response to resources, they may not necessarily be good predictors of species' diets, especially under local environmental conditions (Rosenberg and Cooper 1990).

Although studies of food habits abound, most are single-species studies from single locations that were conducted over a short period of time. Thus little generalization is possible. Further, as noted by Rosenberg and Cooper (1990), one of the reasons that studies of avian diets have been neglected by modern ornithologists is that researchers fear the detail, tedium, and technological expertise thought to be necessary for such studies. Regardless of the reasons, little literature exists on the diets of most species of animals in the world that is useful in describing and especially predicting their patterns of habitat use.

Sampling Techniques

Methods used to study diets of vertebrates can be divided into three basic categories: those involving collection of individual animals, those involving capture or other temporary disturbances of individual animals, and those requiring little or no disturbance of individuals (Rosenberg and Cooper 1990).

Several reviews of dietary assessments are readily available. Although many of these studies directly concern specific groups of animals (such as seabirds), many of the methods also apply to other groups of animals. Rosenberg and Cooper (1990) provided a thorough review of methods used to sample bird diets. Ratti et al. (1982, 765–913) reprinted papers on food habits and

feeding ecology of waterfowl; they included a bibliography of other important references on diet. Each new edition of the Wildlife Society's *Wildlife Management Techniques Manual* includes reviews of methods for birds and mammals (see Haufler and Servello 1994; Litvaitis et al. 1994). Riney (1982, 124–37) summarized studies of mammalian food habits, and many authors in Cooperrider et al. (1986) covered diet studies in wildlife.

The most frequently used method of sampling diets is direct examination of stomach contents. The primary advantage of such sampling is that adequate numbers of stomachs are usually relatively easy to obtain. That is, animals can be collected through trapping or shooting. With shooting, an individual animal can be collected after the researcher has observed its specific foraging behavior; one can then attempt to relate the specific food items in the stomach to those sampled from the foraging substrate and the behavior used to gather food by the animal. For game animals, researchers often take stomach samples from hunter check stations. Another advantage of gut sampling is that the entire contents of the stomach can be obtained. Kill sampling, however, has numerous disadvantages: The animal obviously cannot be resampled at a later date, preventing quantification of temporal (and possibly spatial) changes in food habits. The researcher has a potentially substantial impact on the population under study, negating studies of other aspects of the population's ecology (i.e., abundance, reproductive performance, behavior). Finally, the researcher is often subjected to severe criticism from certain segments of the public when the killing of animals is included in research.

Nondestructive methods of sampling food habits are available for wildlife. Live-caught animals can be forced to regurgitate using a variety of chemical emetics. Although some mortality can occur from the use of emetics, methods are available that will minimize losses (see Rosenberg and Cooper 1990 for review). For many animals, the most easily obtained samples of diets come from their feces, collected either from the environment or during live trapping. In live-trapping studies, droppings can be obtained year-round from animals of any age or any reproductive state, and repeated sampling from known individuals is possible. Ralph et al. (1985) described a technique for collecting and analyzing bird droppings. This and related techniques have been used successfully in many studies of birds (Davies 1976, 1977a, b; Waugh 1979; Waugh and Hails 1983; Tatner 1983; Ormerod 1985) and small mammals (Meserve 1976; Dickman and Huang 1988). Many studies have detailed methods of fecal collection and analysis in ungulates (Riney 1982, 129–31; Haufler and Servello 1994).

Nondestructive methods of determining food habits also involve observation of animal foraging behavior and analysis of food removal rates. Direct observation of food eaten by animals is possible with some species in some vegetation types. Many studies have been designed to quantify the amount of plant material removed by foraging ungulates. Researchers assess the height, weight, and condition of plants over periods of time and then relate the results to the type and amount of food consumed (Dasmann 1949; Severson and May 1967; Willms et al. 1980). Large ungulates can sometimes be observed grazing or browsing: "bite counts," calculated as the number of bites per plant species, are recorded (Willms et al. 1980; Thill 1985). Studies of food removal and bite counts can be combined to develop a picture of food habits. Diurnal birds of prey, such as eagles, vultures, and large hawks, can be observed when they forage in open areas.

Unfortunately, nondestructive methods of assessing food habits do not present a panacea for the researcher. In most cases, foraging animals cannot be observed closely without adversely influencing their behavior, thus negating direct observation of food habits. Because trapping most species is difficult, food in the stomach is often highly digested, one does not know where or when the animal was feeding, and there is little control of the animals captured.

As with most aspects of wildlife research, a combination of carefully selected methods is usually necessary. A useful strategy is to first determine if the species under study shows any foraging behaviors that depart markedly from other closely related species. For example, studies of foraging behavior in the Sierra Nevada showed that many small insectivorous birds (e.g., chickadees, *Parus*) significantly increased their use of bark foraging during winter; a closer examination of food availability and food habits was warranted (Morrison et al. 1989). Preliminary sampling (using direct or indirect methods) of stomach samples also indicates the sample sizes necessary and the time required to analyze those samples, as well as the likely level of resolution possible after a full study is conducted. There is no excuse for simply collecting large numbers of stomachs that will either sit on the laboratory shelf or yield no useful information for use in the management of a species.

Differential digestion rates of food items impose a large potential bias in any study of gut contents. Different kinds of foods at about the same time often are digested at different rates. Further steps must be taken to prevent excessive postmortem digestion of food. For example, small-bodied insects may be gone from the gizzard within 5 minutes, whereas hard seeds may persist for several days (Swanson and Bartonek 1970). Several authors have developed correction factors for the differential rates of digestion shown by animals (Mook and Marshall 1965). Differential digestion of food is not confined to the intertaxonomic level: Rosenberg and Cooper (1990) discussed data that showed that second and third instar moth larvae were digested in less than half the time it took to digest fourth and fifth instars. As noted throughout this book, the goals of a specific study will determine the level of identification required to reach useful conclusions.

Level of Identification

The topic of the proper taxonomic identification of prey items has received little attention in the literature. The taxonomic level selected for diet analysis can have substantial impacts on the ecological interpretation of results. This problem is analogous to the selection of variables for use in habitat models; that is, how finely should we divide categories? If identification to the species level were a simple matter, then this issue would be of minor concern. Not only are many food items difficult to identify when in excellent condition, the mastication and digestion of food complicates the task. The level at which foods are identified is likely to affect similarity measures and conclusions drawn from them.

Greene and Jaksic (1983) studied the influence of prey identification level on measures of niche breadth and niche overlap in raptors, carnivores, and snakes. They calculated niche metrics for the finest prey identification levels (usually specific or generic) reported in diet studies and then recalculated the metrics after combining the prey lists to the ordinal level. They found that niche breadth was consistently larger at the finer prey identification levels than for the ordinal level of classification for all vertebrate groups examined. Calculations using the ordinal levels underestimated niche breadth at higher-resolution levels by 17 to 242% and underestimated niche-breadth scores for single species even more extremely. Food-niche overlaps based on the ordinal level overestimated higher-resolution overlap. Overestimates ranged to infinity when two species did not coincide in the

use of any prey species but appeared to do so because of an ordinal level of identification. Greene and Jaksic clearly showed that using the ordinal level of prey classification can result in serious misinterpretations of some of the potentially most important food-niche and community parameters in assemblages of many animals.

Cooper et al. (1990) offered several guidelines regarding the level of taxonomic identification to choose in a study. Taxonomic levels should be identified that contain enough observations to make analysis meaningful. There are several practical considerations here. First, variables (prey categories) with high numbers of zero counts will not be normally distributed and usually cannot be transformed to normality. Thus multivariate statistical procedures lose validity (see chapter 6). Second, one should decide if dividing a particular order into finer levels will result in any benefit. That is, if ecological and behavioral characteristics of two groups within an order do not differ substantially, then it is unlikely that subdivision will provide much additional information. Food items do not necessarily need to be identified using Linnaean nomenclature. Phenotypically distinguishable taxa (such as grasshopper A, B, and so on) can substitute for Linnaean identification (see also Greene and Jaksic 1983; Wolda 1990). The level of food identification chosen for a study should be made based on the goals of the study, and not a posteriori on the basis of funds, time, or difficulty in identifying the items. Simple preplanning prevents later disappointments, and more importantly, potential waste of animal life when kill sampling is involved.

Analyses

Statistical analyses of behavioral data should be an integral part of the initial study design. A large and varied number of methods of statistical analysis have been used with behavioral data, depending on the goals of the researcher and the form of the data. Clearly, sufficient and appropriate planning should guide the analyses used.

We will not review the many statistical techniques that can be applied to behavioral data. Readers can consult virtually any general statistical text for direction on analyzing behavioral data. Especially useful techniques are those of Siegel (1956); Conover (1980); Snedecor and Cochran (1980); Sokal and Rohlf (1995); and Zar (1999). Texts dealing more specifically with quantitative methods in ethology include Hazlett (1977); Colgan (1978); Siegel and Castellan (1988); Weinberg and Goldberg (1990); Sommer and Sommer (1991); and Lehner (1996). Martin and Bateson (1993, ch. 9) presented a concise but thorough review of fundamental univariate techniques for the study of animal behavior. Our chapter 6 looks at multivariate methods. Although not dealing specifically with statistical methods, Kamil (1988) discussed the application of experimental methods from the perspective of a behavioralist. Although concentrating on applications to foraging studies, many papers in Morrison et al. (1990) can be applied to analysis of any type of behavioral data. Roa (1992) and Manly (1993) discuss interesting analyses for use in experiments of food preference.

Behavioral data are usually recorded as categorical variables, such as various activity types (e.g., walking, running). Continuous data are often later classified into categories for analysis, such as speed of movement. When data are so classified, the result is a contingency table, the cells of which contain frequencies of the various category combinations (e.g., activity type by sex). The null hypothesis of homogeneity of categories is then tested using contingency analysis, such as the chi-square or *G log*–likelihood statistic. When three or more categories are compared, multidimensional contingency table analysis is used (see Colgan and Smith 1978).

Researchers usually record behaviors of animals in some sequential fashion—for example, walk-pause-walk-bite-swallow-walk. Analysis of such data using contingency tables and chi-square and related analyses may not be strictly valid, however, because sequential observations are not likely independent and thus violate the important assumption of independence of most statistical tests (Raphael 1990). The often-sequential nature of data collection can, however, be a benefit in the elucidation of behavioral data. Examining the sequence of behaviors of an individual provides potentially much more information on how the animal exploits its environment than does an overall lumping or averaging of behaviors.

A sequence in which the behavioral pattern always occurs in the same order is termed *deterministic*. Classical behavioralists refer to such sequences as *fixed action patterns*. Vertebrates seldom if ever repeat behaviors in the same order, of course, but they exhibit some amount of variability that may be predictable. These sequences are considered *stochastic* (or probabilistic, to identify the statistical probability that can be assigned to each behavior). Sequences that show no temporal pattern—the component behavior patterns are sequentially independent—are considered *random sequences*. In a random sequence, one behavior or set of behaviors can be followed by any other behavior with equal probability. The conditional probability that one behavior follows another—that is, the probability that behavior B follows behavior A, given that A has occurred—denoted P (B/A), is called the *transitional probability* (Martin and Bateson 1993, 152–54).

There are several methods for analyzing sequential data, such as time-series analyses. Of particular interest to us are analyses involving Markov chains. Markov analysis is a method for distinguishing whether a sequence is random or whether it contains some degree of temporal order. A first-order Markov process is one in which the probability of occurrence for the next event depends only on the immediately preceding event. If the probability depends on the two preceding events, then the process is considered second-order. Higher-order processes are involved as additional events are considered. One analyzes sequences by comparing the observed frequency of each transition with the frequency of transitions that would be expected if the sequences were random (Martin and Bateson 1993, 152). A simple example of Markov analysis was given by Martin and Bateson (1993, 153) and is repeated here in figure 7.4. Raphael (1990)

presented a review and a detailed example of the application of Markov analysis to foraging data; see also Colwell (1973); Riley (1986); Diggle (1990); and Gottman and Roy (1990) for analyses of sequential data.

Analyzing the sequential nature of data can identify changes in the behavior of individuals that might be obscured at least initially by examining only overall averages. Further, such analyses can identify which aspect of an individual's behavior is being impacted by a change (natural or human induced) in its environment. One can also relate each step in a sequence of behaviors to measures of the environment encountered

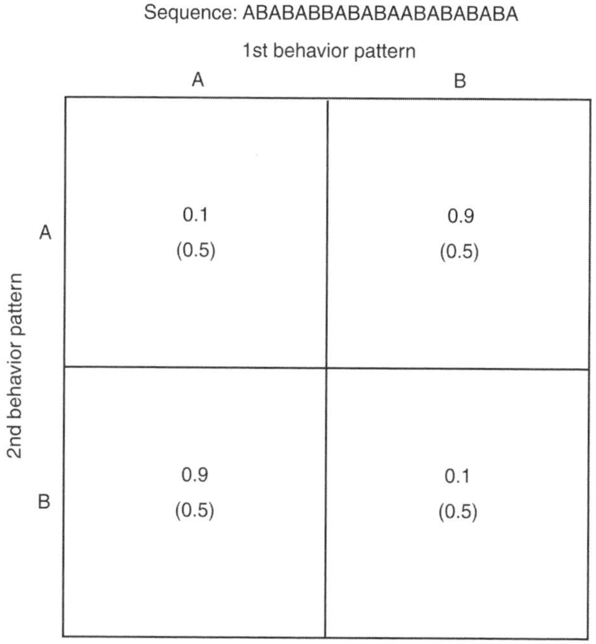

Sequence: ABABABBABABABAABABABABA

Figure 7.4. A highly simplified transition matrix, analyzing the sequence shown above it, which comprises only two different behavior patterns (A and B). The matrix shows the empirical transition probabilities for the four different types of transition (A|A, B|A, A|B, B|B). For example, the lower left cell shows that the conditional probability of B, given the occurrence of A (B|A), is 0.9 (9 out of 10 transitions after A has occurred). For comparison, the transition probabilities under a random model (0.5 for each type in this example) are shown in parentheses. The matrix confirms that A and B tend to alternate (the probabilities of B|A and A|B are high), while repeats are rare (the probabilities of B|B and A|A are low). (From Martin and Bateson 1993, p. 154, fig. 9.5; reprinted with permission of Cambridge University Press.)

during each step. For example, the behavioral sequence of a foraging bird (e.g., hop-hop-glean-probe-hop-hop) can be related to the foraging substrate being encountered (e.g., ground-ground-leaf-bark-ground-ground). Behavioral sequences might lend insight into the response of an animal to treatment effects that might not be evident using averages. Yet, such changes in behavior might explain changes in activity times and perhaps even changes in fecundity. Here again, the goal of the study should determine the detail required.

Indices

Scientists have developed a type of methodology to quantify the use of resources in relation to their availability. Widely known as *preference, selectivity,* or *electivity indices,* these measures seek to compare the frequency of use of a resource with the availability of those items in an animal's environment by representing the data as a single index value; Manly et al. (2002) reviewed the history of the development of these indices. Many of these indices have received a good deal of attention in the wildlife literature, especially with regard to analysis of the food habits of ungulates; they have also been applied to vegetation and other measures of the environment (Morrison 1982). Because of this attention and the potential application of indices to a wide variety of situations, we will discuss some of the important considerations in their use.

The general approach to using electivity indices is to establish a ratio between frequency of an item used by an animal and the amount of that item available for use. For example, Ivelv's electivity index (Ivlev 1961) compares relative availability of an item in the environment (p) with their relative use of that item (r): $E_i = (r_i - p_i)/(r_i + p_i)$. Other indices have similar form, as reviewed by Lechowicz (1982) and Manly et al. (2002). If r and p are equal for all items, then the

animal is choosing the items randomly. If r and p differ, then one usually concludes that the animal is either avoiding (a negative index value) or selecting (positive value) an item.

The most straightforward indices simply consist of the estimated percentages of use of an item divided by the total amount of all items available for use. Values usually range from -1 to 0 for avoidance to 0 to infinity for preference. Such indices have been termed the *forage ratio* (Jacobs 1974) and have been widely used (e.g., Heady and Van Dyne 1965; Chamrad and Box 1968; Petrides 1975; Hobbs and Bowden 1982). The major drawback to these simple indices, however, is their intrinsic asymmetry—that is, their unbounded positive values. A log transformation is used for the forage ratio (Jacobs 1974; Strauss 1979; Lechowicz 1982). Unfortunately, the forage ratio also changes with the relative abundance (p) of items in the environment. Thus this index cannot be used if one wishes to examine the relationship between the relative abundance of items and the animal's preference for those items (Jacobs 1974; Lechowicz 1982).

Readers need only remember our earlier discussions of the great difficulties in accurately quantifying use and availability of resources to identify the overriding factor influencing index values. Clearly, proper determination of the variables to compare in such analyses will largely determine the results. An example modified from Johnson (1980) will illustrate this point. Suppose an investigator collects an animal and finds that its gut contains food items A, B, and C in the percentages shown in the top panel of table 7.3 under the first "Usage" column. A sample of the animal's feeding area reveals that the items were present in the percentages shown under the first "Availability" column. Many investigators would conclude that item A was avoided because use was less than availability, while items B and C were selected. But suppose other investigators do not believe that item A is a valid food item; per-

Table 7.3.   Results of comparing usage and availability data when a commonly available but seldom-used item (A) is included and excluded from consideration

Item	Usage (%)	Availability (%)	Conclusion	Rank Usage	Rank Availability	Rank Difference
			Item A included			
A	2	60	Avoided	3	1	+2
B	43	30	Preferred	2	2	0
C	55	10	Preferred	1	3	−2
			Item A excluded			
A	44	75	Avoided	2	1	+1
B	56	25	Preferred	1	2	−1

Source: Modified from Johnson 1980, table 1. Reprinted with permission of the Ecological Society of America.

haps it was ingested only accidentally while the animal consumed other foods. They would then consider the data in the bottom panel of table 7.3, obtained by deleting item A from the analysis. Now, although item C is still deemed selected for, the assessment of item B has changed from preferred to avoided. Thus we see that conclusions drawn from such analyses will depend markedly on the array of items thought by the investigator to be available to the animal. Naturally, this caution applies whether or not indices are used. The level of prey identification used (as discussed earlier) will also markedly impact conclusions. Here it is critical that the researcher use preliminary studies to identify valid items.

The work of Ivlev (1961), Jacobs (1974), Chesson (1978, 1983), Strauss (1979), Vanderploeg and Scavia (1979a, 1979b), Johnson (1980), and Manly et al. (2002) described the development of the more widely used electivity indices. Lechowicz (1982) compared the characteristics of seven of the electivity indices proposed by these researchers; most of these indices are permutations of Ivlev's original index. Lechowicz found that Ivlev's, Strauss's, and Jacobs's indices could not potentially obtain the full range of index values for all values of r and p.

The index values for intermediate values of r and p depend on the relative abundance of other items in the environment or of those used by the animal. A critical problem with most of the indices is that direct comparisons between indices derived from samples differing in relative abundances are inappropriate (but see Chesson 1983 for exceptions).

To avoid the problems associated with inclusion or exclusion of specific items in the calculation and evaluation of indices, Johnson (1980) developed a procedure based on ranks. He proposed using the difference between the rank of use and the rank of availability. If we use the earlier example with item A included (see table 7.3), the differences in ranks of use and availability are +2, 0, and −2 for items A, B, and C, respectively. Excluding item A from the analysis results in values of +1 and −1 for B and C, respectively. Although the index values themselves change, the difference between B and C remains +2. The loss of information regarding the absolute difference between items realized when using ranks is likely of little consequence. Statistical methods based on ranks are nearly as efficient as methods based on the original data—especially if the assumptions necessary to treat the original data are not met (e.g., assumption of normality). Johnson also provided methods for determining the statistical significance among components of the data. The method of compositional analysis developed in chapter 5 can also be used to analyze diet and activity data.

Literature Cited

Altmann, J. 1974. Observational study of behavior: Sampling methods. *Behaviour* 49:227–67.

Ash, A., M. Coughenour, J. Fryxell, W. Getz, J. Hearne, N. Owen-Smith, D. Ward, and E. A. Laca. 1996. Second international foraging behavior workshop. *Bulletin of the Ecological Society of America* 77:36–38.

Bachman, G. C. 1993. The effect of body condition on the trade-off between vigilance and foraging in Belding's ground squirrels. *Animal Behaviour* 46:233–44.

Baulu, J., and D. E. Redmond Jr. 1978. Some sampling considerations in the quantification of monkey behavior under field and captive conditions. *Primates* 19:391–400.

Berdoy, M., and S. E. Evans. 1990. An automatic recording system for identifying individual small animals. *Animal Behaviour* 39:998–1000.

Bernstein, I. S. 1991. An empirical comparison of focal and ad libitum scoring with commentary on instantaneous scans, all occurrence and one-zero techniques. *Animal Behaviour* 42:721–28.

Block, W. M. 1990. Geographic variation in foraging ecologies of breeding and nonbreeding birds in oak woodlands. *Studies in Avian Biology* 13:264–69.

Brennan, L. A., and M. L. Morrison. 1990. Influence of sample size on interpretation of foraging patterns by chestnut-backed chickadees. *Studies in Avian Biology* 13:187–92.

Caraco, T., S. Martindale, and T. S. Whittham. 1980. An empirical demonstration of risk-sensitive foraging preferences. *Animal Behaviour* 28:820–30.

Carranza, J. 1995. Female attraction by males versus sites in territorial rutting red deer. *Animal Behaviour* 50:445–53.

Chamrad, A. D., and T. W. Box. 1968. Food habits of white-tailed deer in south Texas. *Journal of Range Management* 21:158–64.

Chesson, J. 1978. Measuring preference in selective predation. *Ecology* 59:211–15.

Chesson, J. 1983. The estimation and analysis of preference and its relationship to foraging models. *Ecology* 64:1297–1304.

Colgan, P. W., ed. 1978. *Quantitative ethology.* New York: Wiley.

Colgan, P. W., and J. T. Smith. 1978. Multidimensional contingency table analysis. In *Quantitative ethology,* ed. P. W. Colgan, 145–74. New York: Wiley.

Colwell, R. K. 1973. Competition and coexistence in a simple tropical community. *American Naturalist* 107:737–60.

Conover, W. J. 1980. *Practical nonparametric statistics.* 2nd ed. New York: Wiley.

Cooper, R. J., P. J. Martinat, and R. C. Whitmore. 1990. Dietary similarity among insectivorous birds: Influence of taxonomic versus ecological categorization of prey. *Studies in Avian Biology* 13: 104–9.

Cooperrider, A. Y., R. J. Boyd, and H. R. Stuart, eds. 1986. *Inventory and monitoring of wildlife habitat.* Denver: USDI Bureau of Land Management Service Center.

Courchamp, F., G. S. A. Rasmussen, and D. W. Macdonald. 2002. Small pack size imposes a trade-off between hunting and pup-guarding in the painted hunting dog *Lycaon pictus. Behavioral Ecology* 13: 20–27.

Dasilva, G. L. 1992. The western black-and-white colobus as a low-energy strategist: Activity budgets, energy expenditure and energy intake. *Journal of Animal Ecology* 61:79–91.

Dasmann, W. P. 1949. Deer-livestock forage studies in the interstate winter deer range in California. *Journal of Range Management* 2:206–12.

Davies, N. B. 1976. Food, flicking, and territorial behavior of the pied wagtail (*Motacilla alba yarelli* Gould). *Journal of Animal Ecology* 45:235–52.

Davies, N. B. 1977a. Prey selection and social behavior in wagtails (Aves: Motacillidae). *Journal of Animal Ecology* 46:37–57.

Davies, N. B. 1977b. Prey selection and the search strategy of the spotted flycatcher (*Muscicapa striata*): A field study on optimal foraging. *Animal Behaviour* 25:1016–33.

Davoren, G. K., W. A. Montevecchi, and J. T. Anderson. 2003. Search strategies of a pursuit-diving marine bird and the persistence of prey patches. *Ecological Monographs* 73:463–81.

Defler, T. R. 1995. The time budget of a group of wild

woolly monkeys (*Lagothrix lagotricha*). *International Journal of Primatology* 16:107–20.

de Vries, H., W. J. Netto, and P. L. H. Hanegraaf. 1993. Matman: A program for the analysis of sociometric matrices and behavioural transition matrices. *Behaviour* 125:157–75.

Dewsbury, D. A. 1992. On the problems studied in ethology, comparative psychology, and animal behavior. *Ethology* 92:89–107.

Dickman, C. R., and C. Huang. 1988. The reliability of fecal analysis as a method for determining the diet of insectivorous mammals. *Journal of Mammalogy* 69:108–13.

Diggle, P. J. 1990. *Time series: A biostatistical introduction.* Oxford: Oxford Univ. Press.

Entwistle, A. C., J. R. Speakman, and P. A. Racey. 1995. Effect of using the doubly labelled water technique on long-term recapture in the brown long-eared bat (*Plecotus auritus*). *Canadian Journal of Zoology* 72:783–85.

Finch, D. M. 1984. Parental expenditure of time and energy in the Abert's towhee (*Pipilo aberti*). *Auk* 101:473–86.

Ford, H. A., L. Bridges, and S. Noske. 1990. Interobserver differences in recording foraging behavior of fuscous honeyeaters. *Studies in Avian Biology* 13:199–201.

Glover, T. 2003. Developing operational definitions and measuring interobserver reliability using house crickets (*Acheta domesticus*). In *Exploring animal behavior in laboratory and field,* ed. B. J. Ploger and K. Yasukawa, 31–40. San Diego: Academic Press.

Goldstein, D. L. 1990. Energetics of activity and free living in birds. *Studies in Avian Biology* 13:423–26.

Gottman, J. M., and A. K. Roy. 1990. *Sequential analysis: A guide for behavioral researchers.* Cambridge: Cambridge Univ. Press.

Greene, H. W., and F. M. Jaksic. 1983. Food-niche relationships among sympatric predators: Effects of level of prey identification. *Oikos* 40:151–54.

Hale, A. M., D. A. Williams, and K. N. Rabenold. 2003. Territoriality and neighbor assessment in brown jays (*Cyanocorax morio*) in Costa Rica. *Auk* 120:446–56.

Hall, L. S., P. R. Krausman, and M. L. Morrison. 1997. The habitat concept and a plea for standard terminology. *Wildlife Society Bulletin* 25:173–82.

Hamilton, W. D. 1971. Geometry for the selfish herd. *Journal of Theoretical Biology* 31:293–311.

Harcourt, A. H., and K. J. Stewart. 1984. Gorillas' time feeding: Aspects of methodology, body size, competition and diet. *African Journal of Ecology* 22:207–15.

Haufler, J. B., and F. A. Servello. 1994. Techniques for wildlife nutritional analysis. In *Research and management techniques for wildlife and habitats,* 5th ed., ed. T.A. Bookhout, 307–23. Bethesda, MD: Wildlife Society.

Hazlett, B. A., ed. 1977. *Quantitative methods in the studies of animal behavior.* New York: Academic Press.

Heady, H. F., and G. M. Van Dyne. 1965. Botanical composition of sheep and cattle diets on a mature animal range. *Hilgardia* 36:465–92.

Hejl, S. J., J. Verner, and G. W. Bell. 1990. Sequential versus initial observations in studies of avian foraging. *Studies in Avian Biology* 13:166–73.

Hobbs, N. T. 1989. Linking energy balance to survival in mule deer: Development and test of a simulation model. *Wildlife Monographs* 101:1–39.

Hobbs, N. T., and D. C. Bowden. 1982. Confidence intervals on food preference indices. *Journal of Wildlife Management* 46:505–7.

Hurlbert, S. H. 1984. Pseudoreplication and the design of ecological field experiments. *Ecology* 54:187–211.

Hutto, R. L. 1990. On measuring the availability of food resources. *Studies in Avian Biology* 13:20–28.

Immelmann, K., and C. Beer. 1989. *A dictionary of ethology.* Cambridge: Harvard Univ. Press.

Ivlev, V. S. 1961. *Experimental ecology of the feeding of fishes.* New Haven: Yale Univ. Press.

Jacobs, J. 1974. Quantitative measurement of food selection. *Oecologia* 14:413–17.

Jenkins, S. H. 1982. Management implications of optimal foraging theory: A critique. *Journal of Wildlife Management* 46:255–57.

Johnson, D. H. 1980. The comparison of usage and availability measurements for evaluating resource preference. *Ecology* 61:65–71.

Jones, Z. F., and C. E. Bock. 2003. Relationships between Mexican jays (*Aphelocoma ultramarina*) and northern flickers (*Colaptes auratus*) in an Arizona oak savanna. *Auk* 120:429–32.

Kamil, A. C. 1988. Experimental design in ornithology. In *Current Ornithology,* vol. 5, ed. R. F. Johnston, 312–46. New York: Plenum Press.

Kamil, A. C., and T. D. Sargent, eds. 1981. *Foraging behavior: Ecological, ethological, and psychological approaches.* New York: Garland STPM Press.

Kamil, A. C., J. R. Krebs, and H. R. Pulliam, eds. 1987. *Foraging behavior.* New York: Plenum Press.

Kelly, J. P., and W. W. Weathers. 2002.1 Effects of feeding time constraints on body mass regulation and energy expenditure in wintering dunlin (*Calidris alpine*). *Behavioral Ecology* 13:766–75.

Kenward, R. 1987. *Wildlife radio tagging.* New York: Academic Press.

Kildaw, S. D. 1995. The effect of group size manipulation on the foraging behavior of black-tailed prairie dogs. *Behavioral Ecology* 6:353–58.

Knight, R. L., and K. J. Gutzwiller, eds. 1995. *Wildlife and recreationists: Coexistence through management and research.* Washington, DC: Island Press.

Korschgen, L. J. 1980. Procedures for food-habitat analyses. In *Wildlife management techniques manual,* ed. S. D. Schemnitz. 4th ed. Washington, D.C.: Wildlife Society.

Kotler, B. P., J. E. Gross, and W. A. Mitchell. 1994. Applying patch use to assess aspects of foraging behavior in Nubian ibex. *Journal of Wildlife Management* 58:299–307.

Lechowicz, M. J. 1982. The sampling characteristics of electivity indices. *Oecologia* 52:22–30.

Leger, D. W., and I. A. Didrichsons. 1994. An assessment of data pooling and some alternatives. *Animal Behaviour* 48:823–32.

Lehner, P. N. 1996. *Handbook of ethological methods.* 2nd ed. Cambridge, MA: Cambridge Univ. Press.

L'Heureux, N., M. Lucherini, M. Festa-Bianchet, and J. T. Jorgenson. 1995. Density-dependent mother-yearling association in bighorn sheep. *Animal Behaviour* 49:901–10.

Litvaitis, J. A., K. Titus, and E. M. Anderson. 1994. Measuring vertebrate use of terrestrial habitats and foods. In *Research and management techniques for wildlife and habitats,* 5th ed., ed. T. A. Bookhout, 254–74. Bethesda, MD: Wildlife Society.

Lomnicki, A. 1988. *Population ecology of individuals.* Princeton, NJ: Princeton Univ. Press.

Losito, M. P., R. E. Mirarchi, and G. A. Baldassarre. 1989. New techniques for time-activity studies of avian flocks in view-restricted habitats. *Journal of Field Ornithology* 60:388–96.

Loughry, W. J. 1993. Determinants of time allocation by adult and yearling black-tailed prairie dogs. *Behaviour* 124:23–43.

Machlis, L., P. W. D. Dodd, and J. C. Fentress. 1985. The pooling fallacy: Problems arising when individuals contribute more than one observation to the data set. *Zeitschrift für Tierpsychologie* 68:201–14.

Manly, B. F. J. 1993. Comments on design and analysis of multiple-choice feeding-preference experiments. *Oecologia* 93:149–52.

Manly, B. F. J., L. L. McDonald, D. L. Thomas, T. L. McDonald, and W. P. Erickson. 2002. *Resource selection by animals: Statistical design and analysis for field studies.* 2nd ed. Boston: Kluwer Academic Publishers.

Martin, P., and P. Bateson. 1993. *Measuring behaviour.* 2nd ed. Cambridge: Cambridge Univ. Press.

Masataka, N., and B. Thierry. 1993. Vocal communication of Tonkean macaques in confined environments. *Primates* 34:169–80.

McNay, R. S., J. A. Morgan, and F. L. Bunnell. 1994. Characterizing independence of observations in movements of Columbian black-tailed deer. *Journal of Wildlife Management* 58:422–29.

Mech, L.D. 1983. *Handbook of animal radio-tracking.* Minneapolis: Univ. of Minnesota Press.

Meserve, P. L. 1976. Food relationships of a rodent fauna in a California coastal sage community. *Journal of Mammalogy* 57:300–19.

Mock, P. J. 1991. Daily allocation of time and energy of western bluebirds feeding nestlings. *Condor* 93:598–611.

Moen, A. N. 1973. *Wildlife ecology.* San Francisco: W. H. Freeman.

Mook, L. J., and H. W. Marshall. 1965. Digestion of spruce budworm larvae and pupae in the olive-backed thrush, *Hylocichla ustulata swainsoni* (Tschudi). *Canadian Entomologist* 97:1144–49.

Morrison, M. L. 1982. The structure of western warbler assemblages: ecomorphological analysis of the black-throated gray and hermit warblers. *Auk* 99:503–13.

Morrison, M. L. 1988. On sample sizes and reliable information. *Condor* 90:275–78.

Morrison, M. L., and L. S. Hall. 2002. Standard terminology: Toward a common language to advance ecological understanding and application. In *Predicting species occurrences: Issues of accuracy and scale,* ed. J. M. Scott, P. J. Heglund, M. L. Morrison, J. B. Haufler, M. G. Raphael, W. A. Wall, and F. B. Samson, 43–52. Washington, DC: Island Press.

Morrison, M. L., D. L. Dahlsten, S. M. Tait, R. C. Heald, K. A. Milne, and D. L. Rowney. 1989. *Bird foraging on incense-cedar and incense-cedar scale during winter in California.* Research Paper PSW-

195. Berkeley: USDA Forest Service, Pacific Southwest Forest and Range Experiment Station.

Morrison, M. L., C. J. Ralph, J. Verner, and J. R. Jehl Jr., eds. 1990. Avian foraging: theory, methodology, and applications. *Studies in Avian Biology* no. 13.

Morse, D. H. 1980. *Behavioral mechanisms in ecology.* Cambridge, MA: Harvard Univ. Press.

Nagy, K. A. 1987. Field metabolic rate and food requirement scaling in mammals and birds. *Ecological Monographs* 57:111–28.

Nagy, K. A., and B. S. Obst. 1989. Body size effects on field energy requirements of birds: what determines their field metabolic rates? *International Ornithological Congress* 20:793–99.

Newton, P. 1992. Feeding and ranging patterns of forest hanuman langurs (*Presbytis entellus*). *International Journal of Primatology* 13:245–85.

Nudds, T. D. 1980. Foraging "preference": Theoretical considerations of diet selection by deer. *Journal of Wildlife Management* 44:735–40.

Nudds, T. D. 1982. Theoretical considerations of diet selection by deer: A reply. *Journal of Wildlife Management* 46:257–58.

Ormerod, S. J. 1985. The diet of dippers *Cinclus cinclus* and their nestlings in the catchment of the River Wye, Mid-Wales: A preliminary study of faecal analysis. *Ibis* 127:316–31.

Paterson, J. D., P. Kubicek, and S. Tillekeratne. 1994. Computer data recording and DATAC6, a BASIC program for continuous and interval sampling studies. *International Journal of Primatology* 15:303–15.

Petrides, G. A. 1975. Principal foods versus preferred foods and their relation to stocking rate and range condition. *Biological Conservation* 7:161–69.

Pierce, G. J., and J. G. Ollason. 1987. Eight reasons why optimal foraging theory is a complete waste of time. *Oikos* 49:111–18.

Ploger, B. J. 2003. Learning to describe and quantify animal behavior. In *Exploring animal behavior in laboratory and field,* ed. B. J. Ploger and K. Yasukawa, 11–30. San Diego: Academic Press.

Ploger, B. J., and K. Yasukawa, eds. 2003. *Exploring animal behavior in laboratory and field.* San Diego: Academic Press.

Poysa, H. 1991. Measuring time budgets with instantaneous sampling: A cautionary note. *Animal Behaviour* 42:317–18.

Pulliam, H. R. 1973. On the advantages of flocking. *Journal of Theoretical Biology* 38:419–22.

Quera, V. 1990. A generalized technique to estimate frequency and duration in time sampling. *Behavioral Assessment* 12:409–24.

Ralph, C. P., S. E. Nagata, and C. J. Ralph. 1985. Analysis of droppings to describe diets of small birds. *Journal of Field Ornithology* 56:165–74.

Raphael, M. G. 1988. A portable computer-compatible system for collecting bird count data. *Journal of Field Ornithology* 59:280–85.

Raphael, M. G. 1990. Use of Markov chains in analysis of foraging behavior. *Studies in Avian Biology* 13:288–94.

Ratti, J. T., L. D. Flake, and W. A. Wentz. 1982. *Waterfowl ecology and management: Selected readings.* Bethesda, MD: Wildlife Society.

Remsen, J. V., and S. K. Robinson. 1990. A classification scheme for foraging behavior of birds in terrestrial habitats. *Studies in Avian Biology* 13:144–60.

Riley, C. M. 1986. Foraging behavior and sexual dimorphism in emerald toucanets (*Aulacorhynchus prasinus*) in Costa Rica. M.S. thesis, Univ. of Arkansas, Fayetteville.

Riney, T. 1982. *Study and management of large mammals.* New York: Wiley.

Roa, R. 1992. Design and analysis of multiple-choice feeding-preference experiments. *Oecologia* 89:509–15.

Roberts, G., and P. R. Evans. 1993. Responses of foraging sanderlings to human approaches. *Behaviour* 126:29–43.

Roberts, W. A., and S. Mitchell. 1994. Can a pigeon simultaneously process temporal and numerical information? *Journal of Experimental Psychology: Animal Behavior Processes* 20:66–78.

Rosenberg, K. V., and R. J. Cooper. 1990. Approaches to avian diet analysis. *Studies in Avian Biology* 13:80–90.

Rosenthal, R. 1976. *Experimenter effects in behavioral research.* New York: Irvington.

Ruggiero, L. F., R. S. Holthausen, B. G. Marcot, K. B. Aubry, J. W. Thomas, and E. C. Meslow. 1988. Ecological dependency: The concept and its implications for research and management. *Transactions of the North American Wildlife and Natural Resources Conference* 53:115–26.

Rychlik, L., and E. Jancewicz. 2002. Prey size, prey nutrition, and food handling by shrews of different body sizes. *Behavioral Ecology* 13:216–23.

Sakai, H. F., and B. R. Noon. 1990. Variation in the foraging behaviors of two flycatchers: Associations with stage of breeding cycle. *Studies in Avian Biology* 13:237–44.

Salewski, V., F. Bairlein, and B. Leisler. 2003. Niche partitioning of two Palearctic passerine migrants with Afrotropical residents in their West African winter quarters. *Behavioral Ecology* 14:493–502.

Samuel, M. D., and M. R. Fuller. 1994. Wildlife radiotelemetry. In *Research and management techniques for wildlife and habitats,* 5th ed., ed. T. A. Bookhout, 370–418. Bethesda, MD: Wildlife Society.

Schleidt, W. M., G. Yakalis, M. Donnelly, and J. McGarry. 1984. A proposal for a standard ethogram, exemplified by an ethogram of the bluebreasted quail (*Coturnix chinensis*). *Zeitschrift für Tierpsychologie* 64:193–220.

Schmitz, O. J. 1992. Optimal diet selection by white-tailed deer: Balancing reproduction with starvation risk. *Evolutionary Ecology* 6:125–141.

Schoener, T. W. 1969. Optimal size and specialization in constant and fluctuating environments: An energy-time approach. *Brookhaven Symposium in Biology* 22:103–14.

Severson, K. E., and M. May. 1967. Food preferences of antelope and domestic sheep in Wyoming's Red Desert. *Journal of Range Management* 20:21–25.

Shettleworth, S. J. 1993. Varieties of learning and memory in animals. *Journal of Experimental Psychology: Animal Behavior Processes* 19:5–14.

Siegel, S. 1956. *Nonparametric statistics for the behavioral sciences.* New York: McGraw-Hill.

Siegel, S., and N. J. Castellan. 1988. *Nonparametric statistics for the behavioral sciences.* 2nd ed. New York: McGraw-Hill.

Slater, P. J. B. 1978. Data collection. In *Quantitative ethology,* ed. P. W. Colgan, 7–24. New York: Wiley.

Snedecor, G. W., and W. G. Cochran. 1980. *Statistical methods.* 7th ed. Ames: Iowa State Univ. Press.

Sokal, R. R., and F. J. Rohlf. 1995. *Biometry.* 3rd ed. San Francisco: W. H. Freeman.

Sommer, B., and R. Sommer. 1991. *A practical guide to behavioral research: Tools and techniques.* 3rd ed. New York: Oxford Univ. Press.

Speakman, J. R., and P. A. Racey. 1987. The energetics of pregnancy and lactation in the brown long-eared bat, *Plecotus auritus.* In *Recent advances in the study of bats,* ed. M. B. Fenton, P. A. Racey, and J. M. V. Rayner. Cambridge: Cambridge Univ. Press.

Stamps, J. A., and V. V. Krishnan. 1994. Territory acquisition in lizards. I. First encounters. *Animal Behaviour* 47:1375–85.

Stephens, D. W. 1990. Foraging theory: Up, down, and sideways. *Studies in Avian Biology* 13:444–54.

Stephens, D. W., and J. R. Krebs. 1986. *Foraging theory.* Princeton, NJ: Princeton Univ. Press.

Strauss, R. E. 1979. Reliability estimates for Ivlev's electivity index, the forage ratio, and a proposed linear index of food selection. *Transactions of the American Fisheries Society* 108:344–52.

Swanson, G. A., and J. C. Bartonek. 1970. Bias associated with food analysis in gizzards of blue-winged teal. *Journal of Wildlife Management* 34:739–46.

Swihart, R. K., and N. A. Slade. 1985. Testing for independence of observations in animal movements. *Ecology* 66:1176–84.

Tatner, P. 1983. The diet of urban magpies, *Pica pica. Ibis* 125:90–107.

Thill, R. A. 1985. Cattle and deer compatibility on southern forest range. In *Proceedings of a conference on multispecies grazing,* ed. F. H. Baker and R. K. Jones, 159–77. Morrilton, AR: Winrosk International.

Trainer, J. M., D. B. McDonald, and W. A. Learn. 2002. The development of coordinated singing in cooperatively displaying long-tailed manakins. *Behavioral Ecology* 13:65–69.

Vanderploeg, H. A., and D. Scavia. 1979a. Calculation and use of selectivity coefficients of feeding: Zooplankton grazing. *Ecological Modelling* 7:135–49.

Vanderploeg, H. A., and D. Scavia. 1979b. Two electivity indices for feeding with special reference to zooplankton grazing. *Journal of Fisheries Research Board of Canada* 36:362–65.

Waugh, D. R. 1979. The diet of sand martins in the breeding season. *Bird Study* 26:123–28.

Waugh, D. R., and C. J. Hails. 1983. Foraging ecology of a tropical aerial feeding bird guild. *Ibis* 125:200–217.

Weathers, W. W., W. A. Buttemer, A. M. Hayworth, and K. A. Nagy. 1984. An evaluation of time-budget estimates of daily energy expenditures in birds. *Auk* 101:459–72.

Webb, J. K., and R. Shine. 1993. Prey-size selection, gape limitation and predator vulnerability in Australian blindsnakes (Typhlopidae). *Animal Behaviour* 45:1117–26.

Weckerly, F. W. 1994. Selective feeding by black-tailed deer: Forage quality or abundance? *Journal of Mammalogy* 75:905–13.

Weinberg, S. L., and K. P. Goldberg. 1990. *Statistics for the behavioral sciences.* Cambridge: Cambridge Univ. Press.

White, G. C., and R. A. Garrott. 1990. *Analysis of wildlife radio-tracking data.* San Diego: Academic Press.

Wikelski, M., and F. Trillmich. 1994. Foraging strategies of the Galapagos marine iguana (*Amblyrhynchus cristatus*): Adapting behavioral rules to ontogenetic size change. *Behaviour* 128:255–79.

Williams, P. L. 1990. Use of radiotracking to study foraging in small terrestrial birds. *Studies in Avian Biology* 13:181–86.

Willms, W., A. McLean, R. Tucker, and R. Ritchey. 1980. Deer and cattle diets on summer range in British Columbia. *Journal of Range Management* 33:55–59.

Wilmshurst, J. F., J. M. Fryxell, and R. J. Hudson. 1995. Forage quality and patch choice by wapiti (*Cervus elaphus*). *Behavioral Ecology* 6:209–17.

Wolda, H. 1990. Food availability for an insectivore and how to measure it. *Studies in Avian Biology* 13:38–43.

Yunger, J. A., P. L. Meserve, and J. R. Gutierrez. 2002. Small-mammal foraging behavior: Mechanisms for coexistence and implication for population dynamics. *Ecological Monographs* 72:561–77.

Zar, J. H. 1999. *Biostatistical analysis.* 4th ed. Upper Saddle River, NJ: Prentice Hall.

8 Habitats through Space and Time: Heterogeneity and Disturbance

Nature is an infinite sphere whose center is everywhere and whose circumference is nowhere.

Blaise Pascal

The *habitat* of a species is defined by a location and the set of conditions used by members of the population to meet some or all of their life needs. Components of habitat have been classically viewed as consisting of food, water, and cover (Leopold 1933; Edminster 1938), and more recently as including a wide variety of other environmental factors. The dispersion of all such habitat elements across space and through time in part determines the distribution, density, structure, and productivity of wildlife populations. In this chapter we focus on concepts and measures of habitat at a landscape scale through space and time. Chapter 9 discusses more of the organism and population response to such landscape distributions.

Why Study Habitats and Wildlife in Landscapes?

Landscapes are the great environmental integrators. It is within landscapes that individuals interact as populations, species mix in communities, and communities overlap with abiotic elements of ecosystems. Landscapes integrate all of these factors across spatial scales ranging from as large as the home ranges of individual vertebrate carnivores, large herbivore herds, and mixed foraging flocks of birds to as small as interstices of soil particles in which dwell invertebrate micro- and meiofauna that contribute critically to soil productivity. It is within landscapes that dynamics and disturbance regimes interplay with the ever-changing biota, across scales ranging from sweeping catastrophic fires and volcanic eruptions to soil pits and sunflecks created from single treefalls. It is within landscapes that we are often challenged, as wildlife researchers or managers, to predict the response of environments, organisms, and ecosystems to our activities.

Landscape ecology is the scientific study of species, communities, and ecosystems across geographic areas, typically defined by hydrologic and administrative boundaries. The field of landscape design and architecture (Simonds 1961; McHarg 1969), once a discipline focused on design of entirely human-altered environments, has begun to converge with landscape ecology. This convergence has been propelled by two related disciplines. One is that of environmental psychology, a facet of human ecology that determines the patterns and evolutionary basis of human preference for various natural or altered environments (e.g., Kaplan 1973; Irvine and Kaplan 2001). The other is that of landscape design of environments as applied to seminatural conditions such as managed forests (e.g., Mladenoff et al. 1994; Gustafson and Crow 1998).

Working with landscapes poses special problems for researchers and managers. Researchers must contend with studying subjects that usually have no replicates, no controls, limited baseline (predisturbance) data, and all the environmental noise that complicates data analysis and confounds interpretation. Managers must contend with managing in the face of great variability and uncertainty—including changeable ecological conditions and disturbance events across wide scales of space and time. Landscapes are entirely open systems, as rivers, subsurface water, air, and organisms interchange with adjacent or distant regions. Some solutions to these problems are discussed in this chapter.

Although most wildlife species do not strictly adhere to the cartographer's, hydrographer's, or administrator's boundaries, they often respond to broad landscape patterns as well as to individual patches of resources and environments. Changes in broad patterns of resource patches can insidiously disrupt resource availability and resulting population functions in ways that would not become evident by examining merely local expressions of habitat conditions and occurrence of species. Populations and habitats alike can "unravel" from their overall tapestry within landscapes and within the communities in which they are found. Sometimes such disruptions occur with lag times that mask the true cause unless we examine broader spatial, tempo-

ral, and ecological scales. In addition, humans greatly affect various aspects of environmental processes, as we alter flows of substances, nutrients, energy, fluids, and organisms by changing vegetation conditions, topographies, species pools, and even the soil itself.

A landscape approach to assessment and management of wildlife–habitat relationships is useful also because some ecological processes "emerge" at landscape-scale areas that are not evident or cannot be easily understood within smaller areas. Such landscape processes include the biogeography of species distributions; the hydrology of surface and subsurface flows; the pedology of soil formation, change, and erosion; and the biology of population dynamics and species exchanges, invasions, speciation, and extinction. How humans occupy landscapes and use resources can have drastic effects on each of these processes.

Definition and Classification of Landscapes

A *landscape* is "a part of the space on the earth's surface, consisting of a complex of systems, formed by the activity of rock, water, air, plants, animals and [people] and that by its physiognomy forms a recognizable entity" (Zonneveld 1979). By this general definition, landscapes do not necessarily have a specific size, and might be considered at various scales depending on the area over which unique constellations of ecological processes operate to form "recognizable entities."

More importantly, we should view a "landscape" as pertaining to a specific organism of interest, just as "habitat" is a species- and organism-specific concept. Thus a "landscape" for a large, mobile predator such as a wolf is a geographically larger area than a "landscape" for a mollusk. This difference was brought home to one of us (B. Marcot) while crossing the Ramganga River in Corbett National Park in northern India on elephant-back, where he observed a three-striped roofed terrapin (*Kachuga dhongoka*) nestled within a single soggy footprint of a wild Asian elephant (*Elaphus maximus*) that had passed before. The terrapin's landscape was the footprint and a portion of the immediately adjacent slackwater marsh, whereas the elephant's landscape was a large portion of the entire national park. Thus a *landscape* for a species is that heterogeneous patchwork of specifically selected habitats, resources, and environmental conditions and all biotic and abiotic attributes that influence those conditions, at the appropriate physical scale of geographic extent that an organism or a population encounters over the course of its travels and distributions.

Landscapes contain a common set of elements—including plant and animal communities, ecological processes, and disturbance regimes—that coincide in a particular geographic setting, although each element may extend beyond the delineated landscape area in individual and unique ways. Some landscape factors mediate or exacerbate others. For example, in grasslands, weather conditions, coupled with agricultural development, may alter native fire regimes. This alteration may encourage exotic, pioneer plants to invade and dominate a landscape. Dominance of exotics can then alter the course of future fire regimes. Thus individual components and forces of landscapes do not act in isolation but rather can be mutually determining. We discuss the dynamics of disturbance further below.

The landscape concept may be applied to terrestrial and aquatic environments alike, including marine systems. A coastal, estuarine, lucastrine (lake), or even oceanic or marine environment may contain "recognizable entities" or landscapes if we just look at the right elements. For example, cliff-nesting seabirds may be described as inhabiting coastal landscapes consisting of "recognizable entities" defined by climatic and physiographic components. Even pelagic environments used by albatrosses and whales can be described as marine landscapes (or "waterscapes") defined by current cells, air circulation patterns, upwelling patterns, benthic topography, and weather systems, even if we cannot readily see these elements without measuring devices or help from satellite imagery. Understanding dispersal of organisms in aquatic, particularly marine, environments may aid explanation and management of their ecological communities (Kinlan and Gaines 2003). Much of this chapter, however, will focus discussion on landscapes in terrestrial situations.

Landscape Ecology

Understanding the patterns and reasons for the distribution of species and landscape elements is the challenge of landscape ecology. Thus a functional definition of *landscape ecology* is the study of the response of organisms, species, ecological communities, and ecosystem processes to each other in spatially heterogeneous and temporally changeable environments across a broad area typically delineated by local or regional hydrological boundaries.

Forman and Godron (1986) broadly classified terrestrial landscapes into six general types based on the degree of human disturbance: natural, managed, cultivated, suburban, urban, and megalopolis. *Natural landscapes* are perhaps a special condition because they provide sources or refugia of species sensitive to human disturbance and afford benchmarks from which to measure changes in biodiversity, community structure, population dynamics, and ecosystem processes affected by human intervention. *Managed and cultivated landscapes,* including managed forests, livestock grazing lands, and agricultural lands, collectively constitute what are more frequently being called "seminatural" landscapes (e.g., see Soderstrom and Part 2000). *Suburban and urban landscapes* offer the challenge of providing parks, greenbelts, forest remnants, tree cover, and other elements studied in the discipline of urban wildlife management (Mortberg 2001). *Megalopolis landscapes* offer perhaps the greatest challenge to habitat managers as geographic areas in which much of native biota is lost and in which mostly exotic species thrive.

Scales of Ecological and Management Issues

One of the major advantages of the discipline of landscape ecology is the integration of ecological study across various scales of space and time. However, the term *scale* is often used ambiguously in landscape ecology (and most other) studies and begs clarity of definition (Peterson and Parker 1998; Withers and Meentemeyer 1999; Jenerette and Wu 2000). Table 8.1 lists six dimensions of scale along with suggested guidelines for three levels of magnitude for each dimension.

Geographic extent is one dimension of scale. Typically, the term *large-scale* is used in landscape studies to connote a large geographic extent. This term is unfortunate, because in cartography the same term refers to larger values of map scale ratios—the ratio of a fixed distance on a map, such as 1 inch, to the distance of actual geographical surface it represents—and thus to maps that generally cover a small geographic extent. Mapping ecological entities (e.g., soil map units; Valentine 1981) can vary by map scale. Thus it is important to denote map scale. And unless the context is specifically cartographic, we also suggest replacing the ambiguous term *large-scale* with *large geographic extent,* and *small-scale* with *small geographic extent.*

Another dimension of scale is *spatial resolution,* which can range from images with coarse-grained pixels or large-vector polygons to those with very fine-grained resource patches or point locations of conditions. Whereas geographic ex-

tent and map scale are roughly (inversely) correlated—that is, large-scale maps tend to cover smaller geographic extents (depending on the physical size of the map)—spatial resolution can be a more independent dimension. That is, a small-scale map (e.g., 1:2,000,000 scale) covering a major drainage basin (e.g., the St. Lawrence Seaway watershed) might be represented by coarse-grained spatial resolution (e.g., 1-km^2 pixels), such as is available from many remote-sensing satellite images, or by finer-grained spatial resolution, such as vector polygons <10 ha. Thus it is important to denote the spatial resolution as well as the geographic extent and cartographic map scale of a particular map image (Salajanu and Olson 2001).

Time period is another dimension of scale that is not often made explicit. However, time is vital to interpreting geographic data and landscape simulations (Meentemeyer 1989), particularly for disturbance events and response by organisms and populations. Ideally, to best interpret recurrent patterns and potential influence from human activities, disturbances should be empirically depicted temporally by their duration and frequency. Also, the time period over which data were gathered for geographic analysis (or a particular map) in landscape studies is important to report, as some ecosystem processes, including disturbance regimes, may occur beyond the time period studied. We discuss disturbances further below.

Another dimension of scale, more pertinent to management use, not often explicitly addressed, is administrative hierarchy. *Administrative hierarchy* refers to the breadth of political, social, cultural, or even economic mandates and policies of governments, which may play important roles in some studies or management plans or in interpreting observations (Herzog et al. 2001). Particularly for management, it could be useful to explicate which organization hierarchy levels apply. For example, at the broad-scale

Table 8.1.   Aspects and examples of scale at three levels of magnitude

Aspect of scale	Broad-scale magnitude	Midscale magnitude	Fine-scale magnitude
Geographic extent	Entire major drainage basins, or entire eco-regions ("large-scale" study)	Subbasins or local water-sheds, or more local physiographic prov-inces; large groups of vegetation patches ("medium-scale" study)	Areas smaller than sub-basins or local water-sheds; small groups of individual vegetation patches or substrates ("small-scale" study)
Map scale	"Small-scale" maps, e.g., $\geq$1:1,000,000	"Medium-scale" maps, e.g., 1:100,000	"Large-scale" maps, e.g., 1:24,000
Spatial resolution	Typically coarse-grained, such as for character-ization of vegetation and environmental conditions at 1-km^2 pixel size	Environmental patches within subbasins or local watersheds, e.g., $\geq$10 ha	Fine-grained, e.g., <10-ha patch sizes
Time period	Paleoecological past and evolutionary future	Historic past and approximately one century into the future	Very recent management past and current con-ditions projected only a few years into the future
Administrative hierarchy	International treaties, such as on biodiversity and plant and animal com-merce; also national laws and land manage-ment or resource regu-lations; the "strategic" scale	Individual agency or tribal-specific policies and legal guidelines; state- or provincial-level agency policies, industry policies, and private landholder resource management goals	Local management unit-level operational pol-icies, down to project-level operations; the "tactical" scale
Level of biological organization	General abundance of veg-etation communities, cover types, and struc-tural stages; mapped lo-cations of ecoregions or ecosystems; inference to broad ecological com-munities and species assemblages; "coarse-filter" elements	Distribution and abun-dance of individual species or species groups	Species; gene pools or demes; subspecies or varieties; morphs or ecotypes; "fine-filter" elements

magnitude, studies of marine "landscapes" might pertain to international fishing treaties beyond the coastal sovereign fishing zones of in-dividual nations, and may also involve national policies for protection of marine mammals, fish stocks, or coral reefs. It could be important to

clarify which "scales" of legal mandates and re-source management policies pertain.

Finally, *level of biological organization* refers to the biological dimension of scale, and whether a study or plan pertains to ecosystems, communi-ties, assemblages, or species, or to more finely defined entities such as gene pools or ecotypes. This dimension could also refer to classification levels of vegetation communities or ecosystems, such as plant associations, vegetation types, and ecoregions (e.g., Bailey 2005). Note that a landscape study or management plan might pertain to a fine-scale magnitude of biological organization, such as an inventory of ecotypes, but across a broad geographic extent, such as a drainage basin. In this way, the various dimensions of scale may be applied at different magnitudes for a given purpose.

The biological organization dimension of scale in particular has been abused in much of the conservation and management literature. A rather vast literature has arisen over the past decade on "ecosystem management" (e.g., Armstrong et al. 2003; Butler and Koontz 2005), which is a bit of a misnomer, connoting that land and resource managers understand ecosystems well enough to "manage" and control them with predictable and desired outcomes. Rather, we may understand only a small portion of an ecosystem, such as the response of a few, selected plant or wildlife species, and have little clue how the total system really functions. Peterson et al. (2003) argued, through ecosystem modeling, that apparently rational management approaches can lead to ecological collapse because of our poor understanding of ecosystem dynamics.

Overall, it is important to describe these various dimensions of scale of map-based and remote-sensing image–based products to explain the context in which the information should be used. Applying a GIS model at the wrong scales of geographic extent, resolution, and so forth, can lead to false conclusions. Likewise, understanding the dimensions of scale of a geographically-based analysis, such as mapping of "hot spots" of species richness (also see Chapter 9), helps guide the scales at which such information should be used and not used. For example, species richness

maps depend on the scale (geographic extent and spatial resolution) of the mapping effort (Stoms 1994). In conclusion, we suggest that habitat and landscape studies (and management plans) clearly identify the magnitudes of geographic extent, map scale, spatial resolution, time period, organizational hierarchy (if appropriate), and levels of biological organization addressed and evaluated. In this way, much confusion over terms and methods can be avoided.

Scales of Management Authority and Considerations

Another useful way to consider scale is to think of a "zoning map" showing time and geographic extent pertinent to scientific disciplines. Figure 8.1 maps out "zones" of various scientific disciplines useful in landscape ecology studies. For example, in the upper left corner, one would study biogeography of animals and plants (zoogeography, phytogeography, and "phylogeography" sensu Avise 2000) roughly over a time period of the recent past to the distant geologic past, and over a spatial extent of roughly 10^6 to 10^8 ha. Study of ecological biogeography might extend from the recent past to the near future, and perhaps to smaller spatial extents, say down to 10^5 ha. Researchers of climate change might be interested in projecting global or regional climatic and biome trends from the present to perhaps centuries or a millennium in the future over roughly the same spatial extent.

On a tier down, studies of paleoecology and evolutionary ecology typically pertain to 10,000 years in the past and further back, whereas late prehistoric paleoecology (e.g., of Holocene events) and early historic ecology may peer back only a century to 10 millennia or so, over a spatial extent of perhaps 100 to 1 million ha, depending on the specific study. The analysis of population viability, including stability of metapopulations (see chapter 3), might pertain

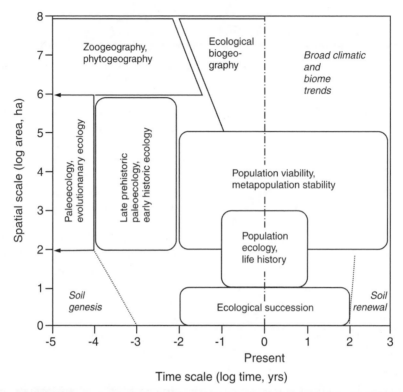

Figure 8.1.   A "zoning map" of ecological scale and species assessment, plotting spatial area against time (note log axes). This figure illustrates the spatial and temporal extents of various disciplines that may be useful in wildlife habitat management and land use planning.

to hindcasting a few decades to a century into the past, and to forecasting decades to a few centuries into the future; the spatial extent here is often less than that of paleoecology studies, again depending on the range of the species and environments studied.

Studies of population ecology and species life history are typically concerned with the recent past and the very near future, perhaps on the order of just a few decades, and on more localized geographic extents. Studies of ecological succession are usually site-specific but can address conditions a century or so in the past and a century or so into the future. Finally, pedogenesis

(soil origin) typically occurs or is studied on a mostly site-specific basis and about a millennium or more into the past, whereas prospects for soil renewal may not begin for centuries into the future. Certainly, other disciplines can also be added to the zoning map.

This type of space-time zoning map may be useful for identifying realms of duration and geographic extent over which managers should address issues of planning effects. Figure 8.2 plots ecosystem management issues according to the disciplines shown in figure 8.1. In the upper-left corner, at distant-past time periods and areas of large geographic extent, the manager might

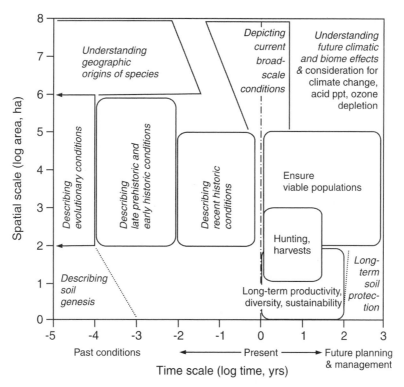

Figure 8.2. A "zoning map" of ecological scale and ecosystem management issues, plotting spatial area against time (note log axes). Compare with figure 8.1; this figure suggests some of the issues that managers may wish to address in wildlife habitat plans.

need to understand geographic origins of species, including centers of origin, centers and routes of species' spread, and the role of refugia. More than for academic interest, such understanding can greatly help identify locations of evolutionary significance deserving conservation consideration. An example can be found in the complex island archipelago of southeast Alaska and northern British Columbia, where unique subspecies and species have evolved and paleoecological conditions have brought large carnivores into sympatry. If ecosystem managers were concerned with representing the range of current broad-scale conditions, they would look to the recent past and near future across wide

geographic areas. And still across a wide geographic extent, the manager might peer into the near future (a century to a millennium) to project climatic or biome effects when considering potential influences of climate change, acid precipitation, and ozone depletion.

Over a lesser geographic extent, the manager might need to describe evolutionary, late prehistoric, early prehistoric, or recent historic conditions. This knowledge would be useful for determining the range of natural conditions in an area, and whether historic conditions truly represent environments in which species persisted over the long term or even evolved (see chapter 2). The goal of ensuring viable populations would prompt the manager to look into the future (on the order of a century to a millennium), if projections allow, and over a geographic extent of perhaps up to half a million hectares or so, depending on the environment and species of interest. Concern for near-term levels of harvest of game animals should be nested within broader concerns for maintaining long-term harvestability and viability of the target species, but these concerns pertain to only the near future and over small geographic areas. Management of long-term productivity, diversity, and sustainability of ecosystems can draw from all of these scales and issues. Description of soil genesis is a site-specific issue that looks back a millennium or more in the past, and describing long-term soil protection should look centuries to millennia into the future.

Such time–space perspectives can aid land management planning and the degree to which plans might study the past and project future cumulative effects. Most land resource management plans are designed to operate for perhaps a decade. Future plan revisions would come later as part of the subsequent planning cycle. But even if the formal duration of a plan is short, we should peer back in time and project into the future to better understand historic conditions and predict future long-term effects lest we continue to chip away at our resource base by what may be termed the long-term tyranny of short-term actions.

Some management issues ought to be addressed at each planning level, and specific scientific disciplines should be brought to bear in informing managers about those issues. For example, in the United States, a plan developed for an entire national forest or grassland may be in formal effect for a decade or so, and pertain to an area on the order of 10^4 to 10^5 ha. Issues that the plan would proximately deal with (see fig. 8.2) might include ensuring viable populations and to some extent representing the full range of eco-

logical or environmental conditions. However, in constructing the plan, managers should peer into the past and project cumulative effects into the future perhaps a millennium or so each way. In so doing, for the past they need to describe late prehistoric, early historic, and recent historic conditions by use of paleoecology and early historic ecology. To ensure future viable populations they might employ population viability analysis.

It is our view that an "ecosystem management" approach should address the full array of policies and land and resource management plans tiered across all geographic extents that account for past and current conditions and future cumulative effects of human activities on biotic and abiotic conditions. Further, such a sweeping "plan" would embody consistent implementation standards, management goals, evaluation criteria, and inventory and monitoring activities across all geographic extents, levels of biological organization, and administrative hierarchies from national policies to local projects. In this way, all issues and conditions, as well as past conditions and future effects, can be consistently treated and monitored across the full breadth of management issues and cumulative effects of past, current, and future activities.

This discourse on planning at multiple scales illustrates that human influence as well as natural disturbance regimes can alter the types, spatial patterns, and development of wildlife habitats. In the next section, we explore specific ways to gauge and interpret heterogeneity of habitats, following which we discuss disturbance dynamics.

Habitat Heterogeneity

Definitions and Concepts

The heterogeneity of resource patches in landscapes has been discussed by different authors in

Table 8.2.   Components of habitat heterogeneity in landscapes

Component	Description	Example	Source
Patch type richness and diversity	Number and relative area of habitat types for a species within a landscape	Number of patches of different types affects species richness, diversity, and numeric dominance of a species in a community	Bascompte et al. 2002; Fitzsimmons 2003
Patch dynamics	Incursion and melding of patches over time as a function of disturbance events and successional growth of vegetation	Distribution of vegetation patches over time as affected by stand-replacing fires and subsequent regrowth	Wu and Loucks 1995
Patch connectivity	Degree of adjacency of patches with similar conditions in a landscape	Connectivity of fencerow edge habitat in an agricultural setting, differentially affecting wildlife with different long-range-dispersal capabilities	Schooley and Wiens 2003; Walker et al. 2003
Patch isolation	Distance from one type of patch to the next (or n^{th}) nearest patch of the same type	Isolated patches less often colonized by species that do not disperse easily through unsuitable environments, such as Bachman's sparrow (*Aimophila aestivalis*) in South Carolina pine woodlands	Bender et al. 2003; Tischendorf et al. 2003
Fragmentation	Breaking up of contiguous environmental or habitat patches into smaller, more disjunct, and more isolated patches of different types	Fragmentation of grasslands differentially affecting bird species with varying sensitivity to habitat areas; differential effects of forest fragmentation on population viability of arboreal marsupials in Australia	Silva et al. 2003; Tallmon et al. 2003
Corridors	More or less linear or constricted arrays of environments or habitats in a landscape serving to connect larger patches	Movement corridors in undisturbed riparian woodland for cougars (*Felis concolor*) in southern California mountains; riparian woodland and shelterbelt corridors in North Dakota supporting populations of migratory birds	Mabry and Barrett 2002; Haddad et al. 2003
Permeability	Degree to which an organism can move among patches within a landscape	Effects of microhabitat and microenvironments within clearcuts on dispersal of red-legged frogs (*Rana aurora*)	Stamps et al. 1987; Chan-McLeod 2003

Table 8.2.   Continued

Component	Description	Example	Source
Edge effect	Incursion of microclimate and vegetation into a patch, typically forested, from a disturbed edge or opening	Clearcuts causing reduced tree stocking density, increased growth, and reproduction of dominant trees, higher tree mortality, and incursion of warmer, drier microclimates into adjacent old-growth forests	Chen et al. 1992, 1995
Edge contrast	Degree of difference in vegetation structure between two adjacent patches	Great contrast in daily average air and soil temperatures, wind velocity, short-wave radiation, and air and soil moisture between clearcuts and old-growth forests	Chen et al. 1993

various ways (table 8.2). Among the aspects of resource patch heterogeneity are patch type richness and diversity, patch dynamics, patch connectivity, patch isolation, fragmentation, corridors, permeability, edge effect, and edge contrast.

We define *habitat heterogeneity* as the degree of discontinuity in environmental conditions across a landscape for a particular species. Remember that *habitat* is a species- or organism-centric term, and that a particular environmental condition or gradient may constitute habitat for one species and a barrier for another.

Environmental conditions can include vegetation composition and structure, as well as dynamic flows of energy, nutrients, resources, and fluids (water and air). Discontinuities in environmental conditions occur as *ecotones,* or relatively sharp breaks in environmental conditions, or as *ecoclines,* or broader gradations in conditions over areas of greater geographic extent. Discontinuities can occur naturally, as with changes in soil type or edges of water bodies, or anthropogenically, as with edges of plowed grasslands or burned forests.

Fragmentation refers to the degree of hetero-geneity of habitats, usually vegetation patches, across a landscape, particularly related to isolation and size of resource patches. Since it refers to habitat, fragmentation is necessarily a species-specific condition. Unfortunately, the term *habitat fragmentation* is engrained in the ecological literature and used with abandon to refer to virtually any sort of heterogeneous condition. Authors thereby equate "fragmentation" with species extirpation (Gu et al. 2002; Solé et al. 2004) in a far too general way. Also, many authors (e.g., Bogaert et al. 2005) refer to "landscape fragmentation," which is strictly incorrect, as it is environments or resources (habitats for specific species) that become fragmented within landscapes, not entire landscapes per se.

Various kinds or degrees of species-specific habitat heterogeneity can be described. In the extreme case, resource or vegetation patches can be isolated into islands surrounded by vastly different and, for denoted species, unsuitable conditions. The response of species and communities to island situations has been the subject of much ecological study and is discussed further below.

Evaluating effects of partial isolation of habitats and environments may pose greater challenges in research and management than do islands. Partial isolation may have a gradient effect on viability of metapopulations (see chapter 3), such as by incrementally lowering a population's crude density (numbers of animals of a population per unit area of unsuitable and suitable environments in a landscape), lowering successful dispersal, and lowering the effective (expected interbreeding) size of populations, even though the absolute numbers of animals in an area may remain the same.

Another kind of heterogeneity is *temporal fragmentation,* sometimes called *ecological continuity.* Ecological continuity refers to the degree to which a particular environment, such as an old forest, occupies a specific area through time. If an old-forest ecosystem is interrupted, such as with widespread forest conversion or cutting, and then allowed to regrow, many of the original species closely associated with such environments may nonetheless be lost. Thus regional and site histories are important to interpreting community composition and species occurrence. Some research on the potential problems of ecological continuity, particularly of old forests, has been done in Europe (Herold and Ulmer 2001), although the concept is still relatively new in land management in North America. European researchers have discovered that vascular plants have differential adaptations to the degree of ecological continuity in taiga habitats of Scandinavia (Delin 1992). Some cryptogams (lichens and bryophytes, including mosses) are adversely affected by temporal disruption of old-forest conditions and thus can serve as indicators of ecological continuity of such environments (Tibell 1992; Selva 1994); similar work has been done on fungi (Selva 2003; Sverdrup-Thygeson and Lindenmayer 2003). Much work remains to determine sensitivity of wildlife species to ecological continuity, both within resource patches and across landscapes, although models of individual response

to habitat patch configurations might provide a useful tool (see chapter 10).

Heterogeneity and *fragmentation* can also refer to subtle discontinuities in environmental conditions rather than to changes in just gross vegetation structure and successional stage. One example is the horizontal separation of vegetation within a stand, such as among canopies of large trees. This kind of fragmentation has sometimes been referred to as *within-stand patchiness* or the *alpha diversity* of vegetation structure (Kitching and Beaver 1990). This kind of fragmentation in forests would likely adversely affect arboreal-dwelling species requiring contiguous canopy structures, such as some primates and rodents that rely on complex, three-dimensional runways through forest canopies. Species likely requiring such conditions include lion-tailed macaques (*Macaca silenus*), a highly endangered primate of wet evergreen *shola* rainforests in southern India; red tree voles (*Arborimus longicaudus*) of western North American conifer forests; and Indian giant squirrels (*Ratufa indica*) of deciduous and moist evergreen forests of peninsular India. Other species requiring dense forest canopy conditions but that may be less vulnerable to forest gaps and canopy separation may include Nilgiri langurs (*Presbytis johni*) and common giant flying squirrels (*Petaurista petaurista*) of south India, and northern flying squirrels (*Glaucomys sabrinus*) of North America. (Doubtless, other species can be added to these lists.) The effect and the number of species that would be included on these lists depend on the degree of facultative or obligate use by the species.

Another poorly studied and subtle aspect of fragmentation is the vertical separation of vegetation layers such as forest canopies and understories. The degree of heterogeneity of vertical forest stand structure is well known to correlate with bird species diversity (see chapter 3; Pearson 1971; Anderson et al. 1979). It may also influence use of stands and landscapes by forest-dwelling raptors that fly and forage below the canopy, such as broad-winged hawks (*Buteo platypterus*), northern goshawks (*Accipiter gentilis*), and some forest owls. Vegetation layer diversity and separation can be greatly affected by vegetation management, such as silvicultural thinning of forests and burning of woodlands and grasslands (Sullivan et al. 2002; Hayes et al. 2003; Patriquin and Barclay 2003; Suzuki and Hayes 2003).

Heterogeneity of environmental conditions can have a positive effect on some community and species assemblage measures. Huston (1994) posited that physical heterogeneity of the environment could interact with species competition and the mobility and size of organisms to determine species diversity (fig. 8.3). In this model, spatial heterogeneity is less effective in preventing competitive exclusion among mobile organisms than among sessile ones. Thus species diversity is highest in environments with high physical heterogeneity, and with species of low mobility or small body size and low competition intensity.

Habitat heterogeneity can also refer to temporal variation in environmental conditions or resource availability. The time dimension is usually modeled as the independent variable in population viability analyses that assess persistence of a population in heterogeneous ("fragmented") habitats. Some authors have found that temporal variation in habitat conditions can mediate whether species can coexist (Billick and Tonkel 2003; Holt and Barfield 2003). Schmidt et al. (2000) modeled populations under various conditions of spatial and temporal variation of habitat conditions related to demographic fitness and concluded that it is important to consider effects of both spatial heterogeneity and temporal variability when analyzing the coexistence of competing, vagile organisms. At broader scales of geographic extent and evolutionary time, Thompson (1999) suggested that coevolution of species is an important ecological condition that continually reshapes interactions across different spatial and temporal scales and that helps structure communities.

Measures of Habitat Heterogeneity

A wealth of mathematical indices has been offered as a way to measure habitat heterogeneity (Li and Wu 2004). Indices of habitat heterogeneity are typically calculated using geographic information system (GIS) and computer analysis software packages such as FRAGSTATS (McGarigal and Marks 1995) and the Indian program BIOCAP (IIRS 1999; Roy and Tomar 2000). Measures derive from geometry, topology, graph theory, matrix algebra, and other fields as ways to represent and analyze spatial arrangements of habitat patches over a surface.

Actual physical heterogeneity of the environment

		LOW	HIGH
Mobility/perception/size of organism	**LOW**	Effective heterogeneity: LOW Competition intensity: HIGH Species diversity: LOW	Effective heterogeneity: HIGH Competition intensity: LOW Species diversity: HIGHEST
	HIGH	Effective heterogeneity: LOW Competition intensity: HIGH Species diversity: LOWEST	Effective heterogeneity: LOW Competition intensity: HIGH Species diversity: LOW

Figure 8.3. Effects of species mobility, physical heterogeneity of the environment, and competition on species diversity. Effective heterogeneity is the combination of the first two elements. Mobility can also be depicted as the ability of organisms to perceive their environment and may be crudely correlated with body size. (Reproduced from M. A. Huston, *Biological Diversity: The Coexistence of Species on Changing Landscapes* [Cambridge: Press Syndicate of Cambridge University (c) 1994], 95, fig. 4.4, with the permission of Cambridge University Press.)

It is a truism, though, that our ability to devise mathematical indices of habitat heterogeneity in theory has outstripped our capability to determine their biological and ecological meanings in reality. We will temper our review of such indices here with several cautions: if the purpose of analyzing the relationships of wildlife to habitat heterogeneity is to add to our ecological understanding, then (1) use only those indices for which empirical evidence shows some direct biological or behavioral response by the organism in question; (2) avoid the temptation to generate a plethora of indices and then claim they are in some sense "important," that they all result from human activities, and that they all negatively affect species, ecological communities, or ecosystems; (3) dissect the components of habitat heterogeneity being analyzed, looking for high correlations among indices—don't just presume that all indices represent "fragmentation"; and, especially, (4) in the spirit of parsimony, use the simplest indices with the greatest explanatory power first.

The last caution is often ignored in studies aimed at "proving" adverse effects of "fragmentation." For example, studies by Askins et al. (1987) of bird populations in 46 forest tracts in Connecticut found that simple measures of total forest area were the best predictors of density and richness of forest-interior bird species associated with small forests, whereas isolation (distance to next nearest forest) was the best predictor for those associated with large forests.

Also, a more general caution is in order. An "index" in general should be thought of as "indexing something"—that is, an index is a *statistical estimator* of some ultimate *parameter*. Most users of indices of habitat heterogeneity fail to realize this relationship or to clearly identify the parameter being indexed. Claiming that "fragmentation" or "habitat patch pattern" or "habitat heterogeneity" is the parameter is inadequate. And defining the parameter in terms of the index—e.g., a "porosity" index is an estimator of the porosity of habitat patches—is tautological. The problem, again, is that we often use indices with little empirical understanding of their biological meaning.

An index of habitat heterogeneity should be expressed as an estimator of some *biological* parameter. Instead of fishing for correlations, the choice of index should match the biological situation. For instance, if some wildlife population is known to be dispersal limited—that is, its population size, density, or rate of increase is most limited by mortality of, or lack of breeding by, dispersing individuals within a landscape—then one should use measures of habitat patch connectivity or permeability (Chan-McLeod 2003) as indices or statistical estimators of the ultimate parameter of (biological) dispersal success within a landscape. An example is the study by St. Clair (2003), who found that forest-dependent songbirds were more reluctant to cross rivers than roads or meadows, and that response to such potential barriers to movement depended on the degree of forest dependence of the species. Thus a landscape with roads and meadows would be more permeable to forest songbirds than would be a landscape with rivers (although crossing roads carries the risk of vehicular collision), so that, following our principle above, these landscape features can serve as indices to (or statistical estimators of) the biological parameter of successful movement by birds.

Such cautions taken, we can now explore a few selected indices of habitat heterogeneity (tables 8.2, 8.3). The simplest measures are total habitat area within a landscape and area of individual habitat patches. Remember that what constitutes habitat and a patch is very much a species-specific question, so the same landscape area—with the same array of substrates, cover types, and vegetation stands—may be mapped and analyzed very differently for different species.

Table 8.3.   Some indices of habitat heterogeneity

Index	Definition	Formula	Variables
Patchiness	Relative size and isolation of vegetation cover patches	$P = \dfrac{\sum_{i=1}^{n} D_i}{N} \cdot 100$	N = number of boundaries between adjacent patches D_i = dissimilarity value for the ith boundary between adjacent patches
Porosity	Number of patches or density of patches within a particular vegetation type regardless of patch size	$PO = \sum_{i=1}^{n} C_{pi}$	n = number of cover classes among all patches C_{pi} = number of closed patches of the ith cover class
Interspersion	A count of dissimilar neighbors of a given patch and the intermixture of vegetation cover types across a landscape	$I = \dfrac{\sum_{i=1}^{n} SF_i}{N}$, where $SF_i = \sum_{i=1}^{n} \dfrac{Edge}{\sqrt{(Area_j \cdot \pi)}}$	$Edge$ = length of edge of the patch in both x and y directions $Area$ = area of the jth patch formed by groups in the ith cover class
Shape	Degree of deviation from a circle	$D = \dfrac{L}{2\sqrt{\pi \cdot A}}$	L = circumference of the path

A = area of the patch

A more complex index is that of *patch shape* (see table 8.3), which was actually published first as an index of the shape of lakes in the physical limnology literature and later reinvented by wildlife biologists (Fried 1975; Patton 1975; Marcot and Meretsky 1983). Moser et al. (2002) suggested a new index of patch shape, which served well to predict vascular plant and bryophyte species richness in rural landscapes. Orrock et al. (2003) found that patch shape, along with habitat corridors, influenced seed predation by invertebrates, rodents, and birds. Ohman and Lamas (2005) used a shape index to reduce edge effects and forest cover fragmentation in long-term forest planning.

More complex indices exist for calculating *patchiness, porosity,* and *interspersion* (see table 8.3). Although it is tempting to refer to all of these as indices of "fragmentation," there is actually a specific fragmentation index that represents the comparative extent of different cover types, such

as forest and nonforest area, and which is calculated as the number of patches of each cover type per unit area. As with other indices, the fragmentation index can be normalized (e.g., in the range of 0 to 10) to compare different landscapes or a single landscape over time.

A *disturbance index* can be calculated from a linear combination of other landscape indices using weighting factors for each index. The determination of weights can be based on adjacency of vegetation types in an analysis of patch juxtaposition. The researcher should be aware that many of these indices of habitat patch pattern are highly correlated (e.g., see fig. 8.4). True to our contention that mathematics may exceed ecological understanding, many additional indices of habitat heterogeneity have been suggested in the literature, including indices of landscape *cohesion* (Opdam et al. 2003); *lacunarity* (McIntyre and Wiens 2000); *landscape division, splitting,* and *effective mesh size* (Jager

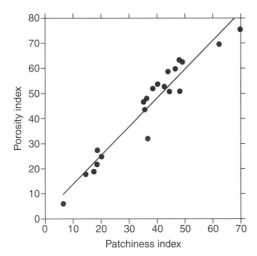

Figure 8.4.   An example of how two indices of habitat patch pattern can be highly correlated. See table 8.3 for formulae. Data are from a study of Asian elephant (*Elephas maximus*)–habitat relations in India (Marcot et al. 2002). The axes represent the percentage of individual elephant census zones that have high levels of either patchiness or porosity. Clearly, the two indices are highly correlated, so that one would do well with choosing just one—that with the greater biological meaning—to represent habitat patch patterns. In this case, however, it is unclear which index has greater biological meaning to Asian elephants; the patchiness index may represent amount of hiding cover or food resources available within a forest patch, whereas the porosity index may represent traversability among patches, and both of these factors influence elephant occurrence and density.

2000); *aggregation* (He et al. 2000); *permeability* (Stamps et al. 1987); and *contagion* (Parresol and McCollum 1997) (also see indices listed by Forman and Godron 1986, 188–189; Wagner and Fortin 2005). Giles and Trani (1999) took indexing one step further and suggested use of a single composite index that collapses six factors: area, class, proportion of dominant class, number of polygons, polygon size variance, and elevation range. Although such composite indices help reduce the dimensionality and complexity of a

problem (cf. the various data reduction methods of principle components analysis), they may mask biological understanding of individual parameters. In fact, the authors warn that identical values of a multifactor index can arise from an array of very different component values.

Many authors (e.g., Nams and Bourgeois 2004; Bolliger et al. 2005) have used *fractal indices*, which essentially measure the degree to which patch edges intrude into patch interiors. A fractal value is a fractional dimension. For example, a fractal value of 2.3 of habitat patches in a landscape may suggest that the geometry of uneven edges of the patches acts as more than just a two-dimensional array of patches but is not quite a three-dimensional habitat structure. Some authors have used fractal indices to describe foraging and movement paths of reindeer (Marell et al. 2002) and the geometric complexity of mole-rat burrows (Le Comber et al. 2002), and also to simulate landscapes (Hargrove et al. 2002).

Indices of habitat heterogeneity have also been based on graph theory (also called "loop analysis"). In this approach, habitat patches are usually represented as nodes or points and adjacent edges between patches as arcs or lines between the points, and various patch pattern indices borrow from graph theory measures of connection complexity (Marcot and Chinn 1982; Cantwell and Forman 1993; Urban and Keitt 2001).

So far we have discussed indices and measures of habitat heterogeneity across space. Other tools are useful for indexing, measuring, and modeling habitat heterogeneity over time. Early models of ecological succession were based on Markov analysis (Horn 1975), a statistical technique that represents transition probabilities among successive states of a system. More general transition matrix and Markovian approaches were later used to model the sensitivity of successional changes to management activities, including harvesting, grazing, fire, and other manipulations (Hann et al. 1998; Rojo and Orois 2005).

Although we discuss modeling in greater detail in chapter 10, we will mention here tools used to track temporal changes in habitat heterogeneity. These include indices and models of ecological succession. For example, Lesica and Cooper (1999) used canonical correspondence analysis on field data from the Centennial Sandhills of southwest Montana to describe shrubland succession from changes in soil organic matter, plant canopy cover, and bare soil area. They modeled succession as a function of the driving variables of fire, ungulate grazing, and burrowing by pocket gophers (*Thomomys talpoides*).

Disturbance Ecology: Dynamics of Habitats in Landscapes

The study of disturbance ecology focuses on dynamics of habitats in landscapes. Disturbance ecology addresses many topics that range widely across scales of space and time (Frelich 2002), including soil dynamics, fire ecology, vegetation succession, meteorology, climatology, and paleoclimatology. To best understand effects on wildlife, disturbances should be depicted according to their frequency, intensity, duration, location, and geographic extent. In this way, disturbances to habitats can be incorporated into spatially explicit models of population demography to project effects on species distribution and viability (Dunning et al. 1995; Boughton and Malvadkar 2002).

Four Types of Disturbances

In a simple classification, four categories of disturbances can be identified according to their intensity and geographic extent (fig. 8.5). *Type I disturbances* are major environmental catastrophes that are relatively short-term and intense

and that affect large areas. They include volcanoes, major fires, floods, and hurricanes (also called typhoons in the Pacific Ocean and cyclones in the tropics).

Hurricanes can have major effects on wildlife habitat and populations, and in some cases, habitat refugia can serve as important protection zones for populations. Hurricanes can affect species and resources differentially. For example, Pierson et al. (1996) found that two severe cyclonic storms on Samoa caused a more severe population decline in the more common and more widely distributed species of flying fox (*Pteropus tonganus*) than in the endemic species (*P. samoensis*). They attributed this difference to the greater susceptibility of *P. tonganus* to hunting mortality in villages, as the species appeared in villages to forage on flowers and fruits after the storms. The greater susceptibility was also due to the greater proportion of foliage in the diet of *P. samoensis,* whereas *P. tonganus* is more highly frugivorous, and, overall, poststorm density of flowers and fruits was depressed. The endemic species was protected from storm effects in rainforest refugia, particularly in areas of high topographic relief, such as volcanic cones and steep valleys protected from wind damage, whereas the more widespread species occurred in the less protected village environments that were more severely damaged.

Using clipping experiments to simulate pruning effects of Hurricane Lili, which in 1996 directly hit the island of Great Exuma, Bahamas, Spiller and Agrawal (2003) concluded that hurricanes can trim shrubs, resulting in new growth that entices up to a 68% increase in herbivory by arthropods. They concluded that such increased sprouting and herbivory following natural or anthropogenic disturbances has wide implications for understanding how storms affect habitats for organisms.

One rather well-studied storm was Hurricane Hugo, which can serve here as an example

		Geographic area affected	
		Widespread (>1000 ha)	Local (1–1000 ha)
Degree of disturbance	High	Type I *Major environmental catastrophe* (volcanoes, major fires, hurricanes)	Type II *Local environmental disturbance* (wind, ice storms, insects, disease)
	Low	Type III *Chronic or systematic change over wide areas* (predators, competition, forestry)	Type IV *Minor environmental change* (local fires, developments)

Figure 8.5.   Four types of disturbance shown by degree, or intensity, and geographic area affected.

of major effects of Type I storm disturbances on wildlife and habitat. Hugo hit South Carolina on September 21, 1989, with sustained winds of 217 km/hr, gusts of 282 km/hr, and a storm surge of 5.8+ m. Damage to forests by Hugo was greater than that from Hurricane Camille, the eruption of Mount St. Helens, and the 1988 Yellowstone fires combined. Hugo damaged forests on over 17,800 sq km, with damage greatest in the coastal Francis Marion National Forest, where about 75% of commercially marketable pine trees were felled by the storm (Ehinger 1991). Many forests of South Carolina hit by Hugo had held some of the densest colonies of the endangered red-cockaded woodpecker found on the U.S. east coast. Prior to Hugo, Francis Marion had supported about 25% of all southeast red-cockaded woodpeckers, including one of the world's largest populations of approximately 500 breeding pairs (West 1989). On Francis Marion, most of the trees damaged were mature pines favored by the woodpeckers; Hugo reduced the woodpecker populations there by 63% and destroyed 87% of known active cavity trees and 50–60% of foraging trees (West 1989).

Recovery of forest conditions for the woodpeckers could take 75 years (Hamrick 1991), and demographic consequences to woodpecker populations will likely be long-lasting (Hooper et al. 1990).

Hugo also passed directly over the 11,330-ha Caribbean National Forest on Puerto Rico and heavily damaged 8900 ha, including breakage and blowdown of many trees and almost complete defoliation of the remaining trees (West 1989). Initial censuses showed that only 23 of the original colony of 46 wild Puerto Rican parrots (*Amazona vittata*), a highly endangered species, remained. However, the captive flock of 53 parrots was unhurt, a prime example of the value of captive breeding of at least some highly endangered species subject to periodic, major disturbance events in the wild. After the storm, many predators were observed in areas of heavy treefall; together with destruction of nest structures caused by Hugo, predation can seriously depress some populations. However, the parrot population has since recovered to population levels approaching that of pre-Hugo, thanks to the surviving and actively reproducing birds, the availability of improved natural nest cavities, and the effectiveness of the enhanced nest management program. Since Hugo, some Puerto Rican parrots have nested in lower elevations and used cavities in tree species never reported since parrot observations began in 1968 (Vilella and Garcia 1995). Thus this major catastrophic event changed the abundance of tree cavities and structure of vegetation such that the availability of suitable sites for this endangered species altered. This alteration resulted in changes in geographic areas used by the birds that in turn helped maintain the population.

Effects of Hurricane Hugo on other forest bird species in Puerto Rico were also studied by Wunderle (1995), who reported that canopy dwellers shifted to foraging in forest understories and openings, and bird assemblages became far less distinct by microhabitat conditions than before the hurricane. Such habitat displacement of birds also included movement of frugivorous birds into preexisting gaps and invasion by forest-edge or shrubby second-growth species into the forest. Wunderle speculated that it may take many years for resources and structures in forest understory and gaps, and associated unique bird assemblages as observed prior to the storm, to again become distinct.

Other examples of Type I disturbances that have been studied in North America include the 1980 eruption of Mount St. Helens (Lawrence and Ripple 2000), the 1988 Yellowstone National Park fires (Turner et al. 2003), and the 1993 Mississippi River floods (Custer et al. 1996; Miller and Nudds 1996). Each kind of Type I disturbance can bring very different changes to environments and species composition, and deserves individual attention in habitat planning that accounts for such infrequent events (see chapter 11). Some Type I disturbance events occur on multiple cycles, such as fire regimes or floods with various repeat frequencies.

Foster et al. (1998) studied five examples of Type I disturbances (which they called LIDs—large, infrequent disturbances), including fire, hurricanes, tornadoes, volcanic eruptions, and floods. They concluded that resulting landscape patterns were controlled by the type of disturbance, topography, and vegetation. They found that different kinds of LIDs produce different edge effects and different extents of enduring legacies that can influence ecosystems for decades or centuries.

Type II disturbances include locally intense environmental changes from events such as windstorms, ice storms, and local outbreaks of defoliating insects. Forest canopy gaps, sometimes called microserules, undergo local succession of plant species and are important contributors to overall vertical forest stand structure and species composition (Moeur 1997; Degen et al. 2005). For example, Lawton and Putz (1988) reported that wind-formed canopy gaps of tropical montane elfin forests in Monteverde, Costa Rica, tended to be small, occur frequently, and annually cover small proportions (≤1%) of the overall forest cover, and that they promoted growth of shade-intolerant plant species, tree saplings on nurse logs and mineral soil disturbed by uprooting trees, and plants from soil seed banks more than from seeds dispersed into the gap. The researchers also found that many saplings in gaps originated from epiphytes in the crowns of the trees that fell. In temperate forests of northern Minnesota, Webb (1989) found that tree damage from thunderstorm winds (25–35 m/s) related to tree size, species, species wood strength, and incidence of species-specific fungal pathogens. He found much within-stand differences in wind-caused mortality and subsequent plant development of the canopy gap, owing to differences in shade tolerance, initial forest structure, gap size, and windfirmness of shade-tolerant understory plants.

The size of forest canopy gaps also influences nitrogen cycling dynamics, as reported by Parsons et al. (1994), in forests of lodgepole pine (*Pinus contorta*) in southwestern Wyoming. In these forests, removal of 15 to 30 tree clusters represented a threshold above which significant losses of available nitrogen to the groundwater occurred.

Outbreaks of defoliating insects such as spruce budworm (*Choristoneura fumiferana*) are another Type II disturbance that can directly or indirectly cause local to extensive changes in forest structure and composition (Alfaro et al. 2001). Complex species interactions can determine control of such insect pests. For example, in spruce–fir forests in northern New Hampshire and western Maine, Crawford and Jennings (1989) reported that the entire bird community showed significant functional responses (increased foraging) to increasing budworm density, whereas only two species, Canada warblers (*Wilsonia pusilla*) and golden-crowned kinglets (*Regulus satrapa*), also showed numerical responses (increased reproduction). The researchers concluded that insectivorous birds are capable of dampening the amplitude of budworm infestations when habitats are suitable for supporting adequate bird populations. This conclusion corroborates other similar findings such as suggested for woodpeckers foraging on bark beetles in northern Colorado (Koplin 1969).

Williams and Liebhold (2000) studied the degree of correlation among spruce budworm outbreaks in eastern North America during 1945–1988 and found that spatial synchrony of outbreaks decreased with distance between local budworm populations and approached zero near 2000 km. They also found geographically blocked clusters of outbreaks along an east–west axis and that synchrony depended in part on temperature and precipitation regimes and dispersal rates and distances of budworms, and

they modeled these findings in a simulation. Their findings suggest that defoliator dynamics are complex, depending on multiple variables.

Like pandemic levels of native pathogens (e.g., root rot) and defoliators (e.g., budworm), other Type II disturbances, especially relatively intense fires in grasslands, shrublands, and forests, often tend to leave spotty footprints rather than entirely denuding large areas of vegetation (Harrison et al. 2003). In this way, as with smaller canopy gaps created by windfall of trees, fires often produce complex vegetation patterns.

As vegetation patches change in time, so do responses by animals as they use patches for breeding, foraging, refuge from predators, resting, and dispersal. Some wildlife species likely evolved in concert with native disturbance regimes and take optimal advantage of resources distributed through space and time in shifting patches. Thus the further that human activities alter native patch disturbance dynamics, the greater may be the discontinuity with the evolved habitat selection behaviors of some species. In some cases, even if a suitable environment is present but is greatly altered in patch distribution pattern and temporal occurrence, an associated wildlife species may be excluded (Ellner and Fussmann 2003; Summerville and Crist 2003). This aspect of coevolution with native disturbance regimes, native resource patch dynamics, habitat selection behaviors of wildlife, and effects of changes in native disturbance regimes on wildlife viability and fitness is little studied and deserves research attention.

Type III disturbances are chronic or systematic changes over wide areas and include slow alteration of native landscapes for human habitations, ecological succession, and long-term climate change. Wildlife relations to Type III disturbances include changes in species abundance and ecosystems from changes in regional climate (Weltzin et al. 2003). Shifts in climate can greatly alter the distribution of vegetation and wildlife over broad areas.

Even climate fluctuations lasting on the order of a few years can greatly affect productivity of some wildlife populations, sometimes far away. One example is the effect of the El Niño warm water cycles in the eastern Pacific Ocean, which interrupted the 1982–1983 breeding chronology and reproductive attempts of seabird communities on Christmas Island in the central Pacific Ocean (Schreiber and Schreiber 1984). The entire seabird community failed reproductively and temporarily disappeared from this equatorial atoll as a result of the distant upwelling effect on food resources. Effects of El Niños may also extend to nonmarine species far from the Pacific (Stapp et al. 2004).

Grant et al. (2000) studied the effect of El Niño on productivity of two species of Darwin's finches on the Galápagos Islands. They found no simple correlation between clutch productivity and El Niño–induced increases in rainfall and temperature. Rather, the clutch productivity varied greatly among El Niño years in response to prior-year weather conditions and the duration of time since the last El Niño event. The authors concluded that wildlife relations to such disturbances can be understood only in a broad temporal context.

Type IV disturbances include minor and local environmental changes. These include low-intensity and local events such as spot fires, low-density rural developments along edges of natural landscapes, and gap dynamics of vegetation canopies (Acevedo et al. 1995). In particular, small vegetation gaps can be caused by natural plant death or by biotic mortality agents such as insect defoliators, plant pathogens, and plant diseases, and also by fire and weather conditions such as ice storms or windstorms. Individual treefalls in forests serve to open canopies, change local microclimates at the forest floor, and afford sun-tolerant plants a foothold (Schnitzer and Carson 2001). Treefalls uproot soil masses and redistribute litter, duff, and upper soil layers along with their associated microfauna and microflora, and can directly affect abundance and distribution of vertebrate wildlife species (Fuller 2000; Greenberg 2002).

Across watersheds and landscapes, disturbances affecting vegetation patch structure and composition tend to alter such ecosystem processes as surface water discharge, nutrient runoff, organic matter input to soils, net productivity, and microclimate. In turn, these changes can influence species composition associated with soil, ground surface, plant canopy, and other substrates. Changes in vegetation patches can alter energy balances of individual organisms, such as by changing food content or values in foraging patches, thereby tipping the balance of foraging efficiency and affecting successful reproduction and fitness or organisms. Managing native vegetation conditions in landscapes that are subject to relatively frequent, intensive Type II and IV disturbances, such as stand-replacing fires, and in which native vegetation occurs only in small, isolated patches, can be a great challenge. In such cases, there may not be sufficient area or number of patches of native vegetation to provide for resilience in the face of intense disturbances, so that high cost and much effort must be expended to prevent changes. In some circumstances, such as the fire-prone interior western United States, disturbances will inevitably occur regardless of, or in this case, because of, efforts to thwart them (Everett et al. 1994).

Often, land use history, such as decades of fire suppression with resulting buildup of fuels, high-grade selection logging of large, old trees, and changes in soil and riparian systems from mining activities, account largely for current or impending dynamics of vegetation. Whether the land can ever return to predisturbance conditions depends on the severity of the disturbance, the incursion by exotic species, the presence of native-species pools, and other factors, such as whether anthropogenic stressors are ongoing and chronic. For example, following forest clearcutting in Hubbard Brook Experimental Forest in New Hampshire, in the absence of further anthropogenic stressors, quick regrowth of dense stands of pin cherry (*Prunus pensylvanica*), a short-lived, early successional tree, was found to reduce loss of nutrients from the ecosystem and to help return much of the nutrient standing crop (N, K, Mg, Na, but not Ca) as quickly as 6 to 8 years postcutting (Marks and Bormann 1972).

Management Lessons

What management lessons can be drawn from this brief review of major categories of disturbances and dynamics of resource patches in landscapes? First, in all but the simplest ecosystems, dynamics of vegetation and environmental factors consist of a complex medley of changes occurring on multiple "schedules." Landscapes must be assessed individually to determine which disturbance types occur, the local site histories, and the likely vegetation responses to any disturbance regimes caused or altered by management activities. For example, fire behavior can be influenced by historical factors such as recent fire occurrence in the area, by proximate factors such as weather and topography, by vegetation factors such as canopy gap openings (which in turn are influenced by windfirmness of the vegetation, orientation to prevailing winds, and other factors), and by many other factors. Fire, like other disturbance regimes, then influences the likelihood of other changes, such as succession, and can greatly alter the suitability of the environment for wildlife.

Second, wildlife can play a major role in affecting disturbance regimes and how habitats respond to disturbances, such as through predation or transportation (phoresis) of disturbance agents (e.g., forest pathogens) and herbivory influence on vegetation otherwise susceptible to disturbances. Such ecological functions of species are often not, but should be, considered in management plans that alter or introduce disturbances.

Third, management may wish to more fully study how activities change native disturbance regimes and how wildlife may respond behaviorally (functionally) and demographically (numerically) to such changes.

Fourth, the specific future responses of most ecosystems that incur disturbances at multiple scales of space, intensity, and time are not very predictable. Rather, what can be better predicted, at least as probabilities or as frequencies, are the disturbance regimes themselves. In this sense, management can then craft a set of desired future dynamics, perhaps to reconstruct or mimic native disturbance regimes in which some species may have evolved optimal habitat selection behaviors. This last point is discussed further in chapter 11.

Fifth, studies of disturbance regimes can help guide land and resource management. For example, one question related to "ecosystem management" of forest resources is, to what extent does management of forest vegetation emulate effects of various kinds of natural disturbances? Do forest clearcutting, partial cutting, and salvage operations have the same effects as wildfires or windstorms? In one comparison, Franklin et al. (2000) reported "immense differences" between even-age silviculture (especially clearcutting) and natural disturbance events such as windthrow, wildfire, and even volcanic eruptions, in terms of the types of forest components ("legacies") retained or created (table 8.4). To meet the challenge of trying to emulate natural disturbances, some researchers have instituted a "variable retention" scheme in which forest openings, patches, and legacy components are

Table 8.4.   Effects of three types of disturbance on forest components

Legacy type	Fire	Wind	Clearcut
Snags	Abundant	Few	None
Logs on forest floor	Common	Abundant	Few or none
Soil disturbance	Low	Patchy	High
Understory community impact	Heavy	Light	Heavy

Source: Franklin et al. 2000. Reprinted with permission of the Society of Conservation Biology.

retained during timber harvest operations so as to more closely match effects of natural disturbance events (Mitchell and Beese 2002). Initial research suggests that such approaches are both ecologically and economically successful.

The implications of the concepts discussed in this chapter to wildlife response and management are addressed in chapter 9.

Literature Cited

Acevedo, M. F., D. L. Urban, and M. Ablan. 1995. Transition and gap models of forest dynamics. *Ecological Applications* 5 (4):1040–55.

Alfaro, R. I., S. Taylor, R. G. Brown, and J. S. Clowater. 2001. Susceptibility of northern British Columbia forests to spruce budworm defoliation. *Forest Ecology and Management* 145 (3):181–90.

Anderson, S. H., H. H. Shugart, and T. M. Smith. 1979. Vertical and temporal habitat utilization within a breeding bird community. In *The role of insectivorous birds in forest ecosystems,* ed. J. G. Dickson, R. N. Conner, R. R. Fleet, and J. C. Kroll, 203–16. New York: Academic Press.

Anderson, S. H., H. H. Shugart, and T. M. Smith. 1979. Vertical and temporal habitat utilization within a breeding bird community. In *The role of insectivorous birds in forest ecosystems,* ed. J. G. Dickson, R. N. Conner, R. R. Fleet, and J. C. Kroll, 203–16. New York: Academic Press.

Armstrong, G. W., W. L. Adamowicz, J. A. Beck, S. G. Cumming, and F. K. A. Schmiegelow. 2003. Coarse filter ecosystem management in a non-equilibrating forest. *Forest Science* 49 (2):209–23.

Askins, R. A., M. J. Philbrick, and D. S. Sugeno. 1987. Relationship between the regional abundance of forest and the composition of forest bird communities. *Biological Conservation* 39:129–52.

Avise, J. C. 2000. *Phylogeography: The history and formation of species.* Cambridge, MA: Harvard Univ. Press.

Bailey, R. G. 2005. Identifying ecoregion boundaries. *Environmental Management* 34 (1):S14–26.

Bascompte, J., H. Possingham, and J. Roughgarden. 2002. Patchy populations in stochastic environments: Critical number of patches for persistence. *American Naturalist* 159:128–37.

Bender, D. J., L. Tischendorf, and L. Fahrig. 2003. Using patch isolation metrics to predict animal movement in binary landscapes. *Landscape Ecology* 18 (1):17–39.

Billick, I., and K. Tonkel. 2003. The relative importance of spatial vs. temporal variability in generating a conditional mutualism. *Ecology* 84 (2):289–95.

Bogaert, J., A. Farina, and R. Ceulemans. 2005. Entropy increase of fragmented habitats: A sign of human impact? *Ecological Indicators* 5 (3):207–12.

Bolliger, J., H. Lischke, and D. G. Green. 2005. Simulating the spatial and temporal dynamics of landscapes using generic and complex models. *Ecological Complexity* 2 (2):107–16.

Boughton, D., and U. Malvadkar. 2002. Extinction risk in successional landscapes subject to catastrophic disturbances. *Conservation Ecology* 6 (2):2; www.consecol.org/vol6/iss2/art2.

Butler, K. F., and T. M. Koontz. 2005. Theory into practice: Implementing ecosystem management objectives in the USDA Forest Service. *Environmental Management* 35 (2):138–50.

Cantwell, M. D., and R. T. T. Forman. 1993. Landscape graphs: Ecological modeling with graph theory to detect configurations common to diverse landscapes. *Landscape Ecology* 8 (4):239–55.

Chan-McLeod, A. C. A. 2003. Factors affecting the permeability of clearcuts to red-legged frogs. *Journal of Wildlife Management* 67 (4):663–71.

Chen, J., J. F. Franklin, and T. A. Spies. 1992. Vegetation responses to edge environments in old-growth Douglas-fir forests. *Ecological Applications* 2 (4):387–96.

Chen, J., J. F. Franklin, and T. A. Spies. 1993. Contrasting microclimates among clearcut, edge, and interior of old-growth Douglas-fir forest. *Agricultural and Forest Meteorology* 63:219–37.

Chen, J., J. F. Franklin, and T. A. Spies. 1995. Growing-season microclimatic gradients from clearcut edges into old-growth Douglas-fir forests. *Ecological Applications* 5 (1):74–86.

Crawford, H. S., and D. T. Jennings. 1989. Predation by birds on spruce budworm *Choristoneura fumiferana:* Functional, numerical, and total responses. *Ecology* 70:152–63.

Custer, T. W., R. K. Hines, and C. M. Custer. 1996. Nest initiation and clutch size of great blue herons on the Mississippi River in relation to the 1993 flood. *Condor* 98 (2):181–88.

Degen, T., F. Devillez, and A.-L. Jacquemart. 2005. Gaps promote plant diversity in beech forests (Luzulo-Fagetum), North Vosges, France. *Annals of Forest Science* 62 (5):429.

Delin, A. 1992. Kärlväxter i taigan i Hälsingland: Deras anpassningar till kontinuitet eller störning. [Vascular plants of the taiga: Adaptations to continuity or to disturbance.] *Svensk Botanisk Tidskrift* 86:147–76.

Dunning, J. B., Jr., D. J. Stewart, B. J. Danielson, B. R. Noon, T. L. Root, R. H. Lamberson, and E. E. Stevens. 1995. Spatially explicit population models: Current forms and future uses. *Ecological Applications* 5 (1):3–11.

Edminster, F. C. 1938. Productivity of the ruffed grouse in New York. *Transactions of the Third North American Wildlife Conference,* 3:825–33.

Ehinger, L. H. 1991. Hurricane Hugo damage. *Journal of Arboriculture* 17 (3):82–83.

Ellner, S. P., and G. Fussmann. 2003. Effects of successional dynamics on metapopulation persistence. *Ecology* 84 (4):882–89.

Everett, R., P. Hessburg, J. Lehmkuhl, M. Jensen, and P. Bourgeron. 1994. Old forests in dynamic landscapes: Dry-site forests of eastern Oregon and Washington. *Journal of Forestry* (January): 22–25.

Fitzsimmons, M. 2003. Effects of deforestation and reforestation on landscape spatial structure in boreal Saskatchewan, Canada. *Forest Ecology and Management* 174 (1–3):577–92.

Forman, R. T. T., and M. Godron. 1986. *Landscape ecology.* New York: Wiley.

Foster, D. R., D. H. Knight, and J. F. Franklin. 1998. Landscape patterns and legacies resulting from large, infrequent forest disturbances. *Ecosystems* 1:497–510.

Franklin, J. F., D. Lindenmayer, J. A. MacMahon, A. McKee, J. Magnuson, D. A. Perry, R. Waide, and D. Foster. 2000. Threads of continuity. *Conservation Biology in Practice* 1 (1):9–16.

Frelich, L. E. 2002. *Forest dynamics and disturbance regimes.* New York: Cambridge Univ. Press.

Fried, E. 1975. A descriptive index of habitat shape irregularity. *New York Fish and Game* 22:166–67.

Fuller, R. J. 2000. Influence of treefall gaps on distributions of breeding birds within interior old-growth stands in Bialowieza Forest, Poland. *Condor* 102 (2):267–74.

Giles, R. H., Jr., and M. K. Trani. 1999. Key elements of landscape pattern measures. *Environmental Management* 23:477–81.

Grant, P. R., B. R. Grant, L. F. Keller, and K. Petren. 2000. Effects of El Niño events on Darwin's finch productivity. *Ecology* 81 (9):2442–57.

Greenberg, C. H. 2002. Response of white-footed mice (*Peromyscus leucopus*) to coarse woody debris and microsite use in southern Appalachian treefall gaps. *Forest Ecology* and Management 164 (1–3):57–66.

Gu, W., R. Heikkila, and I. Hanski. 2002. Estimating the consequences of habitat fragmentation on extinction risk in dynamic landscapes. *Landscape Ecology* 17(8):699–710.

Gustafson, E. J., and T. R. Crow. 1998. Simulating spatial and temporal context of forest management using hypothetical landscapes. *Environmental Management* 22:777–87.

Haddad, N. M., D. R. Bowne, A. Cunningham, B. J. Danielson, D. J. Levey, S. Sargent, and T. Spira. 2003. Corridor use by diverse taxa. *Ecology* 84 (3):609–15.

Hamrick, D. 1991. Assisting homeless woodpeckers. *Birds International* 3 (1):18–27.

Hann, W. J., J. L. Jones, R. E. Keane, P. F. Hessburg, and R. A. Gravenmier. 1998. Landscape dynamics. *Journal of Forestry* 96 (10):10–15.

Hargrove, W. W., F. M. Hoffman, and P. M. Schwartz. 2002. A fractal landscape realizer for generating synthetic maps. *Conservation Ecology* 6 (1):2; www.consecol.org/vol6/iss1/art2.

Harrison, S., B. D. Inouye, and H. D. Safford. 2003. Ecological heterogeneity in the effects of grazing and fire on grassland diversity. *Conservation Biology* 17 (3):837–45.

Hayes, J. P., J. M. Weikel, and M. M. P. Huso. 2003. Response of birds to thinning young Douglas-fir forests. *Ecological Applications* 13 (5):1222–32.

He, H. S., B. E. DeZonia, and D. J. Mladenoff. 2000. An aggregation index (AI) to quantify spatial patterns of landscapes. *Landscape Ecology* 15 (7):591–601.

Herold, A., and U. Ulmer. 2001. Stand stability in the Swiss National Forest Inventory: Assessment technique, reproducibility and relevance. *Forest Ecology and Management* 145 (1–2):29–42.

Herzog, F., A. Lausch, E. Muller, H. Thulke, U. Steinhardt, and S. Lehmann. 2001. Landscape metrics for assessment of landscape destruction and rehabilitation. *Environmental Management* 27:91–107.

Holt, R. D., and M. Barfield. 2003. Impacts of temporal variation on apparent competition and coexistence in open ecosystems. *Oikos* 101 (1):49–58.

Hooper, R. G., J. C. Watson, and R. E. F. Escano. 1990. Hurricane Hugo's initial effects on red-cockaded woodpeckers in the Francis Marion National Forest. *Transactions of the 55th North American Wildlife and Natural Resources Conference* 55:220–24.

Horn, H. S. 1975. Markovian properties of forest succession. In *Ecology and evolution of communities,* ed. M. L. Cody and J. M. Diamond, 196–211. Cambridge, MA: Harvard Univ. Press.

Huston, M. A. 1994. *Biological diversity: The coexistence of species on changing landscapes.* New York: Cambridge Univ. Press.

IIRS. 1999. *BIOCAP user's manual for landscape analysis and modelling biological richness.* Dehra Dun, India: Indian Institute of Remote Sensing, National Remote Sensing Agency, Department of Space, Government of India.

Irvine, K. N., and S. Kaplan. 2001. Coping with change: The small experiment as a strategic approach to environmental sustainability. *Environmental Management* 28:713–25.

Jager, J. A. 2000. Landscape division, splitting index, and effective mesh size: New measures of landscape fragmentation. *Landscape Ecology* 15:115–30.

Jenerette, G. D., and J. Wu. 2000. On the definitions of scale. *Bulletin of the Ecological Society of America* 81 (1):104–5.

Kaplan, S. 1973. Cognitive maps, human needs and the designed environment. In *Environmental design research,* ed. W. F. E. Preiser. Stroudsberg, PA: Dowden, Hutchinson, & Ross.

Kinlan, B. P., and S. D. Gaines. 2003. Propagule dispersal in marine and terrestrial environments: A community perspective. *Ecology* 84 (8):2007–20.

Kitching, R. L., and R. A. Beaver. 1990. Patchiness and community structure. In *Living in a patchy environment,* ed. B. Shorrocks and I. R. Swingland, 147–76. Oxford: Oxford Univ. Press.

Koplin, J. R. 1969. The numerical response of woodpeckers to insect prey in a subalpine forest in Colorado. *Condor* 71 (4):436–38.

Lawrence, R. L., and W. J. Ripple. 2000. Fifteen years of revegetation of Mount St. Helens: A landscape-scale analysis. *Ecology* 81 (10):2742–52.

Lawton, R. O., and F. E. Putz. 1988. Natural disturbance and gap-phase regeneration in a wind-exposed tropical cloud forest. *Ecology* 69:764–78.

Le Comber, S. C., A. C. Spinks, N. C. Bennett, J. U. M. Jarvis, and C. G. Faulkes. 2002. Fractal dimension of African mole-rat burrows. *Canadian Journal of Zoology* 80 (3):436–41.

Leopold, A. 1933. *Game management.* New York: Scribner's.

Lesica, P., and S. V. Cooper. 1999. Succession and disturbance in sandhills vegetation: Constructing models for managing biological diversity. *Conservation Biology* 13 (2):293–302.

Li, H., and J. Wu. 2004. Use and misuse of landscape indices. *Landscape Ecology* 19 (4):389–99.

Mabry, K. E., and G. W. Barrett. 2002. Effects of corridors on home range sizes and interpatch movements of three small mammal species. *Landscape Ecology* 17 (7):629–36.

Marcot, B. G., and P. Z. Chinn. 1982. Use of graph theory measures for assessing diversity of wildlife habitat. In *Mathematical models of renewable resources: Proceedings of the First Pacific Coast Conference on Mathematical Models of Renewable Resources,* ed. R. Lamberson, 69–70. Arcata, CA: Humboldt State Univ.

Marcot, B. G., and V. J. Meretsky. 1983. Shaping stands to enhance habitat diversity. *Journal of Forestry* 81:526–28.

Marcot, B. G., A. Kumar, P. S. Roy, V. B. Sawarkar, A. Gupta, and S. N. Sangama. 2002. Towards a landscape conservation strategy: Analysis of jhum landscape and proposed corridors for managing elephants in South Garo Hills District and Nokrek area, Meghalaya. [English with Hindi summary.] *Indian Forester* (February):207–16.

Marell, A., J. P. Ball, and A. Hofgaard. 2002. Foraging and movement paths of female reindeer: Insights from fractal analysis, correlated random walks, and Levy flights. *Canadian Journal of Zoology* 80 (5):854–65.

Marks, P. L., and F. H. Bormann. 1972. Revegetation following forest cutting: Mechanisms for return to steady-state nutrient cycling. *Science* 176:914–15.

McGarigal, K., and B. J. Marks. 1995. *FRAGSTATS: Spatial pattern analysis program for quantifying landscape structure.* USDA Forest Service General Technical Report PNW-GTR-351. Portland, OR: USDA Forest Service, Pacific Northwest Research Station.

McHarg, I. L. 1969. *Design with nature.* New York: Wiley.

McIntyre, N. E., and J. A. Wiens. 2000. A novel use of the lacunarity index to discern landscape function. *Landscape Ecology* 15 (4):313–21.

Meentemeyer, V. 1989. Geographical perceptions of space, time, and scale. *Landscape Ecology* 3:163–73.

Miller, M. W., and T. D. Nudds. 1996. Prairie landscape change and flooding in the Mississippi River Valley. *Conservation Biology* 10 (3):847–53.

Mitchell, S. J., and W. J. Beese. 2002. The retention system: Reconciling variable retention with the principles of silvicultural systems. *Forestry Chronicle* 78 (3):397–403.

Mladenoff, D. J., M. A. White, T. R. Crow, and J. Pastor. 1994. Applying principles of landscape design and management to integrate old-growth forest enhancement and commodity use. *Conservation Biology* 8 (3):752–62.

Moeur, M. 1997. Spatial models of competition and gap dynamics in old-growth *Tsuga heterophylla/ Thuja plicata* forests. *Forest Ecology and Management* 94:175–86.

Mortberg, U. M. 2001. Resident bird species in urban forest remnants: Landscape and habitat perspectives. *Landscape Ecology* 16 (3):193–203.

Moser, D., H. G. Zechmeister, C. Plutzar, N. Sauberer, T. Wrbka, and G. Grabherr. 2002. Landscape patch shape complexity as an effective measure for plant species richness in rural landscapes. *Landscape Ecology* 17 (7):657–69.

Nams, V. O., and M. Bourgeois. 2004. Fractal analysis measures habitat use at different spatial scales: An example with American marten. *Canadian Journal of Zoology* 82 (11):1738–47.

Ohman, K., and T. Lamas. 2005. Reducing forest fragmentation in long-term forest planning by using the shape index. *Forest Ecology and Management* 212 (1–3):346–57.

Opdam, P., J. Verboom, and R. Pouwels. 2003. Landscape cohesion: An index for the conservation potential of landscapes for biodiversity. *Landscape Ecology* 18 (2):113–26.

Orrock, J. L., B. J. Danielson, M. J. Burns, and D. J. Levey. 2003. Spatial ecology of predator–prey interactions: Corridors and patch shape influence seed predation. *Ecology* 84 (10):2589–99.

Parresol, B. R., and J. McCollum. 1997. Characterizing and comparing landscape diversity using GIS and a contagion index. *Journal of Sustainable Forestry* 5 (1/2):249–61.

Parsons, W. F. J., D. H. Knight, and S. L. Miller. 1994. Root gap dynamics in lodgepole pine forest: Nitrogen transformations in gaps of different size. *Ecological Applications* 4 (2):354–62.

Patriquin, K. J., and R. M. R. Barclay. 2003. Foraging by bats in cleared, thinned and unharvested boreal forest. *Journal of Applied Ecology* 40 (4):646–57.

Patton, D. R. 1975. A diversity index for quantifying habitat "edge." *Wildlife Society Bulletin* 3:171–73.

Pearson, D. L. 1971. Vertical stratification of birds in a tropical dry forest. *Condor* 73:46–55.

Peterson, D. L., and V. T. Parker. 1998. Dimensions of scale in ecology, resource management and society. In *Ecological scale: Theory and applications,* ed. D. L. Peterson and V. T. Parker, 499–522. New York: Columbia Univ. Press.

Peterson, G. D., S. R. Carpenter, and W. A. Brock. 2003. Uncertainty and the management of multistate ecosystems: an apparently rational route to collapse. *Ecology* 84 (6):1403–11.

Pierson, E. D., T. Elmqvist, W. E. Rainey, and P. A. Cox. 1996. Effects of tropical cyclonic storms on flying fox populations on the South Pacific islands of Samoa. *Conservation Biology* 10 (2):438–51.

Rojo, J. M., and S. S. Orois. 2005. A decision support system for optimizing the conversion of rotation forest stands to continuous cover forest stands. *Forest Ecology and Management* 207 (1–2):109–20.

Roy, P. S., and S. Tomar. 2000. Biodiversity characterization at the landscape level using geospatial modelling technique. *Biological Conservation* 95 (1): 95–109.

Salajanu, D., and C. E. Olson. 2001. The significance of spatial resolution: Identifying forest cover from satellite data. *Journal of Forestry* 99 (6):32–38.

Schmidt, K. A., J. M. Earnhardt, J. S. Brown, and R. D. Holt. 2000. Habitat selection under temporal heterogeneity: Exorcizing the Ghost of Competition Past. *Ecology* 81 (9):2622–30.

Schnitzer, S. A., and W. P. Carson. 2001. Treefall gaps and the maintenance of species diversity in a tropical forest. *Ecology* 82 (4):913–19.

Schooley, R. L., and J. A. Wiens. 2003. Finding habitat patches and directional connectivity. *Oikos* 102 (3):559–70.

Selva, S. B. 1994. Lichen diversity and stand continuity in the northern hardwoods and spruce–fir of northern New England and western New Brunswick. *Bryologist* 97 (4):424–29.

Selva, S. B. 2003. Using calicioid lichens and fungi to assess ecological continuity in the Acadian Forest Ecoregion of the Canadian Maritimes. *Forestry Chronicle* 79 (3):550–58.

Silva, M., L. A. Harling, S. A. Field, and K. Teather. 2003. The effects of habitat fragmentation on amphibian species richness of Prince Edward Island. *Canadian Journal of Zoology* 81 (4):563–73.

Simonds, J. O. 1961. *Landscape architecture: The shaping of man's natural environment.* New York: McGraw-Hill.

Soderstrom, B., and T. Part. 2000. Influence of landscape scale on farmland birds breeding in semi-natural pastures. *Conservation Biology* 14 (2):522–33.

Solé, R. V., D. Alonso, and J. Saldaña. 2004. Habitat fragmentation and biodiversity collapse in neutral communities. *Ecological Complexity* 1 (1):65–75.

Spiller, D. A., and A. A. Agrawal. 2003. Intense disturbance enhances plant susceptibility to herbivory: Natural and experimental evidence. *Ecology* 84 (4):890–97.

Stamps, J. A., M. Buechner, and V. V. Krishnan. 1987. The effects of edge permeability and habitat geometry on emigration from patches of habitat. *American Naturalist* 129:533–52.

Stapp, P., M. F. Antolin, and M. Ball. 2004. Patterns of extinction in prairie dog metapopulations: Plague outbreaks follow El Niño events. *Frontiers in Ecology and the Environment* 2 (5):235–40.

St. Clair, C. C. 2003. Comparative permeability of roads, rivers, and meadows to songbirds in Banff National Park. *Conservation Biology* 17 (4):1151–60.

Stoms, D. 1994. Scale dependence of species richness maps. *Professional Geographer* 46:346–58.

Sullivan, T. P., D. S. Sullivan, P. M. Lindgren, and J. O. Boateng. 2002. Influence of conventional and chemical thinning on stand structure and diversity of plant and mammal communities in young lodgepole pine forest. *Forest Ecology and Management* 170 (1–3):173–87.

Summerville, K. S., and T. O. Crist. 2003. Determinants of lepidopteran community composition and species diversity in eastern deciduous forests: Roles of season, ecoregion and patch size. *Oikos* 100 (1):134–48.

Suzuki, N., and J. P. Hayes. 2003. Effects of thinning on small mammals in Oregon coastal forests. *Journal of Wildlife Management* 67 (2):352–71.

Sverdrup-Thygeson, A., and D. B. Lindenmayer. 2003. Ecological continuity and assumed indicator fungi in boreal forest: The importance of the landscape matrix. *Forest Ecology and Management* 174 (1–3):353–63.

Tallmon, D. A., E. S. Jules, N. J. Radke, and L. S. Mills. 2003. Of mice and men and trillium: cascading effects of forest fragmentation. *Ecological Applications* 13 (5):1193–1203.

Thompson, J. N. 1999. Specific hypotheses on the geographic mosaic of coevolution. *American Naturalist* 153:S1–14.

Tibell, L. 1992. Crustose lichens as indicators of forest continuity in boreal coniferous forests. *Norwegian Journal of Botany* 12 (4):427–50.

Tischendorf, L., D. J. Bender, and L. Fahrig. 2003. Evaluation of patch isolation metrics in mosaic landscapes for specialist vs. generalist dispersers. *Landscape Ecology* 18 (1):41–50.

Turner, M. G., W. H. Romme, and D. B. Tucker. 2003. Surprises and lessons from the 1988 Yellowstone fires. *Frontiers in Ecology and the Environment* 1:351–58.

Urban, D., and T. Keitt. 2001. Landscape connectivity: A graph-theoretic perspective. *Ecology* 82 (5): 1205–18.

Valentine, K. W. G. 1981. How soil map units and delineations change with survey intensity and map scale. *Canadian Journal of Soil Science* 61:535–51.

Vilella, F. J., and E. R. García. 1995. Post-hurricane management of the Puerto Rican parrot. In *Integrating people and wildlife for a sustainable future,* ed. J. A. Bissonette and P. R. Krausman, 618–21. Bethesda, MD: Wildlife Society.

Wagner, H. H., and M.-J. Fortin. 2005. Spatial analysis of landscapes: concepts and statistics. *Ecology* 86 (8):1975–87.

Walker, R. S., A. J. Novaro, and L. C. Branch. 2003. Effects of patch attributes, barriers, and distance between patches on the distribution of a rock-dwelling rodent (*Lagidium viscacia*). *Landscape Ecology* 18 (2):185–92.

Webb, S. L. 1989. Contrasting windstorm consequences in two forests, Itasca State Park, Minnesota. *Ecology* 70:1167–80.

Weltzin, J. F., M. E. Loik, S. Schwinning, D. G. Williams, P. A. Fay, B. M. Haddad, J. Harte, T. E. Huxman, A. K. Knapp, and G. Lin. 2003. Assessing the response of terrestrial ecosystems to potential changes in precipitation. *BioScience* 53 (10):941–52.

West, A. J. 1989. Concerning damages to federal, state, and private forest resources and the USDA Forest Service participation in the relief efforts following Hurricane Hugo. Statement of Allan J. West, Deputy Chief for State and Private Forestry, Forest Service, United States Department of Agriculture, before the Subcommittee on Forests, Family Farms and Energy Committee on Agriculture, United States House of Representatives. November 6, 1989, Moncks Corner, South Carolina.

Williams, D. W., and A. W. Liebhold. 2000. Spatial synchrony of spruce budworm outbreaks in eastern North America. *Ecology* 81 (10):2753–66.

Withers, M. A., and V. Meentemeyer. 1999. Concepts of scale in landscape ecology. In *Landscape ecological analysis: Issues and applications,* ed. J. M. Klopatek and R. H. Gardner, 205–52. New York: Springer-Verlag.

Wu, J., and O. L. Loucks. 1995. From balance of nature to hierarchical patch dynamics: A paradigm shift in ecology. *Quarterly Review of Biology* 70:439–66.

Wunderle, J. M., Jr. 1995. Responses of bird populations in a Puerto Rican forest to Hurricane Hugo: The first 18 months. *Condor* 97:879–96.

Zonneveld, I. S. 1979. *Land evaluation and land (scape) science.* Enschede, Netherlands: International Training Center.

9 Wildlife in Landscapes: Populations and Patches

I'm telling you. People come and go in this forest.

EEYORE, *THE HOUSE AT POOH CORNER*

In this chapter we continue the story of habitat patches and landscapes begun in chapter 8. Here we focus on the response of organisms, species, populations, and communities to conditions of patchy environments and habitat dynamics.

Response of Wildlife to Habitat Heterogeneity

Wildlife within landscapes use and select for resource patches that offer food, shelter, and other life needs. The study of how animals use patchy environments has a long tradition in ecology. Studies of *optimal foraging theory* have assessed how organisms select resource patches for foraging and have related return in prey taken to time and energy expended in locating the prey (e.g., Krebs et al. 1974; also see chapter 7). *Foraging energetics* can determine species presence in an area and overall community structure (Oom et al. 2004), and animals can evolve specific behaviors to increase their efficiency in using resource patches. For instance, Morris and Davidson (2000) empirically verified that white-footed mice (*Peromyscus leucopus*) select for foraging patches to optimize their levels of reproductive fitness.

A rich literature exists on the effects of resource heterogeneity on population dynamics and community structure. For example, in the 1970s, much research equated vegetation structural diversity, particularly foliage height diversity (FHD) in forests and woodlands, with bird species diversity (BSD) (Recher 1969; Pearson 1975).

The literature has also amply demonstrated that habitat area and distance from colonizing species pools affect the number of species occurring in specific habitats and the likelihood of occupancy by individual species using the habitat (e.g., Lomolino 1982; Walker et al. 2003). Much of the literature suggests that species' individual and population attributes—mainly dispersal ability, demographics, body size and correlated home range size, and degree of habitat specialization—greatly influence how habitat distance, habitat area, and movement barriers affect individual fitness and population demography.

The influence of habitat heterogeneity depends on the degree of habitat specialization and the use of intervening environments by organisms. Often assumptions are made about species' obligate use of particular environments, such as forest interiors, edges, and old-growth forests, with scant empirical evidence. The real world is often more complicated.

Effects of Edges on Vegetation and Wildlife

One aspect of habitat heterogeneity that can be easily observed and measured is that of edges between vegetation conditions or environmental situations. In the early days of wildlife management, edges in woodlands and forests were viewed as desirable because they enticed game animals such as deer, grouse, and pheasants. Hedgerows in agricultural fields were encouraged as a way to increase huntable game (Edminster 1939; Jones et al. 2005). Over time, however, edges became viewed as less desirable components of landscapes managed for wildlife (see box 9.1).

What is the relation of wildlife communities to edges? *Edges*—sharply defined ecotones and broader ecoclines—are where different communities commingle. Thus plant and animal species richness along edges is often greater than within adjacent homogeneous resource patches. This greater species richness may be more an artifact of the spatial overlap of species assemblages along edges than it is the ecological orientation of organisms to edge environments per se.

For example, species richness is often disproportionately high along riparian areas. Riparian environments likely do offer resources, including cover, water, and food, not found in as high abundance or diversity per unit area in more homogeneous upland situations. However, some proportion of the species richness in riparian environments may be simply due to the fact that riparian environments are linear features that segment a landscape and act as controlling nodes or "cut points" across the quilt of resource patches in a landscape, so that many species must travel through riparian environments as they traverse a landscape. This pattern may inflate species richness of riparian environments disproportionate to their area.

There are few known cases of wildlife species that are absolute obligate users of, and specially adapted to, edge environments, although a large number of species will take advantage of edge conditions as facultative users. Many cases of so-called "edge species" may simply be a case of species using different vegetation communities. Their occurrence near edges may be merely a

---

Box 9.1   Changing Perspectives on Edge Effects

The history of how ecologists and wildlife managers have viewed edges and openings is a lesson in human perspective and scientific method.

In 1933, Aldo Leopold wrote in his classic work *Game Management* that ecotones and edges between vegetation types are areas of particularly high wildlife concentration. Prior to the 1980s, edges and woodland openings were viewed in the literature as largely beneficial to wildlife, with little discussion of potential adverse effects. In 1938, Lay posed the question, How valuable are woodland clearings to wildlife? and concluded that they enhance game populations. Other early studies on the positive effects of edges on wildlife include Petrides' (1942) work on the relation of hedgerows to wintering wildlife in New York, Johnston's (1947) work on breeding birds of forest edges in Illinois, and Sammalisto's (1957) study of birds along woodland–open peatland edges in south Finland. Kelker (1964, 180) reaffirmed Leopold's "accepted ideas" on wildlife management, including "laws of interspersion and dispersion of habitat features."

(box continued on next page)

Box 9.1   Continued

In the1970s, much excellent work was done on determining the use of forest-opening edges for game, largely deer and elk (Ffolliott et al. 1977; Thomas et al. 1979), and on developing guidelines to enhance such conditions. However, also during that decade, some researchers were beginning to determine adverse influences of logging and land use, such as fragmenting environments and causing wildlife populations and communities to be at risk (Whitcomb et al. 1977; Robbins 1979). The 1970s also saw the development of island biogeography theory, and some authors began to use the concepts to describe the adverse effects of fragmenting continental environments on wildlife (Simberloff and Abele 1976).

By the1980s, serious challenges were being made to the assumption that edges and openings are necessarily beneficial to wildlife (Harris 1984), on several fronts. First, there was a growing emphasis on nongame wildlife species (Robbins 1978; Robinson 1988), some of which do not benefit from edges, openings, and fragmentation of environments. Second, studies suggested that habitat fragmentation can serve to reduce habitat patch size and increase isolation, which negatively affects some wildlife species and communities (e.g., Freemark and Merriam 1986; Lehmkuhl 1990). In a sense, this finding was anticipated by Leopold's (1933) "law" of habitat interspersion and dispersion, but now the emphasis was on negative correlations with wildlife populations. Third, the acceleration in anthropogenic changes to native environments, including loss of ancient forests in western North America and rainforests in the tropics, prompted great concern for the survival of area-sensitive habitat specialists. In some cases, too much edge and too many openings led to local overpopulation by game species (Alverson et al. 1988), and researchers began to focus on edges as "predator sinks" of heightened nest predation on songbirds and other wildlife (Andren and Angelstam 1988). Also, some researchers stepped out of the forest and determined that habitat heterogeneity can be a concern for some wildlife species in prairie and grassland environments as well (Wiens 1974).

Research attention turned to the need to conserve remnant patches of native vegetation for area-sensitive and rare species and declining native communities (e.g., Howe et al. 1981). In light of such research, Reese and Ratti (1988) and Guthery and Bingham (1992) raised the need to reevaluate the traditional concept of edge. Soon the conceptual pendulum swung swiftly the other way, and many researchers and managers came to equate fragmentation with loss of native environments, mostly old forests, due to human exploitation, and with the threat of extinction of associated species.

Far from the previous view of benefits to game animals, fragmentation was soon being heralded as the great bearer of species extinction (Wilcox and Murphy 1985; Wilcove 1987). Concern focused on the proportion of landscapes altered by human activities, particularly for urbanization, agriculture, grazing, and timber harvest. Such activities were now being referred to as "stressors" (Barrett 1981). The desirable small woodland openings previously touted by Lay, and the rich habitat edges described by Johnston and Sammalisto were now being seen as dominating native landscapes in boreal, temperate, and tropical regions alike (e.g., Franklin and Forman 1987).

Since the1990s, much research has focused on the mechanisms of environmental fragmentation and species extinction in a wide variety of plant, invertebrate, and vertebrate taxonomic groups, and on how land management causes environmental fragmentation (e.g., Fitzsimmons 2003). Many land and natural resource management plans have addressed means of providing for forest-interior habitats, habitat corridors, and matrix connections, and means of protecting remnant native environments (Lindenmayer and Franklin 2002). In the 1990s, wildlife habitat management embraced a revival of systems ecology in the guise of "ecosystem management," with Aldo Leopold apotheosized as the founder of these precepts (Knight 1996). Gone, however, are the references to the simpler "laws" of the purely positive benefits of habitat edge, interspersion, and dispersion.

This brief review illustrates that conservation issues should be placed in their historical context in order to best understand the evolution of research foci and current management concerns. In historical perspective, it is useful to note that as recently as the1970s, Jack Ward Thomas and his colleagues were battling the then-prevalent adage that good timber management is necessarily good wildlife management (Thomas 1979). Integration of silviculture and wildlife management has come far (e.g., Thompson et al. 2003) and can go even further. It is likely that the early researchers of woodland openings and habitat edges did not discount the evils of excess, but that conditions then did not warrant their being highlighted. Perhaps this account, then, is also a monition on how fast humans have altered native environments, as well as a caveat to future researchers and managers to continue to question what seem to be immutable principles of conservation research and management. In reality, such principles often reside as much in historical and cultural contexts as they do in scientific ones.

result of needing proximity to different environments within their home range areas. Examples include Townsend's solitaires (*Myadestes townsendi*) that establish winter territories along pinyon-juniper-ponderosa pine edges in the southwestern United States (Salomonson and Balda 1977) and hawks (*Buteo* spp.) coexisting along prairie-parkland ecotones (Schmutz et al. 1980). Likewise, cases of adverse effects of edges may simply be a case of overall loss of habitat area (Parker et al. 2005).

Other research on edge environments has revealed complex microclimatic gradients and vegetation response (Chen et al. 1999). Microclimate is an important factor determining foraging site selection by some species. An example is wintering mountain chickadees (*Parus gambeli*), which selected sites with higher air temperatures and lower wind speeds than generally available in forests of south-central Wyoming (Wachob 1996). The productivity of soils and vegetation is sensitive to microclimatic conditions (Zabowski et al. 2000) to the extent that managing vegetation as habitat for wildlife should entail considerations for how opening and changing vegetation structures can influence microclimate (Gray et al. 2002), which, in turn, can affect vegetation response.

Numerous researchers have studied *depth-of-edge influence,* particularly in forest environments. Chen et al. (1993) found that in clearcuts and along clearcut-forest edges, daily average air and soil temperature, wind velocity, and shortwave radiation were consistently higher, and soil and air moister were lower, than within old-growth forest interiors. Local weather patterns affected these differences. Microclimate was most variable temporally along the edge, as compared with clearcuts or forest interiors. In a subsequent study, Chen et al. (1995) determined that microclimate influences extended from 30 to >240 m into the forest, the shallowest influ-ence being from gradients in shortwave radiation and the deepest from humidity.

In oak-chestnut forests of the eastern United States, Matlack (1993) found significant edge effects of light, temperature, litter moisture, vapor pressure deficit, humidity, and shrub cover up to 50 m from the edge. He also reported the influence of aspect on some microclimatic variables, and that a large proportion of forest in small- to medium-sized fragments is climatically altered by edge.

Vegetation response along edges was researched by Young and Mitchell (1994). Siitonen et al. (2005) found complex edge effects related to wood-rotting fungi. Chen et al. (1992) determined that clearcut logging caused reduced tree stocking, increased growth of dominant tree species, and resulted in higher tree mortality along edges of old-growth conifer forests of western North America. Chen et al. (1992) defined a depth-of-edge influence zone as the point along the clearcut–forest gradient at which a given variable returned to a condition representing two-thirds of the interior forest environment, and that this influence zone ranged from 16 to 137 m from the edge, depending on the variable. Canopy cover had shallow depth of edge; tree stem density (stems per ha >6 cm in diameter) and tree diameter at breast height had medium depth of edge; and total basal area had high depth of edge.

Chen et al. (1992) noted that there is no "interior" forest environment in a patch <10 ha if depth-of-edge influence is 137 m, and that edge effects were influenced by topographic position. The concept that small forest patches have no "true" interior environment—that is, their extent is "all edge"—has been used by some managers to discount the ecological value of forest patches and justify their removal. However, studies have shown the high ecological value of retaining remnant patches of native forests within highly managed landscapes (Lindenmayer and Franklin 2002).

Matlack (1993) and Fraver (1994) studied forest edge effects in eastern U.S. forests. Fraver found that, along edges between agricultural lands and mixed-hardwood forests of North Carolina, based on cover of native and exotic plant species, edge effects in forests were up to 50 m on south-facing edges and only 10 to 30 m on north-facing edges. Along Australian wheat-belt edges, Hester and Hobbs (1992) found a greater density and cover of nonnative plant species, decreasing "rapidly" with increasing distance from the edge into native vegetation reserves. Dignan and Bren (2003) modeled light penetration along stream-forest edges as a way to establish guidelines for stream buffers in forests of southeastern Australia. They reported that edge orientation (aspect), distance from the edge, and canopy height most affected depth-of-edge influence for light penetration. On all but south-facing edges, light penetration dropped markedly beyond 10 to 30 m into the forest. Similar results from studies of forest fragmentation and edge influence in tropical environments have been reported (Schlaepfer and Gavin 2001; Laurance et al. 2002).

Edges have also been called predator traps, or predator sinks, by some researchers finding evidence of increased avian predation in such situations. Some researchers have reported increased incidence of avian nest parasitism and predation along edge environments and openings (Rodewald 2002; Lloyd et al. 2005). In forested landscapes of central Vermont, Coker and Capen (1995) reported that brown-headed cowbirds occurred in 46% of openings surveyed, and that cowbird presence was related to opening area, distance to closest chronic disturbance opening, and number of livestock areas within 7 km of the opening. Their work suggested, though, that small (about 4 ha), remote forest openings are unlikely to attract cowbirds in the landscapes studied. Coker and Capen's conclusions contrast with findings by Hahn and Hatfield (1995), who reported that nest parasitism rates by brown-headed cowbirds were significantly higher in forest interiors than old-field communities. They also reported differences in host selection compared with other studies and concluded that cowbirds vary regionally in host and habitat use.

Kolbe and Janzen (2002) found that nests of painted turtles (*Chrysemys picta*) within ~25 to 40 m of water's edge were more apt to be predated than nests farther away, although such effects varied over years. They noted that previous edge predation research had much bias in focusing on birds and at larger spatial scales, and that their findings supported concern for adverse edge effects on wildlife population viability. In other studies, however, Robinson and Robinson (2001) reported that in an Illinois landscape with high avian nest predation, internal edges created by selective logging did not have an additive effect. On the other hand, Jules and Rathcke (1999) reported that *Trillium* plants along edges of old-growth forest fragments had lower fecundity because of decreases in pollinators and increases in seed predation by rodents. Thus, in conclusion, it is clear that effects of edges and habitat fragmentation on plants and animals can vary greatly by species and by the local and landscape contexts.

Boundary Effects

Another topic related to edges and habitat fragmentation that has received much attention in recent years is that of *boundary effects*. The term *boundary* has been used to refer to edges of habitat patches in a landscape and also to administrative borders of protected areas. In the former sense, Matthysen (2002) reported that how individual animals respond to patch boundaries is an important element of models of dispersal and patch occupancy. He evaluated natal dispersal of two species of tit from forest patches 7 to 11 ha in size, and found that one species (the great tit, *Parus major*) tended to emigrate toward the nearest patch border, possibly because these birds were familiar with the forest patch around the natal territory.

In another study, by Tischendorf et al. (2003), whether species are specialist or generalist users of habitat patches in a landscape determined how well measures of patch isolation (see chapter 8) predicted immigration. The authors concluded that generic use of patch isolation measures and patch isolation studies need to be interpreted in the context of structure of the landscape and dispersal distance and behavior of the organism.

Research on boundary effects of protected areas has suggested that isolation of parks and preserves can lead to loss of top predators and wide-ranging terrestrial carnivores. For example, isolation of national parks in North America and Tanzania has contributed to local extinction of large mammals (Newmark 1995, 1996). Exacerbating the effects of simple isolation is the influence of human activities along the boundaries of protected areas (Holland et al. 1991). For instance, in Peru, slash-and-burn agriculture has effectively isolated wild mammals within parks (Naughton-Treves et al. 2003), and many human–animal conflicts occur along ecological and administrative boundaries (Sitati et al. 2003).

Ambrose and Bratton (1990) studied boundary effects in Great Smoky Mountains National Park in the Appalachian Mountains of the eastern United States. They reported that, from 1940 to 1978, the density of forest patches and cleared patches increased, the size of forest patches decreased, and there was little change in the percentage of forest cover. In contrast, inside the park borders, the density of forest patches and cleared patches decreased, and the size of forest patches and the percentage of forest cover increased. Thus, during this time period, land-scape heterogeneity and connectivity of habitat for some forest-dwelling species changed significantly. Furthermore, different sections of the park border varied in these changes, depending on local variations in land use history.

The greater the contrast in environments along administrative borders of parks, the more that parks (or other natural areas) may begin to act as habitat isolates and Freeman's (1986) "genetic islands" for some species. Diligent protection of parks and natural areas from undue or unplanned external threats can help protect wildlife resources (Goggins 1987) within parks and connectivity of populations and habitats across park borders. However, in some cases, habitat isolates for some species may actually require considerable active management of conditions within them if they are to remain useful in contributing to biodiversity. A hands-off approach may not always be the best effective management method.

To offset adverse boundary effects from human settlements and land use along the edges of parks and protected areas, buffer zones can be designated. For example, Willson and Dorcas (2003) and Semlitsch and Bodie (2003) suggested buffering streams and wetlands to protect aquatic salamanders and reptiles. Hylander et al. (2002) suggested using bryophytes to evaluate the efficacy of boreal stream buffers. Ruel et al. (2001) cautioned that windthrow can defeat the use of riparian buffer strips in forests unless wind fetch, tree density, and strip width are taken into account. On a finer scale, buffer areas have been used in wildlife management for many decades to help avoid direct adverse impact of human activities on nesting raptors (e.g., Ward and Salas 2000) and other situations and species of interest.

An interesting review of buffers and boundaries by Martino (2001) concluded that there are two antagonist positions on the use of buffers surrounding national parks and protected areas.

One position argues that buffers should be inviolate areas used as extensions of parks, whereas the other posits that buffers should be used to integrate human presence and activities. The best answer, however, is likely a combination. In some instances, strict protection of some activities such as hunting, slash-and-burn agriculture, and tree felling should be prohibited in buffers along sensitive and small protected areas, whereas in other situations these or other activities may be permitted if effects do not unduly extend into the protected area or disrupt native species and ecosystems therein. Indeed, the Russian system of protected areas—national parks, or *zapovedniks*, and wildlife sanctuaries, or *zakazniks*—provides for such variations in permitted and excluded activities. Martino concluded that there is an urgent need to clearly define the objective of buffer zones in any given situation, and with that we agree.

The Time Dimension: Succession and Climate

Habitat heterogeneity also occurs over time as well as space. We discussed habitat disturbance and temporal changes in chapter 8. Wildlife species respond to temporal habitat changes in many ways, as habitat patches and resource availability flux across landscapes. We will discuss population viability dynamics below, and here will focus on effects of ecological succession and climate change.

Ecological succession occurs following a disturbance to an environment. *Primary succession* occurs when the earth is essentially scraped raw of vegetation, and organisms must colonize the land anew. Examples of primary succession include creation of new terrestrial land area from volcanic eruptions, particularly lava flows such as on the Hawaiian Islands. Another example is glacial activity, which can denude valleys down to bedrock. The classic view of primary succession holds that windswept seeds, spores, and

other disseminules of plants, ferns, lichens, bryophytes, and other hardy organisms invade the newly disturbed area first, taking hold in crevices and sheltered microsites. Such early colonizers are able to withstand extremes of moisture, temperature, and nutrients, and through their biological activity and accretion of detritus begin to transform rock substrates into the first thin soils. Soil formation—pedogenesis—is a critical part of early primary succession upon which other plant colonizers follow that use the litter, duff, and fractured regolith in which to root or otherwise sustain themselves. Later, the early cryptogams (lichens and bryophytes), grasses, and herbs are joined by woody shrubs and, in some environments, eventually trees.

However, nature is usually messier than our models, and primary succession may not necessarily proceed so cleanly. For instance, Fastie (1995) studied multiple pathways of primary succession at Glacier Bay, Alaska, and refuted the assumption used in all side-by-side chronosequence studies that communities at the oldest sites (longest time since exposure from glaciation) developed through stages found in the younger sites. Instead, at Glacier Bay, Fastie found three distinct successional pathways, which depended on distance to seed sources. So again, the landscape perspective proved important in understanding primary successional development.

Secondary succession refers to the sequential development of plant and animal communities following disturbances that leave behind part of the previous community. Examples of secondary succession include secondary forest growth following major disturbances such as fires, clearcutting, and storms; grassland recovery following major droughts or periods of overgrazing; and old-field succession in abandoned farmlands. In each case, the specific sequence of changes in plant and animal species may be very difficult to predict with much certainty. Secondary succession depends in part on seed and propagule sources remaining on-site, such as seed banks within the soil profile.

The huge area of standing forests killed by the eruption of Mount St. Helens in 1980 has undergone what seems to be a combination of primary and secondary succession. Areas blanketed by deep pyroclastic flow and ashfall quickly became invaded by early-seral species, such as lupine (*Lupinus* spp.), that took root in sheltered microsites next to rocks and down tree trunks. However, early colonizers also included tree seedlings that invaded before the substrate was first colonized by grasses and forbs in the traditional model of primary succession. Moreover, pocket gophers and other fossorial wildlife survived the blast in many areas, and their tunneling and digging actions helped turn over the soil profile, incorporating the nutrient-rich ash layer into the underlying soil horizons, exposing buried seed banks, and accelerating secondary succession of plants from on-site (autochthonous) sources. The ecological roles of vertebrates and invertebrates alike greatly influenced vegetation recolonization of the area devastated by the eruption (Bishop 2002).

Turner et al. (1998) studied succession following large, infrequent disturbance events and concluded that the following patterns occur: (1) Initial densities of organisms are lower, then increase. (2) Recovering patches of vegetation serve as refugia or sources of colonization, and their importance increases over time, a process they called *nucleation.* (3) Competitive exclusion of species in the early stages is less important than is chance colonization in determining community composition. (4) Initial and early community composition following disturbances is not very predictable, as species colonization and recovery is a stochastic process. (5) The rate of recovery—that is, the degree of resilience of the ecosystem to return to its initial, predisturbance constitution—will slow

over time. Turner et al. (1998) suggested that succession depends on abundance and spatial arrangement of survivors and residual patches, distance to seed sources, and arrival patterns of colonizers.

To their list we can also add (6) the ecological functions of organisms, such as rock lichens that break down rock surfaces into soil, and digging mammals that overturn soil profiles to enhance soil structure and seed survival. Also, it may be true that functional groups of plants and animals may be more predictable and robust to disturbances than are individual species. For instance, initial research results by one of us (BGM) suggests that all six functional groups of invertebrates inhabiting old-forest remnants in conifer forests of the Cascade Mountains in southern Washington are also present in young clearcuts, in nearly the same numbers of species per functional group, although the specific species may differ markedly. The functional groups include fungivores, predators, omnivores, herbivores, shredders, and parasitoids. Those species inhabiting early successional stages, such as forest clearcuts, may be more generalist in their habitat orientations and resource uses than species inhabiting late successional stages, such as old-growth forests. Similar trends may occur with terrestrial vertebrates.

On a broader time scale, species respond in various ways to local and regional climate change. This response is a topic of great debate and current research focus, particularly as human activity may serve to alter regional or global climates. That humans can greatly alter local climates is indisputable, as evidenced by various forest edge and boundary effects studies (discussed above). Some authors (e.g., Hannah et al. 2005) have suggested that biological reserves and protected areas should be designed with climate change in mind (regardless whether such change is caused by human activities). Other considerations for climate change include better understanding associated changes in fire regimes and crafting appropriate management responses (Whitlock et al. 2003); predicting response of terrestrial ecosystems to changes in precipitation patterns (Weltzin et al. 2003); tracking changes in ecosystem function (Pataki et al. 2003); projecting increases in forest pests (Logan et al. 2003); predicting effects on long-distance migratory wildlife (Jonzen et al. 2002; Lemoine and Bohning-Gaese 2003); predicting changes in host–parasite relations (Jaenike 2002); understanding effects on amphibian survival (Blaustein et al. 2001); and many other factors. Clearly, climate change has the potential of affecting many aspects of species, community, and ecosystem dynamics.

Of Populations and Metapopulations

The concern for unduly fragmenting and attenuating populations across habitat patches and isolates relates to adverse effects on population viability. *Population viability* is defined as the probability of persistence of a population in a defined area, over a defined duration of time. A population with high viability is one with a high probability of persistence. Why talk about probabilities? Because projecting future or potential population size, density, distribution, and rates of change is not an exact process, and results must be couched in terms of probability distributions.

Further, a viable population is self-sustaining on its own—that is, where (birth rate + immigration) ≥ (death rate + emigration), so that the overall rate of change is $\lambda \geq 1.0$. Many of the parameters of "vital rates"—age- or stage-class-specific fecundity and mortality—are usually measured and modeled probabilistically as well, and as density-dependent functions.

However, many, perhaps most, wildlife populations occur in nature not as single panmictic

(fully interbreeding) entities, but rather dispersed through time and space as partially or fully isolated subpopulations. The collective array of all subpopulations is a *metapopulation* (see chapter 3). In a metapopulation, dispersal and breeding are more apt to take place within subpopulations than among them.

Modeling viability of metapopulations is not as simple as just projecting effects of vital rates on a single population structure and size, as with a Leslie matrix life table. Instead, rates of dispersal and breeding among subpopulations must be taken into account, as well as influences of environmental disturbances, degree of synchrony of disturbances among subpopulations, influence of genetic load and variability on vital rates from genetic drift and inbreeding, and other factors. Viability of metapopulations can be expressed in many ways, including the number and size of subpopulations that persist to specified time periods, the spatial dispersion of those nonzero subpopulations, and other ways. In the end, it is the quality, pattern, and extent of habitat patches that influence metapopulation viability.

Dynamics of Colonization: Founders and Rescuers

One of the basic factors that determines occupancy rate of habitats is the dynamic of local colonization and extinction. The now-classic literature on island biogeography (MacArthur and Wilson 1967; Diamond et al. 1976; Simberloff and Abele 1976) has described dynamics of colonization and extinction in both oceanic and continental situations.

Colonization of oceanic islands, and, to some degree, of continental resource patches and landscapes, may occur more frequently with species that are more vagile—that is, are capable of successful, longer-distance movements, such as home range movement, natal dispersal, irruptive movement, and seasonal migration (chapter

3); that produce more "disseminules," or dispersers, per breeding season or unit time; or that are able to withstand unsuitable conditions between islands or resource patches. Two examples of successful long-distance dispersers are exotic aquatic mollusk species dispersing via boaters over long terrestrial distances (Buchan and Padilla 1999) and spiders "ballooning" on airlofted webs across oceans to settle on oceanic islands (Thomas et al. 2003).

Colonization by a species also occurs more frequently when the island or resource patch is larger or contains more area of suitable environmental conditions (Cole 1981); is closer to species source pools; is more heterogeneous, supporting a variety of microhabitats; and, in terrestrial situations, is at least somewhat connected to other resource patches by habitat corridors (Dunning et al. 1995) or habitat remnants or components scattered throughout the intervening matrix lands.

Many researchers and modelers view specific colonization events largely as stochastic incidents dependent on the chances of organisms finding environments with sufficient resources for survival and reproduction. However, Smith and Peacock (1990) proposed that colonization of resource patches by vertebrates in terrestrial systems can be much more deterministic and predictable, and that it is often strongly influenced by the presence of conspecifics, including potential mates, in "recipient" patches. In a simulation modeling study, Ellner and Fussmann (2003) found that the probability of population persistence in a patchy landscape depends in part on the distribution of successional stage durations, and whether patches become suitable for recolonization immediately following a local extinction. Such factors of biological causes of local extinction and patch recovery dynamics can greatly affect whether a species colonizes and persists within a landscape.

One aspect of colonization dynamics of organisms in patchy or insular environments pertains to how local populations get started and their subsequent genetic diversity and viability. Colonizers providing the seeds of "starter" populations are called *founders,* or *founder populations.* Founder populations can begin in newly suitable environments and thus serve to extend the distribution of a species, or they can act to recolonize previously occupied sites left vacant by temporary dips in site suitability or by random local extinction of the species. The contribution of founders to overall metapopulation stability depends on the environmental stability of the site for maintaining the species once it appears, the proportion of valuable "disseminules" or dispersers of the source population that is tapped, and the value of the site as a stepping stone for other patch occupancy, even if the site itself provides suboptimal conditions and is occupied only intermittently.

Also important to founder dynamics is the number of founder organisms, the rate of subsequent interbreeding with adjacent populations, and the rate of immigration contribution from source populations, which are factors that affect the genetic diversity of the local population. If number of founders is low and subsequent breeding exchange or immigration is low, then the population will undergo a genetic bottleneck (see chapter 2). In more extreme cases, it may then suffer deleterious inbreeding effects. Such bottleneck conditions are called the *founder effect.* Paying attention to the founder effect is particularly important for conservation programs of captive propagation or reintroduction of rare species (Backus et al. 1995).

But there are conditions where founders may serve as "nature's experiments" by successfully colonizing newly suitable isolated environments and then evolving unique ecotypes, morphs, subspecies, or, in some island situations, entire suites of new species. Cases of *adaptive radiation* of founders into new species complexes have been documented for many situations, including the honeycreepers of Hawaii, the honeyeaters of Australia, and Darwin's finches in the Galápagos Islands of Ecuador (see box 9.2).

Within each of these groups has evolved mor-

phological specialization. For example, in the Australian honeyeaters, the more nectivorous species are generally longer billed, enabling them to probe a wider variety of flowers. The most specialized of the honeyeaters are the slender-

---

Box 9.2   *Adaptive radiation of species complexes in island and continental settings*

Under the right circumstances of isolation and suitable resources, a founder or original population can evolve into a complex of species. Such adaptive radiation has been documented in island and continental situations alike for a variety of organisms. Some classic examples, well described in the literature, include the following (bird taxonomy follows Sibley and Monroe 1991 unless otherwise noted):

The Hawaiian honeycreepers, of the tribe Drepanidini ("dreps" to the initiated, of family Fringillidae), include more than a dozen extinct species described from subfossils and some 20 or so extant species across the Hawaiian Islands. Many are threatened from habitat alteration.

- The honeyeaters of Australia (family Meliphagidae) constitute a complex of about 67 species (72 if the 5 Australian chats are included in this uncertain taxonomy).
- Darwin's finches (family Fringillidae) in the Galápagos Islands of Ecuador include some 13 species among 4 genera.
- Birds of paradise on New Guinea (family Corvidae) constitute some 38 species, and are a fine example of extreme sexual selection, sexual dimorphism, and adaptive radiation, all in one.

Some less heralded but nonetheless spectacular examples deserving further description include the following:

- On Australia, some 9 species of *Rattus* comprise a set of neoendemic rodents. One account suggests that *Rattus* originated in Southeast Asia and Australia, or at least spread to Australia very early, giving time for radiation of many species.
- In southern Africa, the bird genus *Cisticola* (family Cisticolidae) comprises some 18 species by one account (Sinclair 1997). These include local endemics such as the shortwinged cisticola (*C. brachyptera*) and the singing cisticola (*C. cantans*), both found in tall grass clearings in miombo woodlands in eastern Zimbabwe and Mozambique; and the chirping cisticola (*C. pipiens*), found in reedbeds and papyrus swamps in the Okavango Delta of northern Botswana and neighboring countries. The cisticola species complex, however, does not show a wide array of bill or body morphologies found in other species complexes such as the Hawaiian honeycreepers, in part perhaps because their continental occurrence means competition from other species, or perhaps because they have instead radiated into different geographic areas or habitats but still use the same resources, thereby reducing interspecific competition.
- Vanga shrikes (family Vandigae) are found only on Madagascar. After the original vanga arrived on the island, which has few other bird groups, it evolved into 12 species, each taking advantage of an available niche. Each species developed its own size, color, and shape of bill—all adaptations to the ecological niche it filled. DNA evidence suggests that the vangas are the descendants of a relatively recent immigrant from Africa.
- Flying squirrels on the Indian subcontinent constitute a rather amazing array of 9 genera, 17 species, and 40 subspecies or races. These do not include other squirrel species not adapted to a volant (gliding) style of locomotion. Why so many flying squirrels have evolved in the Indian subcontinent in particular is unclear but may have to do with the great diversity of topography, vegetation, climates, and other species present.
- In sub-Saharan Africa occurs an amazing array of 5 species of ground squirrel, 1 pygmy squirrel, 9 tree squirrels, 10 bush squirrels, and 9 giant and sun squirrels, likely evolved in patterns of adaptive radiation from far fewer original forms. The distribution of these squirrels, along with many primates and small carnivores there, suggests that climatic changes must have been very important, because many of these species live only in restricted regions, often identifiable as refuges, within much more extensive forests (Kingdon 1989, 1997).

beaked species, such as spinebills. Their long, de-curved bills match the equally long, decurved flowers of many native plants (such as the heath *Epacris longiflora*). Short-beaked honeyeaters can extend their tongues to reach nectar at some of the flower bases, but the farther the tongue is extended, the less efficient it becomes. Sometimes the birds pierce the sides of such tubular flowers (and are thus functionally congruent with flowerpiercers of the New World tropics), or instead visit the less specialized flowers (perhaps *Eucalyptus* and *Melaleuca*) where nectar is more readily accessible (Simpson and Day 1999, 376).

The adaptive radiation of bill morphology in Australia's honeyeaters is similar to that in the Hawaiian honeycreepers, and in the more insectivorous and frugivorous Darwin's finches. Although diets vary both within and among these groups and species, the basic principle of adaptive radiation is quite similar.

Note the conservation implications of such adaptations. In Australia, in areas with a year-round nectar or honeydew supply, honeyeaters are usually resident and switch among flowering plant species as each blooms. Where suitable food is available for only part of the year, honeyeaters show regular seasonal movements (such honeyeaters are called "blossom nomads"). For conservation, the full array of flowering plants should be provided for the resident species, and the flowering plants at each of the transient's seasonal ranges. The same can be said for avian and mammalian frugivores of the Neotropics (Graham 2001) and Paleotropics (Kitamura et al. 2002), which follow seasonal shifts in fruiting of trees at fine or gross geographic scales.

How do organisms come to inhabit wildly dispersed and isolated environments? The colonization ability of nonmigratory species across oceanic environments is an extreme case of what has been called "sweepstakes" events, or chance random traverses of widely unsuitable environments to distant habitats. Another interesting

case of *sweepstakes colonization* and subsequent adaptive radiation is found in the little-studied New Zealand lizard fauna (see box 9.3).

New Zealand is an ideal natural laboratory for studying sweepstakes colonization, founder dynamics, adaptive radiation, ecology of sibling species, and reproductive isolating mechanisms among herps, birds, and bats. New Zealand split from the ancient continent of Gondwana about 80 million years ago, spawning the Tasman Sea, and today represents some of the oldest exposed rock on Earth. Much of its flora and fauna is unique and of ancient ancestry. Its colonization history includes the ancient ancestors of the short-tailed bat (*Mystacina tuberculata*), rock wren (*Xenicus gilviventris*), rifleman (*Acanthisitta chloris*), kokako (*Callaeas cinerea*), and saddleback (*Philesturnus carunculatus*), each of which perhaps derived by immigrants from Australia finding their way across the early Tasman Sea. Later, New Zealand rafted farther north, and along with climatic warming, this situation set the stage for colonization by plants and animals from warmer latitudes. For a brief time it was colonized and occupied by tropical biota, including coral reefs, but then returned to more temperate and cooler conditions and biota (Bishop 1992). Colonization of New Zealand by birds has occurred many times over its geologic history and continues today. Many founders likely fell extinct, but those that rooted produced some remarkable descendants. On New Zealand alone, the kakapo (a parrot), takahe (*Porphyrio mantelli*, a rail), kiwis (*Apteryx* spp.), and the late moas all represent the evolution, in at least four distinct families, of both gigantism and flightlessness, traits sometimes occurring in birds in isolated environments with few predators.

Even in recent times, many natural colonizers to New Zealand have been recorded (intentional introductions by humans aside), including the welcome swallow (*Hirundo neoxena*, sometimes treated as a race of the Pacific swallow, *H.*

Box 9.3   *Adaptive radiation of New Zealand lizards*

New Zealand has approximately 37 species of native lizards, 11 of which are rare, vulnerable, or recommended for Red Data Book listing (Towns 1985). All 37 species are endemic and include 16 species (among 3 genera—*Naultinus, Heteropholis,* and *Hoplodactylus*) of geckos and 21 species (among 2 genera—*Cyclodina* and *Leiolopisma*) of skinks. (Taxonomy of all species is not firm, and color variants of some species may yet prove to be new species.) This rather remarkable endemic species diversity likely derived from far fewer ancestral founders, perhaps 1 gecko and 2 skinks, although it is not known from where; geckos and skinks are cosmopolitan and occur widely throughout the Australian and South Pacific regions.

Some evidence suggests that the geckos occurred on New Zealand before it split from the ancient supercontinent Gondwana. Regardless, many of the native lizard taxa of New Zealand currently constitute *sibling species* complexes (also called *cryptic species*)—closely related taxa only recently diverged into separate species that still are nearly identical in appearance. For example, in some of the New Zealand geckos, mouth and tongue color is required for species identification (Towns 1985). These geckos display the tongue and open mouth as aggressive responses, and thereby color might serve as species identification signals to congeners and conspecifics. Many native lizard species have quite limited distributions within New Zealand, but in some parts of the country some sibling species overlap in range. Such overlap suggests that interspecific identification signals and perhaps other reproductive isolating behaviors have evolved sufficiently to permit sympatry without interbreeding and *genetic swamping* (loss of species' genetic identity). Sibling species that occur in sympatry also may use different resources and select different microenvironments, but this needs study for this particular set of lizard species.

---

tahitica) and silvereye (*Zosterops lateralis*). These are two interesting examples of different colonization routes from the same source, Australia, along the "Roaring Forties" westerly winds that have brought so many other dispersers to the New Zealand islands over the millennia. Mapping first-appearance records (from Falla et al. 1991) suggests that the welcome swallow likely colonized New Zealand perhaps three separate times over a period of less than 4 decades. Before 1958, the swallow was considered a rare vagrant, but eventually it spread throughout the two main islands at a rate averaging 50 km/y. By 1975, the welcome swallow appeared 800 km farther east on Chatham Islands (probably a "minor" sweepstakes dispersal event from one or both of the two main New Zealand islands).

In contrast, silvereyes likely first successfully colonized New Zealand, roughly in 1856. On South Island, they spread a distance of 275 km in 4 years, from mid-Canterbury to the Otago Peninsula by 1860, thus at an overland rate of 69 km/y. On North Island, they spread 640 km in 11 years for a rate of 58 km/y, slightly faster than the welcome swallow. Silvereyes are from eastern

and southeastern Australia and Tasmania, where only a portion of their populations migrate. At least several dispersal events occurred before they successfully settled; before 1856 there were scattered records of silvereyes in Otago and Southland, the earliest being in 1832 from Milford Sound in southwestern South Island.

More recently, however, massive compromises to the native New Zealand wildlife have resulted from human-caused habitat changes and introductions of exotic species, including possums, hedgehogs, rats, cats, and other species, much the same story as with Hawaii and many other islands and archipelagoes of the world. In Hawaii, it is estimated that 5 new plant and 18 new exotic arthropod species are introduced annually, many of which have become serious pests (Maxfield 1996).

Some exotic species are introduced for purposes of biological control, soil stabilization, horticulture, or pets but can wreak ecological havoc on native biota. In Hawaii, the rosy snail (*Euglandina rosea*) was introduced from Florida to control populations of another exotic, the African giant snail (*Achatina fulica*), but the rosy snail's diet also includes Hawaii's endemic tree snails. Likewise, the endangered Hawaiian coot (*Fulica alai*) may be outcompeted for food plants, and the endangered Hawaiian stilt (*Himantopus mexicanus knudseni*) for food insects, by the tilapia (*Sarotherodon mossambicus*), an African fish introduced to control growth of algae and weeds in reservoirs and irrigation ditches (Maxfield 1996). Such problems are widespread among islands of the world, making the less disturbed archipelagoes, such as the Galápagos and portions of coastal British Columbia and southeast Alaska, important for scientific study, although these locations, too, are feeling more human pressures over time. And those flightless birds of such evolutionary uniqueness have too often proved to be maladaptive to the onslaught of human pressure and attendant exotic species and often fall extinct, as with moas, or become moribund, as with takahe and kiwis.

Welcome swallows and silvereyes are only two of a relatively recent set of avian colonizers to New Zealand, mostly from Australia. What has induced this recent flurry of successful avian colonization? Perhaps several reasons contribute (F. Schmechel, pers. comm.). Understanding their dynamics sets the stage for better understanding of the dynamics of colonization among continental environments and the potential effects of introducing exotic species. First, recent human land uses in New Zealand have provided new areas of disturbed environments favored by some bird species, such as pastures and suburban areas favored by various introduced finches, and grasslands favored by self-invading spur-winged plovers (*Lobibyx novaehollandiae*) and other shorebirds.

Second, introduced (mostly mammalian) predators have decimated many of the native bird species that otherwise might have outcompeted the avian colonizers. Removal of such key competitors has been implicated in changes in

community composition of other ecosystems (Borer et al. 2005). An example from New Zealand is the pied stilt (*Himantopus leucocephalus*), a recently successful self-invader from Australia. The endemic native black stilts (*H. novaezealandiae*) of New Zealand, which likely derived from an ancient stilt colonizer (perhaps a very early successful invasion of pied stilt), can outcompete pied stilts. But black stilts are highly endangered from introduced mammalian predation and are essentially missing from much of their previous distribution in New Zealand, giving pied stilts a foothold.

Third, habitats and source populations of the invaders themselves may be increasing. In this example, the source populations of bird species that are associated with disturbed environments in eastern Australia and Tasmania may be increasing, thus sending out a greater number of dispersers and potential colonists per unit time. When these three conditions are combined with time and luck, a greater number of intermittent invaders are likely to be successful. Parallels to these three conditions can be found in continental situations to help explain the spread of starlings, egrets, finches, mynas, house mice, Polynesian rats, and many other self-introduced or artificially introduced wildlife species.

Islands often contain interesting endemic biota and are fascinating systems to study. The native biota of Hawaii, for example, consists of many endemics; nearly all insects and terrestrial mollusks have evolved on-site and are endemic, with high endemism rates also of flowering plants, birds, ferns and allies, and other species groups. Distance to species sources affects colonization but also endemism. The rate of successful colonization is inversely related to distance to species pools. But farther distance promotes genetic isolation from subsequent outbreeding or genetic swamping, so that the more isolated the environment, the greater the proportion of endemics, as found on Hawaii. In some cases it is a race between two opposing forces—on the one hand, genetic differentiation and speciation through adaptive radiation, and on the other, local extinction from inbreeding depression, genetic drift, demographic stochasticity, and environmental catastrophes such as major storms.

The mean colonization rate of species onto the Galápagos Islands is on the order of once per 3000 years, whereas Hawaii receives a colonizer once per 70,000 years (some estimates put it at 100,000 years), assuming an initial colonizer biota of about 1000 species in each archipelago (Loope 1989). So, all else being equal, the Galápagos should have 23 times the biodiversity of Hawaii. Both island systems are subtropical or tropical, and both contain tall volcanic mountains. But all things are never equal in ecological systems, and Hawaii is by far the more diverse system for several reasons. Hawaii, at 70 million years, is far older than the more youthful, 3-million-year-old Galápagos and thus has had more time to receive colonizers, develop a wider array of environments, and allow for speciation through adaptive radiation. Hawaii is also twice as large as the Galápagos and thus can support a greater variety of environments and species, and per unit-distance is a bigger target for arriving colonists. Hawaii is also more diverse in climate, whereas the Galápagos are dominated largely by arid scrub habitats and are located equatorially near cold upwelling zones. To continue the comparison, New Zealand in turn dwarfs both the Hawaiian and Galápagos systems in species richness and area, although it is intermediate in distance to species pools.

Not considering developmental history and environmental heterogeneity, one would expect a higher resident biodiversity and more frequent colonization rate in larger islands and in islands that are closer to species pools. Of the four island groups discussed in this section, New Zealand meets these criteria best, followed by the Galápa-

gos and Chathams, and then Hawaii. The ancient age of Hawaii accounts for its biodiversity; through time, it has developed a wide variety of macro- and microenvironments and a high degree of adaptive radiation of founders.

Since colonization and persistence of vegetation and wildlife are at least somewhat allied, it is instructive to know something about dispersal and colonization dynamics of plants as well. Mechanisms of the relatively rare dispersal events of plants onto islands tend to vary significantly from those of rare plants in continental settings. For example, data on the dispersal mode of plant immigrants to the Hawaiian Islands suggest that 75% are dispersed by birds, 23% by oceanic drift, and 1% by air flotation (Carlquist 1974). Plant propagules that are dispersed by birds travel by being eaten and carried internally, or by being mechanically attached, embedded in mud on feet, or embedded in feathers. In contrast, rare vascular plants occurring in a fully continental setting exhibit a fuller array of dispersal means, including gravity, wind, growth or reproduction, and travel by vertebrates (mostly birds and mammals) as the major dispersal means, with water and insects as less common means.

Sweepstakes (and more frequent) dispersal of wildlife can take place in continental as well as oceanic biomes. In some cases, management depends on such intermittent events to afford outbreeding of local populations (that is, interbreeding among subpopulations within a broader metapopulation) and to avoid genetic or demographic problems of small, isolated populations. For example, local colonies of red-cockaded woodpeckers (*Picoides borealis*) in the southeastern United States occur largely in forests of longleaf pine (*Pinus palustris*) on federal lands (mostly national forests). The woodpecker colonies and the forest habitats they use often occur as small patches scattered across landscapes that contain mostly unsuitable environments caused by human activities, such as large-scale reduction or conversion of native forests. It is assumed in some cases that, through random and intermittent dispersal events, the woodpeckers can interchange breeders and genes among some colonies and populations (Walters et al. 2002). In cases of more extreme isolation, biologists have artificially augmented local populations by translocating birds. In other examples, dispersal of organisms among isolated, montane environments may occur only infrequently; successful dispersal may be affected by stochastic events of weather and the chance location of suitable environments (Lomolino et al. 1989; Onorato et al. 2004).

The founder principle is most important for captive breeding situations and zoo populations. A thorough discussion of tracking pedigrees, ensuring correct outbreeding levels, and other related topics is beyond the scope of this chapter. The reader is directed to other, more specialized literature on this subject (e.g., Grier and Barclay 1988; Tudge 1992; Backus et al. 1995).

Colonizers also play a critical role in "rescue" of small local populations potentially imperiled by lack of genetic diversity and too few breeders. Two kinds of rescue can be defined. *Genetic rescue* refers to the supplementing of new genes into a local population that helps avoid problems of fixation of deleterious alleles and increase in genetic homozygosity caused by genetic drift, inbreeding, or the founder effect (also see discussion in chapter 3). *Demographic rescue* is the supplementing of individuals to increase the local effective population size, so that the population does not go extinct through random variations in individual breeding success or number of offspring. As a general rule of thumb, the number of immigrants needed for demographic rescue is approximately an order of magnitude greater than those needed for genetic

rescue, but this depends on generation time, lifespan and age-specific reproductive value, degree of environmental change and stress, and other factors.

Extinction Dynamics

The extinction of local populations is affected by conditions reverse to those promoting colonization. That is, extinction of populations may occur more frequently with species that are less vagile and that are capable of only shorter-distance movements; that produce fewer dispersers per breeding season or unit time; or that are unable to withstand unsuitable conditions between islands or resource patches. Additionally, extinctions may be more frequent when the island or resource patch is smaller or contains less area of suitable environmental conditions; farther from species source pools; less environmentally heterogeneous, supporting a lower variety of microhabitats; and disconnected from other resource patches by the absence of habitat corridors or habitat remnants.

Another condition contributing to extinction is time since isolation of the island or resource patch. In one study, Schoener and Schoener (1983) reported that time to extinction of colonizing anole lizards (*Anolis sagrei*) in Bahamian islands increased monotonically with island area and that rapid colonization occurred above a certain island size, but that number of initial propagules (lizard colonizers) had no effect. In general, the interplay between colonization and extinction dynamics results in specific levels of species richness and diversity in both island and continental situations.

Burkey (1995) found that extinction rates of lizards, birds, and mammals on a set of small islands were greater than on a single large island, and were higher in more fragmented systems. Burkey surmised that some degree of habitat connectivity would help reduce extinction rates in isolated habitat remnants.

One of the more insidious influences of our burgeoning global human population is that of *local extinctions* of wildlife. Much of the literature on population viability, threatened and endangered species, and extinction has focused on the problems of global extinction of species rather than on the incremental effects of local extinctions. Fortunately, this situation is being addressed in recent literature. Barbraud et al. (2003) suggested ways to estimate rates of local extinctions in colonial species.

Further Effects of Isolation

One effect of isolation is the depression of species richness in isolated areas (fig. 9.1). With small populations in insular (island or continental) situations, isolation can increase the proba-

bility of adverse genetic effects, such as fixation of deleterious alleles, increasing homozygosity, and overall decline in allelic diversity of the gene pool as caused by genetic drift. Inbreeding depression—including depressed fertility and fecundity, increased natal mortality, and decreasing age of reproductive senescence—is one manifestation of small effective population size (see chapter 3). Small colonizer populations are subject to founder effects, which set the stage for loss of genetic and phenotypic diversity with subsequent isolation from outbreeding.

At the community level, recent isolation of previously rich environments or those better connected to species source pools can result in a decline in species richness over time. This decline is termed *faunal relaxation* when wildlife species are lost from isolated environments (figs. 9.1, 9.2). Faunal relaxation has been documented for oceanic islands isolated from main-

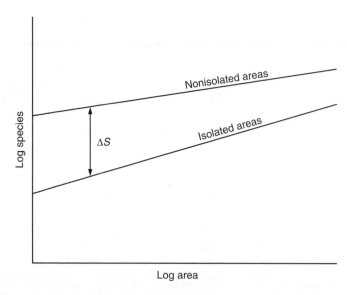

Figure 9.1.   Species–area relations of isolated and nonisolated areas. Species loss, ΔS, represents relaxation effects occurring when land bridges become submerged and isolate islands, or when continental reserves change in status from nonisolated to isolated. (From Boecklen and Simberloff 1986, 255, fig. 2; reprinted by permission of John Wiley and Sons, Inc.)

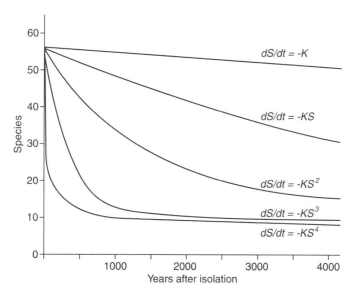

Figure 9.2. Faunal collapse of Nairobi National Park as predicted by five faunal collapse species models. S = species richness (number of species, ordinate); t = time; K = extinction coefficient. K is assumed to be taxon-specific, invariant over time, and inversely related to refuge area. The choice of the model and value of K are arbitrary because no extinctions have been observed to occur. The disparate results among models at longer simulation times strongly suggest the need for taxon-specific field studies of demography, genetics, and population dynamics of individual species. (From Boecklen and Simberloff 1986, 258, fig. 3; reprinted by permission of John Wiley and Sons, Inc.)

lands by submergence of land bridges (Newmark 1987), as well as for patches of rainforest isolated by slash-and-burn or clearcut timber harvesting in the tropics. Eventually, the fauna reaches an equilibrium species richness in which local extinction and emigration of species are equal to immigration and colonization (fig. 9.3).

A classic example of faunal relaxation is the well-studied fauna of Barro Colorado Island (BCI) in Panama, where the creation of the Panama Canal isolated a large patch of native rainforest (Karr 1990). Karr reported that local extinction of birds on BCI after isolation were of species with lower survival rates in the adjacent mainland forest; these species disappeared earlier from the island than did species with higher survival rates. The culprits contributing to extinc-

tion on BCI were species with reduced reproductive success caused by high nest predation and/or altered landscape dynamics, combined with naturally low adult survival rates. Karr did not find a correlation of extinction with population size, although such correlation occurs in native Hawaiian birds.

Isolation of reserves and parks has been a concern for some biologists, who suspect that relaxation effects are causing declines in native wildlife in protected areas. In some cases, legal and ecological boundaries of protected areas, such as national parks, do not necessarily align (Newmark 1995), although it is imperative in such analyses to determine the actual status of the environment and populations in lands adjacent to such protected areas. Newmark (1995, 1996)

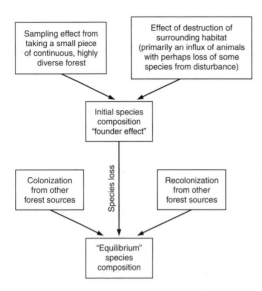

Figure 9.3.   Basic elements of the consequences of habitat isolation and subsequent faunal relaxation (species loss) to new equilibrium levels of diversity. Additional factors (not shown here) of habitat isolation affecting equilibrium levels include changes in habitat patch microclimate, habitat patch edge effects, and effects on foraging efficiency and energetics of organisms. (From Lovejoy et al. 1984, 299, fig. 1; reproduced by permission of the University of Chicago Press.)

analyzed extinction of mammal populations in national parks in western North America and Africa and concluded that extinction rates have exceeded colonization rates, and that extinction rates are greater in smaller park units. He also reported that the major factors affecting greater rates of extinction of mammals were, in decreasing order of importance, small initial population size, occurrence within the order Lagomorpha, and shorter age of maturity (which is positively related to shorter population generation time).

Related to faunal relaxation is the delayed loss of competitive, rare species in islands and environment isolates. The loss can take place over some time following isolation or alteration of

environments, so that extinction eventually "catches up" with the conditions. Thus Tilman et al. (1994) termed the phenomenon *extinction debt,* and Loehle and Li (1996) concluded that the effect is real and should be accounted for in conservation and reserve design. Even dominant and abundant species are not immune to lag delays of extinction debt. Indeed, one of the better uses of population viability analysis (chapter 10) should be to periodically evaluate the status of species thought to be common and stable; unfortunate surprises have indeed happened where abundant species have suddenly fallen locally or globally extinct. In another approach, Hanski and Ovaskainen (2002) noted that landscapes only recently exposed to habitat loss and fragmentation may have an inordinate number of rare species in the short term, which may serve as an indicator of impending extinction debt.

Over millennia, natural selection in extreme isolation can foster amazing new life forms (Latimer et al. 2005). Protection from predators and competitors can result in adaptive radiation of new species complexes in developing environments and ecological vacuums. Adaptive radiation can occur in montane and other isolated terrestrial continental environments as well as on oceanic islands. Additional effects of isolation from competitors or predators include evolution of gigantism, flightlessness, and in some cases polymorphism or, more extreme, adaptive radiation of new species complexes.

Isolation of environmental conditions in continental settings also contributes to development of *relictual faunas,* or ancient species persisting because of suitable environments. By definition, relicts persist in refugia. Relicts can be of any taxonomic group; several were discussed above that occur on New Zealand and southeastern Alaska archipelagoes. Relicts can also occur in continental settings. Welsh (1990) reported that Del Norte salamander (*Plethodon elongatus*), Olympic torrent salamander (*Rhyacotriton olympicus*), and tailed frog (*Ascaphus truei*) are paleoecological relicts that have long been associated with the ancient primeval conifer forests of the Pacific Northwest. Marcot et al. (1998) compared the extant fauna of the interior western United States to Tertiary fossil fauna and reported the persistence of relicts of 7 Tertiary genera (represented by 32 extant species) and 20 Tertiary families (represented by 55 extant genera). Unlike the relict salamanders reported by Welsh, the relict genera and families of the interior West occupy a wide range of environmental conditions, including native grasslands, shrublands, and forests.

In another example of the ecological value of refugia, Cornelius et al. (2000) found that bird species richness in relict old-forest remnants in southern Chile was positively correlated with remnant size, and that the steep regression between bird species richness and forest remnant area resembled a depauperate oceanic archipelago. There likely were little to no rescuer effects from the nearest forest source areas. The authors concluded that large forest remnants or reserves afford better protection for a wider array of bird species than do smaller forest patches.

It may be tempting to use relictual faunas as management or ecological indicators, but some caveats need to be addressed. Relicts tend to occur in odd and disjunct locations so do not necessarily represent zonal or climatic climax conditions. Insofar as relicts are holdovers from earlier environments, their distribution does not necessarily reflect the suitability of current conditions. Thus one should be wary of defining habitat and landscape requirements based on current habitat-use patterns of relicts without knowledge of the paleoecological history of the population and the site (see below). Nonetheless, relicts—including plants and animals—are often of scientific interest and deserve special conservation consideration (Millar and Libby 1991).

Are Isolation and Fragmentation Always Bad?

There are circumstances in which isolation of environments and populations is an advantage for conservation. One advantage is avoiding spread of disease (Hess 1994), parasites, and pathogens. Another advantage is the establishing of several founder populations in sites with different disturbance dynamics and thus different probabilities of success. In this case, overall persistence of the metapopulation is enhanced if there is little correlation of potentially disastrous environmental disturbances among the population centers, and if populations are large enough to avoid genetic problems or if there is occasional gene exchange (outbreeding) among populations. Yet another advantage of isolation is the maintaining of relictual faunas naturally isolated by changes in climate, vegetation, or landform.

In many cases, naturally developed island faunas are best left isolated from exotic species, especially introduced predators and competitors. Species at the edges of their range might occupy some isolated and very different environments through long-distance colonization events or because of relictual distributions in refugia. Peripheral environments may prove important for long-term dispersal of the species and for evolution of unique morphs, subspecies, new species lineages, or entire species complexes (Lesica and Allendorf 1995). Often, it is important to carefully describe the spatial and temporal scale of isolation, as well as the causes, to determine appropriate management actions to maintain the condition or to ameliorate undesirable isolation problems caused by human activities. We disagree with Bunnell et al.'s (2004) conclusion that conservation efforts for contiguously peripheral populations are "doomed to failure" and that conservation should focus instead on disjunct peripheral populations; we assert that both have conservation value and that, with proper partnerships across borders and boundaries, both are feasible.

Fragmentation of environments in landscapes is not necessarily always undesirable. At some scale, nearly all environments and species-specific habitats are fragmented; that is, non-contiguous, if only among different continents or landmasses. Environments or resources are often naturally fragmented through time, as with seasonality of fruits in tropical forests, or through space, as with the heterogeneous occurrence of obligate symbionts, such as some pollinators. In some circumstances, natural fragmentation in space and time of environments or resources can lead to the evolution of new morphs or life forms. As with isolation, it is important to determine the causes and effects of fragmentation of environments to help direct appropriate management action.

The Biogeographic Perspective of Habitat Isolates

In chapter 2 we discussed concepts of evolution and biogeography. Above, we have discussed issues of colonization, adaptive radiation, extinction dynamics, and effects of isolation on island situations. In this section we explore in greater depth the conditions and results of isolation of environments, including kinds of habitat isolates and species–area relations. We compare resource patches with oceanic islands for the purpose applying principles of island biogeography and results of environmental fragmentation and isolation to continental landscape conditions.

Classic island biogeography compares species richness and dynamics of colonization and extinction with island area and distance to source species pools (see below). In recent decades, island biogeography has been extended to describe such relations in continental settings in which environments and habitats for species be-

come fragmented and isolated. The aim is to predict the occurrence and persistence of organisms in habitat patches and protected areas under anthropogenic stress.

Of Paleohistory and Phylogeography

Before we try to explain the dynamics of isolation, it may be helpful first to describe how the situation of isolation originates and why some populations are isolated. Understanding the basis of isolation is vital for knowing what, if anything, can or should be done to ameliorate the adverse (anthropogenic) effects of isolation, including rarity and vulnerability to disturbances.

We should also be clear what we mean by "isolation," which can play out differently across spatial and temporal scales. Entire populations or metapopulations of a species can be isolated from others across vast expanses of environments hostile to successful dispersal, such as great gray owls (*Strix nebulosa*) occurring in Nearctic and Palearctic zoogeographic regions being isolated by arctic and oceanic environments across which the owls would not cross. Such broad geographic isolation is sometimes the setting for eventual character divergence and the arising of new species in allopatry (disjunct distributions).

If the span of time of such isolation has not been long, or if selective pressures and other forces causing divergence of gene pools is not severe, then the resulting species are closely related, and secondary contact may result in hybridization. Apparently, this is the case with co-occurring northern spotted owls (*Strix occidentalis caurina*) and barred owls (*Strix varia*) in western North America, where the latter species has only recently invaded. These two species may have originally evolved from a common ancestral form that was split in geographic occurrence on a broad continental scale, isolating what eventually became the spotted owl in western North America. (Interestingly, a comparable example can be found in southern South America with the similar rusty-barred owl *Strix hylophila* and rufous-legged owl *Strix rufipes,* whose ranges adjoin but do not overlap, although the evolutionary history of these species apparently has also not been studied.) In this instance, the spotted/barred owl hybrids (Hamer et al. 1994)—called "sparred owls"—may be a sign that the two parent species have not diverged sufficient reproductive or behavioral isolating mechanisms to ensure their separate coexistence in sympatry. (As well, the more aggressive barred owls often directly outcompete spotted owls for nesting sites.) This effect from barred owls has important and difficult implications for recovery of the federally threatened spotted owl (Peterson and Robins 2003; Haig et al. 2004).

If the span of time is long or if selection forces differ significantly, then such species in secondary contact may have diverged enough to retain their identities and not suffer "genetic swamping" through hybridization. It is hypothesized that such has happened in Amazonia, where multiple species of antbirds had evolved during multiple periods of environmental change but can now coexist upon secondary contact because their foraging specialization is an ancient trait (Rosenberg 1997).

Species can become regionally or locally endemic—found nowhere else—in two main ways (Gentry 1986). *Paleoendemics* are species that currently occur in refugia of suitable environments that have become isolated because of large-scale and long-term changes to environmental conditions. Some paleoendemics may be archaic taxa with primitive traits, and some may be close to natural extinction. An example is western pond turtle (*Clemmys marmorata*), which is broadly endemic to western North America; the modern species is from an ancient lineage and is now found only in subtracted, isolated aquatic environments. In fact, many herps, including ranid frogs and plethodontid salamanders, having isolated and disjunct distributions tend to be paleoendemics.

Neoendemics are species that have only recently evolved to species status, usually in disjunct or isolated environments. An example may be the complex of four (or five, depending on recent taxonomy) torrent salamanders (*Rhyocotriton* spp.) occurring within western North American mountain ranges (Good and Wake 1992).

Paleoendemics tend to have more than one disjunct population, whereas endemics confined to just one population can be either paleo- or neoendemics (Kruckeberg and Rabinwitz 1985). However, the converse cannot be asserted; that is, species with more than one disjunct population cannot be assumed to be a paleoendemic because patterns of disjunct populations can arise from a number of other factors, such as recent colonization of newly suitable environments, subtraction of ranges by human alteration of habitats, and elimination of organisms from hunting and other direct persecution.

Garza (1996) coined the term *semiendemic* to refer to species that migrate and spend part of their year elsewhere but that restrict their distribution during one part of the year to a specific geographic area. Semiendemics can include migratory species, including many nonresident tropical birds. Garza argued that, for purposes of habitat conservation, species that are confined to a relatively small geographic area for a portion of the year should still be considered as restricted to that area. A salient example of a semiendemic species is the endangered Kirtland's warbler (*Dendroica kirtlandii*), which migrates to the Bahama archipelago in the Caribbean during the nonbreeding season but concentrates in jack pine forests of the Great Lakes states for breeding. Actually, because this species concentrates geographically on both breeding and wintering grounds, it should be considered a semiendemic in both locations and both parts of the year.

Other kinds of isolation can occur at narrower scales of geographic extent, such as from effects of habitat fragmentation. We discuss implications of this scale next.

Kinds and Effects of Habitat Isolates

Habitat isolates can take many forms (see box 9.4). Few studies have empirically demonstrated specific demographic effects of resource patch isolation, or the improvement of habitat connectivity, on dynamics of small populations. In one study of Bachman's sparrow (*Aimophila aestivalis*) populations in landscapes with linear patches of pine woodlands in South Carolina, Dunning et al. (1995) found that isolated woodland patches were less likely to be colonized than were nonisolated patches. They also reported that woodland corridor configurations aided

successful colonization of newly created woodland patches. They concluded that occupancy of woodland habitats at a regional scale could be aided by design of habitat corridors, particularly for species that do not disperse well through inhospitable environments. We further discuss habitat corridors below.

In southwestern Oregon, Mills (1995) found that populations of California red-backed voles (*Clethrionomys californicus*) were virtually confined to isolated patches of forest remnants, seldom using the intervening clearcuts. Also, he found that trapping rates of voles were six times greater in remnant interiors than on the edge, likely corresponding to a greater density of the vole's preferred food, hypogeous (underground fruiting) sporocarps of mycorrhizal fungi. Thus, in this landscape, forest habitat isolates offered the only refuge for this species, which also

---

Box 9.4. Some interesting kinds of habitat isolates

A *habitat isolate* is a set of isolated environmental conditions used by specific species. The term *isolate* may refer to oceanic islands, continental habitat islands (such as patches of montane forest), and individual patches of resources or environments occurring in a continental landscape setting. Habitat isolates may also include a variety of fascinating *azonal* conditions (that is, environmental conditions atypical to a particular region). Some interesting examples are

- *polinias,* open-water areas used by bearded seals (*Erignathus barbatus*) and walruses (*Odobenus rosmarus*) of the high Arctic
- *'gator holes,* water-holding depressions created by American alligators (*Alligator mississippiensis*) and used by a variety of wildlife in the Everglades and Big Cypress Swamp in southern Florida (Bondavalli and Ulanowicz 1999)
- *kipukas,* remnant native forests persisting on mounds in a landscape of relatively recent lava flows (Tisdale et al. 1965), such as on shield volcanoes of Hawaii above about 1000 m elevation, in which persist Hawaiian honeycreepers and other native birds such as the 'oma'o, or Hawaiian thrush (*Myadestes obscurus*)
- *tapuis,* ancient colossal erosional-remnant limestone pillars rising from Venezuelan rainforests, upon which have evolved unique plants, insects, and amphibians
- *blue holes* (Caribbean) or *cenotes* (Central America), steep, water-filled sinkholes occurring in karst landscapes, used for watering by many wildlife species and for nesting by some, including the turquoise-browed motmot (*Eumomota superciliosa*) in Yucatán, Mexico
- *nunataks,* tundra islands surrounded by coalescent glaciers that support relict plants derived from interglacial periods, such as in southeast Alaska (Heusser 1955);
- *kopjes,* naturally occurring insular rock outcrops, such as used by unique bird guilds, particularly rare frugivores and nectivores, in Serengeti National Park in Tanzania (Trager and Mistry 2003)
- *dambos,* permanently water-logged drainages that support many locally endemic plants (ferns, herbs, and shrubs) and animals in Nyika National Park in Malawi

---

showed evidence of being a forest-interior specialist. The long-term viability of such isolated vole colonies, however, is unknown.

Island faunas can be particularly vulnerable to isolation effects. Overall, many of the 40 or more bird species native to Hawaii, including honeycreepers and other species groups known only from the fossil record, became extinct after Polynesians originally settled the islands. In addition, since later European settlement of the islands 2 centuries ago, another 22 native bird species or subspecies of all taxonomic groups have become extinct. Causes of extinction have been attributed to hunting by humans; introduction of disease, rats, cats, mongooses, grazing livestock, exotic birds, and game mammals; felling of native forests; urbanization; and spread of exotic vegetation (Freed and Cann 1989).

Species–Area Relations

Community composition within isolated environments is dictated in part by relations of individual species to habitat area. Species–area relations have formed one of the hearts of classic island biogeography, and a great deal has been published on this topic in many geographic areas and among many taxonomic groups.

At its simplest, the number of species, *S*, is affected by resource patch area, *A*, according to $S = CA^Z$, where C = a scaling constant that varies by taxon and location, and z = the rate at which the number of species increases with increasing area. When plotted on a log-log relationship, where $\log S = \log C + z \log A$, the function appears as a straight line, and z becomes the slope of the line. From a number of studies of species richness on oceanic archipelagoes, z varies from approximately 0.24 for breeding land and freshwater birds in the West Indies and land vertebrates in islands of Lake Michigan, to 0.49 for breeding land and freshwater birds in islands of the Gulf of Guinea (MacArthur and Willson

1967). In consequence of the z factor, as a general rule of thumb, twice the number of species seems to require 10 times the area (Darlington 1957; see also Harris 1984).

In the mid-1970s, researchers were using species–area relations to predict extinction of species (Gilpin and Diamond 1976) and extending island biography to continental situations. The relationship between the presence or frequency of occurrence of a species in replicate habitat patches to habitat patch size has been called an *incidence function,* which has been calculated for terrestrial continental birds (Cornelius et al 2000), seabirds (Whittam and Siegel-Causey 1981), butterflies (Hanski et al. 1996), and many other taxa.

Howe et al. (1981) reported that area of small rainforest remnants in New South Wales, Australia, was the best single predictor of bird species richness. Their species–area relationship suggested a z value of 0.50. Patch isolation, disturbance by livestock, and distance to water, however, all tended to reduce the number of resident species found. Lomolino (1984) reported that species–area and species-isolation relations of terrestrial mammals on 19 archipelagoes were consistent with basic predictions from the equilibrium theory of island biogeography (where species richness can be predicted, given island or habitat isolate area, distance from species colonization pool, and rates of immigration, emigration, and local extinction). Also, Lomolino noted that the species–area relationship weakened (the z value declined) with more isolated islands and with less vagile species.

Many cautions have been leveled at interpreting species–area relations and incidence functions of individual species. Perhaps the foremost caution is that of *sampling effect,* or the chance inclusion of a given species in a large area simply due to more area "sampled." A parsimonious approach to interpreting species–area and incidence functions would demand first assuming no causal relation or underlying mechanism, even if correlations are statistically significant. However, much anecdotal and some experimental evidence suggests strong ecological causes of such functions, although in some cases they are poorly understood.

Another caution is to remember that most studies provide only a snapshot of systems in transition. Species richness and occurrence of individual species are often in flux, particularly in continental habitat isolates. The snapshot does not reveal the *trends* of richness levels or occupancy rates, which may be influenced by seasonal changes in habitat use, or longer-term dynamics such as faunal relaxation (loss of species) or tension (the increase in richness in newly suitable environments).

Environmental isolates in continental landscapes differ from oceanic islands in several functions: (1) Movements of organisms among habitat patches within landscapes may be more complex than among islands. This complexity is because intervening environments in landscapes may provide varying degrees of required resources and thus may serve as stepping-stone or sink areas. (2) Colonization and extinction processes may be more complex in landscapes than on islands. (3) Smaller environmental patches in landscapes may cease to function as habitat for a species if patch size and area of resources are small in relation to key life-history requirements. Also, as we have seen, edge and boundary effects can render small patches unusable for some interior-dwelling species, and the context of the patch in the landscape matrix can greatly affect its use. (4) Species-specific dispersal habitat in landscapes may be suboptimal but usable, whereas in the situation of true islands, ocean stretches seldom provide useful dispersal habitat for terrestrial organisms. All four of these conditions can contribute to greater species richness and less pronounced faunal relaxation effects in environmental patches occurring in a continental setting than on oceanic islands having the same area.

Further, in the short term, species may still persist in an area despite loss of habitat quality or fragmentation of habitats, because some species, particularly birds, have high site fidelity and return to sites altered beyond suitability. However, Hoover (2003) demonstrated that prothonotary warblers (*Protonotaria citrea*) in Illinois changed their site fidelity when they experienced low reproductive rates. Also, local redistribution of animals—a functional response—may result in abnormally and temporarily high densities in remaining suitable environments. Nonreproductive "floater" individuals can also continue to colonize remaining suitable environments after a disturbance event and thereby keep population densities abnormally high until the lag catches up and the population crashes. For these reasons, one should take great care when interpreting data on population density in disturbed and isolated habitats unless one knows something about the organisms' site fidelity and the population's structure and dynamics.

Conservation of Wildlife in Heterogeneous Environments

In this section we discuss the conservation of wildlife and habitats in mixed environments, including the value of remnant habitat patches and habitat corridors, and planning for habitat heterogeneity at the landscape scale.

The Conservation Value of Remnant Patches of Natural Environments

Our discussion of potential adverse effects of isolation and fragmentation of environments should not be taken to mean that remnant patches of natural environments have little or no conservation value. Quite the contrary, remnant patches may be all that is left in a landscape, watershed, or geographic region from which to rebuild a more natural biota or ecosystem (Pardini et al. 2005). Thus remnants can have disproportionately high conservation value *per unit area* than do large natural areas, depending on the landscape context and management needs.

Often, remnant old-growth forests may be the final local bastion for species closely associated with such environments, including species of fungi, lichens, bryophytes, vascular plants, and invertebrates, as well as vertebrates. Protection of small, isolated old-forest remnants may be worthwhile. In some situations, such conservation measures may be highly efficient in that they would entail only a small land area but would have great benefits. Small old-growth forest remnants would not support all life history requirements of mid- and large-sized species, such as some mustelids and other wide-ranging carnivores. However, these forest remnants can nonetheless provide valuable species source pools for propagules and inocula of many plants and small animals vital to the function of native ecosystems and soil productivity (Amaranthus et al. 1994). And forest fragments can provide at least some resources used by native vertebrates (Howe et al. 1981) and can provide vital ecosystem services, such as pollination of crop plants by bees (Ricketts 2004).

In some countries, small remnants of native forests are maintained for their value as environmental benchmarks, for scenic interest, or as sites of rich floral or faunal diversity. In India, old forest remnants in landscapes of managed teak (*Tectona grandis*) plantations are termed "preservation plots" and serve, in part, as restoration benchmarks. In Asia, Africa, and elsewhere, remnant forest patches are preserved as "sacred groves" for their spiritual values (Decher 1997; Tiwari et al. 1998; Bhagwat et al. 2005). Sacred groves often have great ecological value as well, and harbor many plants and animals natu-

rally rare or made rare from surrounding anthropogenic changes to their habitats. In many cases, remnants of natural environments can serve as connectors between larger patches, as we discuss next.

Remnant patches of native environments offer vital service to conserving sources of associated plants and animals and providing stepping-stone connectivity of a habitat for a species throughout a landscape. Beyond this, native remnants can serve a purely utilitarian and anthropocentric need, such as by providing sources of valuable plants, pharmaceuticals, and foods (Schelhas 1995). Remnants can also provide valuable learning experiences for restoration management. We encourage any efforts to consider the conservation of remnant native environments in the full context of these and other values.

Utility of Habitat Corridors and Related Connectors

Use of habitat corridors for linking populations of a species was discussed in chapter 3. A number of summary papers on habitat corridors are available (Haddad et al. 2003).

Corridors can include linear strips of environments that link larger habitat blocks for particular species. One type of a linear habitat corridor is a riparian forest buffer. Riparian forest buffers have long been part of many silvicultural management prescriptions for providing movement corridors for wildlife species in managed forest landscapes (e.g., Ruel et al. 2001), and they have been proposed as a means for maintaining regional biodiversity (Hagar 1999; Vesely and McComb 2002), although such roles have been little studied.

In boreal balsam fir (*Abies balsamea*) stands in Québec Province, Canada, Darveau et al. (1995) reported that bird densities increased 30% to 70% in riparian forest strips the year following timber harvest, and decreased thereafter to approximately pretreatment levels, representing an initial enticement of disturbance-tolerant birds and a subsequent faunal relaxation. Forest-dwelling bird species were less abundant than ubiquitous species. Four forest birds—golden-crowned kinglet (*Regulus satrapa*), Swainson's thrush (*Catharus ustulatus*), blackpoll warbler (*Dendroica striata*), and black-throated green warbler (*D. virens*)—were virtually absent in 20-m-wide strips but present in 60-m-wide strips. This incidence function suggests an effective riparian forest buffer size for forest bird conservation, but data are needed on longer-term demographic response.

Some biologists have posed that riparian buffer strips can serve to attract predators, perhaps at undesirable density and diversity. Vander Haegen and DeGraaf (1996) found higher predation rates on both ground and shrub bird nests in riparian buffer strips created by commercial clearcutting than those in intact forests. Predation rates were similar in mainstem and tributary buffer strips. The predators consisted of six mostly forest-dwelling species that used the buffers to forage and perhaps travel. The authors recommended buffer strips >150 m wide along riparian zones to reduce edge-related nest predation, especially in landscapes where buffers comprise a significant portion of the existing remnant forest. However, the effect of higher predation on bird species fitness in the clearcut-created buffers is unknown and needs empirical study. Evidence of higher predation, as with competition, does not necessarily mean that populations of target species are no longer viable.

Corridors can also include specific habitat components more or less linearly arranged across a landscape, such as perch poles in the desert. In the Mojave Desert of California, Knight and Kawashima (1993) found higher densities of common ravens (*Corvus corax*) along highway and powerline transects than in

control areas with no highways or powerlines within 3.2 km, and raven nests more abundant along powerlines. Ravens may have been attracted to highways for road-kill carrion. Red-tailed hawks (*Buteo jamaicensis*) and their nests were more abundant along powerline than along highway or control transects. The authors recommended that land managers evaluate possible effects on vertebrate populations and species interactions when assessing future linear right-of-way projects.

Another form of corridor is transmission-line cuts, which can open forest or woodland canopies; provide lush grass, forb, or shrub cover; and offer much linear edge across a landscape that serves to intersect other resource patches. Effects of transmission-line corridors on wildlife were studied by King and Byers (2002), Forrester et al. (2005), and many others. Chasko and Gates (1982) found that, in a Maryland oak–hickory forest, the corridor was dominated by mixed-habitat bird species rather than by grassland bird species. The authors defined mixed-habitat bird species as species that use two or more vegetation conditions, such as grasslands and shrubs. They also found that the few isolated shrub patches occurring in the grassy corridor provided "habitat islands" where nest density and fledging success were high. Predators apparently were not able to exploit patchily distributed shrub nests in the corridor; therefore, the authors recommended managing for increased vegetation heterogeneity within transmission-line corridors to increase nest density and success of mixed-habitat bird species.

Corridors and habitat connections have been posed for large mammals as well. Silvicultural prescriptions designed to provide deer habitat corridors have been popular for some time (e.g., Wallmo 1969). Beier's (1993) simulation study of cougar habitat merged consideration of minimum habitat area and corridor use. He concluded that habitat areas as small as 2200 sq km could support cougars with low extinction risk if demographic rescue rates along habitat corridors were on the order of one to four immigrants per decade.

Studying arboreal marsupials in southeastern Australia, Lindenmayer and Nix (1993) found that designing wildlife habitat corridors based on suitable habitat, species home range, and predictions of minimum corridor width alone was insufficient. They posed that additional design criteria should include site context and connectivity (also see Sverdrup-Thygeson and Lindenmayer 2003) and the social structure, diet, and foraging patterns of desired species. In particular, they found that the key variable affecting use of linear strips of vegetation by arboreal marsupials was the number of trees with hollows as potential nest sites.

Habitat isolates, corridors, and connectors do not, however, exist in static form. In most ecosystems, they are subject to systematic and planned changes and to stochastic disturbance events, which often render their long-term conservation problematic and a major challenge.

The Landscape as the Planning Area

Perhaps the most prudent approach to planning for wildlife habitat is to consider the landscape as a whole and then decide which elements within it best contribute to specific conservation goals. The first challenge, of course, is to articulate those goals unambiguously and precisely, so it is clear which species, communities, habitats, and vegetation and environmental conditions are of conservation value and interest, which usually include those that are scarce, isolated, disjunct, or declining.

A true landscape planning approach then entails integrating management goals across ownerships and land allocation categories. This may entail recognizing reasonable expectations for the conservation contribution by individual owners and lands. For example, private tree farms cannot be expected to maintain a full array of species, habitats, and ecological functions on their own, but they may provide some degree of young-forest environments for some dispersing species.

Lindenmayer and Franklin (2002) emphasized the value of habitat elements for older-forest species in the *matrix* of forests occurring outside reserve systems. They suggested that matrix forests can play four key roles in biodiversity conservation: (1) supporting populations of some species; (2) regulating and even affording the movement and dispersal of some organisms; (3) buffering sensitive areas and reserves; and (4) maintaining the integrity of aquatic systems. In conjunction with corridors and reserve areas, management of the forest matrix can contribute to a landscape approach of conservation.

As another example, Marcot et al. (2001) suggested a landscape approach to tropical forest conservation that would explicitly identify lands contributing to four spatial scales of a protected area network (PAN): (1) large, existing or potential national parks and other major reserves—these may include protected land allocations such as botanical waysides, wilderness areas, wildlife sanctuaries, and research natural areas; (2) corridors and lands serving to connect the first two areas; (3) small patches of key wildlife resources within lands otherwise allocated to intensive resource extraction such as timber concessions; and (4) individual components, substrates, or elements of high wildlife value within intensive resource-use areas. Their PAN elements 2, 3, and 4 might include matrix lands discussed by Lindenmayer and Franklin (2002).

The landscape planning concepts and approaches of Lindenmayer and Franklin (2002) and Marcot et al. (2001) can also be extended beyond forest environments. For example, Tscharntke et al. (2002) reported on the value of small fragments of grasslands in cropland landscapes for conserving insect communities.

Another consideration in landscape planning for wildlife habitat is to account for disturbance dynamics and how they can create, destroy, or change connectivity of desired habitat conditions. The challenge is to calculate the area needed in various cover types, given the expected frequency, intensity, location, and distribution of disturbances, be they wildfire, logging, or other events. This calculation is particularly problematic when conserving rare, isolated habitats and populations in highly disturbance-prone environments, such as conserving old-forest groves in fire-prone forest types. The approach should entail calculating (1) the area of specific habitat conditions required for meeting conservation goals; (2) the frequency and area affected by disturbance events; and (3) the post-disturbance recovery time for producing desired conditions again. Geographic information system (GIS) simulation and analysis may be useful (e.g., Herzog et al. 2001; Schelhaas et al. 2002). One important consideration for such landscape planning for habitat is to include contingency for surprise changes in environmental conditions because of unexpected disturbance events, changes in social values, and increases in economic value of undervalued resources.

Monitoring Habitats and Providing for Wildlife in Landscapes

For decades, if not centuries, managers have successfully provided some aspects of wildlife habitat at the landscape scale. These include use of salt licks and watering holes for ungulates, nest boxes for waterfowl, and many other such activities that implicitly account for landscape patterns of vegetation and habitat selection behaviors of animals. More recent, however, is the focus on ecosystem processes and multiple species, and the need to monitor environments and wildlife response at landscape scales. Monitoring landscape effects on wildlife populations can include tracking trends in populations and habitat occupancy, and determining habitat selection behaviors and numerical and functional responses of organisms.

Understanding how populations and organisms occur in resource patches and use fragmented environments, and the reasonable goals set by various landowners throughout a landscape, can help managers set realistic expectations for wildlife presence and population density. Since not all resource patches may be occupied by a species simultaneously, those patches that are small, that contain less than the full complement of resources for the species, and that are located peripherally, could be monitored among seasons and years to determine true trends in occupancy patterns and to track age and reproductive status of associated organisms. Such monitoring should occur before changing management direction that would severely reduce habitat quality, particularly for at-risk species. At the very least, preactivity monitoring sets the baseline from which changes can be gauged (see chapter 7).

In general, to help ensure viable populations, management guidelines should provide for the full range of habitat conditions and habitat quality needed to maintain a well-distributed population. Certainly, not all species are equally well distributed, nor is a given species equally abundant across its distributional range. Thus, results of monitoring and implementing management guidelines should be tempered with knowledge of the natural abundance patterns of species. Providing a full array of resources and environments to help ensure persistence of a population should entail providing seasonal and range movement habitats; habitats for migration, dispersal, and resting (including waterfowl loafing); and habitats to maintain genetic diversity by ensuring a large effective population and connected subpopulations where natural conditions provide. In addition, providing for peripheral habitat, suboptimal or secondary habitats, sinks, corridors, key links, and other environments critical to maintaining metapopulations (see chapter 3) may be in order. Such management guidelines are best developed from local empirical knowledge of the life history and ecology of the species in the field. No model or generalized set of guidelines can replace basic field zoology and autecological studies.

Monitoring would best serve its purpose when keyed to "early warning signals" of impending changes in populations. However, many long-lived vertebrate wildlife species or species with high site fidelity may not respond quickly to initial changes in environmental and habitat conditions (see also the discussion above on faunal relaxation and extinction debt). Population responses—functional or numerical—might show lag effects. By the time the population abundance has responded, it may be too late to alter the course of environmental change.

This problem of lag effect can be solved by selecting (1) species with short life-spans, high reproductive rates, high motility, and high habitat specificity, such as some invertebrates (e.g., Read and Andersen 2000); and/or (2) population parameters that respond quickly to habitat changes. Such population parameters might include changes in breeding attempts, foraging behaviors, selection of foraging substrates, juvenile sex ratios and population age structure, or proportion of nonreproductive and nonterritorial floaters in the population. Also, functional responses of populations might occur more quickly than numerical responses. Thus, foraging activities, nest-site switching, or congregations of breeding units (e.g., pairs, colonies, herds, or leks) may serve as better early warning signals than would breeding rates and overall population density. Also, some abnormal onto-

genic developmental traits might precede decreases in survival and reproduction, such as found by Lens et al. (2002) in asymmetries in tarsus length of endangered Taita thrushes (*Turdus helleri*). Ultimately, the purpose is to index and ensure long-term high fitness of individuals—that is, the reproductive vitality of offspring.

Other early warning signals might include dysfunctions in social or individual behaviors, such as increases in "divorce rates" of otherwise tightly pair-bound species such as cranes and some owls, or increases in the rate at which highly territorial organisms such as raptors abandon seemingly suitable areas. Increases in the nonbreeding segment of the population (including floaters or helpers) and declines in individual health (e.g., percentage body fat) may also signal decline in environmental suitability.

At a broader landscape scale, monitoring of various indicators of ecosystem health may help reveal early warning signs of impending loss of ecological integrity. In two examples, Whittier et al. (2002) suggested monitoring a range of physical and chemical indicators of stress to lake ecosystems in the northeastern United States, and Hausner et al. (2003) suggested monitoring specific bird guilds in northern coastal birch forests as indicators of overall land use impacts.

The trick in successful monitoring is not only to detect adverse changes in time to correct them, but to determine the cause (e.g., Cane and Tepedino 2001). Attribution of cause is deceptively difficult and is one reason why passive adaptive management (learning from management activities not designed as experiments) often fails to serve as good monitoring or learning tools. Causes can be proximate, such as woodland caribou (*Rangifer tarandus*) starving during winter because of low density of lichens fallen from tree canopies that provide food, or ultimate, such as widespread decline in air quality caused by industrial and automobile emissions that adversely affect the growth and reproduction of lichens and increase their mortality. Hansson (2002) found that geographic variations in populations of two vole species were caused by small-scale movements and interspecific interactions as well as by large-scale patterns of food availability, weather, and predator density.

Monitoring for proximate or ultimate causes—in this chapter, of habitat and population isolation and attendant conservation concerns—means beginning with an understanding of how the system works, representing that understanding in a conceptual and diagrammatic model (see chapter 10), and then crafting a statistically correct sampling design to determine trends of the parameters of interest and the factors that most likely influence such trends. Statistical considerations can be integrated into an active adaptive management approach (learning from management activities that are designed as experiments). Only by determining the relative contribution of causal factors can we determine what aspects of management guidelines to support or amend.

Literature Cited

Alverson, W. S., D. M. Waller, and S. L. Solheim. 1988. Forests too deer: Edge effects in northern Wisconsin. *Conservation Biology* 2:348–58.

Amaranthus, M., J. M. Trappe, L. Bednar, and D. Arthur. 1994. Hypogeous fungal production in mature Douglas-fir forest fragments and surrounding plantations and its relation to coarse woody debris and animal mycophagy. *Canadian Journal of Forest Research* 24 (11):2157–65.

Ambrose, J. P., and S. P. Bratton. 1990. Trends in landscape heterogeneity along the borders of Great Smoky Mountains National Park. *Conservation Biology* 4:135–43.

Andren, H. P., and P. Angelstam. 1988. Elevated predation rates as an edge effect in habitat islands: Experimental evidence. *Ecology* 69:544–47.

Backus, V. L., E. H. Bryant, C. R. Hughes, and L. M. Meffert. 1995. Effect of migration or inbreeding followed by selection on low-founder-number populations: Implications for captive breeding programs. *Conservation Biology* 9 (5):1216–24.

Barbraud, C., J. D. Nichols, J. E. Hines, and H. Hafner. 2003. Estimating rates of local extinction and colonization in colonial species and an extension to the metapopulation and community levels. *Oikos* 101 (1):113–26.

Barrett, G. W. 1981. Stress ecology: An integrative approach. In *Stress effects on natural ecosystems*, ed. G. W. Barrett and R. Rosenberg. New York: Wiley.

Beier, P. 1993. Determining minimum habitat areas and habitat corridors for cougars. *Conservation Biology* 7 (1):94–108.

Bhagwat, S. A., C. G. Kushalappa, P. H. Williams, and N. D. Brown. 2005. The role of informal protected areas in maintaining biodiversity in the Western Ghats of India. *Ecology and Society* 10 (1):8 [online]; www.ecologyandsociety.org/vol10/iss1/art8.

Bishop, J. G. 2002. Early primary succession on Mount St. Helens: Impact of insect herbivores on colonizing lupines. *Ecology* 83 (1):191–202.

Bishop, N. 1992. *Natural history of New Zealand.* Auckland, NZ: Hodder & Stoughton.

Blaustein, A. R., L. K. Belden, D. H. Olson, D. M. Green, T. L. Root, and J. M. Kiesecker. 2001. Amphibian breeding and climate change. *Conservation Biology* 15 (6):1804–9.

Boecklen, W. J., and D. Simberloff. 1986. Area-based extinction models in conservation. In *Dynamics of extinction,* ed. D. K. Elliott, 247–76. New York: Wiley.

Bondavalli, C., and R. E. Ulanowicz. 1999. Unexpected effects of predators upon their prey: The case of the American alligator. *Ecosystems* 2:49–63.

Borer, E. T., E. W. Seabloom, J. B. Shurin, K. E. Anderson, C. A. Blanchette, B. Broitman, S. D. Cooper, and B. S. Halpern. 2005. What determines the strength of a trophic cascade? *Ecology* 86 (2):528–37.

Buchan, L. A. J., and D. K. Padilla. 1999. Estimating the probability of long-distance overland dispersal of invading aquatic species. *Ecological Applications* 9 (1):254–65.

Bunnell, F. L., R. W. Campbell, and K. A. Squires. 2004. Conservation priorities for peripheral species: The example of British Columbia. *Canadian Journal of Forest Research* 34:2240–47.

Burkey, T. V. 1995. Extinction rates in archipelagoes: Implications for populations in fragmented habitats. *Conservation Biology* 9 (3):527–41.

Cane, J. H., and V. J. Tepedino. 2001. Causes and extent of declines among native North American invertebrate pollinators: Detection, evidence, and consequences. *Conservation Ecology* 5 (1):1 [online]; www.consecol.org/vol5/iss1/art1.

Carlquist, S. 1974. *Island biology.* New York: Columbia Univ. Press.

Chasko, G. G., and J. E. Gates. 1982. Avian habitat suitability along a transmission-line corridor in an oak–hickory forest region. *Wildlife Monographs* 82:1–41.

Chen, J., J. F. Franklin, and T. A. Spies. 1992. Vegetation responses to edge environments in old-growth Douglas-fir forests. *Ecological Applications* 2 (4):387–96.

Chen, J., J. F. Franklin, and T. A. Spies. 1993. Contrasting microclimates among clearcut, edge, and interior of old-growth Douglas-fir forest. *Agricultural and Forest Meteorology* 63:219–37.

Chen, J., J. F. Franklin, and T. A. Spies. 1995. Growing-season microclimatic gradients from clearcut edges into old-growth Douglas-fir forests. *Ecological Applications* 5 (1):74–86.

Chen, J., S. C. Saunders, T. R. Crow, R. J. Naiman, K. D. Brosofska, G. D. Mroz, B. L. Brookshire, and J. F. Franklin. 1999. Microclimate in forest ecosystem and landscape ecology. *BioScience* 49 (4):288–97.

Coker, D. R., and D. E. Capen. 1995. Landscape-level habitat use by brown-headed cowbirds in Vermont. *Journal of Wildlife Management* 59 (4):631–37.

Cole, B. J. 1981. Colonizing abilities, island size, and the number of species on archipelagoes. *American Naturalist* 117:629–38.

Cornelius, C., H. Cofre, and P. A. Marquet. 2000. Effects of habitat fragmentation on bird species in a relict temperate forest in semiarid Chile. *Conservation Biology* 14 (2):534–43.

Darlington, P. J. 1957. *Zoogeography: The geographic distribution of animals.* New York: Wiley.

Darveau, M., P. Beauchesne, L. Belanger, J. Huot, and P. Larue. 1995. Riparian forest strips as habitat for breeding birds in boreal forest. *Journal of Wildlife Management* 59 (1):67–78.

Decher, J. 1997. Conservation, small mammals, and the future of sacred groves in West Africa. *Biodiversity and Conservation* 6:1007–26.

Diamond, J. M., J. Terborgh, R. F. Whitcomb, J. F. Lynch, P. A. Opler, C. S. Robbins, D. S. Simberloff, and L. G. Abele. 1976. Island biogeography and conservation: Strategy and limitations. *Science* 193:1027–32.

Dignan, P., and L. Bren. 2003. Modelling light penetration edge effects for stream buffer design in mountain ash forest in southeastern Australia. *Forest Ecology and Management* 179 (1–3):95–106.

Dunning, J. B., Jr., R. Borgella Jr., K. Clements, and G. K. Meffe. 1995. Patch isolation, corridor effects, and colonization by a resident sparrow in a managed pine woodland. *Conservation Biology* 9 (3):542–50.

Edminster, F. C. 1939. Hedge plantings for erosion control and wildlife management. *Transactions of the North American Wildlife and Natural Resources Conference* 4:534–41.

Ellner, S. P., and G. Fussmann. 2003. Effects of successional dynamics on metapopulation persistence. *Ecology* 84 (4):882–89.

Falla, R. A., R. B. Sibson, and E. G. Turbott. 1991. *Collins guide to the birds of New Zealand and outlying islands.* Auckland, NZ: HarperCollins.

Fastie, C. L. 1995. Causes and ecosystem consequences of multiple pathways of primary succession at Glacier Bay, Alaska. *Ecology* 76 (6):1899–1916.

Ffolliott, P. F., R. E. Thill, W. P. Clary, and F. R. Larsen. 1977. Animal use of ponderosa pine forest openings. *Journal of Wildlife Management* 41:782–84.

Fitzsimmons, M. 2003. Effects of deforestation and reforestation on landscape spatial structure in boreal Saskatchewan, Canada. *Forest Ecology and Management* 174 (1–3):577–92.

Forrester, J. A., D. J. Leopold, and S. D. Hafner. 2005. Maintaining critical habitat in a heavily managed landscape: Effects of power line corridor management on Karner blue butterfly (*Lycaeides melissa samuelis*) habitat. *Restoration Ecology* 13 (3):488–98.

Franklin, J. F., and R. T. T. Forman. 1987. Creating landscape patterns by forest cutting: Ecological consequences and principles. *Landscape Ecology* 1:5–18.

Fraver, S. 1994. Vegetation responses along edge-to-interior gradients in the mixed hardwood forests of the Roanoke River Basin, North Carolina. *Conservation Biology* 8 (3):822–32.

Freed, L. A., and R. L. Cann. 1989. Integrated conservation strategy for Hawaiian forest birds. *BioScience* 39 (7):475–79.

Freeman, J. 1986. The parks as genetic islands. *National Parks* 60:12–17.

Freemark, K. E., and H. G. Merriam. 1986. Importance of area and habitat heterogeneity to bird assemblages in temperate forest fragments. *Biological Conservation* 36:115–41.

Garza, H. G. 1996. The conservation importance of semiendemic species. *Conservation Biology* 10 (2):674–75.

Gentry, A. H. 1986. Endemism in tropical versus temperate plant communities. In *Conservation biology: The science of scarcity and diversity,* ed. M. E. Soulé, 153–81. Sunderland, MA: Sinauer Associates.

Gilpin, M. E., and J. M. Diamond. 1976. Calculation of immigration and extinction curves from the species-area-distance relation. *Proceedings of the National Academy of Science* 73 (11):4130–34.

Goggins, G. C. 1987. Protecting the wildlife resources of national parks from external threats. *Land and Water Law Review* 22:1–27.

Good, D. A., and D. B. Wake. 1992. Geographic variation and speciation in the torrent salamanders of the genus *Rhyacotriton* (Caudata: Rhyacotritonidae). *University of California Publications in Zoology* 126:1–91.

Graham, C. H. 2001. Factors influencing movement patterns of keel-billed toucans in a fragmented tropical landscape in southern Mexico. *Conservation Biology* 15 (6):1789–98.

Gray, A. N., T. A. Spies, and M. J. Easter. 2002. Microclimatic and soil moisture responses to gap formation in coastal Douglas-fir forests. *Canadian Journal of Forest Research* 32:332–43.

Grier, J. W., and J. H. Barclay. 1988. Dynamics of founder populations established by reintroduction. In *Peregrine falcon populations,* ed. T. J. Cade, J. H. Enderson, C. G. Thelander, and C. M. White, 698–700. Boise, Idaho: The Peregrine Fund.

Guthery, F. S., and R. L. Bingham. 1992. On Leopold's principle of edge. *Wildlife Society Bulletin* 20:340–44.

Haddad, N. M., D. R. Bowne, A. Cunningham, B. J. Danielson, D. J. Levey, S. Sargent, and T. Spira. 2003. Corridor use by diverse taxa. *Ecology* 84 (3):609–15.

Hagar, J. C. 1999. Influence of riparian buffer width on bird assemblages in western Oregon. *Journal of Wildlife Management* 63 (2):484–96.

Hahn, D. C., and J. S. Hatfield. 1995. Parasitism at the landscape scale: Cowbirds prefer forests. *Conservation Biology* 9 (6):1415–24.

Haig, S. M., T. D. Mullins, E. D. Forsman, P. W. Trail, and L. Wennerberg. 2004. Genetic identification of spotted owls, barred owls, and their hybrids: Legal implications of hybrid identity. *Conservation Biology* 18 (5):1347–57.

Hamer, T. E., E. D. Forsman, A. D. Fuchs, and M. L. Walters. 1994. Hybridization between barred and spotted owls. *Auk* 111:487–92.

Hannah, L., G. Midgley, G. Hughes, and B. Bomhard. 2005. The view from the Cape: Extinction risk, protected areas, and climate change. *BioScience* 55 (3):231–42.

Hanski, I., A. Moilanen, T. Pakkala, and M. Kuussaari. 1996. The quantitative incidence function model and persistence of an endangered butterfly metapopulation. *Conservation Biology* 10 (2):578–90.

Hanski, I., and O. Ovaskainen. 2002. Extinction debt at extinction threshold. *Conservation Biology* 16 (3):666–73.

Hansson, L. 2002. Dynamics and trophic interactions of small rodents: Landscape or regional effects on spatial variation? *Oecologia* 130 (2):259–66.

Harris, L. D. 1984. *The fragmented forest.* Chicago: Univ. of Chicago Press.

Hausner, V. H., N. G. Yoccoz, and R. A. Ims. 2003. Selecting indicator traits for monitoring land use impacts: Birds in northern coastal birch forests. *Ecological Applications* 13 (4):999–1012.

Herzog, F., A. Lausch, E. Muller, H. Thulke, U. Steinhardt, and S. Lehmann. 2001. Landscape metrics for assessment of landscape destruction and rehabilitation. *Environmental Management* 27:91–107.

Heusser, C. J. 1955. Pollen profiles from Prince William Sound and southeastern Kenai Peninsula, Alaska. *Ecology* 36 (2):185–202.

Hess, G. R. 1994. Conservation corridors and contagious disease: A cautionary note. *Conservation Biology* 8 (1):256–62.

Hester, A. J., and R. J. Hobbs. 1992. Influence of fire and soil nutrients on native and non-native annuals at remnant vegetation edges in the western Australian wheatbelt. *Journal of Vegetation Science* 3 (1):101–8.

Holland, M. M., P. G. Risser, and R. J. Naiman, eds. 1991. *Ecotones: The role of landscape boundaries in the management and restoration of changing environments.* New York: Chapman & Hall.

Hoover, J. P. 2003. Decision rules for site fidelity in a migratory bird, the prothonotary warbler. *Ecology* 84 (2):416–30.

Howe, R. W., T. D. Howe, and H. A. Ford. 1981. Bird

distributions on small rainforest remnants in New South Wales. *Australian Wildlife Research* 8:637–51.

Hylander, K., B. G. Jonsson, and C. Nilsson. 2002. Evaluating buffer strips along boreal streams using bryophytes as indicators. *Ecological Applications* 12 (3):797–806.

Jaenike, J. 2002. Time-delayed effects of climate variation on host–parasite dynamics. *Ecology* 83 (4): 917–24.

Johnston, V. R. 1947. Breeding birds of the forest edge in Illinois. *Condor* 49:45–53.

Jones, G. A., K. E. Sieving, and S. K. Jacobson. 2005. Avian diversity and functional insectivory on north-central Florida farmlands. *Conservation Biology* 19 (4):1234–45.

Jonzen, N., A. Hedenstrom, C. Hjort, A. Lindstrom, P. Lundberg, and A. Andersson. 2002. Climate patterns and the stochastic dynamics of migratory birds. *Oikos* 97 (3):329–36.

Jules, E. S., and B. J. Rathcke. 1999. Mechanisms of reduced Trillium recruitment along edges of old-growth forest fragments. *Conservation Biology* 13 (4):784–93.

Karr, J. R. 1990. Avian survival rates and the extinction process on Barro Colorado Island, Panama. *Conservation Biology* 4:391–97.

Kelker, G. H. 1964. Appraisal of ideas advanced by Aldo Leopold thirty years ago. *Journal of Wildlife Management* 28:180–85.

King, D. I., and B. E. Byers. 2002. An evaluation of powerline rights-of-way as habitat for early-successional shrubland birds. *Wildlife Society Bulletin* 30 (3):868–74.

Kingdon, J. 1989. *Island Africa: The evolution of Africa's rare animals and plants*. Princeton, N.J.: Princeton Univ. Press.

Kingdon, J. 1997. *The Kingdon field guide to African mammals*. San Diego: Academic Press.

Kitamura, S., T. Yumoto, P. Poonswad, P. Chuailua, K. Plongmai, T. Maruhashi, and N. Noma. 2002. Interactions between fleshy fruits and frugivores in a tropical seasonal forest in Thailand. *Oecologia* 133 (4):559–72.

Knight, R. L. 1996. Aldo Leopold, the land ethic, and ecosystem management. *Journal of Wildlife Management* 60 (3):471–74.

Knight, R. L., and J. Y. Kawashima. 1993. Responses of raven and red-tailed hawk populations to linear right-of-ways. *Journal of Wildlife Management* 57 (2):266–71.

Kolbe, J. J., and F. J. Janzen. 2002. Spatial and temporal dynamics of turtle nest predation: Edge effects. *Oikos* 99 (3):538–44.

Krebs, J. R., J. C. Ryan, and E. L. Charnov. 1974. Hunting by expectation or optimal foraging? A study of patch use by chickadees. *Animal Behavior* 22:953–64.

Kruckeberg, A. R., and D. Rabinwitz. 1985. Biological aspects of endemism in higher plants. *Annual Review of Ecology and Systematics* 16:447–79.

Latimer, A. M., J. A. Silander, and R. M. Cowling. 2005. Neutral ecological theory reveals isolation and rapid speciation in a biodiversity hot spot. *Science* 5741:1722–24.

Laurance, W. F., T. E. Lovejoy, H. L. Vasconcelos, E. M. Bruna, R. K. Didham, P. C. Stouffer, C. Gascon, R. O. Bierregaard, S. G. Laurance, and E. Sampaio. 2002. Ecosystem decay of Amazonian forest fragments: A 22-year investigation. *Conservation Biology* 16 (3):605–18.

Lay, D. W. 1938. How valuable are woodland clearings to wildlife? *Wilson Bulletin* 50:254–56.

Lehmkuhl, J. F. 1990. The effects of forest fragmentation on vertebrate communities in western Oregon and Washington. *Northwest Environmental Journal* 6:433–34.

Lemoine, N., and K. Bohning-Gaese. 2003. Potential impact of global climate change on species richness of long-distance migrants. *Conservation Biology* 17 (2):577–86.

Lens, L., S. Van Dongen, and E. Matthysen. 2002. Fluctuating asymmetry as an early warning system in the critically endangered Taita thrush. *Conservation Biology* 16 (2):479–87.

Leopold, A. 1933. *Game management*. New York: Scribner's.

Lesica, P., and F. W. Allendorf. 1995. When are peripheral populations valuable for conservation? *Conservation Biology* 9 (4):753–60.

Lindenmayer, D. B., and J. F. Franklin. 2002. *Conserving forest biodiversity: A comprehensive multiscaled approach*. Washington, DC: Island Press.

Lindenmayer, D. B., and H. A. Nix. 1993. Ecological principles for the design of wildlife corridors. *Conservation Biology* 7 (3):627–30.

Lloyd, P., T. E. Martin, R. L. Redmond, U. Langner, and M. M. Hart. 2005. Linking demographic effects of habitat fragmentation across landscapes to continental source–sink dynamics. *Ecological Applications* 15 (5):1504–14.

Loehle, C., and B. Li. 1996. Habitat destruction and

the extinction debt revisited. *Ecological Applications* 6 (3):784–89.

Logan, J. A., J. Régni?re, and J. A. Powell. 2003. Assessing the impacts of global warming on forest pest dynamics. *Frontiers in Ecology and the Environment* 1 (3):130–37.

Lomolino, M. V. 1982. Species–area and species–distance relationships of terrestrial mammals in the Thousand Island region. *Oecologia* 54:72–75.

Lomolino, M. V. 1984. Mammalian island biogeography: Effects of area, isolation and vagility. *Oecologia* 61:376–82.

Lomolino, M. V., J. H. Brown, and R. Davis. 1989. Island biogeography of montane forest mammals in the American Southwest. *Ecology* 70:180–94.

Loope, L. L. 1989. Island ecosystems. In *Conservation biology in Hawai'i,* ed. C. P. Stone and D. B. Stone, 3–6. Honolulu: University of Hawaii.

Lovejoy, T. E., J. M. Rankin, R. O. Bierregaard Jr., K. S. Brown Jr., L. H. Emmons, and M. E. Van der Voort. 1984. Ecosystem decay of Amazon forest remnants. In *Extinctions,* ed. M. H. Nitecki, 295–325. Chicago: Univ. of Chicago Press.

MacArthur, R. H., and E. O. Wilson. 1967. *The theory of island biogeography.* Princeton, NJ: Princeton Univ. Press.

Marcot, B. G., L. K. Croft, J. F. Lehmkuhl, R. H. Naney, C. G. Niwa, W. R. Owen, and R. E. Sandquist. 1998. *Macroecology, paleoecology, and ecological integrity of terrestrial species and communities of the interior Columbia River Basin and portions of the Klamath and Great Basins.* General Technical Report PNW-GTR-410. Portland, OR: USDA Forest Service, Pacific Northwest Research Station.

Marcot, B. G., R. E. Gullison, and J. R. Barborak. 2001. Protecting habitat elements and natural areas in the managed forest matrix. In *The cutting edge: Conserving wildlife in logged tropical forests,* ed. R. A. Fimbel, A. Grajal, and J. G. Robinson, 523–58. New York: Columbia Univ. Press.

Martino, D. 2001. Buffer zones around protected areas: A brief literature review. *Electronic Green Journal* 15 [online]; http://egj.lib.uidaho.edu/egj15/martino1.html.

Matlack, G. R. 1993. Microenvironment variation within and among forest edge sites in the eastern United States. *Biological Conservation* 66:185–94.

Matthysen, E. 2002. Boundary effects on dispersal between habitat patches by forest birds (*Parus major, P. caeruleus*). *Landscape Ecology* 17 (6):509–15.

Maxfield, B. A. 1996. The Hawaiian Islands, 20 years later. *Endangered Species* Bulletin 21 (4):18–21.

Millar, C. I., and W. J. Libby. 1991. Strategies for conserving clinal, ecotypic, and disjunct population diversity in widespread species. In *Genetics and conservation of rare plants,* ed. D. A. Falk and K. E. Holsinger, 149–70. New York: Oxford Univ. Press.

Mills, L. S. 1995. Edge effects and isolation: Red-backed voles on forest remnants. *Conservation Biology* 9 (2):395–403.

Morris, D. W., and D. L. Davidson. 2000. Optimally foraging mice match patch use with habitat differences in fitness. *Ecology* 81 (8):2061–66.

Naughton-Treves, L., J. L. Mena, A. Treves, N. Alvarez, and V. C. Radeloff. 2003. Wildlife survival beyond park boundaries: The impact of slash-and-burn agriculture and hunting on mammals in Tambopata, Peru. *Conservation Biology* 17 (4):1106–17.

Newmark, W. D. 1987. A land-bridge island perspective on mammalian extinctions in western North American parks. *Nature* 325:430–32.

Newmark, W. D. 1995. Extinction of mammal populations in western North American national parks. *Conservation Biology* 9 (3):512–26.

Newmark, W. D. 1996. Insularization of Tanzanian parks and the local extinction of large mammals. *Conservation Biology* 10 (6):1549–56.

Onorato, D. P., E. C. Hellgren, R. A. Van Den Bussche, and J. R. Skiles. 2004. Paternity and relatedness of American black bears recolonizing a desert montane island. *Canadian Journal of Zoology* 82 (8):1201–10.

Oom, S. P., J. A. Beecham, C. J. Legg, and A. J. Hester. 2004. Foraging in a complex environment: From foraging strategies to emergent spatial properties. *Ecological Complexity* 1 (2):299–327.

Pardini, R., S. M. de Souza, R. Braga-Neto, and J. P. Metzger. 2005. The role of forest structure, fragment size and corridors in maintaining small mammal abundance and diversity in an Atlantic forest landscape. *Biological Conservation* 124 (2):253–66.

Parker, T. H., B. M. Stansberry, C. D. Becker, and P. S. Gipson. 2005. Edge and area effects on the occurrence of migrant forest songbirds. *Conservation Biology* 19 (4):1157–67.

Pataki, D. E., D. S. Ellsworth, R. D. Evans, M. Gonzalez-Meler, J. King, S. W. Leavitt, G. Lin, R. Matamala, E. Pendall, and R. Siegwolf. 2003. Tracing changes in ecosystem function under elevated carbon dioxide conditions. *BioScience* 53 (9):805–18.

Pearson, D. L. 1975. The relation of foliage complexity to ecological diversity of three Amazonian bird communities. *Condor* 77:453–66.

Peterson, A. T., and C. R. Robins. 2003. Using ecological-niche modeling to predict barred owl invasions with implications for spotted owl conservation. *Conservation Biology* 17 (4):1161–65.

Petrides, G. A. 1942. Relation of hedgerows in winter to wildlife in central New York. *Journal of Wildlife Management* 6 (4):261–80.

Read, J. L., and A. N. Andersen. 2000. The value of ants as early warning bioindicators: Responses to pulsed cattle grazing at an Australian arid zone locality. *Journal of Arid Environments* 45 (3):231–51.

Recher, H. F. 1969. Bird species diversity and habitat diversity in Australia and North America. *American Naturalist* 103:75–80.

Reese, K. P., and J. T. Ratti. 1988. Edge effect: a concept under scrutiny. *Transactions of the North American Wildlife and Natural Resources Conference* 53:127–36.

Ricketts, T. H. 2004. Tropical forest fragments enhance pollinator activity in nearby coffee crops. *Conservation Biology* 18 (5):1262–71.

Robbins, C. S. 1978. Determining habitat requirements of nongame species. *Transactions of the North American Wildlife and Natural Resources Conference* 43:57–68.

Robbins, C. S. 1979. Effect of forest fragmentation on bird populations In *Management of north central and northeastern forests for nongame birds,* ed. R. M. DeGraaf and K. E. Evans, 198–212. General Technical Report NC-51. St. Paul, MN: USDA Forest Service, North Central Forest Experiment Station.

Robinson, S. K. 1988. Reappraisal of the costs and benefits of habitat heterogeneity for nongame wildlife. *Transactions of the North American Wildlife and Natural Resources Conference* 53:145–55.

Robinson, S. K., and W. D. Robinson. 2001. Avian nesting success in a selectively harvested north temperate deciduous forest. *Conservation Biology* 15 (6):1763–71.

Rodewald, A. D. 2002. Nest predation in forested regions: Landscape and edge effects. *Journal of Wildlife Management* 66 (3):634–40.

Rosenberg, K. V. 1997. Ecology of dead-leaf foraging specialists and their contribution to Amazonian bird diversity. *Ornithological Monographs* 48:673–700.

Ruel, J. C., D. Pin, and K. Cooper. 2001. Windthrow in riparian buffer strips: Effect of wind exposure, thinning and strip width. *Forest Ecology and Management* 143 (1–3):105–13.

Salomonson, M. G., and R. P. Balda. 1977. Winter territoriality of Townsend's solitaires (*Myadestes townsendi*) in a piñon-juniper-ponderosa pine ecotone. *Condor* 79:148–61.

Sammalisto, L. 1957. The effect of the woodland–open peatland edge on some peatland birds in south Finland. *Ornis Fennica* 34:81–89.

Schelhaas, M. J., G. J. Nabuurs, M. Sonntag, and A. Pussinen. 2002. Adding natural disturbances to a large-scale forest scenario model and a case study for Switzerland. *Forest Ecology and Management* 167 (1–3):13–26.

Schelhas, J. 1995. Conserving the biological and human benefits of forest remnants in the tropical landscape: Research needs and policy recommendations. In *Integrating people and wildlife for a sustainable future,* ed. J. A. Bissonette and P. R. Krausman, 53–56. Bethesda, MD: Wildlife Society.

Schlaepfer, M. A., and T. A. Gavin. 2001. Edge effects on lizards and frogs in tropical forest fragments. *Conservation Biology* 15 (4):1079–90.

Schmutz, J. K., S. M. Schmutz, and D. A. Boag. 1980. Coexistence of three species of hawks (*Buteo* spp.) in the prairie-parkland ecotone. *Canadian Journal of Zoology* 58:1075–89.

Schoener, T. W., and A. Schoener. 1983. The time to extinction of a colonizing propagule of lizards increases with island area. *Nature* 302:332–34.

Semlitsch, R. D., and J. R. Bodie. 2003. Biological criteria for buffer zones around wetlands and riparian habitats for amphibians and reptiles. *Conservation Biology* 17 (5):1219–28.

Sibley, C. G., and B. L. Monroe Jr. 1991. *Distribution and taxonomy of the birds of the world.* New Haven: Yale Univ. Press.

Siitonen, P., A. Lehtinen, and M. Siitonen. 2005. Effects of forest edges on the distribution, abundance, and regional persistence of wood-rotting fungi. *Conservation Biology* 19 (1):250–60.

Simberloff, D. S., and L. G. Abele. 1976. Island biogeography theory and conservation practice. *Science* 191:285–86.

Simpson, K., and N. Day. 1999. *Field guide to the birds of Australia.* 6th ed. Victoria, Australia: Penguin.

Sinclair, I. 1997. *Field guide to the birds of southern Africa.* Cape Town, S. Africa: Struik.

Sitati, N. W., M. J. Walpole, R. J. Smith, and N. Leader-Williams. 2003. Predicting spatial aspects of human–elephant conflict. *Journal of Applied Ecology* 40 (4):667–77.

Smith, A. T., and M. M. Peacock. 1990. Conspecific attraction and the determination of metapopulation colonization rates. *Conservation Biology* 4:320–23.

Sverdrup-Thygeson, A., and D. B. Lindenmayer. 2003. Ecological continuity and assumed indicator fungi in boreal forest: The importance of the landscape matrix. *Forest Ecology and Management* 174 (1–3): 353–63.

Thomas, C. F., P. Brain, and P. C. Jepson. 2003. Aerial activity of linyphiid spiders: Modelling dispersal distances from meteorology and behaviour. *Journal of Applied Ecology* 40 (5):912–27.

Thomas, J. W., ed. 1979. *Wildlife habitats in managed forests: The Blue Mountains of Oregon and Washington.* USDA Forest Service Agriculture Handbook No. 553. Portland, OR: USDA Forest Service.

Thomas, J. W., C. Maser, and J. E. Rodiek. 1979. Edges. In *Wildlife habitats in managed forests,* ed. J. W. Thomas, 48–59. USDA Forest Service Agriculture Handbook No. 553. Portland, OR: USDA Forest Service.

Thompson, I. D., J. A. Baker, and M. Ter-Mikaelian. 2003. A review of the long-term effects of post-harvest silviculture on vertebrate wildlife, and predictive models, with an emphasis on boreal forests in Ontario, Canada. *Forest Ecology and Management* 177 (1–3):441–69.

Tilman, D., R. M. May, C. L. Lehman, and M. A. Nowak. 1994. Habitat destruction and the extinction debt. *Nature* 371:65–66.

Tischendorf, L., D. J. Bender, and L. Fahrig. 2003. Evaluation of patch isolation metrics in mosaic landscapes for specialist vs. generalist dispersers. *Landscape Ecology* 18 (1):41–50.

Tisdale, E. W., M. Hironaka, and M. A. Fosberg. 1965. An area of pristine vegetation in Craters of the Moon National Monument, Idaho. *Ecology* 46 (3):349–52.

Tiwari, B. K., S. K. Barik, and R. S. Tripathi. 1998. Biodiversity value, status, and strategies for conservation of sacred groves of Meghalaya, India. *Ecosystem Health* 4 (1):20.

Towns, D. R. 1985. *A field guide to the lizards of New Zealand.* Occasional Publication No. 7. Wellington, NZ: New Zealand Wildlife Service.

Trager, M., and S. Mistry. 2003. Avian community composition of kopjes in a heterogeneous landscape. *Oecologia* 135 (3):458–68.

Tscharntke, T., I. Steffan-Dewenter, A. Kruess, and C. Thies. 2002. Contribution of small habitat fragments to conservation of insect communities of grassland-cropland landscapes. *Ecological Applications* 12 (2):354–63.

Tudge, C. 1992. *Last animals at the zoo: How mass extinction can be stopped.* Washington, DC: Island Press.

Turner, M. G., W. L. Baker, C. J. Peterson, and R. K. Peet. 1998. Factors influencing succession: Lessons from large, infrequent natural disturbances. *Ecosystems* 1:511–23.

Vander Haegen, W. M., and R. M. Degraaf. 1996. Predation on artificial nests in forested riparian buffer strips. *Journal of Wildlife Management* 60 (3):542–50.

Vesely, D. G., and W. C. McComb. 2002. Salamander abundance and amphibian species richness in riparian buffer strips in the Oregon Coast Range. *Forest Science* 48 (2):291–98.

Wachob, D. G. 1996. The effect of thermal microclimate on foraging site selection by wintering mountain chickadees. *Condor* 98:114–22.

Walker, R. S., A. J. Novaro, and L. C. Branch. 2003. Effects of patch attributes, barriers, and distance between patches on the distribution of a rock-dwelling rodent (*Lagidium viscacia*). *Landscape Ecology* 18 (2):185–192.

Wallmo, O. C. 1969. Response of deer to alternate-strip clearcutting of lodgepole pine and spruce–fir timber in Colorado. USDA Forest Service Research Note RM-141. Fort Collins, CO: USDA Forest Service, Rocky Mountain Forest and Range Experiment Station.

Walters, J. R., L. B. Crowder, and J. A. Priddy. 2002. Population viability analysis for red-cockaded woodpeckers using an individual-based model. *Ecological Applications* 12 (1):249–60.

Ward, J. P., Jr., and D. Salas. 2000. Adequacy of roost locations for defining buffers around Mexican spotted owl nests. *Wildlife Society Bulletin* 28 (3):688–98.

Welsh, H. H. 1990. Relictual amphibians and old-growth forest. *Conservation Biology* 4:309–19.

Weltzin, J. F., M. E. Loik, S. Schwinning, D. G. Williams, P. A. Fay, B. M. Haddad, J. Harte, T. E. Huxman, A. K. Knapp, and G. Lin. 2003. Assessing the response of terrestrial ecosystems to potential changes in precipitation. *BioScience* 53 (10):941–52.

Whitcomb, B. L., R. F. Whitcomb, and D. Bystrak. 1977. Long-term turnover and effects of selective logging on the avifauna of forest fragments. *American Birds* 31:17–23.

Whitlock, C., S. L. Shafer, and J. Marlon. 2003. The role of climate and vegetation change in shaping past and future fire regimes in the northwestern US and the implications for ecosystem management. *Forest Ecology and Management* 178 (1–2): 5–21.

Whittam, T. S., and D. Siegel-Causey. 1981. Species incidence functions and Alaskan seabird colonies. *Journal of Biogeography* 8:421–25.

Whittier, T., S. G. Paulsen, D. P. Larsen, S. A. Peterson, A. Herlihy, and P. Kaufmann. 2002. Indicators of ecological stress and their extent in the population of Northeastern Lakes: A regional scale assessment. *BioScience* 52 (3):235–47.

Wiens, J. A. 1974. Habitat heterogeneity and avian community structure in North American grasslands. *American Midland Naturalist* 91:195–13.

Wilcove, D. S. 1987. From fragmentation to extinction. *Natural Areas Journal* 7:23–29.

Wilcox, B. A., and D. D. Murphy. 1985. Conservation strategy: The effects of fragmentation on extinction. *American Naturalist* 125:879–87.

Willson, J. D., and M. E. Dorcas. 2003. Effects of habitat disturbance on stream salamanders: Implications for buffer zones and watershed management. *Conservation Biology* 17 (3):763–71.

Young, A., and N. Mitchell. 1994. Microclimate and vegetation edge effects in a fragmented podocarp-broadleaf forest in New Zealand. *Biological Conservation* 67:63–72.

Zabowski, D., B. Java, G. Scherer, R. L. Everett, and R. Ottmar. 2000. Timber harvesting residue treatment: Part 1. Responses of conifer seedlings, soils and microclimate. *Forest Ecology and Management* 126 (1):25–34.

10   Modeling Wildlife–Habitat Relationships

Models are not like religion. You can have more than one . . . and you don't have to be-
lieve them.

DANIEL PAULY AND VILLY CHRISTENSEN*

In this chapter we explore the basis and use of models of wildlife–habitat relationships. First, we discuss the use, types, and objectives for modeling wildlife–habitat relationships. Then, we discuss how scientific uncertainty affects wildlife modeling and management, and how models should be used in light of uncertainties. We then review general types of model structures; traditional and new model forms used in research and management of wildlife–habitat relationships; and how models can be used in habitat planning and conservation. We end with a discussion of model validation.

Use and Types of Models

In this section we define models and discuss objectives for modeling types of predictions, model selection, and accounting for correlation and causation.

What Is a Model?

In its broadest sense, a model—from the Latin *modus,* meaning mode or measure—is any for-

mal representation of some part of the real world. Hall and Day (1977) suggested that a model can be conceptual, diagrammatic, mathematical, or computational. These forms can also be viewed as stages in a logical model-building process.

Developing a *conceptual model* may entail synthesizing current scientific understanding, field observations, and professional judgment of a particular species or habitat, and proposing a few hypotheses to explain the species' distribution and abundance. Even (especially) at the conceptual stage, it is vital to explicitly state assumptions and simplifications necessary for the model to be true or useful. The *diagrammatic stage* takes a conceptual model one step further by explicitly showing interrelationships among various environmental parameters and species' behaviors. The *mathematical stage* quantifies these relationships by applying coefficients of change and formulae of correlation or causality. Finally, the *computational stage* aids in exploring or solving the mathematical relationships by analyzing the behavior of formulae on computers.

The conceptual and diagrammatic stages of modeling are often the most difficult, and the most revealing, stages of building ecological theories and enhancing understanding. They must derive from a well-shaped statement of modeling goals and objectives and from basic understanding and articulation of the system being represented.

Objectives for Modeling

The main objectives for developing models of wildlife–habitat relationships are (1) to *formalize* or describe our current understanding about a species or an ecological system; (2) to *understand* which environmental factors affect distribution and abundance of a species; (3) to *predict* future distribution and abundance of a species; (4) to *identify* weaknesses in and improve our understanding; and (5) to *generate testable hypotheses* about the species or system of interest.

Not all of these goals are mutually reachable. For example, many observational field studies may result in statistical descriptions of wildlife-habitat relationships. Such observational descriptions are pertinent to specific locations, environmental conditions, and time periods, and help to explain observed patterns. They should not be assumed to necessarily also provide much power to predict conditions beyond those contexts with any reliability, but such studies are often used this way. At best, they can be used to generate hypotheses. Typically, though, most interest in modeling wildlife–habitat relationships does deal with prediction. In this book, *predictive modeling* refers to estimating the historic, future, or potential presence, distribution, or abundance of a wildlife species or group of species, given information on actual or possible environmental and habitat conditions. We include historic conditions under prediction because retrospective studies are so important.

Types of Predictions

There are two main types of predictions that may be made from models. One is *hindcasting,* which identifies key environmental variables, typically those of vegetation structure or environmental attributes, that account for observed variation in species variables such as abundance. Hindcasting is used to explain historic patterns observed in species occurrence and abundance and is pertinent, strictly speaking, only to the time and place at which the original data were gathered. Hindcasting is typically done from retrospective studies that try to tease out main correlations or causes from conditions or changes that have already occurred. Retrospective studies are vitally helpful in many fields—for example, as used by Louda et al. (2003) to reduce risks in biocontrol programs. Retrospective approaches have also been used to reconstruct historic vegetation (Schulte and Mladenoff 2001), to study the effects of habitat fragmentation on birds (Manolis et al. 2001), and in many other areas.

The other class of prediction is *forecasting.* Forecasting is an explicit attempt to predict future or potential species conditions, given environmental conditions at a time or place not represented by the field data used to generate the model in the first place. Many workers use results of hindcasting, such as obtained with use of correlation, regression, or multivariate statistics, to predict future species conditions, typically under alternative habitat management scenarios (e.g., McCune et al. 2003). However, without proper description of the initial investigation and without validation studies, predictions from hindcasting may be quite unreliable because environmental, demographic, and ecological conditions may vary significantly among locations or over time. At best, using hindcasting models for prediction in new situations entails the assumption that factors not accounted for in the prediction model are insignificant or are unchanged. This assumption should be explicitly stated.

What is the best means of predicting species responses to environmental conditions? Proper forecasting techniques account for autocorrelation of a variable over some time series or over spatial (such as environmental) gradients (Lichstein et al. 2002, Diniz-Filho and Telles 2002). More fundamentally, forecasting should be based on an understanding of the causes of the distribution and abundance of a species, rather than simply correlations as with hindcasting.

Types of predictions can be *deterministic,* as with point estimates of some future or expected population density; *statistical,* as with estimates of central tendency values and some measure of variation of a parameter, such as a mean population density plus or minus some standard deviation or tolerance interval; or purely *probabilistic,* as with estimates of the likelihoods of population persistence to future time periods.

Selecting Models

The manager should be wary of models that produce purely deterministic predictions because there is no measure of uncertainty of the prediction or variation around the outcome. Without knowledge of uncertainty and variation, the manager has no way to judge the relative risk associated with alternative courses of action. As we will suggest below, one of the best uses of wildlife–habitat relationships models is to aid risk analysis in decision making.

If the purpose of modeling is to assist management, one might consider a multiscale approach to model development and selection. A first step might be to clearly identify the management scales (see Table 8.1) in terms of geographic extent, map scale, spatial resolution, time period, administrative hierarchy, and levels of biological organization for which the management activity or plan is directed. Next, one

might identify the key areas of scientific inquiry (fig. 8.1) and management issues (fig. 8.2) that pertain to a particular management activity or plan, along with the expected duration of the activity or plan and the need for assessing prehistoric or historic conditions and cumulative effects. Finally, one might consider models by their purpose and function and select the kinds of models most pertinent to the questions, issues, and scales. Of course, other factors will guide model selection as well, such as availability and the need to integrate with other management activities or plans.

Of Correlates and Causes

When we build a model, including a statistical evaluation (hindcast) based on empirical observations, it is not always evident which factors are correlates and which are true causes. This is an old problem (Wright 1921). Often it is important to identify true causes in order to know which management activities to change or reaffirm to better meet objectives. But, frequently, teasing apart the "causal web" of wildlife–habitat relationships with any degree of predictive confidence is immensely difficult.

Consider four progressively more complex situations, as depicted in figure 10.1. In the first instance (fig. 10.1A), some species response, S, such as population presence or abundance, is assumed to be directly caused and explained by some environmental variable, E. There may also be some degree of unexplained variation in species response (shown as *?* in the figure). The unexplained variation (*?*) is due to measurement error, experimental error, or the effect of other environmental factors not included in the study. It can be quantified by such means as calculating residuals in a regression analysis or partitioning error terms in an analysis of variance.

However, as depicted in figure 10.1B, in the real world we may be measuring one environ-

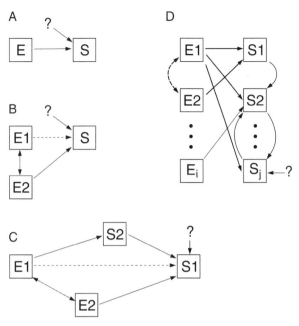

Figure 10.1.   Causes and correlates: four increasingly complex and realistic scenarios of wildlife–habitat relationships. S = wildlife species response; E = environmental factors; ? = unexplained variation; solid arrows = causal relations; dashed-line arrows = correlational relations that may or may not be causal; large dots = a sequence of other E and S variables. See text for explanation. (From Marcot 1998, 137.)

mental variable, E1, when the real causal factor is another, unmeasured environmental variable, E2. In this case, the two environmental variables are themselves correlated and there may or may not be a causal relation between them. E1 and E2 may be vastly different kinds of environmental factors and may operate at different spatial or temporal scales as well. We think we have explained the biological response of the species S by the observed correlation with E1, and we may even be able to predict S from E1 to a limited degree (in which case E1 is termed the "latent variable"), but we may be greatly mistaken to presume that management of E1 will necessarily affect the species as we wish. Further, the unexplained variation (?) in S is then due to the less than perfect correlation between E1 and E2 as well as to measurement and experimental error,

and to influential environmental factors beyond E1 and E2 not addressed in the study. Such a situation may be analyzed using univariate multiple regression techniques.

As an example, we may find that mean fecundity rates (S) in a population of Townsend's vole (*Microtus townsendii*) are negatively correlated with food abundance during the previous season (E1), and thereby infer that high food abundance leads to high population density, which in turn suppresses mean fecundity levels as a population regulation mechanism. However, on closer inspection, it may turn out that the real culprit, E2, causing lower mean fecundity of voles is parasitism by botflies (*Cuterebra grisea*) (e.g., Boonstra et al. 1980). It may turn out that both food resources and botfly incidence are affected by weather, so that environmental factors

E1 and E2 are correlated, but there is no direct causal link between them. If our study focused only on food resources and vole fecundity, we would conclude that food abundance is the cause and that managing for higher vole densities could be afforded by managing for more consistent food resource levels. This conclusion would be in error. Also, the unexplained variation in vole fecundity would be caused by the less than perfect correlation between food resources and botfly incidence, as well as by additional factors beyond food or botflies not addressed in the study.

However, the real world is often even more complicated. As shown in figure 10.1c, another species, S2—potentially a competitor, predator, or symbiont—may also play a role in affecting the species of interest, S1. (We included the parasite as an environmental factor in the last example.) Following our example above, mean vole fecundity might be influenced by some (hypothetical) predator, S2, that selectively removes high-fecundity individuals from the vole population. S2 itself may share some environmental factor, E1, that correlates with (but does not cause) S1. In an even more complicated but increasingly realistic schematic, as in figure 10.1d, S2 may also be influenced by other environmental factors beyond E1. And so the causal web expands.

It is important to recognize such *causal webs* of organisms and environmental factors, and to focus attention on the main influences (instead of the entire possible causal web). We should at least challenge ourselves to draw a causal diagram (the diagrammatic phase of modeling as outlined above)—often called an influence diagram—so that we can hypothesize which are major causal factors, which are minor causal factors, and which are merely correlative. Differentiating between causes and correlates is critical for guiding costly habitat management activities to respond to complex environmental issues

such as changes in air quality or regional climate, and for establishing an appropriate monitoring scheme, including identifying and tracking key indicators.

Path regression analysis is a statistical technique that can aid in quantifying the relative contribution of causal factors (Shipley 2002). Path regression is used to determine and display the partial correlation coefficients of individual factors that can influence a species population or, in some instances, a management objective, such as used by Howe and Brown (2000) to model effects of rodent foraging on vegetation communities, and by Johnson et al. (2001) to model foraging behavior of woodland caribou (*Rangifer tarandus caribou*). An example of a path regression analysis is shown in figure 10.2. Other techniques useful for teasing out the relative contributions of environmental factors, species factors, and uncertainty are those of multivariate multiple regression and two-stage regression.

Uncertainty and Unknowns in Wildlife–Habitat Relationship Models

Biological models do not predict species distribution and abundance without error. Rather, modeling wildlife–habitat relationships, like managing species habitats, typically entails dealing with the following kinds of obstacles:

- Imprecise data
- Uncertain inferences
- Limiting and fallacious assumptions
- Unforeseen environmental, administrative, and social circumstances
- Risks of failure

Imperfections are often present in habitat analyses and management decisions but are especially

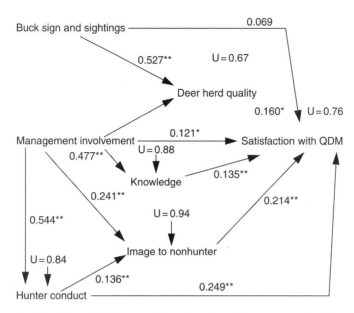

Figure 10.2.   Example of a path analysis that partitions the various factors accounting for variation in public satisfaction with quality deer management (QDM). The values are partial correlations (i.e., the correlation of each factor once the contribution of all other factors is accounted for). U = unexplained variance, calculated as $(1 - R^2)^{1/2}$; * $P = < 0.05$, ** $P = < 0.0001$. (Reproduced from Woods et al. 1996 [fig. 1], by permission of the Wildlife Society.)

important when there is risk of reducing a wildlife population or eliminating a species. Uncertainty may be encountered when analyzing biological data, when making inferences about species' responses to environmental conditions, and when selecting and instituting a management plan.

Types of uncertainty may be classified as scientific uncertainty and decision-making uncertainty. Just as analyzing species and habitats entails a different process from that used to make decisions on resource management, so too are the kinds and implications of uncertainties from the analysis process distinct from those in a decision-making process. Results of a technical study, such as a risk analysis of population viability, may be part (but only part) of the infor-

mation used by a decision maker in developing a habitat management plan.

Types of Uncertainty

Scientific uncertainty in habitat modeling refers to the nature of the data and the ways in which information on species and habitats is represented and applied. Scientific uncertainty essentially means that our predictions of how species respond to environmental conditions are not perfect. Uncertainty may occur because (1) the system itself is naturally variable and very complex, and thus difficult to predict; (2) the process of estimating values of parameters in the habitat model entails a degree of error; (3) models used to generate predictions are in some sense invalid;

or (4) the scientific question being asked is ambiguous or incorrect.

VARIABILITY OF NATURAL SYSTEMS— NOISE IN THE MESSAGE

Many aspects of natural systems vary over time. Predicting attributes of the system—the "message" we are trying to interpret—may often involve observing and modeling traits that are influenced from outside factors (fig. 10.1); that is, "noise" inherent in the message. Such noise introduces variation in measurements and uncertainty in estimating and predicting attributes of the system. In statistical models of habitat relationships, noise is typically depicted as unexplained variation in the occurrence or abundance of a species. One kind of unexplained statistical variation is the value of "residuals" in regression models. Sometimes this kind of noise in the system can be a useful source of information itself (e.g., Motta 2003).

In a most revealing review of models in ecology and evolution, Moller and Jennions (2002) determined that the mean amount of variance (r^2) explained was a dismal 2.51 to 5.42%, and that the mean effect size reported was between Pearson $r = 0.180$ and 0.193. They concluded that, because of so much noise in the data, most studies had inadequate sample sizes to determine the absence of a particular relationship with a power of 80% and $\alpha = 0.05$. This conclusion is not encouraging and suggests the need to conduct initial pilot studies to estimate the degree of variation in parameters so sample sizes can be adjusted accordingly (see chapter 4). However, Peek et al. (2003) viewed this same study from the perspective that about *half* of the variation in statistical models is *not* explained (the unexplained residual or random error), suggesting a far higher success rate in ecological modeling. They concluded that the success of ecological models should be judged not by single-factor relationships, but by overall model performance. Still, in

a risk management framework, and to set realistic expectations for results of their actions, decision makers should know that such models may explain or predict only half the variation in whatever conditions they wish to manage.

De Valpine and Hastings (2002) suggested a numerical method for incorporating noise and observation error when fitting population models, and their tests of fit suggested that such an approach works best with stable-point Ricker recruitment models and worst with Beverton-Holt models. Hewitt et al. (2001) suggested treating temporal variability as a parameter instead of as noise, thereby increasing the detection of treatment effects in before–after/control–impact (BACI) studies.

UNCERTAINTY OF EMPIRICAL INFORMATION—ERRORS OF ESTIMATION

Values of environmental parameters are typically estimated from a sample set of observations. A parameter, for example, may be the mean number of tree stems per hectare or the variance of litter sizes of black-footed ferrets (*Mustela nigripes*) to the extent that these can be attributed to individual and environmental variation. When a parameter is estimated from a sample set of observations, from a statistical viewpoint, uncertainty or errors in estimation occur. The estimations are *biased* if each of the values of the observations are consistently lesser or greater than actual (unknown) values; *inaccurate* if the estimated value of the parameter of interest (such as a mean or a variance) is substantially different from the true value; or *imprecise* if values of individual observations vary widely among each other.

Each of these errors in estimating the value of a parameter constitutes a different kind of scientific or statistical uncertainty. Bias is estimated by the difference between the mean of observed values and the true parameter value; accuracy is measured by mean square error of observed values; and precision is estimated by variance in the observed values. Statistical estimators always have these properties of bias, accuracy, and precision, as well as others, particularly *consistency*. Bias tells you how far you missed the bull's-eye on average; accuracy tells you the spread of misses among individual trials; and precision tells you how tightly grouped your trials are, whether they missed the bull's-eye or not. Then, *consistency* tells you if you're even shooting at the right target. An estimator can be consistent, but biased, and adjusting the sampling methodology can reduce bias; likewise, an estimator can be consistent, but imprecise or inaccurate, and increasing sample size can increase precision and accuracy.

Such errors of estimation can arise from a number of sampling problems, including inadequate sample size, observations taken from disparate times or places, and samples taken nonrandomly or nonsystematically, depending on the assumptions of the estimator being used. Errors of estimating the value of parameters may also arise from applying the wrong kind of estimator, such as in applying a formula for calculating variance. If correct use of the formula assumes that observations were made independently and randomly—when they were actually made over a time series or systematically, such as at even intervals over a transect—then an error of applying the wrong kind of estimator has been made. (With systematic samples, the appropriate variance estimator should instead use the mean square of *successive differences* in values measured along the time or spatial series of samples.)

MODEL VALIDITY AND UNCERTAINTY OF MODEL STRUCTURE

Model validity refers to a broad spectrum of performance standards and criteria. Examples are model credibility, realism, generality, precision, breadth, and depth (Marcot et al. 1983). The various criteria refer to such attributes of models as the number of parameters in a model and their interactions, the context within which a model was developed or should be used, and the underlying and simplifying assumptions of the model structure (see table 10.1 for definitions of model validation criteria).

A parameter that is estimated precisely, accurately, and without bias may still be used inappropriately, as in a model that is applied to the wrong environment, location, season, or species.

APPROPRIATENESS OF THE PROBLEM— ASKING THE RIGHT QUESTION

The context in which a theory is applied or a model is used may introduce yet another source of uncertainty. Even given that a model has been validated—that is, shown to be a useful tool and to generate acceptable predictions according to particular criteria—it still may be applied to the wrong problem. (This conclusion is analogous to the property of consistency in statistical estimation.) In some cases, this problem of inappropriate application may be unavoidable if no other models are available.

For example, a life table model that assumes equal sex ratios and that adults breed each year may generate acceptable predictions for use with Dall sheep (*Ovis dalli*) but may generate grossly inaccurate predictions when used for species with variable or quite different social breeding organizations, such as pronghorn (*Antilocapra americana*). This application would call into question the reliability of the model when used with some species or under some circumstances.

Further, the hypothesis or problem being addressed by using a particular model may be ambiguous or even unanswerable. For example, a model of species–habitat relationships that describes vegetation types may not provide a particularly useful foundation for answering questions about landscape dynamics necessary for maintaining viable populations of the species.

Table 10.1.   Criteria useful for validating wildlife–habitat relationship models

Criterion	Explanation
Precision	The capability of the model to replicate particular system parameters
Generality	The capability of the model to represent a broad range of similar systems
Realism	The capability of the model for relevant variables and relations
Precision	The number of significant figures in a prediction or simulation
Accuracy	The degree to which a simulation reflects reality
Robustness	Conclusions that are not particularly sensitive to model structure
Validity	The capability of the model to produce all empirically correct models.
Usefulness	The existence of some empirically correct model.
Reliability	The fraction of models that are empirically correct
Adequacy	The fraction of pertinent empirical observations that can be simulated
Resolution	The number of parameters of a system that the model attempts to mimic
Wholeness	The number of biological processes and interactions reflected in the model
Heurism	The degree to which the model usefully furthers empirical and theoretical investigations
Adaptability	The future development and application
Availability	The existence of other, simpler, validated models that perform the same function
Appeal	The degree to which model results match our intuition and stimulating thought, and practicable.
Breadth, Depth	The number and kinds of variables chosen to describe each (habitat) component
Face validity	The credibility of the model
Sensitivity	The match of model variables and parameters with real-world counterparts, and their variation causing outputs that match historical data; also, the dependence of model output on specific variations of variables
Hypothesis validity	The realism with which subsystem models interact
Technical and operational validity	The identification and importance of all divergence in model assumptions from reality, as well as the identification and importance of the validity of the data
Dynamic validity	The analysis of provisions for application to be modified in light of new circumstances

Source: Based on Marcot et al. 1983; reproduced by permission of the Wildlife Management Institute.

Accounting for Error in Modeling

One of the major problems in using models of wildlife–habitat relationships is that of *propagation of error.* Error can arise from model structure, missing data, mismatched scales of geographic extent and spatial resolution, and other systematic sources, as well as from measurement error and the stochastic nature of biological systems. How do all such errors compound in a particular model? The problem of error propagation has been poorly addressed in the statistical and modeling literature and needs much work. One approach to depicting the compounding of error is to partition the variance associated with model output into additive factors, each representing the major sources of error. This method is analogous to methods used in analysis of variance, in which mean square errors are partitioned into sampling error and experimental error. This approach may entail an analytic formulation for summing variance and covariance terms (box 10.1). Other approaches may invite use of model sensitivity analysis. We also further discuss errors in modeling below under "Validating Wildlife–Habitat Relationship Models."

Box 10.1   *Propagation of error in modeling wildlife–habitat relationships*

How does error compound in wildlife–habitat models, and why should we worry about it?

Every variable and function in a wildlife–habitat model can have several kinds of associated error, including measurement error, experimental error, and random error. Although it may be feasible to estimate such error for each variable or simple relation in a univariate sense, it is the compounding of error among variables and in complex functions that combine variables that may seriously affect final model output. The result of such *error propagation* may be model output that is significantly biased, imprecise, or inaccurate. Thus it may be critical to understand how error terms compound.

The biostatistical literature continues to poorly address or even to ignore the estimation of error propagation, in large part because it is such a wicked analytic problem. The classic approach to the problem is to dissect the variance of some variable *y* into its Taylor series expansion terms of component measured quantities $F(x_1, x_2, x_3, \ldots)$ (Kotz et al. 1982):

$$\text{var}(y) = \sum \left(\frac{\partial F}{\partial x_i} \right)^2 \text{var}(x_i) + 2 \sum_{ij} \left(\frac{\partial F}{\partial x_j} \right) \left(\frac{\partial F}{\partial x_j} \right) \text{cov}(x_i x_j).$$

The wicked part of this problem is not in estimating the first variance term, but in the covariance terms $\text{cov}(x_i x_j)$. In the simplest case, if x_i and x_j are uncorrelated, then $\text{cov}(x_i x_j) = 0$, and the entire second term for var(*y*) drops out. Ecological variables are quite often at least partially correlated (figs. 10.1, 10.2), so that $\text{cov}(x_i x_j) > 0$ and the second term in the above equation is nonzero and needs to be calculated.

In the case of estimating covariance among means, the parent covariance term $\text{cov}(x_i x_j)$, or $\sigma_{\mu\upsilon}^2$, can be calculated as (Eadie 1983):

$$\sigma_{\mu\upsilon}^2 = \lim_{N \to} \frac{1}{N} \sum \left[(\mu_i - \bar{\mu})(\upsilon_i - \bar{\upsilon}) \right].$$

This covariance term may be extremely difficult to impossible to measure from empirical data. This difficulty is especially true with real-world studies of landscapes, ecosystems, and populations in conditions that are poorly, or not, replicable, and for which control conditions are not feasible.

However, it is the covariance between ecological variables that may often be a major source of variation and error in model output. Even in the simple case of two interacting variables, *x* and *y*, the appropriate variance estimator involves the wicked covariance term: for the interaction term *xy*, variance is calculated as $y^2 \text{var}(x) + x^2 \text{var}(y) + 2xy \text{cov}(x,y)$ (Kotz et al. 1982, 549). Empirically, covariance among all such key variables in a wildlife–habitat model may be impossible to estimate empirically and to calculate analytically.

Propagation of error also means that initial errors may have a fatal effect on the final results—that is, that small changes in initial data may produce large changes in final results. Such problems are called *ill-conditioned* and may include population models that exhibit chaotic or initially unpredictable behavior (Morris 1990; Hassell et al. 1991). Additional sources of error, particularly in computer models and that can propagate across functions, include rounding errors and truncation errors.

What is a modeler to do? One tractable approach to evaluating error propagation is to conduct sensitivity analyses of the model, whereby changes in outputs are plotted as a function of incremental changes in input variables (Nelson 2003). This approach can help identify the domains over which models exhibit chaotic behavior, such as when populations exhibit irregular cycles or suddenly crash or expand with only minor changes in the input variables. Sensitivity analysis can be one phase of model validation, discussed further in the text. Another analytic approach to dissecting propagated errors is forward and backward analysis (see Fröberg 1969, 3–9).

Decision Making under Uncertainty and What to Do about It

Probably all management decisions dealing with wildlife habitat are made under some uncertainty of current conditions or future effects. Decision-making uncertainty can arise from imprecise data, uncertain inferences, limiting and fallacious assumptions, and unforeseen environmental, administrative, and social circumstances. Each of these factors can contribute to risks of failure in meeting desired management goals.

How might the manager or decision maker proceed under such uncertainties? A host of decision-analysis techniques are available that aid in assessing the value of perfect information, the value of sample information, the credibility of information, and quantitative measures of the state of knowledge (e.g., Clemen and Reilly 2001; McDonald and McDonald 2003; Pielke and Conant 2003). We review some decision-aiding approaches and models below. Using these approaches to identify areas and degrees of decision-making uncertainty may also be useful for establishing management activities as adaptive management experiments. In a sense, uncertainty is an opportunity for testing management hypotheses about outcomes of actions, as long as basic tenets of adaptive management are not violated (see chapter 11).

Balancing Theory with Empiricism

We have dedicated several chapters in this book to reviewing study design and measures of habitat and wildlife behavior. Empirical field studies—whether observational descriptions or experimental tests—can be used to develop models of wildlife–habitat relationships. Theory, however, often plays an important role in model development. Theoretical models may tend to be more robust and general, whereas empirical models may be more locally accurate and precise. Each complements the other. Ultimately, empirical models can be used to induce more general theoretical ones, and theoretical models can help guide the specific development of empirical models.

Using Models to Generate Research Hypotheses

The modeling process should be a means by which we challenge ourselves to explicitly articulate what we think we know about some system. This challenge occurs in the conceptual and diagrammatic phases of model development as discussed previously. To this end, model output can be used to generate research hypotheses. In an adaptive management context, management activities can be crafted to test the more important assumptions or provide information on the key unknowns. When management activities are applied on the ground, these assumptions and unknowns may be termed *management hypotheses*.

To scientifically test management hypotheses, management activities should be crafted to follow the guidelines for correct study design, including evaluation of baseline conditions, provision of controls, adequate study and treatment duration, appropriate spatial scale, and adequate replication of controls and treatments. Such considerations pertain to ensuring consistency, reducing bias, and providing for appropriate levels of accuracy and precision of estimates. Then, effects of the management activities can be analyzed to validate the original assumptions and provide new information to revise or reaffirm the management hypotheses and guidelines. In turn, the models used to suggest the original management activities would be updated and new activities suggested, if warranted by the findings. Thus the ideal adaptive management process is cyclic, not linear, and entails strict adherence to correct experimental design. That is, management trials on the ground are crafted as scientific experiments, and their results are used to pose new management hypotheses and to further craft and test new management trials.

As models become more complex, and as management objectives broaden to include landscapes and ecosystems, use of models for generating research hypotheses should become more salient in the decision-making process. Real-world constraints of uncertain future research budgets, changing management goals, and balancing the need for short-term publications and long-term studies must be addressed in this use of models.

Types of Model Structures

The array of models in wildlife ecology and conservation is bewildering. There has been an exponential increase in the number of publications on modeling in the ecological and wildlife literature since the middle of the last century (fig. 10.3), with no asymptote in sight. Clearly, mod-

eling continues to play central roles in many facets of wildlife–habitat relationships (WHR) research and management.

In this section we briefly review some general types of WHR model structures as a way to provide a classification of types of models. In the next section we discuss specific models that are directly or potentially useful to WHR research and management.

Statistical Empirical Models

Statistical empirical models provide statistical analyses or summaries of empirical data. These models are usually based on statistical correlations among variables. They include many kinds of bivariate and multivariate analyses that describe relations among variables (see chapter 6). Recently, meta-analysis methods have come into vogue that provide a means of combining information, and making inference across, multiple studies (Pena 1997; Gurevitch et al. 2001). Meta-analysis has been used in a variety of research questions, including understanding the effects of

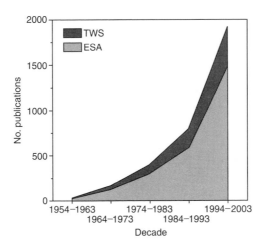

Figure 10.3.   Number of articles on models or modeling published in all journals by The Wildlife Society (TWS) and the Ecological Society of America (ESA), by decade, 1954–2003.

competition on predator–prey interactions (Bolnick and Preisser 2005), determining the influence of organic agriculture on biodiversity (Bengtsson et al. 2005), and estimating vital rates of raptors (Boyce et al. 2005).

Statistical empirical models also include statistical approaches that derive relations among variables in a post hoc manner. Some of these newer approaches are termed *knowledge discovery, data mining,* and *rule induction* (Jeffers 1991; Hastie et al. 2001). These approaches entail discovering relations among variables in typically large data sets or from a series of examples. These newer approaches deviate significantly from the traditional "frequentist" statistical methods. Some entail use of Bayesian statistics, and others depart entirely from statistical considerations of hypothesis-driven analyses altogether. Some classical, frequentist statisticians despair of such post hoc approaches as "fishing expeditions" that may yield spurious results with unknown degrees of confidence (Anderson et al. 2001). However, the methods seem to be here to stay, and when used judiciously and thoughtfully, they can be very helpful in further hypothesis creation. We discuss some tools under these new approaches below.

Habitat Relationship Models

A variety of traditional WHR models have been used for many years. These include wildlife-habitat matrix models, gap analysis models, habitat suitability index (HSI) models, habitat effectiveness (HE) models, habitat evaluation procedures (HEP), and others. The basis for these models can be a combination of expert opinion, literature, and field data. These models may be structured as simple look-up tables (WHR matrix models), as geometric mean equations incorporating key or limiting factors (HSI, HE models), or as hybrids of equation-based analysis within additional assessment procedures (HEP models). They may also predict wildlife species presence and distribution based on vegetation and land cover conditions (WHR matrix models, gap analysis models).

Analytic and Numerical Population Models

Analytic and numerical population models include more traditional population demographic and genetic models, such as Leslie matrix life tables (e.g., Henny et al. 1970; Taylor and Carley 1988; Miller et al. 2002) and calculations of rates of inbreeding and genetic drift (e.g., Perrin and Mazalov 1999; Edmands and Timmerman 2003).

Simulation Models

Simulation models include traditional dynamic models that simulate time-based interactions of variables of populations and ecosystems. Most simulation models are discrete and based on difference equations or other representations of time-step functions, and as such derive largely from "queueing theory" (Gordon 1978). A queue is a sequence of events waiting to happen (for example, dispersal of an organism from a particular habitat patch); how they happen is determined by the structure of the model (such as equations that describe the direction and distance of dispersal), and when they happen is usually based on a probability distribution of event frequency (such as the probability of dispersal events per time increment). Time-dynamic simulation models are used widely in ecology (e.g., Bolliger et al. 2005).

Geographic Information System–Based Models

Simulation models also include, more recently, geographic information system (GIS)–based, geographically referenced ("spatially explicit") models of landscape patterns and disturbance events (see chapter 8) and simulations of individual-based movement throughout a landscape and use of habitat patches (see chapter 9). GIS models can be very flexible and incorporate other modeling constructs, such as population viability analysis. For example, Zhu et al. (1998) integrated GIS with a decision support structure using a knowledge-based information system, and illustrated its utility for strategic planning for development of the island of Islay off Scotland. Reynolds et al. (2000) used the Ecosystem Management Development System (EMDS) model (Reynolds et al. 1997), which integrates a fuzzy logic modeling framework within a GIS system, to evaluate watershed conditions. Raphael et al. (2001), Wisdom et al. (2002), and Rowland et al. (2003) integrated Bayesian belief networks with GIS to evaluate broad-scale management effects on terrestrial wildlife in the interior western United States. Zhang et al. (1997) integrated a neural network model (see below) into GIS to classify vegetation types from remote sensing imagery. Many other examples of GIS modeling, including hybrid models of GIS with other modeling constructs, are available (O'Neil et al. 2005).

Knowledge-Based (Expert) Models

Knowledge-based (expert) models represent the experience and judgment of human experts rather than being based strictly on empirical data, simulations, or theoretical constructs. These models include expert systems of various types, as well as models in which the state variables and relations are determined by expert judgment. Knowledge-based models are often devised by use of an intermediary "knowledge engineer" or someone who quizzes the expert and puts the gained expertise into formulae, functions, and computer code. Such expert-based models can pose particular challenges for verification and validation.

A Review of Specific Types of Wildlife–Habitat Models

In this section we will review specific examples of various types of models that are useful for assessing wildlife–habitat relationships. We split our review here into traditional WHR models and a set of newer, more avant-garde types of models.

Traditional Models of Wildlife–Habitat Relationships

In this section we review types of wildlife-habitat relationship models that have been commonly used in many studies of management situations.

MODELS OF VEGETATION COMPOSITION AND STRUCTURE

A number of models have been developed that display current and future composition and structure of vegetation stands. These models include *forest stand growth and yield models* of many kinds, which are typically based on growth and yield information generally available for a variety of commercial forest types.

It is common practice to use forest growth models designed for silviculture to infer amount of wildlife habitat (e.g., Thompson et al. 2003). However, this use assumes that the vegetation variables output by the model are directly pertinent to wildlife—that is, that they are the vegetation conditions selected by specific wildlife species (see fig. 10.1). In many cases, however, the representation of habitat for specific wildlife species may be uncertain, at best. Thus the user should interpret actual wildlife response with a fair degree of caution and skepticism. In other words, a portion of the uncertainty (variation) in species' response may be due to simply not tracking the most appropriate habitat variables. This uncertainty also means that if successful wildlife management is dependent on models built for other purposes, the models should be evaluated for how precisely and fully they represent key habitat attributes for specific species of interest.

Models of ecological succession are useful for predicting changes in vegetation composition and structure over time. Such models are used in conjunction with WHR matrix models to depict changes in macrohabitat for wildlife species. As a systems model, a model of successional changes can also incorporate the ecological functions of wildlife as influencing vegetation. For example, the models of Lesica and Cooper (1999) of shrub vegetation succession in the Centennial Sandhills of southwest Montana incorporated the salient effects of pocket gopher (*Thomomys talpoides*) burrowing and ungulate browsing, as well as fire, on maintaining early seral vegetation.

DISTURBANCE MODELS

Disturbance models can include a variety of process, simulation, or even analytic models to determine how environmental conditions, especially vegetation and river systems, are affected by intermittent disturbance events such as fire, radical climate change, drought, and floods. Disturbances are being considered more frequently in models used to assess wildlife–habitat relationships and other kinds of land use (Schelhaas et al. 2002). Some disturbance models pertain to fine-grain features of environments, such as simulating forest canopy gaps (Lundquist and Beatty 2002). Other disturbance models address broad geographic areas, such as the effects of fires on the amount and distribution of old-growth forests across a broad area—for example, the Pacific Northwest of the United States (Wimberly 2002).

Disturbance models can be immensely useful for conservation of rare species or recovery of threatened species, by understanding the roles—negative and positive—played by disturbance

events. For example, Root (1998) determined that the long-term viability of the threatened Florida scrub jay (*Aphelocoma coerulescens*) requires fires at least every 30 years—fires serving to maintain a diversity of habitat patches used by the species. Probst and Weinrich (1993) found a similar, positive relation between the endangered Kirtland's warbler (*Dendroica kirtlandii*) and frequent fires in its jack pine (*Pinus banksiana*) forest habitat in Michigan.

Recently, there has been a return to modeling disturbances by using *stable-state analysis* (Beisner et al. 2003; Didham and Watts 2005). Popular in the 1960s, this modeling approach views specific patterns of composition and structure of ecosystems and ecological communities as semi-stable conditions that can be perturbed to varying degrees by disturbance events. If the perturbation is sufficiently strong, the system may enter a new configuration and a different stable state. Examples include repeated fires, livestock grazing, and agricultural development in native grasslands that allow invasive species to gain an irreversible foothold and basically destroy native grass and forb communities (Harrison et al. 2003); the creation of new stable states in forests of the eastern United States under immense herbivory pressure from overabundant deer herds (Stromayer and Warren 1997); grasslands maintained by herbivore foraging (Seabloom and Richards 2003); and, in a case of the *absence* of a natural disturbance event, the exclusion of wildfire in sagebrush-steppe or arid savanna communities that leads to invasion ("encroachment") by conifer trees or other woody plants (e.g., Skarpe 1991).

The degree to which systems, including wildlife species, can rebound following a disturbance event is a measure of their *resilience* (Gunderson et al. 2002), and the degree to which systems do not change is their *resistance* (Knapp et al. 2001; Byers and Noonburg 2003). One implication of this rebounding for managing resource production, including wildlife habitat, is that more re-silient and resistant systems are better able to withstand the onslaught of disturbance events. And by implication, more resilient and resistant systems can produce more predictable and sustainable levels of renewable natural resources. Another implication of multistate modeling is that some systems can attain several stable states, or may take a long time to recover to original or desired states (Ludwig et al. 1997), or, if pushed too far, may never be able to be restored to a previous state. These outcomes can have important implications for guiding conservation and restoration actions, particularly if recovery times following disturbances do not match expectations (Paine et al. 1998).

Another type of disturbance model pertains to *climate change*. Much work is being done to understand climate trends as disturbance events themselves; the influence of climate shifts on other disturbances, such as fire frequency (Whitlock et al. 2003) and pests and parasites (Jaenike 2002; Logan et al. 2003); and the implications of these influences overall changes in ecosystems (Beckage et al. 2005), ecological communities (Weltzin et al. 2003), and individual wildlife taxa (Torti and Dunn 2005). For example, analyzing data on temperature and precipitation trends in Europe, Lemoine and Bohning-Gaese (2003) predicted that climate change would result in a decrease in the proportion of long-distance migratory birds and an increase in the proportion and number of short-distance migrants and residents. They concluded, with some empirical corroboration, that increasingly warm winters will disproportionately threaten long-distance migrants (also see Jonzen et al. 2002).

WHR MATRIX MODELS

WHR matrix models have been around since the 1970s—and probably far earlier in more primitive forms. They are essentially static tables that relate wildlife species to habitat types and com-ponents (Verner and Boss 1980) and that depict other features of their life history (Johnson and O'Neil 2001). Often the data are categorical and qualitative, and are derived from a combination of field studies and (largely) professional judgment. Such matrices or information bases may be useful for predicting the potential presence of wildlife species associated with specific environmental conditions, although they tend to err on the side of commission (predicting more species to be present than actually are). This error is because factors not specified in a query of WHR matrix databases are tacitly presumed to be optimal or do not affect species absence. Still, such matrices are of great value to explore general trends and patterns of potential species presence and community composition.

GAP ANALYSIS MODELS

Gap analysis is a method of identifying "gaps" in conservation area networks in which high species richness or locations of particular species or taxa of interest are unprotected. Gap analysis maps are typically produced in GIS by overlaying land ownership and management categories with species' range maps or distribution maps of vegetation and land cover categories that represent species' habitats. Then, "hot spots" of high species richness, centers of endemism or rarity, or other such features of conservation interest are delineated, and it is then determined if such sites are unprotected (Jennings 2000).

Gap analysis has been used widely as part of reserve design tools (also see discussion below on other methods of reserve design). As examples, Clark and Slusher (2000) used gap analysis to design a national wildlife refuge in the Kankakee River/Grand Marsh area in Indiana and Illinois. Caicco et al. (1995) used gap analysis to determine the conservation and management status of vegetation communities in Idaho. Allen et al. (2001) used gap analysis to model viable mammal populations, and suggested that defining minimum critical areas via gap analysis was a useful way to produce better maps of critical unprotected areas for mammal species.

Several cautions pertain to gap analysis. Hot spots of high species richness may represent disturbed or ecotonal environments rather than cores of species' ranges and optimal habitat conditions, so some care needs to be taken in interpreting results and in understanding the underlying distributional data. Church et al. (2000) suggested a method to balance species richness objectives with assessment of habitat quality. Flather et al. (1997) suggested that several key assumptions of gap analysis should bear critical evaluation, including (1) that a subset of taxa can represent overall diversity patterns, and (2) that accurate reserve design can be affected by data uncertainty and error propagation in the underlying distributional data. Other concerns for accuracy of mapping gaps and species richness hot spots were expressed by Dean et al. (1997) and NCASI (1996), who found that boundaries of species-rich areas vary greatly depending on the accuracy of the underlying habitat and species distribution maps.

HABITAT SUITABILITY AND EFFECTIVENESS MODELS

One of the more popular and simpler approaches to modeling WHR has been the use of *habitat suitability index (HSI) models*. HSI models are used extensively by USDI Fish and Wildlife Service (Schamberger et al. 1982) and other federal resource management agencies. These models typically denote habitat suitability of a species as the geometric mean of n environmental variables deemed to most affect species presence, distribution, or abundance. The general model form is: $HSI = (V_1 \cdot V_2 \cdot \ldots V_n)^{1/n}$, where the V's represent n key environmental variables. Each variable and the resulting HSI values are scaled from 0 to 1. The overall HSI value assumedly represents the final response of the species to the combination of the values of the environmental parameters. A geometric mean is used so that when one variable goes to zero, HSI = 0.

For example, the three environmental variables denoted in an HSI model for yellow warbler (*Dendroica petechia*) are percentage of deciduous shrub crown cover, average height of deciduous shrub canopy, and percentage of shrub canopy consisting of hydrophytic shrubs (Schroeder 1983). The resulting suitability index in the yellow warbler model represents relative habitat values for reproduction. HSI models have been constructed for a wide variety of species in the United States.

HSI models are useful for representing in a simple and understandable form the major environmental factors thought to most influence occurrence and abundance of a wildlife species. However, HSI models are best viewed as hypotheses of species–habitat relationships rather than as causal functions (Schamberger et al. 1982). Their value lies in documenting a repeatable assessment procedure and in providing an index to a very few, and easily evaluated, environmental characteristics that can be compared among alternative management plans.

However, HSI models do not provide information on population size, trend, or behavioral response by individuals to shifts in resource conditions, and seldom include interaction or error terms. In fact, Bart et al. (1984) found that HSI models performed poorly because they were not based on field data and did not sufficiently account for interactions between the predictor (habitat) variables. Cole and Smith (1983) voiced similar concerns. Also, as we have discussed elsewhere, users should be wary of models that provide purely deterministic predictions with no statements of uncertainty or variability. At best, such models should be viewed as providing potentially testable hypotheses (although the 0–1 indices of HSI models are not particularly interpretable as empirical parameters) and interpreted as average conditions at a relatively broad geographic extent.

Comparable to habitat suitability index models are habitat capability and habitat effectiveness models. These models essentially perform the same function as HSI models but may vary slightly in structure. *Habitat capability (HC) models* typically provide an estimate of the total area within which resources for a particular species can be found, or rank a given area for the relative capability of supporting a species, given a few key environmental factors. *Habitat effectiveness (HE) models* rank resources in an area according to the degree to which maximum use or carrying capacity can be met.

An HE model was constructed for assessing habitat effectiveness for Rocky Mountain elk (*Cervus elaphus nelsonii*) winter range in the Blue Mountains of eastern Oregon and Washington (Wisdom et al. 1986). This model calculates an elk habitat effectiveness index as the geometric mean of four environmental variables, including distance from cover-forage edge, miles of road open to motorized traffic per square mile of habitat, habitat types and successional stage, and type of management treatment. The model was evaluated by Holthausen et al. (1994) by use of expert opinion.

With HSI, HC, and HE models, it is difficult to interpret if the resulting index value is intended to represent environmental conditions or population response. Also, the sensitivity of the resulting habitat index values to any one environmental variable is diminished as more variables are added to the model. This behavior is a function of the mathematics of a geometric mean model, and may not accurately reflect actual habitat use or population response. Finally, as with HSI models, HC and HE models should be used to represent relative environmental conditions and as a means of generating hypotheses about species–habitat relationships rather than

as evidence of causal relations or as reliable predictions of actual species response. Bender et al. (1996) provided a procedure for evaluating confidence intervals for HSI models; this approach may be extended to HC and HE models as well.

A related modeling approach is *habitat evaluation procedures (HEP)*. The USDI Fish and Wildlife Service has used HEP models extensively to assess environmental conditions at the species level (Flood et al. 1977). The procedure is based on *habitat units (HUs)*, which are defined as the product of habitat quality (on a 0–1 index, as from a habitat suitability index) and habitat quantity. HEP models have typically been based on HSI models that serve to estimate habitat unit scores (e.g., Cole and Smith 1983). HEP models may require much field data on specific environmental attributes, such as forage quality or quantity. However, the procedure provides a structured way to document a repeatable assessment of environmental conditions. HEP is often used to evaluate impacts of, and mitigations for, proposed projects on environmental conditions for species of special interest. Roberts and O'Neil (1985) provided a procedure for selecting species for HEP assessments. Rewa and Michael (1984) provided a way of evaluating environmental quality for ecological guilds by using a HEP approach.

Wakeley and O'Neil (1988) presented methods to increase efficiency in applying HEP. Their suggestions included delineating cover types by using remote imagery and combining types; choosing wildlife species to model for which there is available inventory information; choosing model forms that make best use of available inventory data and that focus on the most important life history components; designing field sampling for environmental conditions to be cost-effective and tailored to the range of modeled conditions; and using computers to aid in collecting and analyzing field inventory data and conducting model analysis.

POPULATION AND METAPOPULATION DEMOGRAPHY AND SIMULATION MODELS

A rather vast literature has developed in the past two decades on population viability concepts, models, and conservation (for entry, see McCullough 1996; Morris et al. 1999; Beissinger and McCullough 2002; see also chapter 3). *Population viability analysis (PVA)* usually entails using stochastic models of population demography and perhaps population genetics (usually for analyzing effects of inbreeding and genetic drift in small or isolated populations). Some popular PVA modeling packages include the RAMAS series (Ferson 2002) and VORTEX (Lacy and Kreeger 1992).

PVA can also entail individual-based, geographically referenced simulation models designed to analyze habitat patch occupancy by organisms and effects of environmental changes and disturbances on habitat patterns. In these models, individual demography is linked to location-specific environmental conditions, including quality and extent of habitat (Kareiva and Wennergren 1995; Mooij and DeAngelis 2003). Spatially explicit models predict occupancy of habitat patches in heterogeneous landscapes by breeding individuals, as well as various population and metapopulation trends. As examples, Akçakaya et al. (1995) assessed the effect of spatial patterns of habitats on viability of populations of helmeted honeyeaters (*Lichenostomus melanops cassidix*), and Raphael and Holthausen (2002) analyzed effects of habitat management alternatives on populations of northern spotted owls (*Strix occidentalis caurina*). It should be remembered, however, that such models usually do not explicitly include factors that can have strong influences on population dynamics in patchy environments, such as density-dependent survival, fecundity, and dispersal.

Brook et al. (1999) evaluated four PVA modeling packages and concluded that subtle differences among the models can affect results and conclusions. This outcome is not unexpected but is a wake-up call to those who use only one model or who rely heavily on models for decision making. It suggests that model users should fully understand their models—the state variables, functions, and especially the implications of what real-world attributes are *not* represented well or at all. Reviews of spatially explicit population models were provided by Dunning et al. (1995), Holt et al. (1995), Turner et al. (1995), and Seppelt (2005). Their general conclusions were that such models can fundamentally aid basic ecological knowledge of landscape phenomena and the application of landscape ecology to conservation and management. However, one caveat worth remembering when using spatially explicit population models is that many factors usually not explicitly included in the model—such as density-dependent demographic effects, competitors, predators, and effects of harvest—often have strong influences on population dynamics.

A number of models are available for analyzing data on population demography and census results. Examples include several models for analyzing the Breeding Bird Survey (BBS) data (Sauer et al. 2003), such as COMDYN (Hines et al. 1999).

LANDSCAPE MODELS

Landscape models include a wide variety of tools to help depict and predict habitat patch patterns across watersheds, basins, and beyond. For example, Westphal et al. (2003) used stochastic dynamic programming to design optimal landscape patterns for persistence of metapopulations of the Mount Lofty Ranges southern emu-wren (*Stipiturus malachurus intermedius*), a critically endangered bird of Australia. Knapp et al. (2003) used semiparametric logistic regression and spatial autocorrelation to estimate likelihoods that mountain yellow-legged frogs

(*Rana muscosa*) would occupy habitat patches across a landscape. Many other modeling techniques have been devised, including a wide array of metrics of habitat patch patterns used in GIS (see table 8.3; O'Neill et al. 1988; McGarigal and Marks 1995).

As with using forest stand growth and yield models to infer wildlife habitat, models designed to guide spatial scheduling of activities at the landscape scale for forest harvest, transportation infrastructures, and related pursuits are sometimes used to interpret wildlife habitat at the landscape scale. Their use carries the same caution—that is, that landscape-scale scheduling models may not produce the most useful or pertinent variables for predicting wildlife species' response with relative certainty. Examples include using models of timber harvest scheduling across multiple stands of varying ages to infer extent and connectivity of wildlife habitat (Rempel and Kaufmann 2003; Taylor et al. 2003).

A number of algorithms have been devised by which optimal or satisfactory scheduling of resource-use activities at the landscape scale are calculated. An example is from Falcão and Borghes (2002) who combined three random and systematic search heuristic algorithms to calculate the best forest management schedule under spatial constraints (also see Boston and Bettinger 2002). Again, the pertinence of such models, if used to design and evaluate landscape characteristics for wildlife presumed to be associated with represented forest conditions, needs to be evaluated with caution.

COMMUNITY STRUCTURE AND ECOSYSTEM
PROCESS MODELS

Of increasing pertinence to modeling and managing wildlife and its habitat are ecosystem process models. These models can be quite valuable for understanding the ecological roles of wildlife in structuring their communities and food webs, and in regulating flow of energy and cy-

cling of nutrients and substances within their ecosystems.

Systems modeling was developed in the 1950s under the field of operations research and was adopted by ecologists in succeeding decades (e.g., Holling 1966; Patten 1971; Grant et al. 1997). A vast literature is available on modeling food webs and trophic dynamics of ecological communities and ecosystems (e.g., Montoya and Sole 2003), with current debates in ecology focused on relations between ecosystem complexity and stability (e.g., Wardle and Grime 2003) and such dynamics of ecosystems as resilience, resistance, elasticity, and restoration (Brang 2001; Redman and Kinzig 2003). Such models are useful for managing wildlife when they reveal the specific dynamic roles played by wildlife species and diverse communities in regulating community and ecosystem function and structure (Kinzig et al. 2001). For example, Eichner and Pethig (2003) found that modeled population dynamics varied according to food chain equilibria far more than would be predicted from traditional Lotka-Volterra equations or predator–prey relations.

The ECOPATH, ECOSIM, and ECOSPACE modeling shells (Christensen and Pauly 1992; Walters et al. 1999; Pauly et al. 2000) have been used to study equilibrium conditions and trophic networks in aquatic systems. ECOPATH has been used to model ecosystem dynamics of a coral reef in French Frigate Shoals (Polovina 1984), a sandy barrier lagoon in Taiwan (Lin et al. 1999), the Newfoundland–Labrador continental shelf (Bundy et al. 2000), and other aquatic ecosystems. It may also be useful if adapted for terrestrial wildlife community and ecosystem analysis.

Other modeling tools useful for simulating time-dynamics relations of predator–prey relations, energy flow, and other aspects of community and ecosystem functioning include use of discrete difference equations in computer simulation programs such as STELLA (High

Performance Systems, Inc., Lebanon, New Hampshire). STELLA was used by Costanza et al. (1990) to model dynamics of coastal landscapes, and by Hudson (1995) to depict relations of people and wildlife (fig. 10.4).

A major manifestation of ecosystem modeling in wildlife management appears in the guise of so-called ecosystem management, which essentially is the management of selected components of a system, such as specific species and their habitats, under the assumption of benefits for the entire ecosystem (see chapter 11).

Decision Support and Knowledge-Based Models of Wildlife–Habitat Relationships

In this section we review a host of model types that deviate from the traditional population, habitat, landscape, and WHR approaches surveyed above. Decision support and knowledge-based models include decision modeling approaches used for some time now in wildlife habitat management, as well as a number of new approaches just being applied to WHR assessment and management. This field is developing remarkably fast.

DECISION SUPPORT MODELS

A major challenge in conservation is managing species and ecosystems when scientific knowledge is scant and uncertainty is great. Decision support models (DSMs) can aid this challenge by (1) evaluating the implications of uncertainty in meeting management goals, (2) combining empirical data with expert judgment, and (3) identifying key habitat elements, through sensitivity testing and validation, as a basis for prioritizing inventory and monitoring. DSMs are tools to aid decision makers; they should *not* be used as a blanket final decision, as a replacement for unclear thinking, or as an unexplainable black box.

DSMs include a wide range of tools, such as Bayesian analyses and belief networks, data and text mining, decision modeling such as decision tree analysis, expert systems, fuzzy logic and fuzzy set theory models, genetic algorithms, rule and network induction, neural networks, reliability analyses, quantitative (environmental) risk analysis, simulation and scenario modeling, and other approaches.

Successful use of DSMs in risk analysis and risk management for plant and animal conservation depends largely on the availability of data or experts, and the willingness of decision makers to articulate their risk attitudes and decision criteria. These are no small hurdles. Many DSMs can aid in merging scientific data with expert knowledge, although no model can replace empirical field studies in basic zoology, taxonomy, demography, and population genetics. DSMs can neither substitute for such work nor create knowledge and understanding where such information is initially lacking.

The appropriate role of DSMs is as a support structure. DSMs can be used to test the value and cost of additional information, the importance and influence of missing data, and the potential utility of alternative management activities. They can help evaluate the likelihood of various outcomes and the social and ecological costs and values of those outcomes. They can also provide a framework for combining information from multiple, disparate sources, including expert judgment. They can provide a structure—call it a working hypothesis—for depicting environmental correlates of species in a "causal web" of factors that affect species' distribution, abundance, and persistence.

SELECTING DSMS

Selecting decision support models for use in species-environment planning and management should follow several criteria. A good decision model should be able to combine disparate data

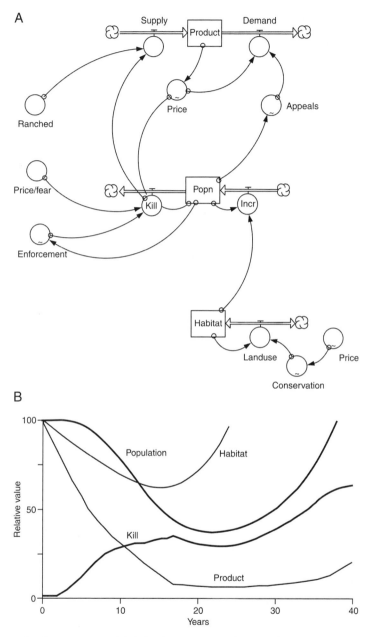

Figure 10.4. Example of a STELLA model. (A) Relations among factors influencing the dynamics of a commercially exploited wildlife population (boxes = stocks; arrows = flows; valves = rates; circles = auxiliary variables). (B) Model run illustrating recovery after an initial phase of overexploitation. (Based on Hudson 1995, 320–21.)

sources and handle missing data, combine different kinds of variables, clearly display influences, handle uncertainties and represent outcomes as probabilities, and identify major factors that can be influenced by management. A number of DSMs, such as Bayesian belief networks, provide for combining empirical data with expert judgment (Heckerman et al. 1994), often from multiple experts.

DSMs should be able to clearly display major influences on (the "causal web" of) wildlife population viability or quality of habitat, including the values and interactions of such key environmental correlates. In this way, DSMs should provide a lucid communication medium to clearly show the intuitive effects of management activities, and to express predicted outcomes as likelihoods as a basis for risk analysis and risk management. And, lastly, DSMs should be able to provide a means to test the sensitivity of management activities and to determine which key environmental correlates have the greatest influence on the organism to help prioritize inventory, monitoring, or restoration and conservation actions.

BUILDING DSMS

Part or all of most DSMs are based on expert judgment. There are specific methods for extracting and representing knowledge from experts, and this procedure has been called "knowledge engineering" in the artificial intelligence literature. It entails the process of interviewing experts, extracting their expert knowledge, quizzing them on their degree of certainty, and accurately representing that knowledge in computer models, such as in expert system control rules (Hink and Woods 1987).

The acceptability of using expert opinion for analyzing WHR has been established in a number of modeling efforts and the ecological literature. Examples include using expert opinion to interpret or supplement empirical evidence of

population trends or habitat effectiveness (e.g., Cohen 1988; Blaustein et al. 1994; Holthausen et al. 1994); to evaluate effects of environmental pollution on forest systems (de Steiguer 1990); to determine species richness of Diptera (Petersen and Meier 2003); to model vegetation and faunal distribution (Pearce et al. 2001); and to evaluate population viability of a broad array of wildlife species and taxa under alternative forest management plans (FEMAT 1993). However, use of expert judgment should be approached with due caution (Kahneman et al. 1985; Seoane et al. 2005), as it may be broader in scope but more biased and less accurate than empirical data. Using expert opinion needs an established methodology to ensure credibility and avoid bias, such as the guidelines suggested by Meyer and Booker (1990) and Cleaves (1994). When used correctly, however, expert knowledge can provide a cost-effective means of providing predictions on effects of management on biodiversity and wildlife (Martin et al. 2005).

Review of New Approaches in the DSM Arena

Following is a brief review of a number of new or more resent approaches in the DSM arena.

BAYESIAN ANALYSES AND BELIEF
NETWORK MODELING

Once the poor stepchild of traditional "frequentist" statistics, Bayesian statistical approaches to modeling WHR have been used widely to evaluate wildlife populations, effects of habitat management, and other aspects of wildlife ecology and management (Dorazio and Johnson 2003). A straightforward application of Bayes' theorem was used in the PATREC, or pattern recognition, WHR models of the 1970s, which still provide a valuable structure for analyzing species–habitat relations (see box 10.2). For example, Grubb et al. (2003) used the PATREC approach of Bayesian inference to assess nesting habitat suitability for

---

Box 10.2 *Bayesian modeling and wildlife habitat*

What to do when sample sizes are few—as with unreplicable landscapes or with threatened species with tiny populations? What to do when controls do not exist, when baseline conditions cannot be established, and when unforeseen disturbances wreck the experiment? These and other nightmares haunt many real-world wildlife studies, in which funding levels and the pace of human activities and natural perturbations seldom allow for perfect experimental designs.

One answer may lie in the use of Bayesian statistics, which provide a useful complement to other traditional approaches.

Bayesian approaches entail first describing a priori probabilities of outcomes given specific conditions, such as the specific environmental states E present with populations of specific sizes S. Priors are thus denoted as the conditional probability of the environmental condition given a specific population size, or P (E|S). For various population sizes, S, a *likelihood function* of priors can then be plotted. We will return to likelihood functions in a moment.

For a particular study area, the unconditional probabilities (overall frequency distribution) of population sizes P (S) and of environmental conditions P(E) are additional factors. Then, the posterior probability P(S|E) of predicting a population size given an environmental state can be calculated by using Bayes' theorem: P(S|E) = [P(E|S) P(S)] / P(E). A graph can then be plotted showing posterior probabilities of population size for various environmental states.

Another way of expressing Bayes' theorem (Reckhow 1990, 2056) more explicitly displays the role of null hypotheses H_O and competing alternative hypotheses H_A in relation to data x:

$$P(H_o \mid x_1, \ldots, x_n) = \frac{P(x_1, \ldots, x_n \mid H_o) \cdot P(H_o)}{P(x_1, \ldots, x_n) \cdot [P(H_o) + P(x_1, \ldots, x_n \mid H_A) \cdot P(H_A)]}.$$

In this formulation, the odds for H_o against H_A can be calculated as the ratio of the likelihood function of conditions x given the null hypothesis, to the likelihood function of conditions x given the alternative hypothesis. This odds ratio is roughly analogous to the P value in classical statistics. The basic formula can also be extended to accommodate more than one alternative hypothesis.

Major advantages of the Bayesian approach are (1) it makes use of existing knowledge or expert judgment in the estimation of the prior probabilities, and (2) it produces a useful formula that predicts outcomes in terms of likelihoods or odds.

Major complaints against the Bayesian approach are (1) prior probabilities can be biased when based on best guesses rather than on empirical research; (2) prior probability values often greatly influence the posterior probabilities, so that even minor bias or inaccuracy will change outcomes; and (3) the environmental states must be depicted in only a few, oversimplified categories.

In more complex Bayesian approaches, the "independent variable" E can be partitioned into multiple components and the unconditional and conditional likelihoods for each component evaluated separately. Other variants to the approach provide for sequential estimation of the posteriors so that biases can be reduced by successive approximations or by continuous data collection. Some authors have revised the formulae for calculating posterior probabilities under various assumptions of statistical distributions of the prior probabilities.

---

bald eagles (*Haliaeetus leucocephalus*) in the western United States. Wikle (2003) demonstrated use of a Bayesian approach to assess the probability of spread of invasive species in an analytic diffusion model of house sparrows (*Carpodacus mexicanus*) in the eastern United States. Bayesian approaches to analyzing metapopulations have been used by Goodman (2002), O'Hara et al. (2002), Wade (2002), ter Braak and

Etienne (2003), and others. Jonsen et al. (2003) used meta-analysis to combine data on animal movement from several sources in a Bayesian statistical framework.

Other uses of Bayesian statistics include sequential and hierarchical empirical Bayesian approaches (Gazey and Staley 1986; Ver Hoef 1996; Oman 2000), which incrementally incorporate new data to refine estimates of population response or other prediction variables and probabilistic relations. Empirical Bayesian approaches to learning may be useful for studying landscapes or ecosystems, which tend to be unique, meaning there is only a sample size of one (Schindler 1998), and also for studying populations of threatened or endangered species that cannot be subjected to experimental treatments and replication. Such dynamic approaches differ fundamentally from more traditional statistical correlation models that produce fixed results and that avoid use of prior information.

In fact, one of the hallmarks of Bayesian approaches in general is the use of prior knowledge to structure probabilistic relations among variables for predicting outcomes. That is, a Bayesian approach incorporates prior knowledge such as by setting *a priori probabilities* that in turn influence the calculation of new states and outcomes as *posterior probabilities*. In wildlife and environmental management, this Bayesian approach means that outcomes are represented as likelihoods, which in turn are statements of uncertainty (Toivonen et al. 2001) and which can feed directly into risk analyses. Bayesian models, however, can be rather arbitrary, in that values of priors can greatly affect posteriors, which means that unless the prior probabilities are sound and unless the model structure itself is in some way validated, the analysis can lead to faulty predictions with unknown accuracy and bias.

One aspect of Bayesian modeling that has become popular is *Bayesian belief networks (BBNs)*. In its best form, a BBN is essentially a causal model representing the major factors of some system (Marcot et al. 2001). Such causal models are also called *dependency networks* or *influence diagrams,* and they display probabilistic relations among stressor and wildlife variables (e.g., fig. 10.5). BBNs have been used for a wide variety of ecological problems, including aspen management (Haas 1991), wildlife assessment in

the interior western United States (Raphael et al. 2001), and participatory resource management (Cain et al. 1999).

DATA AND TEXT MINING

Data and text mining constitute one form of "knowledge discovery," or statistical learning approaches in which systematic patterns and correlations among variables are discerned from large data bases or documents (Hastie et al. 2001). The objective is to produce predictions. The main techniques of data and text mining are *bagging* (in which alternative data classification and regression analyses are considered); *boosting* (in which multiple models or classifiers are generated and weighted for prediction or classification); and *stacking* and *metalearning* (in which predictions from multiple models that may be very different in structure are combined).

Whereas data mining is used on numeric data, text mining is based on analysis of multiple text documents by extracting and analyzing cooccurrence of key phrases, words, and concepts. Text mining may have great utility in wildlife management for analyzing anecdotal written descriptions of historic habitat conditions, stories of wildlife encounters, and other nonquantitative information sources. We know of no such use to date.

Examples of data and text mining modeling tools are CBA (classification based on association, developed by the School of Computing at the National University of Singapore) and Weka (a collection of machine learning algorithms written as open-source software in Java code developed by the University of Waikato).

DECISION TREE ANALYSIS

Decision tree analysis is a more traditional approach than most of the other techniques discussed in this section. Decision trees typically depict a series of decisions with various chance outcomes denoted by probabilities. Fully speci-

A

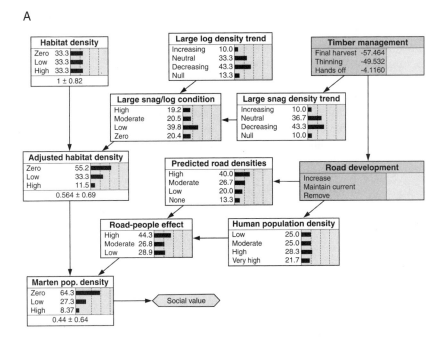

B

Adjusted habitat density	Road-people effect	Zero	Low	High
Zero	High	100.00	0.000	0.000
Zero	Moderate	100.00	0.000	0.000
Zero	Low	100.00	0.000	0.000
Low	High	50.000	50.000	0.000
Low	Moderate	20.000	80.000	0.000
Low	Low	0.000	100.00	0.000
High	High	0.000	50.000	50.000
High	Moderate	0.000	20.000	80.000
High	High	0.000	0.000	100.00

Figure 10.5. Hypothetical example of a Bayesian belief network model. (A) The model includes alternative management decisions (nodes "Timber management" and "Road development"); the intermediate habitat attributes they would affect; the final influence on a wildlife species of interest; and utilities ("social value," incorporating social benefits and economic costs) of wildlife outcomes. The numbers in the timber management node reflect overall costs of each decision, given the probability structure of the model, and the costs and benefits associated with each outcome. (B) The conditional probability table (CPT) underlying the node "Marten pop. density," illustrating the combined effects of adjusted habitat density, which accounts for management influence on snags and down logs (denning habitat for marten) and road-people effects (representing disturbance of marten). For example, if adjusted habitat density is high but road-people effects are also high, there is only a 50% likelihood of marten populations being high. Numbers in CPTs can be derived from empirical data, best professional estimates, or a combination of the two. (Model constructed by B. Marcot using the program Netica, by Norsys, Inc.)

fied decision trees also show costs or benefits of each decision, effects of chance outcomes on those values, overall utilities of final outcomes, and expected values of each decision pathway. The best decision is the one with the lowest expected cost or highest benefit. Expected values are calculated as the sum of the products of probabilities and values along a given decision pathway. An example of a hypothetical decision pathway is shown in figure 10.6 and discussed in box 10.3.

Decision trees can be crafted from expert opinion (Failing et al. 2004) or induced from data analysis (Kampichler and Platen 2004). They can be a very useful way to explore potential costs and benefits of alternative conservation actions, such as explored by Maguire et al. (1988) for conservation of black-footed ferrets (*Mustela nigripes*) in Montana. New uses of decision tree modeling include induction of decision

structures from empirical data. An example is from Stockwell et al. (1990), who induced decision trees to predict density of greater gliders (*Petauroides volans*) in Australia. However, various problems with decision tree analysis include the difficulty of representing a full decision pathway, identifying all major chance responses, estimating probabilities of outcomes, and estimating future costs and benefits of current and future decision actions.

CLASSIFICATION AND REGRESSION TREES

Classification and regression trees (CARTs) are diagrams that depict prediction variables found to have the greatest explanatory power for some response variable (Breiman et al. 1984). Classification trees are based on categorical response variables (such as categories of pelage color), whereas regression trees are based on continuous response variables (such as body length).

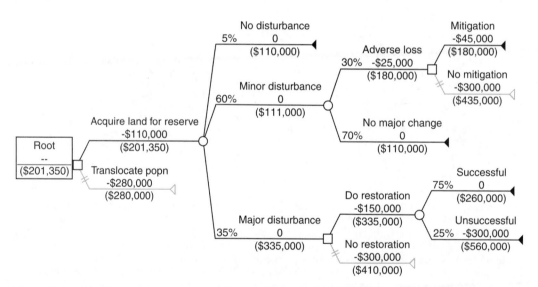

Figure 10.6. Hypothetical example of a decision tree designed to evaluate whether to translocate a threatened wildlife population or acquire land for a reserve. Boxes = decision nodes, circles = response nodes, with probabilities of various chance outcomes. Values above the lines are (hypothetical) dollar costs for each activity; and values below the lines are dollar expected values, given probabilities of chance outcomes. (Model constructed by B. Marcot by using the program DecisionPro, by Vanguard Software Corporation.)

Box 10.3   Using decision trees for conservation planning

Decision trees are used widely in many kinds of decision making. In decision tress, decision points are shown as boxes and responses are shown as circles. To illustrate, the hypothetical scenario shown in figure 10.6 is explained as follows. You have the option of translocating a threatened population for $280,000 or acquiring the land for establishing a reserve for only $110,000. If you acquire the reserve, management of the land will change to allow natural disturbance regimes to resume. You estimate the chance of no disturbance events in the near term to be low, about 5%; the chance of a minor disturbance in the near term to be about 60%; and the chance of a major disturbance to be 35%. If there is a minor disturbance, there is a further 30% chance of significant adverse loss of habitat and a 70% chance of no significant loss. The adverse loss will cost $25,000 to curtail (e.g., fighting a fire, stopping flooding), and then you will be faced with a decision of mitigating the loss for $45,000, or not mitigating it and thereby losing the population. In the case of a major disturbance event, restoration will cost $150,000 and has a 75% chance of being successful, but if you choose not to conduct restoration, it will be an unrecoverable loss, and further mitigation will cost $300,000. If you choose to do restoration and it is unsuccessful, then the population is lost and further mitigation will also cost $300,000.

The question in this scenario is, What to do?

The central initial question is, Should you purchase the land for the reserve or translocate the population? Which decision has the lowest expected cost? Which has the highest probability of population persistence regardless of the cost? and what is that associated cost?

These questions are answered by solving the decision tree, which means calculating expected values for each decision node. This solution is done by backwards calculation—that is, starting at the end of each branch of the tree and working backwards. Expected values are the running sum of the products of probabilities of chance outcomes times the cost (or benefit, if so displayed) of each event or decision. For example, in figure 10.6, on the bottom outcome of "major disturbance," the expected value (cost) of doing restoration, not knowing initially if it will be successful, is $335,000, which is (75%)($260,000) + (25%)($560,000). Note that the values $260,000 (successful outcome) and $560,000 (unsuccessful outcome) are the sums of expected values for each respective pathway through the decision tree. The best decision to be made at any given decision node in the tree (the boxes in fig. 10.6) would be the one that has the lower expected value of cost. So if there is a major disturbance, you should do restoration (expected cost = $335,000) rather than not (expected cost = $410,000).

Calculating backwards through the rest of the tree, it turns out that the expected value of acquiring the land for a reserve, given all the probabilities and costs of subsequent disturbances, losses, and mitigation, is about $201,000, whereas the expected value of translocating the population is a significantly higher $280,000 (again, this is purely a hypothetical example). So the prudent—less expensive—decision is to purchase the land.

However, the expected values can be sensitive to the accuracy of the future cost estimates, accuracy of the probabilities of outcomes, how the tree is structured in the first place, and the difference in risk among alternative decision pathways. On this last point, one should consider a decision's risk in addition to its expected value. This is told by assessing all possible outcomes that could result from an initial decision—that is, the range of expected values shown along each decision pathway in the decision tree. One aspect of risk is denoted by the joint probability of outcomes. For example, the probability of a minor disturbance *and* an adverse loss is (60%)(30%) = 18%, and the probability of a major disturbance and an unsuccessful mitigation is (35%)(25%) = 9%. These may be low-enough odds to risk the land acquisition decision if cost is less of a factor.

Another systematic way to account for risk is to incorporate a risk attitude in calculations of the expected values. The *risk attitude* of the decision maker—that is, whether the decision maker is risk-averse, risk-neutral, or risk seeking—can greatly alter expected values and best-decision pathways. Risk attitude can be determined by posing hypothetical betting scenarios to the decision maker—for example, asking how much the decision maker would be willing to receive for certain as an equivalent to a 50% chance of receiving $100,000. In our example, if the decision maker says a sure receipt of $40,000 is just as good as a 50% chance of receiving $100,000, this response defines the decision maker's risk attitude for the project. Depicting such trade-offs in a utility function can be used to modify the expected values in the decision tree. In such an example, the new expected values are now approximately $302,000 to acquire the land and $280,000 to translocate the population; clearly, translocation now costs less, so translocation is the preferred decision if cost is the main factor. If absolute numbers are hard to come by, one can also simply determine what risk attitude would be needed to equalize the expected values of both decisions and then determine if one's attitude is more or less risk seeking than that; this determination would tip the balance in favor of one decision over another.

Much has been written on variations and statistical methods and considerations in CART modeling, and CART has been used rather extensively in ecological modeling (De'ath and Fabricius 2000). As examples, Lehmkuhl et al. (2001) used a regression tree to determine which life history attributes best predict level of viability risk for a sample set of 60 wildlife species in the Pacific Northwest of the United States (fig. 10.7). Munger et al. (1998) used CART modeling to predict occurrence of Co-lumbia spotted frogs (*Rana luteiventris*) and Pacific treefrogs (*Hyla* [=*Pseudacris*] *regilla*) from U.S. National Wetland Inventory data. Grubb and King (1991) predicted effects of human disturbance on bald eagles using classification trees. Andersen et al. (2000) used regression trees to model desert tortoise (*Gopherus agassizii*) habitat in the Mojave Desert. Kintsch and Urban (2002) used classification and regression trees to determine the degree to which biotic communities could be predicted by parameters

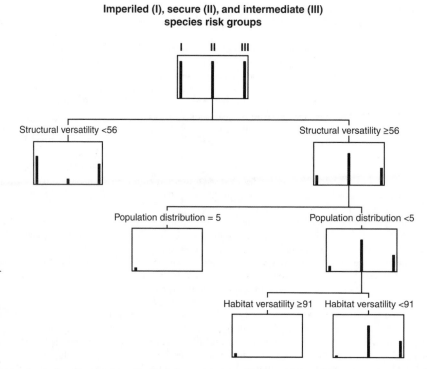

Figure 10.7. Example of regression tree modeling of three categories of species viability risk levels predicted from life history and habitat use attributes, based on 60 wildlife species in the Pacific Northwest of the United States. Structural versatility is an index of the percentage of structural condition classes of vegetation types used by each species; population distribution is an index of the geographic range of each species; and habitat versatility is an index of the percentage of vegetation types used by each species. In this model, none of the life history attributes of species, including reproductive rate, mean number of progeny, and other factors, were statistically significant predictors. (From Lehmkuhl et al. 2001, 484 [fig. 4] reprinted with permission of Oregon State University Press Copyright 2001.)

of their physical environment, and to identify focal indicator species.

De'ath (2002) proposed a method of devising multivariate regression trees to model species–environment relations, which may hold promise in much WHR work. CART analyses can be done in many general statistical software packages, although there are also specific programs developed for this purpose, such as CART and MARS (Salford Systems) and SIPINA (D. A. Zighed and R. Rakotomalala, University of Lyon).

EXPERT SYSTEMS

Once the shining future of artificial intelligence research, the use of expert systems seems to have faded in recent years. The classic framework of expert systems is in the form of if-then-else control rules that guide classification, diagnosis, or other evaluation of some condition. Control rules are based on a human expert's understanding in some narrowly defined field, are typically couched in terms of degrees of confidence or probabilities, and when chained together form an interactive system by which some problem can be solved by querying the expert's knowledge. Classic expert systems suffered from the difficulty of combining opinions from multiple experts (although procedures were eventually developed for this purpose; e.g., Clarke et al. 1990) but excelled in combining qualitative and quantitative information, in handling unknowns, and in explaining reasoning behind queries. Some early examples of the use of expert systems in WHR include Marcot (1986) and Robertson et al. (1991).

Current forms of expert systems have deviated from the classic control rule format, where expertise is now integrated into other model structures. While these are not true expert systems in the classic sense, they have advanced the field by incorporating expert knowledge into other modeling approaches, such as GIS, fuzzy

logic models, and decision analysis. Such hybrid models—perhaps now call them "expert-based systems"—have been devised for such problems as conservation of rivers (Pedroli et al. 2002) and coral reefs (O'Connor 2000). Commercially available hybrid expert system shells include DXpress (Knowledge Industries), which is used in a Bayesian-based modeling shell; CORVID (EXSYS Inc.), for building online expert advisory systems; and others.

FUZZY LOGIC AND FUZZY SET MODELS

Fuzzy logic or fuzzy set models describe the logical relations among factors that affect the degree to which some entity (such as a species) belongs to a particular set or outcome (such as having a particular level of viability). Confusing to many users, a fuzzy logic value is not the same as a probability. Values in fuzzy logic (often expressed between 0 and 1 or between −1 and +1) refer to the strength of evidence that would put some entity into some set (McNeill and Freiberger 1993)—for example, evidence that a particular wildlife population belongs to the set of threatened species (Regan and Colyvan 2000). A popular fuzzy logic modeling shell is NetWeaver (M. C. Saunders and B. J. Miller, Pennsylvania State University).

Fuzzy logic models have been devised for a variety of classification and evaluation problems, such as mapping historic forests in Michigan (Brown 1998), predicting coral reef development (Meesters et al. 1998), and prioritizing habitat management for a salamander (Pyke 2005). Fuzzy logic models have also been used in procedures for evaluating suitability of lands for conservation (Stoms et al. 2002) and have been integrated with GIS in risk assessment tools (Reynolds et al. 1997; Bojorquez-Tapia et al. 2002) (also see fig. 10.8). Advantages of a fuzzy logic model are that it is relatively easy to build; the relation between some environmental factor and a species response, for example, can be easily

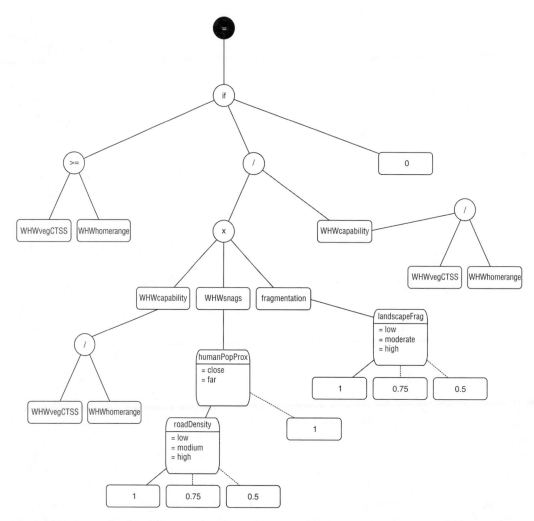

Figure 10.8. Example of the structure of a fuzzy logic model predicting density of white-headed woodpecker (WHW; *Picoides albolarvatus*) territories, using the NetWeaver fuzzy logic modeling shell (Reynolds et al. 1997). WHWvegCTSS = total area of vegetation cover types and seral stages usable by WHWs; WHWhomerange = mean home range size of a WHW territorial pair; WHWcapability = effective area of habitat divided by home range size; WHWsnags = influence of snag density on WHWcapability; humanPopProx = relative proximity of human habitats to WHW habitat; roadDensity = relative density of roads within WHW habitat; fragmentation = influence of WHW habitat fragmentation on WHWcapability; landscapeFrag = relative degree of fragmentation of WHW vegetation conditions. The functions refer to fuzzy arithmetic. (Model constructed by B. Marcot by using the program NetWeaver, by Penn State University.)

depicted in user-defined curves expressing strength of evidence. Fuzzy models are useful for quickly building knowledge-based systems to compare relative outcomes among alternative conditions. A disadvantage of fuzzy models is that it is unclear how to interpret and, especially, validate the final results, such as total strength of all evidence for some outcome, as a fuzzy value is neither a statement of probability nor a directly measurable empirical variable.

GENETIC ALGORITHMS

Genetic algorithms are computer programs designed to mimic the evolution and adaptation of genetically based populations by retaining characteristics denoted as having "adaptive advantage" and discarding maladaptive behaviors. Genetic programs include the "game of life" (also called artificial life; Stein 1991), cellular automata, and related programs (Goldberg 1988). As examples, genetic modeling has been used successfully to solve problems of scheduling optimal spatial forest harvest (Boston and Bettinger 2002), to model site selection by species (Moilanen and Cabeza 2002), and to test models that use indicator species to predict species richness (Thomson et al. 2005).

RULE AND NETWORK INDUCTION MODELS

These modeling techniques take a fully or partially specified database and induce an expert rule set that optimally (or otherwise) describes the (known) outcomes based on the descriptor variables in the examples (Jeffers 1991). For instance, a database might consist of a series of observations where a species is present or absent, along with variables representing the environmental conditions at each site. A rule induction model would produce a "key" of sorts that identifies which environmental variables best account for presence and absence of the species. In this way, rule induction models are similar to classification and regression tree models but can handle more kinds of data (categorical, continuous, binary, and ordinal) and, most usefully, missing data. A variation of the rule induction approach is used to induce a network of factors that can then be developed into an influence diagram, Bayesian belief network, loop model, decision tree, or other tools. An example of a rule induction modeling shell is See5/C5 from Rulequest Research (Australia), and a network induction modeling shell is BKD (Bayesian Knowledge Discovery) from The Open University, United Kingdom.

Rule and network induction models are very powerful for helping to make sense of a database of known examples. In this way, they are similar to data mining models but produce specific tools for prediction as well as hindcasting (explanation). A common method used in rule and network induction models is the ID3 algorithm (Quinlan 1986; also see Shapiro 1987). More recently developed is the EM algorithm used in Bayesian networks, which is applied to learning from new data to update model probabilities and to induce network structures (Lauritzen 1995; Bauer et al. 1997). As an example, Uhrmacher et al. (1997) used a fuzzy-based rule induction approach to analyze the dynamic behavior of an ecological system by generalizing system behavior and dynamics from specific cases. Lehmkuhl et al. (2001) presented an optimally generated rule to predict level of viability risk based on species' life history characteristics, induced from 60 examples species in the Pacific Northwest of the United States by using the ID3 rule induction algorithm (fig. 10.9). The rule induction approach tends to include predictor variables that best match given examples even if the correlation among variables is not statistically significant (as with regression trees). Thus the user needs to evaluate whether results are still biologically meaningful and useful, even if they are not statistically significant in the classic parametric sense.

A. Population is contiguous.
  B. The organism forages underwater or aerially, or foraging substrate is unknown.
    C. The upper elevation range of typical or regular occurrence is up to 1000 ft. (no identification)
    CC. The upper elevation range of typical or regular occurrence is up to 3000 ft. (no identification)
    CCC. The upper elevation of typical or regular occurrence is up to 5000 ft. Group III
    CCCC. The upper elevation range of typical or regular occurrence is >5000 ft. Group I
  BB. The organism does not forage underwater or aerially, and foraging substrate is not unknown.
    C. The average age at first breeding (females) is <6 months. (no identification)
    CC. The average age at first breeding (females) is 1 year. Group II
    CCC. The average age at first breeding (females) is 2 years.
      D. The structural versatility of the species is <99. Group II
      DD. The structural versatility of the species is >=99. Group III
    CCCC. The average age at first breeding (females) is 3 years. Group III
    CCCCC. The average age at first breeding (females) is 4+ years. Group II

AA. Population distribution consists of gaps.
  B. The taxonomic order is Caudata.
    C. The structural versatility of the species is <90.50. Group II
    CC. The structural versatility of the species is >90.50. Group III
  BB. The taxonomic order is Anura. (no identification)
  BBB. The taxonomic order is Squamata. Group III
  BBBB. The taxonomic order is Falconiformes.
    C. The average age at first breeding (females) is <6 months. (no identification)
    CC. The average age at first breeding (females) is 1 year. Group II
    CCC. The average age at first breeding (females) is 2 years. Group III
    CCCC. The average age at first breeding (females) is 3 years. Group I
    CCCCC. The average age at first breeding (females) is 4+ years. (no identification)
  BBBBB. The taxonomic order is Charadriiformes. Group II
  BBBBBB. The taxonomic order is Strigiformes.
    C. The habitat versatility of the species is <34.50. Group I
    CC. The habitat versatility of the species is >=34.50. and <50.00. Group III
    CC. The habitat versatility of the species is >=50.00. Group II
  BBBBBBB. The taxonomic order is Apodiformes.
    C. The habitat versatility of the species is <53.50. Group II
    CC. The habitat versatility of the species is >=53.50. Group III

Figure 10.9.   Example of a rule induction model called SARA (Species at Risk Advisor), using the ID3 rule induction algorithm (see text for explanation), of species viability risk levels predicted from life history and habitat use attributes. This model was derived from the same data set used in figure 10.7. (From Lehmkuhl et al. 2001, 486–87 [fig 6]; reprinted with permission of Oregon State University Press Copyright 2001.)

NEURAL NETWORKS

Neural networks are models that predict the value of some response variable (such as population density of some wildlife species) from a set of predictor variables (such as habitat attributes) by generating intermediate dummy variables and juggling their weights and relationships. Neural networks are constructed from a set of examples and are "trained" by iteratively readjusting weights and functions among the variables to produce a set of equations that best fit the examples. The dummy variables ("perceptrons" in neural networking parlance), created by training and solving a neural network, are intermediate combinations of the predictor variables (Kosko 1990).

Overall, neural network models do not produce explainable networks, particularly the intermediate network strata (the dummy variables), which are the critical parts of the networks that most influence the predicted outcomes. Instead, these models result in a "black

BBBBBBBB. The taxonomic order is Piciformes.
   C. The habitat versatility of the species is <50.00. Group III
   CC. The habitat versatility of the species is >+50.00. Group II
BBBBBBBBB. The taxonomic order is Passeriformes.
   C. It is a "patch" species, likely using only 1 homogenous habitat patch during the life cycle. Group III
   CC. It is a "mosaic" species, likely using an aggregate of habitat patches but 1 structural stage.
      D. The migration or seasonal movement is <100 km. (no identification)
      DD. The migration or seasonal movement is 100 – 100 km. Group II
      DDD. The migration or seasonal movement is >1000 km. Group II
      DDDD. The species is non-migratory. Group II
   CCC. It is a "generalist" species, likely using all or many patch types, & >1 structural stage. Group III
   CCCC. It is a "contrast" species, likely requiring contrast between 2 structural stages in close proximity.
   (no identification)
BBBBBBBBBB. The taxonomic order is Rodentia.
   C. The structural versatility of the species is <28.50. Group I
   CC. The structural versatility of the species is >=28.50. Group II
BBBBBBBBBBB. The taxonomic order is Carnivora. Group II

AAA. Population distribution consists of patchily distributed populations.
   B. The average age at first breeding (females) is <6 months. Group II
   BB. The average age at first breeding (females) is 1 year. Group II
   BBB. The average age at first breeding (females) is 2 years. Group III
   BBBB. The average age at first breeding (females) is 3 years.
      C. The habitat versatility of the species is <34.50. Group I
      CC. The habitat versatility of the species is >=34.50. Group II
   BBBBB. The average age at first breeding (females) is 4+ years. Group III

AAAA. Population distribution consists of isolated population(s).
   B. The migration or seasonal movement is <100 km. Group III
   BB. The migration or seasonal movement is 100 – 100 km. Group II
   BBB. The migration or seasonal movement is >1000 km. Group I
   BBBB. The species is nonmigratory. Group I

AAAAA. Population distribution is scarce.
   B. The habitat versatility of the species is <16.00. Group III
   BB. The habitat versatility of the species is >16.00. Group I

Figure 10.9. Continued

box" that may nonetheless be useful as a tool to calculate outcomes for some tasks. Neural networks are more useful for interpolating values of some continuous variables, such as topographic relief or some climate variable (Cook and Wolfe 1991; Derr and Slutz 1994; Christopherson 1997), than for extrapolating values beyond the domain of the predictor variables. They are not particularly useful for aiding understanding of the underlying ecological causal web.

Neural networks have been used in natural resource and wildlife management in a variety of problems, such as developing vegetation management plans (Deadman and Gimblett 1997),

monitoring and simulating changes in forest resources (Gimblett and Ball 1995), forecasting recreational use of wilderness areas (Pattie 1992), classifying land cover types from Landsat imagery (Skirvin and Dryden 1997), and other areas. Yeh and Li (2003) devised interesting models of urban planning by integrating GIS, cellular automata, and neural networks. Monteil et al. (2005) used neural networks to determine correlates to bird species richness in forest patches. Olden (2003) used a neural network model to predict species composition of temperate lake fish communities from habitat attributes, and suggested that such predictive models are powerful ways for explicitly considering and conserving species membership and their functional roles in the community (fig. 10.10).

Liu et al. (2003) compared two neural network modeling approaches and three traditional statistical methods for classifying forest inventory data into ecological types. They reported that accuracy of the neural network models was at least 90% and did as well as the kth-nearest-neighbor statistical classification method. However, as with all neural network models, theirs failed to explain the underlying causal web structure that accounted for the accurate results.

Still, neural networks can be rather flexible and can be combined with other modeling approaches (such as GIS; see "GIS-Based Models" section above). Ejrnaes et al. (2002) combined ordination techniques with neural network modeling to successfully predict habitat quality (in terms of species richness, nativeness, rarity, and beta diversity) of vascular plant species.

Examples of commercially available neural network modeling shells include the Neural-Works series (NeuralWare Inc., Carnegie Pennsylvania) and SNNS (Stuttgart Neural Network Simulator, University of Stuttgart).

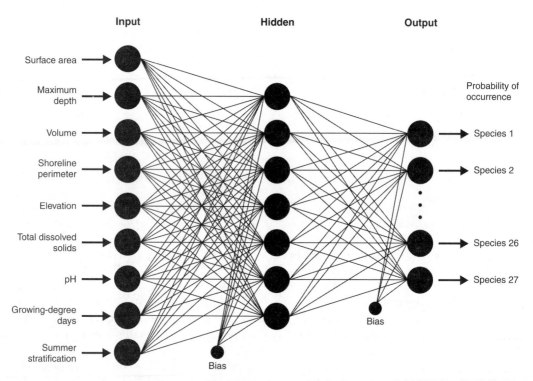

Figure 10.10.   Example of a neural network model of the presence and absence of 27 fish species as a function of nine lake habitat variables. Nodes marked "Hidden" are used to determine the best combination of weights among all input variables. Note the use of "Bias" variables that account for unexplained variance (similar to the nodes marked *?* in fig. 10.1). (From Olden 2003, 857 [fig. 1].)

RELIABILITY MODELS

Reliability analysis refers to equations that depict the probability of failure of a system component (such as a species within an ecosystem) over time. Also called survival analysis and failure time analysis, reliability modeling describes and depicts system trends and patterns more than it represents and explains the underlying influences. As used in reliability engineering, failure models are typically based on negative exponential or other probabilistic distributions of MTBF (mean time between failures) of some element (such as an electronic component; Barlow 1998). Reliability analysis is inherently probabilistically based, and in that sense corresponds well with how population viability is viewed as a probabilistic phenomenon. However, reliability models have not been used much in WHR modeling, although the probability distributions may be useful for calculating expected persistence of species or habitat conditions, given effects of stressors.

QUANTITATIVE (ENVIRONMENTAL) RISK ANALYSIS

Quantitative or environmental risk analysis, as a general topic, includes a very wide variety of models and tools used to estimate likelihoods of outcomes, given alternative management actions and subsequent environmental conditions, and to clearly display and interpret uncertainty in such decisions (Morgan and Henrion 1990). Quantitative environmental or ecological risk assessment is generally used to collect, organize, analyze, and present scientific information to improve decision making (Serveiss 2002). Tools used in risk analysis can include all of the models discussed in this chapter. Additionally, the RAMAS Risk Calc modeling shell (Ferson 2002) was specifically designed to help analyze species viability in a risk assessment framework.

There is also a formal quantitative risk analysis (QRA) methodology developed by the U.S. Environmental Protection Agency (EPA 1996). QRA has been used mostly to evaluate toxicological human health risks. QRA has also been used successfully with laboratory toxicity tests on single species to help predict impacts on aquatic ecological communities (de Vlaming and Norberg-King 1999).

LOOP ANALYSIS MODELS

Loop analysis models are used to represent the patterns of relationships among variables. Loop analysis draws from graph theory, in which variables are shown as nodes and their relations as paths or arcs between the nodes (Puccia and Levins 1985). A loop diagram can be a useful starting point to depict a causal web, as in the diagrammatic phase of modeling. A simple loop diagram can be evaluated for its qualitative patterns (Dambacher et al. 2003a, 2003b), such as degree of connectivity and stability of food webs. More sophisticated loop diagrams can also include quantitative causal modeling or path analysis. The patterns of loop diagrams can be evaluated using what is called structural equation modeling, which has been used in ecology (Iriondo et al. 2003; Pugesek et al. 2003).

An early application of loop analysis was demonstrated by Marcot and Chinn (1982), who devised graph theory measures of habitat connectivity. Examples of habitat maps represented as graphs and nearest-neighbor adjacency matrices are shown in figure 10.11. The adjacency matrix of a "habitat graph" can be mathematically analyzed for various connectivity features, such as the average juxtaposition of each habitat patch; presence of habitat patch "cut points" whose disturbance might disconnect the graph (e.g., node 5 in fig. 10.11C); overall connectivity of the graph (e.g., mean number of arcs per node); and distance (number of intervening habitat patches) between any two nodes (patches). Their analysis demonstrated that linear features such as roads and riparian areas

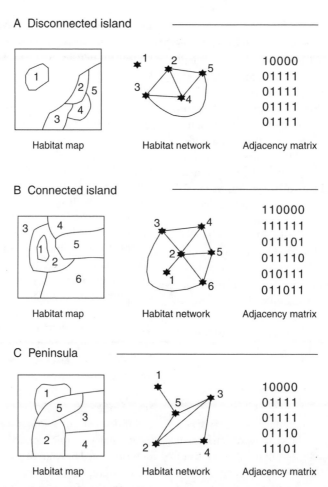

Figure 10.11. Example of a habitat map and its graph theory analog. Note how each patch on the habitat map is represented by a node on the habitat network, and patches that share a common edge are shown by a *1* in the adjacency matrix. Matrix algebra calculations can reveal further information about connectivity of individual nodes and of the overall map.

usually create cut points in a graph and serve as nodes with the highest local connectivity to other nodes. This condition may help explain the high species richness of native linear features, such as riparian areas, and the potential for disproportionately disrupting a landscape from anthropogenic linear features, such as roads and transmission line corridors.

Other authors later suggested various graph theory measures of habitat patch patterns across a landscape (e.g., Cantwell and Forman 1993; Ricotta et al. 2000; Urban and Keitt 2001). Bunn et al. (2000) demonstrated the use of graph theory to analyze the very different degrees of connectivity of a given landscape for American mink (*Mustela vison*) and prothonotary warblers (*Protonotaria citrea*). They analyzed, for each species, the functional distance between reachable habitat patches by using least-cost path modeling and discovered that habitats within their test landscape were fundamentally connected for mink and unconnected for warblers. In general, a graph theory and loop analysis approach to assessing habitat patch patterns and connectivity may have advantages over more complicated mathematical spatial indexes (see table 8.3; O'Neill et al. 1988; McGarigal and Marks 1995) by being simpler to explain and calculate, and by having more direct pertinence to movement patterns of animals.

This overview of various traditional and newer model constructs does not include a number of other approaches with potential for WHR analysis and management, such as rough set theory, a mathematical tool similar to fuzzy set theory to manage uncertainty and vagueness in data sets (Berger 2004, Tan 2005). Doubtless, many new modeling concepts and tools will continue to be developed that push the envelope of traditional statistical analysis and that may find utility for WHR assessment and management. This is an exciting time for wildlife modelers.

Models for Habitat Conservation

In this section, we briefly review two topics related to modeling habitat conservation for wildlife: models that aid optimal allocation of lands for designating conservation areas, and scenario models that aid overall wildlife habitat and land use planning.

Models for Optimizing Land Allocations

A host of modeling approaches has appeared in the past decade to aid optimal design of conservation areas. Kingsland (2002) suggested that the science of nature reserve design has bene-

fited from optimality modeling derived from operations research from the 1970s and 1980s, in a decision analysis framework. More recently, use of simulation modeling of vegetation growth and use of new optimization algorithms have provided the field of reserve design with many practical tools.

Many tools used for design of reserves and conservation areas draw from some of the modeling approaches discussed above. For example, the approach used by Strange et al. (2002) uses genetic algorithms of "evolutionary self-organization" to find the optimal solution. Guisse and Gimblett (1997) used neural networks to help map conflicting recreational impacts on state parks. Other hybrid design models were suggested by Nalle et al. (2002).

A popular model for weighing costs and values of land areas for conservation area planning is SITES, which produces alternative maps of boundaries of potential reserves to meet stated conservation objectives, such as protection of undisturbed ecosystems. SITES is complicated to run and understand, but it has utility in designing reserves and regional conservation plans (Carroll et al. 2002). An advantage of SITES is that one can specify some selected land allocation boundaries to be inviolate so as to aid conservation planning within a landscape already partitioned under existing land use plans.

Many other models of conservation area planning have been proposed, such as using a focal species (e.g., moose, *Alces alces*) to design reserves in Nova Scotia, Canada (Snaith and Beazley 2002). Rothley (2002) proposed a method of applying multiobjective integer programming (MOIP) to surrogate design criteria, such as reserve size or connectedness. McDonnell et al. (2002) used nonlinear integer programming in their algorithm to design reserves in Northern Territory, Australia. Cumming (2002) used habitat shape geometry to design reserves for staving off invasive species. Many other examples and approaches based on concepts of reserve complementarity (Williams et al. 2000), redundancy (ReVelle et al. 2002), representativeness (Powell et al. 2000), and other criteria are available in the literature.

The lessons to be drawn from such a variety of approaches are that (1) there is no one single approach to reserve design that meets all conservation area planning objectives; (2) thus it is critical to articulate those planning objectives first and not let the design model dictate them; and (3) a combination of models or trying alternative design algorithms may be a useful way to test the practicality and utility of their results.

Scenario Modeling

In one sense, most wildlife habitat planning involves exploring alternative scenarios for land allocation, habitat modification, and natural resource use. There are also more formal approaches to scenario planning useful for natural resource and wildlife conservation (Bennett et al. 2003). In this sense, a "scenario" can be a specific level of expected use of a resource or land area, and even a worldview of how resources ought to be used or conserved, such as under utilitarian, humanistic, ecological, or other ethical stances (Callicott et al. 1999). In fact, scenario modeling can even test implications of alternative worldviews (Marcot et al. 2002).

A scenario model is a type of systems model designed to understand social and ecological systems (Bennett et al. 2003). In this sense, scenario modeling of social-economic-ecological systems has been around since the 1970s (e.g., Jantsch 1972). In scenario modeling, wildlife management is viewed as part of a broader social, economic, and ecological system (Walker et al. 2002). Scenario planning can use specific kinds of models discussed above to assess effects of alternative social direction and resource use interests on wildlife conservation, ecosystem services, and human communities. Irwin and Freeman (2002) used such an approach to evaluate social and ecological implications of conservation options for the Tallapoosa River in Alabama under various adaptive management scenarios. Sheppard and Meitner (2005) evaluated forest planning scenarios with stakeholder groups by using multicriteria analysis and visualization methods.

Validating Wildlife–Habitat Relationship Models

Most of the types of models discussed in this chapter are difficult to validate. In part, this difficulty exists because the models consist of many relations and entities that confound simple prediction (Gentiol and Blake 1981). Many also represent outcomes in abstruse terms, such as unitless indexes, fuzzy set membership, Bayesian posterior likelihoods, or influence webs, which sometimes defy simple interpretation and easy comparison with empirical observation. The problem of validation of the more avant-garde modeling approaches discussed above has received limited attention in the wildlife literature.

The concept of model validation should be addressed in any modeling exercise. Analyses that cannot be falsified or otherwise rigorously tested are essentially little more than belief systems and have little to no value in science. It is therefore essential in the model-building process to represent variables and their relations as empirically quantifiable entities. For example, it is easy to build an expert system, a Bayesian belief network, or an influence diagram with vaguely worded variables such as "habitat quality" and "species response." It is more difficult, but essential, to craft such models using empirically verifiable variables, such as area of a specific vegetation condition or density of reproductive individuals.

Species-habitat models built using various kinds of complicated DSM tools often approximate habitat evaluation or suitability models (e.g., Adamus 1996) and could be evaluated by validating their specific ecological predictions. Other approaches may include using Bayesian statistics to use new information to test, calibrate, and refine certain kinds of network models. Also see Reinhardt et al. (1992), Preece (1994), and Sequeira et al. (1996) for other examples of validating decision support- and knowledge-based models.

Purposes of Validation

Validating wildlife-habitat models should be part of each step in building and using such tools (Marcot et al. 1983). Model validation is best viewed as a general approach to developing, calibrating, and testing models, and should be conducted in a variety of ways throughout the model development and application process.

Aspects of validating a model (see table 10.1) include:

1. *Verifying the model*—verifying that mathematical equations are correct or that the computer program code has been written without bugs
2. *Testing the audience*—ensuring that the audience for whom the model is intended will accept and use the tool
3. *Running the model*—confirming that the model can be run with available or obtainable data
4. *Assessing purpose and context*—ensuring that the purpose of the model and the conditions in which it is to be used have been clearly stated and adhered to in its use
5. *Testing the output*—assessing whether the output of the model matches real-world biological conditions

Model validation is typically associated with just the first and final items on this list. Each item, however, contributes to successful development and application of a wildlife–habitat model.

VERIFYING THE MODEL

Ensuring that formulae and computer code are written correctly is a simple but important aspect of model validation. A similar task is documentation. *Documentation* refers to explicitly explaining the development procedure used to create the model; writing down major assumptions and uncertainties inherent in the model; disclosing sources of information and analyses used to develop variables and their relationships in the model; and annotating any computer code. The more a model is verified, the more open it is to understanding—and critique.

Verification is an important aspect of modeling where meeting legal mandates is a concern, as in developing models for use in National Environmental Policy Act (NEPA) documents, such as environmental impact statements. In this case, keeping careful records in process documents is paramount.

TESTING THE AUDIENCE

The best model in the world may fail to be used if it is too complex or too esoteric. It will also be ignored if existing administrative organizations or policies do not provide for its use, or if it is, for some reason, not credible. In an operational sense, a model is valid, in part, if it is accepted (has face validity) and is usable in the intended work setting. Thus developing models with teams combining managers and researchers helps enhance the utility of such tools (Bunnell 1989). A team approach would help ensure that the model addresses the correct question, is based on data available from existing databases, is credible, and can be used in the everyday course of work.

In a typical management situation, a wildlife-habitat model is used to predict response of wildlife species to potential environmental conditions created from alternative management activities. For such a model to be used, it must run from information available from existing or easily obtainable inventories of vegetation and environments. This requirement may limit accuracy of model predictions, however, if inventories are dated or incomplete, or fail to include pertinent variables. In such a case, the models may be more useful for suggesting changes to inventory procedures.

At best, proxy variables might be used to represent the missing variables (Marcot et al. 2001), such as using size of a forest opening to represent a more complex representation of habitat patch juxtaposition, or using road density available as a GIS data layer to represent hunter access. When proxy variables are used, their degree of correlation with the intended predictor variable should be evaluated.

ASSESSING PURPOSE AND CONTEXT

The purpose of a model is often incompletely stated, and clarifying the intended purpose should guide how the model is used. If a model is truly intended to predict real-world environments and populations, it should be evaluated using a different set of criteria than if its purpose is to formalize our knowledge and understanding.

Also, the context of a model should be specified by the model builder. Context includes the range of environmental conditions (e.g., weather), types of environments, and seasons in which the model was built and tested. It is the onus of the model builder to document these conditions, and of the model user to adhere to those conditions. When a model is used outside its intended context, its accuracy and reliability are essentially unknown unless they have been formally evaluated.

TESTING THE OUTPUT

Too often, models created for prediction are untested against real-world situations. This problem occurs for a variety of reasons. Models are often developed without regard to validation until after they are built, and postconstruction validation costs too much time and money. Or, models are built mostly from theory and are difficult or impossible to test, even if they are used for prediction. Or, it is unclear or unspecified what the model output represents, such as with models that calculate some relative index of habitat value.

Depending on the purpose, not all models require rigorous field testing. However, the validity (and uncertainty) of models used for helping make decisions about irreversible or expensive losses of environments and populations should be known. In this case, the accuracy, bias, precision, and reliability of the model should be evaluated (e.g., see guidelines by Golbeck 1987).

A Type I error in prediction occurs when a model predicts species presence (or some other measure) and the species is actually absent. This error could occur because of inadequate or incorrect sampling for the species, because the field study was conducted during the wrong season, because the species is inherently rare and does not maximally occupy all suitable environments, or because the model was wrong and overstated the value of environmental parameters or failed to account for an environmental condition that deducts the presence of the species. The degree to which a model avoids Type I errors is given by the confidence coefficient P (where $P = 1 - \alpha$, where α is the significance level).

Contrariwise, a Type II error in prediction occurs at rate β when the model predicts absence and the species is actually present. This error could occur because the animals detected were wandering or their presence is not indicative of actual environmental quality; because of sampling design; or because the model is wrong and does not include a vital parameter that affects presence of the species. The degree to which a model avoids Type II errors is given by the power of the model, $1 - \beta$ (Steidl et al. 1997). Power provides a means for selecting models for trend analysis (Gerrodette 1987), detecting environmental impacts (Osenberg et al. 1994), and determining population declines (Strayer 1999). Tyre et al. (2003) suggested an extension of logistic modeling using a zero-inflated binomial model to better estimate Type II error rates.

Typically, the modeler must make a trade-off between reducing Type I errors and Type II errors (Roback and Askins 2005; Verhoeven et al. 2005). The ramifications of each type of error of model prediction depend on how the model is to be used. If the objective is to identify needs for mitigation, such as with purchasing or trading habitats with high opportunity costs, or by restoring or enhancing environmental conditions, then the model must accurately predict species presence. That is, frequencies of Type I errors should be minimized because costs of actions based on model predictions are high. On the other hand, if the model is to be used for predicting impacts, especially on rare or vulnerable species, then errors in predicting species presence or positive responses may be tolerable, but false predictions of species absence or negative responses might be of greater concern than in the case of mitigation. In this case, the power of a model and its ability to avoid Type II errors is critical.

In general, most habitat models can be expected to account for less than half the variation in species density or abundance. On-site environmental conditions generally account for even less variation in population density when considering migratory species, especially migratory

birds. At first it might seem that low correlations in a habitat model are not very useful, but considering the large array of other factors that affect populations, even relatively low correlations may provide useful insights.

This characteristic of low correlation is also a lesson for the manager who will use the model for maintaining environmental conditions. The manager must understand that most models that predict species presence, population density, or species richness from environmental characteristics are likely to capture only a portion, typically half or less, of the variation in those species' parameters. This low explanatory power does not mean that habitat is unimportant; it is usually critical. It means that one cannot manage for environmental conditions alone and expect with high confidence that the population will show a direct response. Another way of interpreting this low explanatory power is that by managing for (readily measurable) environmental conditions, we control only a portion of the factors that affect the occurrence and abundance of species.

Given these validation results, the appropriate use of habitat models then appears to be to help us recognize the degree (correlation) to which we can provide for species presence and abundance, and thus which environmental parameters under consideration are the more critical. Such models can also be used to assess potential (hypothetical) effects on species from alternative management scenarios. However, if such models are used specifically to predict population size, the predictions should be treated as hypotheses. Such predictions would assume that all factors not considered by the model—that is, the 50% or more of unexplained variation in occurrence or abundance—are unimportant or are at optimal values. This assumption is invariably false.

Overall, validating models is a many-faceted problem and should be done routinely as models are built and used. Validation should address the appropriateness of objectives and structure of the model; the utility, reliability, accuracy, and completeness of the model; and its credibility.

Some Methods of Model Validation

We have given some tips above about model validation. Specific methods used to test the performance of models are many, and the modeling literature about testing and validation is rich and deep. In brief, methods of model validation can be categorized as (1) those that empirically test specific model predictions, (2) those that evaluate the appropriateness of the underlying model structure, and (3) those that assess the usability of the model for its intended audience and purpose.

Empirical tests of model predictions can use various statistical methods, such as cross-validation of a data set (building the model with a randomly chosen portion of the data set, and testing it against the other half) using techniques of data bootstrapping (with replacement of the data after random selection) or jackknifing (without replacement) (see Meyer et al. 1986, Lillegard et al. 2005). Recently, use of the Akaike's information criterion (AIC) (Anderson et al. 1994) has come into favor. AIC is used to guide selection of best-fitting statistical model structures (e.g., Spendelow et al. 1995; Halley and Inchausti 2002; Rushton et al. 2004).

Evaluating the underlying model structure can be notoriously difficult, particularly if the model hides its structure well, such as with neural networks. The state variables and relations, however, should be articulated through the model verification and documentation phase. The veracity of how they are represented in the model, and how well they in turn represent accepted ecological theory, can then be evaluated

through peer review. Lastly, assessing the usability of a model for its intended purpose and audience can be done by use of questionnaires and interviews.

Another method of model validation is use of sensitivity testing, particularly for the model-building stage. By evaluating the degree of sensitivity of model output to its structure and underlying state relations (e.g., Pacala et al. 1996), one can determine whether the model is performing as expected and desired. Further validation may be necessary once the model is built and performing to specifications, however, or else the model is simply a representation of one's understanding and biases. For some modeling exercises, such as evaluating the performance of a model, this level of variation may be enough. But for use in real-world conservation, further validation may be needed to ensure that the model is calibrated correctly and is adequately representing reality. Other validation methods suggested in the literature include use of diffusion approximation to validate time series data on population counts to validate population viability models (Holmes and Fagan 2002) and use of logistic regression to analyze sensitivity of population viability analysis models (Cross and Beissinger 2001).

Examples of Validation

The past decade has seen a number of studies aimed at validating various kinds of WHR models. This increase in validation studies is most encouraging, because validation is difficult but often essential. In each of these tests, different criteria were used to test various aspects of model prediction, including the robustness of the models to being used in various ways, sensitivity of predictions to precision of input variables, and accuracy of predictions of species' abundances as compared among different seral stages. Following are examples of recent or important validation studies.

Raphael and Marcot (1986) validated for errors of omission and commission a WHR model of amphibians, reptiles, birds, and mammals in a Douglas-fir sere in California. Results suggested that, because of errors of commission (predicting more species than actually occur), WHR matrix models are probably best used to predict the occurrence of species in general vegetation types and environmental conditions across broad regions rather than at the individual-stand scale. A shortcoming of such models is that they do not quantify population response. Thus such models cannot be used to gauge population density or to quantify population trend.

Edwards et al. (1996) tested predictions from a gap analysis of terrestrial vertebrate species in national parks in Utah using long-term species lists and found error rates of omission of species to range from 0% to 25%, and those of commission to range from 4% to 33%; error rates were highest with amphibians and reptiles, lowest with birds and mammals, and lower with larger-sized parks. They concluded that the gap analysis WHR models were adequate for aiding conservation planning at the ecoregional level. Overall, errors of commission in WHR matrix models can be explained, in part, by the fact that fine-scale environmental data are typically lacking with which to "trim" predicted lists of species, which also causes errors when predicting spatial distributions of species at levels of fine-scale resolution and small geographic areas (Mackey and Lindenmayer 2001).

Stephens et al. (2002) compared a simple species matrix model of alpine marmot (*Marmota marmota*) in southern Germany with detailed empirical data from 13 years' study on behavior

and demography and found that the simplest matrix model adequately predicted population size and density under equilibrium conditions, but not under dynamically changing conditions. For the latter, it was necessary to have data on behavior—in particular, Allee effects on population dynamics.

Dettmers et al. (2002) tested the predictions of presence of forest bird species by habitats in the southern United States as modeled by a combination of published census and natural history data, field experience, and expert opinion. Using point-count survey data from three states, they found that 23% to 52% of the models correctly predicted ranks of positive associations between predictions and observation data. The models performed better in predicting species restricted to mid-aged to mature deciduous forests or to high elevations, and poorer for species in mature deciduous forests because those species were also observed to use some early-aged deciduous forest sites.

A study by Fleishman et al. (2001) of 10 data sets revealed that umbrella species (selected a priori using a recently developed index) were no more effective than randomly selected species in protecting unrelated species. Such use of umbrella species may be useful, but their work suggests caution in unbridled use of the concept without testing.

Several recent studies have focused on comparing and testing models used in population demography and viability analysis. Mills et al. (1996) evaluated four population viability analysis programs and found that idiosyncrasies among the programs of their input format, and whether and how the models handled density dependence, led to great differences in estimates of extinction rates and expected population size.

They concluded that PVAs should include at least one scenario run without density dependence and to exercise caution when interpreting PVA modeling results.

Likewise, Lindenmayer et al. (2000) tested VORTEX to predict abundance of three species of arboreal marsupials in southeastern Australia and found that model runs based only on patch area and home range size overpredicted number of occupied patches and total abundance of animals. These authors also suggested caution when using such tools to predict dynamics and response of populations in fragmented environments.

Coulson et al. (2001) cautioned that PVA models are generally unable to perform well because of the inability of precisely predicting catastrophic disturbances. However, despite their conclusion that PVA models have limited utility as absolute predictors, it is our experience that PVAs can be useful for comparing relative effects of alternative habitat management scenarios.

VALIDATION OF SINGLE-SPECIES MODELS

Fielding and Haworth (1995) tested predictions from a range of models developed using discriminant analysis and logistic regression of nest location and occupancy of golden eagles (*Aquila chrysaetos*), ravens (*Corvus corax*), and buzzards (*Buteo buteo*) in northwest Scotland. They found that the models' predictions were correct 6% to 100% of the time. They attributed this widely variable outcome in prediction success to methodological and ecological processes and cautioned that great care must be taken about making predictions from such studies and that such systems may be inherently *unpredictable.*

Cross and Beissinger (2001) compared logistic regression of demographic data with results of stochastic and deterministic demographic models of African wild dogs (*Lycaon pictus*). The logistic model results suggested that pup survival explained the most variation in the proba-

bility of extinction regardless of density dependence, whereas the standard demographic models suggested that adult survival was more important. The authors concluded that logistic regression is a useful tool for exploring sensitivity of extinction probability from vital rate parameters, although it is clear that model structure, analysis methods, and variance of the vital rate parameters all affected results.

Fecske et al. (2002) reported on a test of a habitat-relation model for American marten (*Martes americana*) in the Black Hills of South Dakota. They used stepwise logistic regression to test their model predictions of presence of the species, and found that their model correctly predicted presence in 60% of the 46 10.2-km^2 quadrats surveyed. Roloff et al. (2001) tested a spatially explicit habitat effectiveness model of Rocky Mountain elk, using two-way ANOVA to compare observed use with predicted habitat quality. They found that the elk model performed more consistently during fall than other seasons, and for subherds unaffected by recent fire. Their results suggested the need to model elk herds by season and disturbance influences, as affecting forage dynamics, topographic usage, and road effects.

Other validation studies include tests of three habitat suitability models for Kirtland's warbler in Michigan (Nelson and Buech 1996).

Conclusions

WHR models are essential partners on our journey to better understand, manage, and monitor wildlife species and communities and their environments. As natural environments on this globe become increasingly stressed and scarce under bourgeoning human populations and natural resource use, we are coming to rely more on technologies such as remote sensing imagery, on expert judgment, and on the various kinds of

WHR models to help us find acceptable balances between conservation and exploitation.

No single model is fully general, completely accurate and precise, and completely devoid of bias. As with human understanding, models will always be wrong—perhaps a better term is "not useful"—in some context. This does not invalidate models, and we should avoid the urge to cry "fatal flaw" and engage in a "battle of the models" when a prediction goes awry or when personal interests contrast with model results. Instead, we should strive to understand in which contexts a model may be most useful and the degree of that utility, and then view with due caution any application of the model outside those circumstances. Only then can managers best understand when to use a model and what confidence to place in it.

Models based on expert judgment can provide valuable information, but they should be peer reviewed and, where possible, empirically validated or at least calibrated, else they constitute little more than belief systems. We also advocate use of multiple models because models tell us as much about the modeler's biases and knowledge and the specific structure of the model as they do about the real world.

Literature Cited

*Epigraph taken from a presentation about ecosystem modeling using Ecopath with Ecosim, given at the United Nations University Fisheries Training Programme, Institute of Marine Science, Reykjavik, Iceland.December 17, 2002.

Adamus, P. R. 1996. Validating a habitat evaluation method for predicting avian richness. *Wildlife Society Bulletin* 23 (4):743–49.

Akçakaya, H. R., M. A. McCarthy, and J. L. Pearce. 1995. Linking landscape data with population viability analysis: Management options for the helmeted honeyeater. *Biological Conservation* 73:169–76.

Allen, C. R., L. G. Pearlstine, and W. M. Kitchens.

2001. Modeling viable mammal populations in gap analysis. *Biological Conservation* 99:135–44.

Andersen, M. C., J. M. Watts, J. E. Frelich, S. R. Yool, G. I. Wakefield, J. F. McCauley, and P. B. Fahnestock. 2000. Regression tree modeling of desert tortoise habitat in the central Mojave Desert. *Ecological Applications* 10 (3):890–900.

Anderson, D. R., K. P. Burnham, and G. C. White. 1994. AIC model selection in overdispersed capture-recapture data. *Ecology* 75 (6):1780–93.

Anderson, D. R., K. P. Burnham, W. R. Gould, and S. Cherry. 2001. Concerns about finding effects that are actually spurious. *Wildlife Society Bulletin* 29 (1):311–16.

Barlow, R. E. 1998. *Engineering reliability*. Philadelphia: Society for Industrial and Applied Mathematics.

Bart, J., D. R. Petit, and G. Linscombe. 1984. Field evaluation of two models developed following the habitat evaluation procedures. *Transactions of the North American Wildlife and Natural Resources Conference* 49:489–99.

Bauer, E., D. Koller, and Y. Singer. 1997. Update rules for parameter estimation in Bayesian networks. In *Proceedings of the 13th Annual Conference on Uncertainty in AI (UAI)*, 3–13. Providence, RI.

Beckage, B., W. J. Platt, and B. Panko. 2005. A climate-based approach to the restoration of fire-dependent ecosystems. *Restoration Ecology* 13 (3): 429–31.

Beisner, B. E., D. T. Haydon, and K. Cuddington. 2003. Alternative stable states in ecology. *Frontiers in Ecology and the Environment* 1 (7):376–82.

Beissinger, S. R., and D. R. McCullough, eds. 2002. *Population viability analysis*. Chicago: Univ. of Chicago Press.

Bender, L. C., G. J. Roloff, and J. B. Haufler. 1996. Evaluating confidence intervals for habitat suitability models. *Wildlife Society Bulletin* 24 (2):347–52.

Bengtsson, J., J. Ahnstrom, and A. C. Weibull. 2005. The effects of organic agriculture on biodiversity and abundance: A meta-analysis. *Journal of Applied Ecology* 42 (2):261–69.

Bennett, E. M., S. R. Carpenter, G. D. Peterson, G. S. Cumming, M. Zurek, and P. Pingali. 2003. Why global scenarios need ecology. *Frontiers in Ecology and the Environment* 1 (6):322–29.

Berger, P. A. 2004. Rough set rule induction for suitability assessment. *Environmental Management* 34 (4):546–58.

Blaustein, A. R., D. B. Wake, and W. P. Sousa. 1994. Amphibian declines: Judging stability, persistence, and susceptibility of populations to local and global extinctions. *Conservation Biology* 8 (1):60–71.

Bojorquez-Tapia, L. A., L. Juarez, and G. Cruz-Bello. 2002. Integrating fuzzy logic, optimization, and GIS for ecological impact assessments. *Environmental Management* 30 (3):418–22.

Bolnick, D. I., and E. L. Preisser. 2005. Resource competition modifies the strength of trait-mediated predator–prey interactions: A meta-analysis. *Ecology* 86 (10):2771–79.

Boonstra, R., C. J. Krebs, and T. D. Beacham. 1980. Impact of botfly parasitism on *Microtus townsendii* populations. *Canadian Journal of Zoology* 58: 1683–92.

Boston, K., and P. Bettinger. 2002. Combining tabu search and genetic algorithm heuristic techniques to solve spatial harvest scheduling problems. *Forest Science* 48 (1):35–46.

Boyce, M. S., L. L. Irwin, and R. Barker. 2005. Demographic meta-analysis: Synthesizing vital rates for spotted owls. *Journal of Applied Ecology* 42 (1):38–49.

Brang, P. 2001. Resistance and elasticity: Promising concepts for the management of protection forests in the European Alps. *Forest Ecology and Management* 145 (1–2):107–19.

Breiman, L., J. H. Friedman, and R. A. Olshein. 1984. *Classification and regression trees.* New York: Chapman & Hall.

Brook, B. W., J. R. Cannon, R. C. Lacy, C. Mirande, and R. Frankham. 1999. Comparison of the population viability analysis packages GAPPS, INMAT, RAMAS, and VORTEX for the whooping crane (*Grus americana*). *Animal Conservation* 2:23–31.

Brown, D. G. 1998. Mapping historical forest types in Baraga County Michigan, USA, as fuzzy sets. *Plant Ecology* 134:97–111.

Bundy, A., G. R. Lilly, and P. A. Shelton. 2000. *A mass balance model for the Newfoundland-Labrador Shelf.* Canadian Technical Report of Fisheries and Aquatic Sciences, 2310. Department of Fisheries and Oceans.

Bunnell, F. L. 1989. *Alchemy and uncertainty: What good are models?* General Technical Report PNW-GTR-232. Portland OR: USDA Forest Service, Pacific Northwest Research Station.

Byers, J. E., and E. G. Noonburg. 2003. Scale dependent effects of biotic resistance to biological invasion. *Ecology* 84 (6):1428–33.

Caicco, S. L., J. M. Scott, B. Butterfield, and B. Csuti. 1995. A gap analysis of the management status of the vegetation of Idaho (U.S.A.). *Conservation Biology* 9 (3):498–511.

Cain, J. D., C. H. Batchelor, and D. K. N. Waughray. 1999. Belief networks: A framework for the participatory development of natural resource management strategies. *Environment, Development and Sustainability* 1:123–33.

Callicott, J. B., L. B. Crowder, and K. Mumford. 1999. Current normative concepts in conservation. *Conservation Biology* 13 (1):22–35.

Cantwell, M. D., and R. T. T. Forman. 1993. Landscape graphs: Ecological modeling with graph theory to detect configurations common to diverse landscapes. *Landscape Ecology* 8 (4):239–55.

Carroll, C., R. F. Noss, P. C. Paquet, and N. H. Schumaker. 2003. Use of population viability analysis and reserve selection algorithms in regional conservation plans. *Ecological Applications* 13 (6): 1773–89.

Christensen, V., and D. Pauly. 1992. ECOPATH II: A software for balancing steady-state ecosystem models and calculating network characteristics. *Ecological Modelling* 61:169–85.

Christopherson, D. 1997. Artificial intelligence and weather forecasting: An update. *AI Applications* 11 (2):81–93.

Church, R., R. Gerrard, A. Hollander, and D. Stoms. 2000. Understanding the tradeoffs between site quality and species presence in reserve site selection. *Forest Science* 46 (2):157–67.

Clark, F. S., and R. B. Slusher. 2000. Using spatial analysis to drive reserve design: A case study of a national wildlife refuge in Indiana and Illinois (USA). *Landscape Ecology* 15:75–84.

Clarke, N. D., J. A. Stone, and T. J. Vyn. 1990. Conservation tillage expert system for southwestern Ontario: Multiple experts and decision techniques. *AI Applications in Natural Resource Management* 4:78–84.

Cleaves, D. A. 1994. *Assessing uncertainty in expert judgments about natural resources.* General Technical Report SO-110. New Orleans: USDA Forest Service, Southern Forest Experiment Station.

Clemen, R. T., and T. Reilly. 2001. *Making hard decisions with decision tools.* Pacific Grove, CA: Duxbury Thomson Learning.

Cohen, Y. 1988. Bayesian estimation of clutch size for scientific and management purposes. *Journal of Wildlife Management* 52 (4):787–93.

Cole, C. A., and R. L. Smith. 1983. Habitat suitability

indices for monitoring wildlife populations: An evaluation. *Transactions of the North American Wildlife and Natural Resources Conference* 48:367–75.

Cook, D. F., and M. L. Wolfe. 1991. A back-propagation neural network to predict average air temperatures. *AI Applications* 5:40–46.

Costanza, R., F. H. Sklar, and M. L. White. 1990. Modeling coastal landscape dynamics. *BioScience* 40 (2):91–107.

Coulson, T., G. M. Mace, E. Hudson, and H. Possingham. 2001. The use and abuse of population viability analysis. *Trends in Ecology and Evolution* 16:219–21.

Cross, P. C., and S. R. Beissinger. 2001. Using logistic regression to analyze the sensitivity of PVA models: A comparison of methods based on African wild dog models. *Conservation Biology* 15 (5):1335–46.

Cumming, G. 2002. Habitat shape, species invasions, and reserve design: Insights from simple models. *Conservation Ecology* 6 (1):3 [online]; www.consecol.org/vol6/iss1/art3.

Dambacher, J. M., H. W. Li, and P. A. Rossignol. 2003a. Qualitative predictions in model ecosystems. *Ecological Modelling* 161:79–93.

Dambacher, J. M., H. Luh, H. W. Li, and P. A. Rossignol. 2003b. Qualitative stability and ambiguity in model ecosystems. *American Naturalist* 161 (6):876–88.

Deadman, P. J., and H. R. Gimblett. 1997. Applying neural networks to vegetation management plan development. *AI Applications* 11 (3):107–12.

Dean, D. J., K. R. Wilson, and C. H. Flather. 1997. Spatial error analysis of species richness for a gap analysis map. *Photogrammetric Engineering and Remote Sensing* 63:1211–17.

De'ath, G. 2002. Multivariate regression trees: A new technique for modeling species-environment relationships. *Ecology* 83 (4):1105–17.

De'ath, G., and K. E. Fabricius. 2000. Classification and regression trees: A powerful yet simple technique for ecological data analysis. *Ecology* 81 (11):3178–92.

Derr, V. E., and R. J. Slutz. 1994. Prediction of El Niño events in the Pacific by means of neural networks. *AI Applications* 8 (2):51–63.

de Steiguer, J. E. 1990. Using subjective judgment to assess air pollution effects on forests. In *XIX World Congress, 5–11 August 1990: Science in forestry, IUFRO's second century*, 131–40. Quebec: Canadian IUFRO World Congress Organizing Committee.

Dettmers, R., D. A. Buehler, and K. E. Franzreb. 2002. Testing habitat-relationship models for forest birds of the southeastern United States. *Journal of Wildlife Management* 66 (2):417–24.

de Valpine, P., and A. Hastings. 2002. Fitting population models incorporating process noise and observation error. *Ecological Monographs* 72 (1):57–76.

de Vlaming, V., and T. J. Norberg-King. 1999. *A review of single species toxicity tests: Are the rests reliable predictors of aquatic ecosystem community responses?* EPA/600/R-97/114. Washington, DC: U.S. Environmental Protection Agency, Office of Research and Development.

Didham, R. K., and C. H. Watts. 2005. Are systems with strong underlying abiotic regimes more likely to exhibit alternative stable states? *Oikos* 110 (2):409–16.

Diniz-Filho, J. A. F., and M. P. D. Telles. 2002. Spatial autocorrelation analysis and the identification of operational units for conservation in continuous populations. *Conservation Biology* 16 (4):924–35.

Dorazio, R. M., and F. A. Johnson. 2003. Bayesian inference and decision theory: A framework for decision making in natural resource management. *Ecological Applications* 13 (2):556–63.

Dunning, J. B., Jr., D. J. Stewart, B. J. Danielson, B. R. Noon, T. L. Root, R. H. Lamberson, and E. E. Stevens. 1995. Spatially explicit population models: Current forms and future uses. *Ecological Applications* 5 (1):3–11.

Eadie, W. T. 1983. *Statistical methods in experimental physics*. 2nd repr. ed. Amsterdam: Elsevier Science.

Edmands, S., and C. C. Timmerman. 2003. Modeling factors affecting the severity of outbreeding depression. *Conservation Biology* 17 (3):883–92.

Edwards, T. C., Jr., E. T. Deshler, D. Foster, and G. G. Moisen. 1996. Adequacy of wildlife–habitat relation models for estimating spatial distributions of terrestrial vertebrates. *Conservation Biology* 10 (1):263–70.

Eichner, T., and R. Pethig. 2003. The impact of scarcity and abundance in food chains on species population dynamics. *Natural Resource Modeling* 16 (3): 259–303.

Ejrnaes, R., E. Aude, B. Nygaard, and B. Munier. 2002. Prediction of habitat quality using ordination and neural networks. *Ecological Applications* 12 (4): 1180–87.

EPA. 1996. Proposed guidelines for ecological risk assessment. Risk Assessment Forum, U.S. Environmental Protection Agency, Job Number 1547. Unpublished draft report EPA/630/R-95/002B. Washington, DC.

Failing, L., G. Horn, and P. Higgins. 2004. Using expert judgment and stakeholder values to evaluate adaptive management options. *Ecology and Society* 9 (1):13; www.ecologyandsociety.org/vol9/iss1/art13.

Falcão, A. O., and J. G. Borghes. 2002. Combining random and systematic search heuristic procedures for solving spatially constrained forest management scheduling models. *Forest Science* 48 (3): 608–21.

Fecske, D. M., J. A. Jenks, and V. J. Smith. 2002. Field evaluation of a habitat-relation model for the American marten. *Wildlife Society Bulletin* 30 (3):775–82.

FEMAT. 1993. *Forest ecosystem management: An ecological, economic, and social assessment.* Report of the Forest Ecosystem Management Assessment Team. Washington, DC: U.S. Government Printing Office.

Ferson, S. 2002. *RAMAS Risk Calc 4.0: Risk assessment with uncertain numbers.* Boca Raton, FL: Lewis Press.

Fielding, A. H., and P. F. Haworth. 1995. Testing the generality of bird-habitat models. *Conservation Biology* 9 (6):1466–81.

Flather, C. H., K. R. Wilson, D. J. Dean, and W. C. Mc-Comb. 1997. Identifying gaps in conservation networks: Of indicators and uncertainty in geographic-based analyses. *Ecological Applications* 7:531–42.

Fleishman, E., R. B. Blair, and D. D. Murphy. 2001. Empirical validation of a method for umbrella species selection. *Ecological Applications* 11 (5):1489–1501.

Flood, B. S., M. E. Sangster, R. S. Sparrow, and T. S. Baskett. 1977. *A handbook for habitat evaluation procedure.* Research Publication 132. Washington, DC: USDI Fish and Wildlife Service.

Fröberg, C. 1969. *Introduction to numerical analysis.* 2nd ed. Reading, MA: Addison-Wesley.

Gazey, W. J., and M. J. Staley. 1986. Population estimation from mark-recapture experiments using a sequential Bayes algorithm. *Ecology* 67 (4):941–51.

Gentiol, S., and G. Blake. 1981. Validation of complex ecosystem models. *Ecological Modelling* 14:21–38.

Gerrodette, T. 1987. A power analysis for detecting trends. *Ecology* 68:1364–72.

Gimblett, R. H., and G. L. Ball. 1995. Neural network architectures for monitoring and simulating changes in forest resource management. *AI Applications* 9 (2):103–23.

Golbeck, A. L. 1987. *Evaluating statistical validity of research reports: A guide for managers, planners, and researchers.* General Technical Report PSW-87. Berkeley, CA: USDA Forest Service, Pacific Southwest Forest and Range Experiment Station.

Goldberg, D. E. 1988. *Genetic algorithms in search, optimization, and machine learning.* Reading, MA: Addison-Wesley.

Goodman, D. 2002. Predictive Bayesian population viability analysis: A logic for listing criteria, delisting criteria, and recovery plans. In *Population viability analysis,* ed. S. R. Beissinger and D. R. McCullough, 447–69. Chicago: Univ. of Chicago Press.

Gordon, G. 1978. *System simulation.* 2nd ed. Englewood Cliffs, NJ: Prentice Hall.

Grant, W. E., E. K. Pedersen, and S. L. Marin. 1997. *Ecology and natural resource management: Systems analysis and simulation.* New York: Wiley.

Grubb, T. G., W. W. Bowerman, A. J. Bath, J. P. Giesy, and D. V. C. Weseloh. 2003. *Evaluating Great Lakes bald eagle nesting habitat with Bayesian inference.* Research Paper RMRS-RP-45. Fort Collins, CO: USDA Forest Service, Rocky Mountain Research Station.

Grubb, T. G., and R. M. King. 1991. Assessing human disturbance of breeding bald eagles with classification tree models. *Journal of Wildlife Management* 53:500–511.

Guisse, A. W., and H. R. Gimblett. 1997. Assessing and mapping conflicting recreation values in state park settings using neural networks. *AI Applications* 11 (3):79–89.

Gunderson, L. H., C. S. Holling, L. Pritchard Jr., and G. D. Peterson. 2002. Resilience of large-scale resource systems. In *Resilience and the behavior of large-scale systems,* ed. L. H. Gunderson and L. Pritchard Jr., 3–20. Washington, DC: Island Press.

Gurevitch, J., P. S. Curtis, and M. H. Jones. 2001. Meta-analysis in ecology. *Advances in Ecological Research* 32:200–247.

Haas, T. C. 1991. A Bayesian belief network advisory system for aspen regeneration. *Forest Science* 37 (2):627–54.

Hall, C. A. S., and J. W. Day. 1977. Systems and models: terms and basic principles. In *Ecosystem modeling in theory and practice,* ed. C. A. S. Hall and J. W. Day, 6–36. New York: Wiley Interscience.

Halley, J., and P. Inchausti. 2002. Lognormality in ecological time series. *Oikos* 99 (3):518–30.

Harrison, S., B. D. Inouye, and H. D. Safford. 2003. Ecological heterogeneity in the effects of grazing and fire on grassland diversity. *Conservation Biology* 17 (3):837–45.

Hassell, M. P., H. N. Comins, and R. M. May. 1991. Spatial structure and chaos in insect population dynamics. *Nature* 353:255–58.

Hastie, T., R. Tibshirani, and J. Friedman. 2001. *The elements of statistical learning: Data mining, inference, and prediction.* New York: Springer.

Heckerman, D., D. Geiger, and D. M. Chickering. 1994. Learning Bayesian networks: The combination of knowledge and statistical data. In *Uncertainty in artificial intelligence: Proceedings of the Tenth Conference,* ed. R. L. de Mantaras and D. Poole, 293–301. San Francisco: Morgan Kaufmann.

Henny, C. J., W. S. Overton, and H. M. Wight. 1970. Determining parameters for populations by using structural models. *Journal of Wildlife Management* 34:690–703.

Hewitt, J. E., S. E. Thrush, and V. J. Cummings. 2001. Assessing environmental impacts: Effects of spatial and temporal variability at likely impact scales. *Ecological Applications* 11 (5):1502–16.

Hines, J. E., T. Boulinier, J. D. Nichols, J. R. Sauer, and K. H. Pollock. 1999. COMDYN: Software to study the dynamics of animal communities using a capture-recapture approach. *Bird Study* 46 (suppl.): S209–17.

Hink, R. F., and D. L. Woods. 1987. How humans process uncertain knowledge: An introduction for knowledge engineers. *AI Magazine* 8 (3):41–53.

Holling, C. S. 1966. Strategy of building models of complex ecological systems. In *Systems analysis in ecology,* ed. K. Watts, 195–214. New York: Academic Press.

Holmes, E. E., and W. F. Fagan. 2002. Validating population viability analysis for corrupted data sets. *Ecology* 83 (9):2379–86.

Holt, R. D., S. W. Pacala, T. W. Smith, and J. Liu. 1995. Linking contemporary vegetation models with spatially explicit animal population models. *Ecological Applications* 5 (1):20–27.

Holthausen, R. S., M. J. Wisdom, J. Pierce, D. K. Edwards, and M. M. Rowland. 1994. *Using expert opinion to evaluate a habitat effectiveness model for elk in western Oregon and Washington.* Research Paper PNW-RP-479. Portland, OR: USDA Forest Service, Pacific Northwest Research Station.

Howe, H. F., and J. S. Brown. 2000. Early effects of rodent granivory on experimental forb communities. *Ecological Applications* 10 (3):917–24.

Hudson, R. J. 1995. Paths to conservation. In *Integrating people and wildlife for a sustainable future,* ed. J. A. Bissonette and P. R. Krausman, 318–22. Bethesda, MD: Wildlife Society.

Iriondo, J. M., M. J. Albert, and A. Escudero. 2003. Structural equation modelling: An alternative for assessing causal relationships in threatened plant populations. *Biological Conservation* 113 (3):367–77.

Irwin, E. R., and M. C. Freeman. 2002. Proposal for adaptive management to conserve biotic integrity in a regulated segment of the Tallapoosa River, Alabama, U.S.A. *Conservation Biology* 16 (5):1212–22.

Jaenike, J. 2002. Time-delayed effects of climate variation on host–parasite dynamics. *Ecology* 83 (4): 917–24.

Jantsch, E. 1972. *Technological planning and social futures.* London: Casell/Associated Business Programmes.

Jeffers, J. N. R. 1991. Rule induction methods in forestry research. *AI Applications* 5:37–44.

Jennings, M. D. 2000. Gap analysis: Concepts, methods, and recent results. *Landscape Ecology* 15:5–20.

Johnson, C. J., K. L. Parker, and D. C. Heard. 2001. Foraging across a variable landscape: Behavioral decisions made by woodland caribou at multiple spatial scales. *Oecologia* 128 (1):590–602.

Johnson, D. H., and T. O'Neil, eds. 2001. *Wildlife-habitat relationships in Oregon and Washington.* Corvallis: Oregon State Univ. Press.

Jonsen, I. D., R. A. Myers, and J. M. Flemming. 2003. Meta-analysis of animal movement using state-space models. *Ecology* 84 (11):3055–63.

Jonzen, N., A. Hedenstrom, C. Hjort, A. Lindstrom, P. Lundberg, and A. Andersson. 2002. Climate patterns and the stochastic dynamics of migratory birds. *Oikos* 97 (3):329–36.

Kahneman, D., P. Slovic, A. Tversky, and editors. 1985. *Judgment under uncertainty: Heuristics and biases.* Cambridge: Cambridge Univ. Press.

Kampichler, C., and R. Platen. 2004. Ground beetle occurrence and moor degradation: Modelling a bioindication system by automated decision-tree induction and fuzzy logic. *Ecological Indicators* 4:99–109.

Kareiva, P., and U. Wennergren. 1995. Connecting landscape patterns to ecosystem and population processes. *Nature* 373:299–302.

Kingsland, S. E. 2002. Creating a science of nature reserve design: Perspectives from history. *Environmental Modeling and Assessment* 7 (2):61–69.

Kintsch, J. A., and D. L. Urban. 2002. Focal species, community representation, and physical proxies as conservation strategies: A case study in the Amphibolite Mountains, North Carolina, U.S.A. *Conservation Biology* 16 (4):936–47.

Kinzig, A. P., S. W. Pacala, and D. Tilman, eds. 2001. *The functional consequences of biodiversity: Empirical progress and theoretical extensions.* Princeton, NJ: Princeton Univ. Press.

Knapp, R. A., K. R. Matthews, and O. Sarnelle. 2001. Resistance and resilience of alpine lake fauna to fish introductions. *Ecological Monographs* 71 (3): 401–21.

Knapp, R. A., K. R. Matthews, H. K. Preisler, and R. Jellison. 2003. Developing probabilistic models to predict amphibian site occupancy in a patchy landscape. *Ecological Applications* 13 (4):1069–82.

Kosko, B. 1990. *Neural networks and fuzzy systems: A dynamical systems approach to machine intelligence.* Englewood Cliffs, NJ: Prentice Hall.

Kotz, S., N. L. Johnson, and C. B. Read, eds. 1982. *Encyclopedia of statistical sciences.* Vol. 2, *Classification: Eye estimate.* New York: Wiley.

Lacy, R. C., and T. Kreeger. 1992. *Vortex users manual: A stochastic simulation of the extinction process.* Chicago: Chicago Zoological Society.

Lauritzen, S. L. 1995. The EM algorithm for graphical association models with missing data. *Computational Statistics and Data Analysis* 19 (2):191–201.

Lehmkuhl, J. F., B. G. Marcot, and T. Quinn. 2001. Characterizing species at risk. In *Wildlife–habitat relationships in Oregon and Washington,* ed. D. H. Johnson and T. A. O'Neil, 474–500. Corvallis: Oregon State Univ. Press.

Lemoine, N., and K. Bohning-Gaese. 2003. Potential impact of global climate change on species richness of long-distance migrants. *Conservation Biology* 17 (2):577–86.

Lesica, P., and S. V. Cooper. 1999. Succession and disturbance in sandhills vegetation: Constructing models for managing biological diversity. *Conservation Biology* 13 (2):293–302.

Lichstein, J. W., T. R. Simons, S. A. Shriner, and K. E. Franzreb. 2002. Spatial autocorrelation and autoregressive models in ecology. *Ecological Monographs* 72 (3):445–63.

Lillegard, M., S. Engen, and B. E. Saether. 2005. Bootstrap methods for estimating spatial synchrony of fluctuating populations. *Oikos* 109 (2):342–50.

Lin, H. J., K. T. Shao, S. R. Kuo, H. L. Hsieh, S. L. Wong, I. M. Chen, W. T. Lo, and J. J. Hung. 1999. A trophic model of a sandy barrier lagoon at Chiku in southwestern Taiwan. *Estuarine, Coastal and Shelf Science* 48 (5):575–88.

Lindenmayer, D. B., R. C. Lacy, and M. L. Pope. 2000. Testing a simulation model for population viability analysis. *Ecological Applications* 10 (2):580–97.

Liu, C., L. Zhang, C. J. Davis, D. S. Solomon, T. B. Brann, and L. E. Caldwell. 2003. Comparison of neural networks and statistical methods in classification of ecological habitats using FIA data. *Forest Science* 49 (4):619–31.

Logan, J. A., J. Régnière, and J. A. Powell. 2003. Assessing the impacts of global warming on forest pest dynamics. *Frontiers in Ecology and the Environment* 1 (3):130–37.

Louda, S. M., R. W. Pemberton, M. T. Johnson, and P. A. Follett. 2003. Nontarget effects: The Achilles' heel of biological control? Retrospective analyses to reduce risk associated with biocontrol introductions. *Annual Review of Entomology* 48:365–96.

Ludwig, D., B. Walker, and C. S. Holling. 1997. Sustainability, stability, and resilience. *Conservation Ecology* 1 (1):7 [online]; www.consecol.org/vol1/iss1/art7.

Lundquist, J. E., and J. S. Beatty. 2002. A method for characterizing and mimicking forest canopy gaps caused by different disturbances. *Forest Science* 48 (3):582–94.

Mackey, B. G., and D. B. Lindenmayer. 2001. Towards a hierarchical framework for modelling the spatial distribution of animals. *Journal of Biogeography* 28:1147–66.

Manolis, J. C., D. E. Andersen, and F. J. Cuthbert. 2001. Patterns in clearcut edge and fragmentation effect studies in northern hardwood-conifer landscapes: Retrospective power analysis and Minnesota results. *Wildlife Society Bulletin* 28 (4):1088–1101.

Maguire, L. A., T. W. Clark, R. Crete, J. Cada, C. Groves, M. L. Shaffer, and U. S. Seal. 1988. Black-footed ferret recovery in Montana: A decision analysis. *Wildlife Society Bulletin* 16:111–20.

Marcot, B. G. 1986. Use of expert systems in wildlife-habitat modeling. In *Wildlife 2000: Modeling habi-*

tat relationships of terrestrial vertebrates, ed. J. Verner, M. L. Morrison, and C. J. Ralph, 145–50. Madison: Univ. of Wisconsin Press.

Marcot, B. G. 1998. Selecting appropriate statistical procedures and asking the right questions: A synthesis. In Statistical methods for adaptive management studies, ed. V. Sit and B. Taylor, 129–42. Victoria, BC: B.C. Ministry of Forests Research Branch;   www.for.gov.bc.ca/hfd/pubs/docs/lmh/lmh42.htm.

Marcot, B. G., and P. Z. Chinn. 1982. Use of graph theory measures for assessing diversity of wildlife habitat. In Mathematical models of renewable resources. Proceedings of the First Pacific Coast Conference on Mathematical Models of Renewable Resources, ed. R. Lamberson, 69–70. Arcata, CA: Humboldt State University Mathematical Modeling Group.

Marcot, B. G., M. G. Raphael, and K. H. Berry. 1983. Monitoring wildlife habitat and validation of wildlife–habitat relationships models. Transactions of the North American Wildlife and Natural Resources Conference 48:315–29.

Marcot, B. G., R. S. Holthausen, M. G. Raphael, M. M. Rowland, and M. J. Wisdom. 2001. Using Bayesian belief networks to evaluate fish and wildlife population viability under land management alternatives from an environmental impact statement. Forest Ecology and Management 153 (1–3):29–42.

Marcot, B. G., W. E. McConnaha, P. H. Whitney, T. A. O'Neil, P. J. Paquet, L. E. Mobrand, G. R. Blair, L. C. Lestelle, K. M. Malone, and K. I. Jenkins. 2002. A multi-species framework approach for the Columbia River Basin: Integrating fish, wildlife, and ecological functions. Report prepared for the Northwest Power and Conservation Council, Portland, OR; www.edthome.org/framework.

Martin, T. G., P. M. Kuhnert, K. Mengersen, and H. P. Possingham. 2005. The power of expert opinion in ecological models using Bayesian methods: Impact of grazing on birds. Ecological Applications 15 (1):266–80.

McCullough, D. R. 1996. Metapopulations and wildlife conservation. Washington, DC: Island Press.

McCune, B., S. D. Berryman, J. H. Cissel, and A. I. Gitelman. 2003. Use of a smoother to forecast occurrence of epiphytic lichens under alternative forest management plans. Ecological Applications 13 (4):1110–23.

McDonald, T. L., and L. L. McDonald. 2003. A new ecological risk assessment procedure using resource selection models and geographic information systems. Wildlife Society Bulletin 30 (4):1015–21.

McDonnell, M. D., H. P. Possingham, I. R. Ball, and E. A. Cousins. 2002. Mathematical models for spatially cohesive reserve design. Environmental Modeling and Assessment 7 (2):107–14.

McGarigal, K., and B. J. Marks. 1995. FRAGSTATS: Spatial pattern analysis program for quantifying landscape structure. General Technical Report PNW-GTR-351. Portland, OR: USDA Forest Service, Pacific Northwest Research Station.

McNeill, D., and P. Freiberger. 1993. Fuzzy logic. New York: Simon & Schuster.

Meesters, E. H., R. P. M. Bak, S. Westmacott, M. Ridgley, and S. Dollar. 1998. A fuzzy logic model to predict coral reef development under nutrient and sediment stress. Conservation Biology 12 (5):957–65.

Meyer, J. S., C. G. Ingersoll, L. L. McDonald, and M. S. Boyce. 1986. Estimating uncertainty in population growth rates: Jackknife vs. bootstrap techniques. Ecology 67:1156–66.

Meyer, M. A., and J. M. Booker. 1990. Eliciting and analyzing expert judgment: A practical guide. Washington, DC: U.S. Nuclear Regulatory Commission, Office of Nuclear Regulatory Research, Division of Systems Research.

Miller, D. H., A. L. Jensen, and J. H. Hammill. 2002. Density dependent matrix model for gray wolf population projection. Ecological Modelling 15: 271–78.

Mills, L. S., S. G. Hayes, C. Baldwin, M. J. Wisdom, J. Citta, D. J. Mattson, and K. Murphy. 1996. Factors leading to different viability predictions for a grizzly bear data set. Conservation Biology 10 (3): 863–73.

Moilanen, A., and M. Cabeza. 2002. Single-species dynamic site selection. Ecological Applications 2 (3):913–26.

Moller, A. P., and M. D. Jennions. 2002. How much variance can be explained by ecologists and evolutionary biologists? Oecologia 132 (4):492–500.

Monteil, C., M. Deconchat, and G. Balent. 2005. Simple neural network reveals unexpected patterns of bird species richness in forest fragments. Landscape Ecology 20 (5):513–27.

Montoya, J. M., and R. V. Sole. 2003. Topological properties of food webs: From real data to community assembly models. Oikos 102 (3):614–22.

Mooij, W. M., and D. L. DeAngelis. 2003. Uncertainty

in spatially explicit animal dispersal models. *Ecological Applications* 13 (3):794–805.

Morgan, M. G., and M. Henrion. 1990. *Uncertainty: A guide to dealing with uncertainty in quantitative risk and policy analysis.* Cambridge: Cambridge Univ. Press.

Morris, W., D. Doak, M. Groom, P. Kareiva, J. Fieberg, L. Gerber, P. Murphy, and D. Thomson. 1999. *A practical handbook for population viability analysis.* Washington, DC: The Nature Conservancy.

Morris, W. F. 1990. Problems in detecting chaotic behavior in natural populations by fitting simple discrete models. *Ecology* 71:1849–162.

Motta, R. 2003. Ungulate impact on rowan (*Sorbus aucuparia* L.) and Norway spruce (*Picea abies* (L.) Karst.) height structure in mountain forests in the eastern Italian Alps. *Forest Ecology and Management* 181(1–2):139–50.

Munger, J. C., M. Gerber, K. Madrid, M. Carroll, W. Petersen, and L. Heberger. 1998. U.S. National Wetland Inventory classifications as predictors of the occurrence of Columbia spotted frogs (*Rana luteiventris*) and Pacific treefrogs (*Hyla regilla*). *Conservation Biology* 12 (2):320–30.

Nalle, D. J., J. L. Arthur, and J. Sessions. 2002. Designing compact and contiguous reserve networks with a hybrid heuristic algorithm. *Forest Science* 48 (1):59–68.

NCASI. 1996. *The National Gap Analysis Program: Ecological assumptions and sensitivity to uncertainty.* Technical Bulletin No. 720. Research Triangle Park, NC: National Council of the Paper Industry for Air and Stream Improvement.

Nelson, J. 2003. Forest-level models and challenges for their successful application. *Canadian Journal of Forest Research* 33:422–29.

Nelson, M. D., and R. R. Buech. 1996. A test of 3 models of Kirtland's warbler habitat suitability. *Wildlife Society Bulletin* 24 (1):89–97.

O'Connor, R. J. 2000. Expert systems, fuzzy logic, and coral reef development under environmental stress. *Conservation Biology* 14 (3):904–6.

O'Hara, R. B., E. Arjas, H. T. T. Toivonen, and I. Hanski. 2002. Bayesian analysis of metapopulation data. *Ecology* 83 (9):2408–15.

Olden, J. D. 2003. A species-specific approach to modeling biological communities and its potential for conservation. *Conservation Biology* 17 (3):854–63.

Oman, S. D. 2000. Minimax hierarchical empirical Bayes estimation in multivariate regression. *Journal of Multivariate Analysis* 80 (2):285–301.

O'Neil, T. A., P. Bettinger, B. G. Marcot, W. Luscombe, G. Koeln, H. Bruner, C. Barrett, J. Gaines, and S. Bernatas. 2005. In *Wildlife techniques manual*, 6th ed., ed. C. E. Braun, 418–47. Washington, DC: Wildlife Society. O'Neill, R. V., J. R. Krummel, R. H. Gardner, G. Sugihara, B. Jackson, D. L. DeAngelis, B. T. Milne, M. G. Turner, B. Zygmunt, S. W. Christensen, V. H. Dale, and R. L. Graham. 1988. Indices of landscape pattern. *Landscape Ecology* 1 (3):153–62.

Osenberg, C. W., R. J. Schmitt, S. J. Holbrook, K. E. Abu-Saba, and A. R. Flegal. 1994. Detection of environmental impacts: Natural variability, effect size, and power analysis. *Ecological Applications* 4 (1):16–30.

Pacala, S. W., C. D. Canham, J. Saponara, J. A. Silander Jr., R. K. Kobe, and E. Ribbens. 1996. Forest models defined by field measurements: Estimation, error analysis and dynamics. *Ecological Monographs* 66 (1):1–43.

Paine, R. T., M. J. Tegner, and E. A. Johnson. 1998. Compounded perturbations yield ecological surprises. *Ecosystems* 1:535–45.

Patten, B. C. 1971. *Systems analysis and simulation in ecology.* 3 vols. New York: Academic Press.

Pattie, D. C. 1992. Using neural networks to forecast recreation in wilderness areas. *AI Applications* 6 (2):57–59.

Pauly, D., V. Christensen, and C. Walters. 2000. Ecopath, ecosim, and ecospace as tools for evaluating ecosystem impact of fisheries. *ICES Journal of Marine Science* 57:697–706.

Pearce, J. L., K. Cherry, M. Drielsma, S. Fierrier, and G. Whish. 2001. Incorporating expert opinion and fine-scale vegetation mapping into statistical models of faunal distribution. *Journal of Applied Ecology* 38 (2):412–24 .

Pedroli, B., G. de Blust, K. van Looy, and S. van Rooij. 2002. Setting targets in strategies for river restoration. *Landscape Ecology* 17 (1 suppl.):5–18.

Peek, M. S., A. J. Leffler, S. D. Flint, and R. J. Ryel. 2003. How much variance is explained by ecologists? Additional perspectives. *Oecologia* 137 (2):161–70.

Pena, D. 1997. Combining information in statistical modeling. *American Statistician* 51 (4):326–32.

Perrin, N., and V. Mazalov. 1999. Dispersal and inbreeding avoidance. *American Naturalist* 154:282–92.

Petersen, F. T., and R. Meier. 2003. Testing species-richness estimation methods on single-sample collection data using the Danish Diptera. *Biodiversity and Conservation* 12 (4):667–86.

Pielke, R. A., and R. T. Conant. 2003. Best practices in prediction for decision-making: lessons from the atmospheric and earth sciences. *Ecology* 84 (6):1351–58.

Polovina, J. J. 1984. Model of a coral reef ecosystem. The ECOPATH model and its application to French Frigate Shoals. *Coral Reefs* 3:1–11.

Powell, G. V. N., J. Barborak, and M. Rodriguez-S. 2000. Assessing representativeness of protected natural areas in Costa Rica for conserving biodiversity: A preliminary gap analysis. *Biological Conservation* 93:35–41.

Preece, A. D. 1994. Validation and verification of knowledge-based systems. *AI Magazine* 15 (1):65–66.

Probst, J. R., and J. Weinrich. 1993. Relating Kirtland's warbler population to changing landscape composition and structure. *Landscape Ecology* 8 (4): 257–71.

Puccia, C. J., and R. Levins. 1985. *Qualitative modeling of complex systems: an introduction to loop analysis and time averaging.* Cambridge, MA: Harvard Univ. Press.

Pugesek, B.H., A. Tomer, and A. von Eye. 2003. *Structural equation modeling: Applications in ecological and evolutionary biology.* Cambridge: Cambridge Univ. Press.

Pyke, C. R. 2005. Assessing suitability for conservation action: Prioritizing interpond linkages for the California tiger salamander. *Conservation Biology* 19(2):492–503.

Quinlan, J. R. 1986. Induction of decision trees. *Machine Learning* 1 (1):81–106.

Raphael, M. G., and R. S. Holthausen. 2002. The use of demographic data and a spatially explicit population model to analyze effects of habitat management on northern spotted owls. In *Predicting species occurrences: Issues of scale and accuracy,* ed. J. M. Scott, P. J. Heglund, M. L. Morrison, M. Raphael, J. Haufler, and B. Wall, 701–712. Washington, DC: Island Press.

Raphael, M. G., and B. G. Marcot. 1986. Validation of a wildlife–habitat-relationships model: Vertebrates in a Douglas-fir sere. In *Wildlife 2000: Modeling habitat relationships of terrestrial vertebrates,* ed. J. Verner, M. L. Morrison, and C. J. Ralph, 129–38. Madison: Univ. of Wisconsin Press.

Raphael, M. G., M. J. Wisdom, M. M. Rowland, R. S. Holthausen, B. C. Wales, B. G. Marcot, and T. D.

Rich. 2001. Status and trends of habitats of terrestrial vertebrates in relation to land management in the interior Columbia River Basin. *Forest Ecology and Management* 153 (1–3):63–87.

Reckhow, K. H. 1990. Bayesian inference in non-replicated ecological studies. *Ecology* 71:2053–59.

Regan, H. M., and M. Colyvan. 2000. Fuzzy sets and threatened species classification. *Conservation Biology* 14 (4):1197–99.

Redman, C. L., and A. P. Kinzig. 2003. Resilience of past landscapes: Resilience theory, society, and the Longue Durée. *Conservation Ecology* 7 (1) [online]; www.consecol.org/vol7/iss1/art14/index.html.

Reinhardt, E. D., A. H. Wright, and D. H. Jackson. 1992. Development and validation of a knowledge-based system to design fire prescriptions. *AI Applications* 6 (4):3–14.

Rempel, R. S., and C. K. Kaufmann. 2003. Spatial modeling of harvest constraints on wood supply versus wildlife habitat objectives. *Environmental Management* 32 (3):334–47.

ReVelle, C. S., J. C. Williams, and J. J. Boland. 2002. Counterpart models in facility location science and reserve selection science. *Environmental Modeling and Assessment* 7 (2):71–80.

Rewa, C. A., and E. D. Michael. 1984. Use of habitat evaluation procedures (HEP) in assessing guild habitat value. *Transactions of the Northeast Section of the Wildlife Society* 41:122–29.

Reynolds, K., J. Slade, M. Saunders, and B. Miller. 1997. *NetWeaver for EMDS version 1.0 user guide: A knowledge base development system.* Portland, OR: USDA Forest Service, Pacific Northwest Research Station.

Reynolds, K. M., M. Jensen, J. Andreasen, and I. Goodman. 2000. Knowledge-based assessment of watershed condition. *Computers and Electronics in Agriculture* 27:315–33.

Ricotta, C., A. Stanisci, G. C. Avena, and C. Blasi. 2000. Quantifying the network connectivity of landscape mosaics: A graph-theoretical approach. *Community Ecology* 1 (1):89–94.

Roback, P. J., and R. A. Askins. 2005. Judicious use of multiple hypothesis tests. *Conservation Biology* 19 (1):261–67.

Roberts, R. H., and L. J. O'Neil. 1985. Species selection for habitat assessments. *Transactions of the North American Wildlife and Natural Resources Conference* 50:352–62.

Robertson, D., A. Bundy, R. Muetzelfeldt, M. Haggith, and M. Uschold. 1991. *Eco-logic: Logic-based ap-*

proaches to ecological modelling. Cambridge, MA: Massachusetts Institute of Technology Press.

Roloff, G. J., J. J. Millspaugh, R. A. Gitzen, and G. C. Brundige. 2001. Validation tests of a spatially explicit habitat effectiveness model for Rocky Mountain elk. Journal of Wildlife Management 65 (4):899–914.

Root, K. V. 1998. Evaluating the effects of habitat quality, connectivity, and catastrophes on a threatened species. Ecological Applications 8 (3):854–65.

Rothley, K. D. 2002. Dynamically-based criteria for the identification of optimal bioreserve networks. Environmental Modeling and Assessment 7 (2):123–28.

Rowland, M. M., M. J. Wisdom, D. H. Johnson, B. C. Wales, J. P. Copeland, and F. B. Edelmann. 2003. Evaluation of landscape models for wolverines in the interior northwest, United States of America. Journal of Mammalogy 84 (1):92–105.

Rushton, S. P., S. J. Ormerod, and G. Kerby. 2004. New paradigms for modelling species distributions? Journal of Applied Ecology 41:193–200.

Sauer, J. R., J. E. Hines, and J. Fallon. 2003. The North American Breeding Bird Survey: Results and analysis 1966–2002. Version 2003.1. Laurel, MD: USGS Patuxent Wildlife Research Center; www.pwrc.usgs.gov.

Schamberger, M., A. H. Farmer, and J. W. Terrell. 1982. Habitat suitability index models: Introduction. FWS/OBS-82/10. Washington, DC: USDI Fish and Wildlife Service.

Schelhaas, M. J., G. J. Nabuurs, M. Sonntag, and A. Pussinen. 2002. Adding natural disturbances to a large-scale forest scenario model and a case study for Switzerland. Forest Ecology and Management 167 (1–3):13–26.

Schindler, D. W. 1998. Whole-ecosystem experiments: Replication versus realism: The need for ecosystem-scale experiments. Ecosystems 1:323–34.

Schroeder, R. L. 1983. Habitat suitability index models: Yellow warbler. FWS/OBS-82/10.27. Washington, DC: USDI Fish and Wildlife Service.

Schulte, L. A., and D. J. Mladenoff. 2001. The original US Public Land Survey Records: Their use and limitations in reconstructing resettlement vegetation. Journal of Forestry 99 (10):5–11.

Seabloom, E. W., and S. A. Richards. 2003. Multiple stable equilibria in grasslands mediated by herbivore population dynamics and foraging behavior. Ecology 84 (11):2891–2904.

Seoane, J., J. Bustamante, and R. Diaz-Delgado. 2005.

Effect of expert opinion on the predictive ability of environmental models of bird distribution. Conservation Biology 19 (2):512–22.

Seppelt, R. 2005. Simulating invasions in fragmented habitats: Theoretical considerations, a simple example and some general implications. Ecological Complexity 2 (3):219–31.

Sequeira, R. A., J. L. Willers, and R. L. Olson. 1996. Validation of a deterministic model-based decision support system. AI Applications 10 (1):25–40.

Serveiss, V. B. 2002. Applying ecological risk principles to watershed assessment and management. Environmental Management 29:145–54.

Shapiro, A. D. 1987. Structured induction in expert systems. Reading, MA: Addison-Wesley.

Sheppard, S. R., and M. Meitner. 2005. Using multicriteria analysis and visualization for sustainable forest management planning with stakeholder groups. Forest Ecology and Management 207 (1–2): 171–87.

Shipley, B. 2002. Cause and correlation in biology: A user's guide to path analysis, structural equations and causal inference. New York: Cambridge Univ. Press.

Skarpe, C. 1991. Spatial patterns and dynamics of woody vegetation in an arid savanna. Journal of Vegetation Science 2 (4):565–72.

Skirvin, S. M., and G. Dryden. 1997. Classification of Landsat thematic mapper image data, Chiricahua National Monument, Arizona. AI Applications 11 (3):90–98.

Snaith, T. V., and K. F. Beazley. 2002. Moose (Alces alces americana [Gray Linnaeus Clinton] Peterson) as a focal species for reserve design in Nova Scotia, Canada. Natural Areas Journal 22 (3):235–40.

Spendelow, J. A., J. D. Nichols, I. C. T. Nisbet, H. Hays, G. D. Cormons, J. Burger, C. Safina, J. E. Hines, and M. Gochfeld. 1995. Estimating annual survival and movement rates of adults within a metapopulation of roseate terns. Ecology 76 (8):2415–28.

Steidl, R. J., J. P. Haynes, and E. Schauber. 1997. Statistical power analysis in wildlife research. Journal of Wildlife Management 61 (2):270–79.

Stein, R. M. 1991. Real artificial life. Byte 16 (1): 289–98.

Stephens, P. A., F. Frey-roos, W. Arnold, and W. J. Sutherland. 2002. Model complexity and population predictions: The alpine marmot as a case study. Journal of Animal Ecology 71 (2):343–61.

Stockwell, D. R. B., S. M. Davey, J. R. Davis, and I. R. Noble. 1990. Using induction of decision trees to predict greater glider density. *AI Applications in Natural Resource Management* 4 (4):33–43 .

Stoms, D. M., J. M. McDonald, and F. W. Davis. 2002. Fuzzy assessment of land suitability for scientific research reserves. *Environmental Management* 29 (4):545–58.

Strange, N., H. Meilby, and B. J. Thorsen. 2002. Optimization of land use in afforestation areas using evolutionary self-organization. *Forest Science* 48 (3):543–55.

Strayer, D. L. 1999. Statistical power of presence-absence data to detect population declines. *Conservation Biology* 13 (5):1034–38.

Stromayer, K. A. K., and R. J. Warren. 1997. Are overabundant deer herds in the eastern United States creating alternate stable states in forest plant communities? *Wildlife Society Bulletin* 25 (2):227–34.

Tan, R. R. 2005. Rule-based life cycle impact assessment using modified rough set induction methodology. *Environmental Modelling & Software* 20 (5):509–13.

Taylor, M., and J. S. Carley. 1988. Life table analysis of age structured populations in seasonal environments. *Journal of Wildlife Management* 52:366–73.

Taylor, R. J., T. Regan, H. Regan, M. Burgman, and K. Bonham. 2003. Impacts of plantation development, harvesting schedules and rotation lengths on the rare snail *Tasmaphena lamproides* in northwest Tasmania: A population viability analysis. *Forest Ecology and Management* 175 (1–3):455–66.

ter Braak, C. J. F., and R. S. Etienne. 2003. Improved Bayesian analysis of metapopulation data with an application to a tree frog metapopulation. *Ecology* 84 (1):231–41.

Thompson, I. D., J. A. Baker, and M. Ter-Mikaelian. 2003. A review of the long-term effects of post-harvest silviculture on vertebrate wildlife, and predictive models, with an emphasis on boreal forests in Ontario, Canada. *Forest Ecology and Management* 177 (1–3):441–69.

Thomson, J. R., E. Fleishman, R. Mac Nally, and D. S. Dobkin. 2005. Influence of the temporal resolution of data on the success of indicator species models of species richness across multiple taxonomic groups. *Biological Conservation* 124 (4):503–18.

Toivonen, H. T. T., H. Mannila, A. Korhala, and H. Olander. 2001. Applying Bayesian statistics to organism-based environmental reconstruction. *Ecological Applications* 11 (2):618–30.

Torti, V. M., and P. O. Dunn. 2005. Variable effects of climate change on six species of North American birds. *Oecologia* 145 (3):486–95.

Turner, M. B., G. J. Arthaud, R. T. Engstrom, S. J. Hejl, J. Liu, S. Loeb, and K. McKelvey. 1995. The usefulness of spatially explicit population models in land management. *Ecological Applications* 5 (1):12–16.

Tyre, A. J., B. Tenhumberg, S. A. Field, D. Niejalke, K. Parris, and H. P. Possingham. 2003. Improving precision and reducing bias in biological surveys: Estimating false-negative error rates. *Ecological Applications* 13 (6):1790–1801.

Uhrmacher, A. M., F. E. Cellier, and R. J. Frye. 1997. Applying fuzzy-based inductive reasoning to analyze qualitatively the dynamic behavior of an ecological system. *AI Applications* 11 (2):1–10.

Urban, D., and T. Keitt. 2001. Landscape connectivity: A graph-theoretic perspective. *Ecology* 82 (5):1205–18.

Ver Hoef, J. M. 1996. Parametric empirical Bayes methods for ecological applications. *Ecological Applications* 6 (4):1047–55.

Verhoeven, K. J., K. L. Simonsen, and L. M. McIntyre. 2005. Implementing false discovery rate control: Increasing your power. *Oikos* 108 (3):653–47.

Verner, J., and A. S. Boss. 1980. *California wildlife and their habitats: Western Sierra Nevada*. General Technical Report PSW-37. Berkeley: USDA Forest Service, Pacific Southwest Forest and Range Experiment Station.

Wade, P. R. 2002. Bayesian population viability analysis. In *Population viability analysis*, ed. S. R. Beissinger and D. R. McCullough, 213–38. Chicago: Univ. of Chicago Press.

Wakeley, J. S., and L. J. O'Neil. 1988. *Techniques to increase efficiency and reduce effort in applications of the habitat evaluation procedures (HEP)*. Environmental Impact Research Program Technical Report EL-88-13. Vicksburg, MS: U.S. Army Corps of Engineers.

Walker, B., S. Carpenter, J. Anderies, N. Abel, G. Cumming, M. Janssen, L. Lebel, J. Norberg, G. D. Peterson, and R. Pritchard. 2002. Resilience management in social-ecological systems: A working hypothesis for a participatory approach. *Conservation Ecology* 6 (1):14 [online]; www.consecol.org/vol6/iss1/art14.

Walters, C., D. Pauly, and V. Christensen. 1999. Ecospace: Prediction of mesoscale spatial patterns in trophic relationships of exploited ecosystems, with emphasis on the impacts of marine protected areas. *Ecosystems* 2:539–54.

Wardle, D. A., and J. P. Grime. 2003. Biodiversity and stability of grassland ecosystem functioning. *Oikos* 100 (3):622–23.

Weltzin, J. F., M. E. Loik, S. Schwinning, D. G. Williams, P. A. Fay, B. M. Haddad, J. Harte, T. E. Huxman, A. K. Knapp, and G. Lin. 2003. Assessing the response of terrestrial ecosystems to potential changes in precipitation. *BioScience* 53 (10):941–52.

Whitlock, C., S. L. Shafer, and J. Marlon. 2003. The role of climate and vegetation change in shaping past and future fire regimes in the northwestern US and the implications for ecosystem management. *Forest Ecology and Management* 178 (1–2):5–21.

Wikle, C. K. 2003. Hierarchical Bayesian models for predicting the spread of ecological processes. *Ecology* 84 (6):1382–94.

Wilcox, B. A. 1986. Extinction models and conservation. *Trends in Evolution and Ecology* 1:46–48.

Williams, P. H., N. D. Burgess, and C. Rahbek. 2000. Flagship species, ecological complementarity and conserving the diversity of mammals and birds in sub-Sahara Africa. Animal Conservation 3:249–60.

Wimberly, M. C. 2002. Spatial simulation of historical landscape patterns in coastal forests of the Pacific Northwest. *Canadian Journal of Forest Research* 21:1316–28.

Wisdom, M. J., L. R. Bright, C. G. Carey, W. W. Hines, R. J. Pedersen, D. A. Smithey, J. W. Thomas, and G. W. Witmer. 1986. *A model to evaluate elk habitat in western Oregon.* Publication No. R6-F&WL-216-1986. Portland OR: USDA Forest Service, Pacific Northwest Research Station.

Wisdom, M. J., M. M. Rowland, B. C. Wales, M. A. Hemstrom, W. J. Hann, M. G. Raphael, R. S. Holthausen, R. A. Gravenmier, and T. D. Rich. 2002. Modeled effects of sagebrush-steppe restoration on greater sage-grouse in the interior Columbia Basin, U.S.A. *Conservation Biology* 16 (5):1223–31.

Woods, G. R., D. C. Guynn, W. E. Hammitt, and M. E. Patterson. 1996. Determinants of participant satisfaction with quality deer management. *Wildlife Society Bulletin* 24 (2):318–24.

Wright, S. 1921. Correlation and causation. *Journal of Agricultural Research* 20:557–85.

Yeh, A. G.-O., and X. Li. 2003. Simulation of development alternatives using neural networks, cellular automata, and GIS for urban planning. *Photogrammetric Engineering and Remote Sensing* 69 (9):1043–52.

Zhang, X., C. Li, and Y. Yuan. 1997. Application of neural networks to identifying vegetation types from satellite images. *AI Applications* 11 (3):99–106.

Zhu, X., R. G. Healey, and R. J. Aspinall. 1998. A knowledge-based systems approach to design of spatial decision support systems for environmental management. *Environmental Management* 22:35–48.

PART III

The Management of Wildlife Habitat

Managing Habitat for Animals in an Evolutionary and Ecosystem Context

11

Indiscriminant cutting of trees is calamitous. It is the citizen's duty to guard our forest wealth.

EMPEROR SHIVAJI*

Concepts and theories, statistical analysis tools, and modeling technologies for habitat assessment have come a long way since the early days. The tools of habitat management have continued to evolve in response to new environmental problems and to new scientific concepts and findings. In some cases, what seemed to be utterly sufficient and immutable axioms of habitat management—such as the value of forest openings and edges for benefiting wildlife (e.g., Lay 1938)—have given way to different perspectives and new knowledge gleaned from changing landscapes, such as the more recent concern for excessive fragmentation of old- or native-forest cover (e.g., Perault and Lomolino 2000; also see box 9.1).

Learning from traditional approaches to wildlife management, including their successes and failures, helps us prepare for management challenges to come. The basic tenets of habitat management provided by early wildlife biologists such as Leopold (1933) still prove useful today: wild animals need essentials of cover, food, and water for survival of individuals and

for a chance of continuation of the population. But to this foundation, we provide some additional tenets to aid future assessments and management.

We first discuss managing wildlife in situ in an ecosystem context. Borrowing from the old German concept of Umwelt, we view wildlife organisms, populations, species, and communities as a function not just of cover, food, and water ("habitat" in the traditional sense). They also respond to numerous other biotic and abiotic factors. In this regard, wildlife is a function of (1) the full set of environmental factors—including but extending beyond those essential three traditional habitat factors of cover, food, and water—which together influence realized fitness of organisms; (2) the ecological roles of other species; and (3) abiotic conditions and events, including what we may term systematic or chronic changes and acute environmental disturbances or perturbations. In this fuller context, wildlife managers should attend not just to conserving taxonomic species, and not just to maintaining viability of populations (particularly threatened

379

ones), but also to providing the rich complex of organisms' ecological functions; the full set of ecosystem processes; the genetic diversity within and among species, demes, morphs, ecotypes, and other subspecific entities; and all the Umwelt conditions collectively required for their persistence and development.

We raise the question, What is wildlife? and propose that the term and the approaches currently taken for their management be extended to consider the full array of all biota present in an ecosystem as well as their ecological functions. We explore the question from a variety of dimensions, all affecting conservation policy and management success on the ground. We propose an enhanced approach to depicting, modeling, and predicting the status and condition of wildlife organisms in an ecosystem, by extending beyond the traditional wildlife–habitat relationships approach to a fuller species–environment relations framework.

Ultimately, we hope that successful wildlife management will provide global opportunities to maintain current native biodiversity as well as conditions to support long-term evolutionary processes of species lineages. In this vein, we propose some checkpoints and tenets for managing wildlife in contexts of ecological domains and evolutionary time frames.

Since much of land management in the 21st century will likely entail unprecedented changes in environmental conditions, we advocate an aggressive and rigorous use of adaptive habitat management. In this approach—at heart, a fundamental philosophy of land use and personal resource use habits—we should learn from experiences, including planned and unplanned natural experiments, and then modify our management actions and resource use behaviors accordingly to better meet wildlife conservation goals. Much has been written on ideal procedures to adaptive management, but more needs to be said about real-world problems in imple-

menting them. In the spirit of learning from history and experience, we state some basic tenets of the adaptive management approach and cite cases of both success and failure. Instead of prognosticating future declines of scarce habitat and endangered species, which can be found in a variety of other sources, we focus on exploring a more optimistic future of the science and sociology of habitat management.

Managing Wildlife in an Ecosystem Context

Managing wildlife in context of their ecosystems entails clarifying the taxa and conditions that qualify, and focusing management activities on ecosystem components.

What Is Wildlife?

More than just a means of focusing the work of agencies and the subject of this book, the question, What is wildlife? has recently been evolving along scientific, management, social, cultural, legal, and even ethical and esthetic dimensions (box 11.1). Traditional views have focused on wildlife solely as terrestrial vertebrates—initially, game birds and mammals, and later also organisms of conservation concern, principally threatened and endangered species. Although much of wildlife management today still focuses on these elements, as they indeed are still important concerns for conservation action, the term *wildlife* is being broadened along several fronts.

For example, thanks to increasing interest in cumulative effects analysis and integrating evolutionary perspectives into wildlife science, the question, What is wildlife? in turn prompts asking which conditions of conservation interest, evolutionary significance, and ecological function to include. Should relatively scarce subspecies, particularly increasingly uncommon local endemics such as the Hawaiian coot, or 'alae ke

Box 11.1   What is wildlife?

Following are different dimensions of the question that wildlife biologists, managers, lawyers, judges, economists, social scientists, political scientists, politicians, indigenous peoples, hunters, conservationists, educators, and artists need to ponder: What is wildlife? The answers lie not in specific responses to each question (that's the easy way out), but rather in how the needs expressed in the multiple dimensions of all the questions can be met at the same time.

What is wildlife?
- The scientific dimension
  Which taxonomic classes pertain?
  Which conditions of evolutionary significance pertain?
  Which ecological functional groups of organisms pertain?
- The management dimension
  What is a population?
  Which population and species deserves specific management attention, and which are assumed to derive secondary benefit?
  Which taxa have agency or landholder status for management focus (sensitive, rare, game, etc.)?
  Is the focus on populations, species, habitats, or ecosystems?
- The legal dimension
  Which populations or species are threatened or endangered (or rare, sensu Red Data Book listings used outside the United States)?
  Which population or species are candidates for listing?
  What is mandated for ensuring viability, biodiversity, and persistence (nonextinction)?
  Which population or species are legally hunted or gathered, and is there a mandate for sustained harvest above and beyond minimal viability?
  What are obligations to tribes and indigenous people?
  What are the paralegal obligations for representing the rights of nonhuman living entities in judicial and land use arenas?
- The economic dimension
  How should use of wildlife resources help support local private economies, such as through fur trapping and game ranching?
  What role should questions of economic impact play in determining wildlife habitat management mandates on private and on public lands?
- The cultural dimension
  What is necessary to provide for the rights and persistence of indigenous peoples?
  How should patterns of other traditional use of wildlife resources be managed?
  What are the obligations for providing for the future generations of all peoples?
- The ethical dimension
  What takes precedence—individual organisms, populations, species, or systems?
  Which populations or species should fall first in triage approach?
  How should human habits of land use and resource consumption be met or modified?
- The aesthetic dimension
  What should science, management, politics, and publics learn from artistic perspectives?
  How should the needs and interests of nature artists and educators be met?

In pluralistic societies designed to resolve conflicting interests through confrontational means, such as courts of law, the process of finding the difficult answers that integrate across these dimensions is often tedious and expensive, and is sometimes imperfect. Rarely do litigative compromises satisfy everyone fully. Newer approaches to conflict resolution (e.g., Maguire 1991) are meant to circumvent purely judicial and litigative solutions. In a world of increasing scarcity of wildlife resources, it is probably wise to begin to explore and refine such new approaches to provide better for future generations, if only the next one.

'oke'o (*Fulica americana alai*); the Hawaiian gallinule, or 'alae 'ula (*Gallinula chloropus sandvicensis*); and the black-crowned night-heron, or 'auku'u (*Nycticorax nycticorax hoactli*), of the Hawaiian islands, be considered in equal footing as full species, for scientific, management, and legal priorities? Similar questions pertain to considering ecological functional groups of organisms as subjects of wildlife investigations and regulatory mandates.

The question, What is wildlife? raises important management concerns. Wildlife management—whether on private lands for game ranching; on state or federal public lands; on lands of indigenous peoples; or on resource industry lands, including farms and commercial forests—has been mostly defined in terms of economic impacts and legal mandates. Empirically, much of wildlife management focuses on meeting regulatory edicts and litigative exigencies. The focus is often on the lists of threatened, endangered, and rare or sensitive species. While other species are not excluded from the realm of wildlife management, they often receive diminished or no formal management attention. Arguments have been raised that it is too complicated to address the full ecological community and that selected signposts or indicators, largely species of legal concern or consumptive interest, must be chosen.

As such, some federal land management agencies and other landholders bank on "coarse-filter" approaches. The basic tenet of this approach is that managing generally defined or broad-scale habitat conditions—for example, managing forests within historic ranges of natural disturbances such as wildfire (Armstrong et al. 2003)—will provide for the needs of all associated native species (Hunter 1991). This tenet has seldom been formally tested. Our experience suggests that, more often than not, it is likely to be proven dangerously wrong. In the coarse-filter approach, "wildlife" is operationally de-fined as the often-unspecified wildlife community that is associated with some general macrohabitat condition, as defined for one or a few species (the indicator species approach—e.g., Abate 1992), or defined as in some other general way, such as by reconstructing historic conditions (the range of natural variations approach—e.g., Morgan et al. 1994, Fulé et al. 1997). However, at least in forest management in the western United States, the environmental requirements of many species—particularly nonvertebrate taxa—are not necessarily met by this approach (Landres et al. 1988; Niemi et al. 1997). At best, it is wise to check the validity of the coarse-filter approach to biodiversity conservation, lest some elements or species be excluded. More recently, Hunter (2005) introduced the concept of a "mesofilter conservation strategy" to straddle the coarse- and fine-filter scales, a strategy that focuses on providing habitat elements of ecosystems important to many species. We concur with this refinement but again levy our charge not to casually presume that coarse- and mesofilter approaches will necessarily and always provide for all species.

The legal dimension raises other questions on what is wildlife. Legal listings can include wildlife listed as rare, threatened, or endangered, such as in international Red Data Books on rare species (e.g., King and Warren 1981) or by government agencies such as USDI Fish and Wildlife Service (FWS) following the Endangered Species Act (ESA). Additional listings with potential litigative implications may include those suggested by the International Union for Conservation of Nature and Natural Resources (IUCN) and associated organizations (McNeely et al. 1990). Typically included in such compendia are species, geographically defined subspecies, and species in portions of their range ("populations" in the loose sense). Most of these lists, and thus what qualifies as wildlife in the legal or quasi-legal sense, change often. For instance, FWS has drastically reduced its formal list of candidate species by virtually doing away with the candidate 2 and 3 categories in the listing status and trimming species remaining in the candidate 1 category. Quite literally, then, many of the taxa now excluded from these candidate lists have no federal legal status as potentially threatened or endangered wildlife under current ESA operations. However, at least in the United States, individual states can and often do offer some additional protection.

Other federal legislation for management of wildlife in the United States has put further legal boundaries on the definition of wildlife, such as by specifying organisms for conservation in sundry acts (e.g., the Marine Mammal Protection Act, the Migratory Bird Treaty Act, and others). Federal regulations implementing the National Forest Management Act of 1976 (36 CFR 219.19) mandated that USDA Forest Service provide for viability of all native and desired nonnative vertebrates on Forest Service lands, but recent changes have replaced this mandate with more general (coarse-filter) guidelines for maintaining ecosystem diversity. The regulations provide other mandates for maintaining biodiversity and therein refer to plant and animal species and communities, but they do not further specify such organisms by taxonomic group (e.g., vertebrates). Political pressure to drastically abbreviate regulations pertaining to conservation of endangered species and viable populations continues. Thus it has been a point of ongoing legal challenge and internal debate as to which wildlife species, if any specifically, qualify under changing regulations.

Additional questions of legal definitions of wildlife pertain to organisms hunted and fished by the general public, as well as to obligations for providing native tribes and indigenous peoples with specific plants and animals for hunting and gathering purposes (Alcorn 1993). In some judicial systems, organisms have been represented in court cases by interest groups, which has raised the question of the legal standing of nonhuman living entities (Stone 1974).

A number of other questions as to what qualifies as wildlife pertain to economic, cultural, ethical, and even esthetic dimensions (see box 11.1). Economic issues focus on questions of crop depredation, disease and pathogen organisms, forest insect pests, and similar direct assaults (or benefits) to natural or food resources and human health. They also focus on the indirect effects of providing habitat for threatened or endangered organisms—provisions that are then seen to have adverse economic effects on human communities and resource use interests. Cultural issues focus on rights of indigenous peoples to use particular lands and habitats and to hunt or gather plants and animals. Other cultural issues deal with traditional, although not necessarily indigenous, use of lands and wildlife populations for hunting purposes. Ethical issues can pertain to general use of resources or to specific animal groups. Animal rights interest groups tend to focus on species that seem closer to humans in their degree of sentience (and in their expression of pain) and tend to exclude most fish, herps, invertebrates, and plants. Artistic questions may deal with a sense of place or with organisms of specific esthetic interest, and thereby exclude habitats and organisms not meeting such standards (for example, some dangerous or venomous organisms, snakes, spiders, and microorganisms).

It is clear that such a simple question as, What is wildlife? has deceptively vast ramifications for guiding social desires, legal decisions, directions for management, public education, and ultimately the future of organisms and habitats across the land. It is equally clear that no one definition of wildlife satisfies all such interests. We propose that the science of wildlife ecology and management of wildlife habitat expand to fully encompass all organisms collectively addressed by these dimensions in an ecosystem context. Effectively, wildlife should ultimately be defined as no less than the full array of living organisms of all taxonomic groups in terrestrial, riparian, and aquatic environments, including marine, estuarine, and belowground ecosystems.

How Can Ecosystems Be Managed?

Our new era of "ecosystem management" (e.g., Zorn et al. 2001) presumes much (Butler and Koontz 2005). It presumes that we know enough about how ecosystems work—including the roles and interactions of "wildlife"—that we can effectively depict, predict, and control effects of our activities. The Wildlife Society has recently suggested a set of performance measures for tracking effects of human activity on ecological sustainability (Haufler et al. 2002). However, much of the science underlying an ability to predict ecosystem response is wanting. At best, we can be sure that many of our activities will indeed affect ecosystems, but to what end is mostly a best guess or pertinent to only a small portion of what constitutes an ecosystem, such as abundance of game or production of rangelands.

A first step toward ecosystem management should entail definitions (see Glossary). What is an ecosystem, and how can ecosystems be mapped for management purposes? Much of the controversy surrounding ecosystems as a unit of study and management is really about an ecosystem *as a definable unit.* Here we will briefly review this discussion, because it is important that students understand the ongoing controversy prior to using the term *ecosystem,* and especially before designing an "ecosystem" study.

Defining, and choosing indicators for, ecosystems would be possible if the science of ecology were able to provide us with simple, rigorous models for describing and predicting the status of ecosystems. However, at this point our knowl-

edge is largely inadequate to identify the most important variables (Keddy et al. 1993). Likewise, Tracy and Brussard (1994) argued that we do not yet know which few ecosystem processes are the most important to study, to index, and to preserve. They noted that ecosystems are arbitrarily defined study units that range from water droplets to the entire biosphere! And as summarized by Peters (1991, 91), *ecosystem,* as currently used, is a "multidimensional, unlimited, relativistic" entity representing the environment.

Ecologists have not yet adopted a single, general classification system for (nonsystematic) ecological units above the species (the Linnaean classification system), except perhaps for the biome, which is too large and too crude to be a useful scale of study and management (Orians 1993). Any classification system of ecological units is certain to be highly contentious, both within the biological community and among policymakers. The controversy surrounding identification of old-growth forest in the Pacific Northwest, and delineation of wetlands across the United States, are recent examples. Other examples that address particular management issues are classifications of ecosystems for fish and forest health (Rieman et al. 2000), and use of remote-sensing data to classify forest "ecosystems" (Treitz and Howarth 2000). However, none of these examples begins with a causal model of how ecosystems work in all their parts and relations, and from that builds a classification of ecosystem types. Thus we are led to classifying ecosystems in a more operational and issue-driven way.

But do these limitations mean that the ecosystem concept has no value in our analyses of wildlife habitat? We think not, because this concept embodies all the interactions that influence what an animal does and the context in which lineages evolve. It thus reminds us that the animal lives in a complex situation, and that changes in individual factors can have a cascading effect on the individual as well as the population and community. Any advancement in our understanding of these interactions thus advances our ability to manage the species. It forces us to view animals in a broader context of space and time than previously witnessed in wildlife management.

How do we move beyond the standard ecological rhetoric—espoused in the preceding paragraph—and actually implement the ecosystem concept in studies of wildlife habitat? Although in preceding chapters we have not explicitly identified the ecosystem concept as such, much of this book has emphasized the conducting of studies within clearly elucidated spatial scales. We did not previously adduce the ecosystem concept because we feared it would distract from the sampling and analytical issues at hand, but below we make this attempt. Further, much of our wildlife research will, and should, remain at relatively microscales (e.g., James and Coskey 2002), although ultimately both should be integrated (e.g., Gehrt and Chelsvig 2003; Spencer and Thompson 2003).

Regardless of the scale of study, organisms are influenced by factors that operate across broader geographic areas, but which cannot be effectively measured at the relatively smaller geographic area of study. For example, studies of microhabitat seldom measure the abundance of predators, although predators can influence the presence or absence of the animal at a specific site and time, and the distribution or range of predators may extend well beyond the geographic area covered by the study. Likewise, microhabitat studies are seldom able to measure habitat quality (e.g., some measure of fecundity; see chapter 6), including effects of climate, although quality is probably the most appropriate measure of habitat.

Thus, viewing a habitat study in light of the larger ecosystem should help determine what factors to measure and at what scale to measure

them. A rather traditional principle first advanced in the general systems theory literature of the 1960s (e.g., Lance and Williams 1967) and later echoed in hierarchy theory (Peterson and Parker 1998), and which applies to studies of wildlife in an ecosystem context, is to look at least two levels of spatial scale or organization smaller than that of the entity of central interest, to determine the specific mechanisms responsible for observed patterns, and two levels larger, to describe the environmental context and emergent behaviors or patterns.

If the goal is to determine why an animal uses a specific site, the ecosystem concept tells us that it will usually be necessary to measure things at many scales. The ecosystem concept thus places a habitat study within a hierarchy (Kolasa 1989), not unlike the overall hierarchical concept of habitat selection developed in chapter 7. Yes, it would be helpful to draw a line on a map around an ecosystem; we do not think that is likely to happen with full consensus. However, it is possible to draw lines between many of the factors that we know influence an animal, such as specific vegetation conditions, presence of cliffs and talus, and areas of moderate winter climate. Food web theory, predator–prey relationships, competitive interactions, and so on have well-developed bodies of theory and empirical studies that lend insight into how animals perceive their surroundings. All of these interactions are part of what we perceive as "ecosystem functions." Our discussion of the debate surrounding keystone species in chapter 2 exemplifies how autecology supports understanding of ecosystem functions.

The days of measuring some vegetation plots and calling the measurements a habitat study are rapidly falling behind us. There is still a need for such work, especially on little-known species and habitats. Such work has formed a solid foundation upon which we can now build a better understanding of ecological relationships by expanding beyond just vegetation as descriptors

Box 11.2   *Thinking green in an urban environment: City greenbelts as natural environments in urban landscapes.*

Each kind of landscape, from natural to megalopolis, can provide some aspects of wildlife habitat to help maintain regional biodiversity. One facet of landscape planning for wildlife can involve providing greenbelts near major cities.

Examples of planning for city greenbelts can be found in the Russian Far East, where vast forests still cover much of the land, although many of the forests are now being harvested for much-needed revenue. Russian cities are sometimes designed with a mandatory nearby greenbelt of vegetation cover. Specific formulae are used to determine the size of greenbelt forests within a specific distance of the city and according to the city's population.

The purpose of greenbelts is to protect forest environments for wildlife, water sources, recreation, and other uses. One such site is Khekhtsir Zapovednik, a scientific nature reserve located south of Khabarovsk city, in Khabarovski Krai (province), in the Russian Far East near the border with China. This nature reserve was established in 1963 and once housed kozaks who prevented cutting of its old forests. It is situated on an ancient, low-lying montane island along the east side of the Ussuri River and is ecologically unique in its collection of rare plants and primary forests. The isolated mountain range on which it is located is slowly being submerged over geologic time by the alluvium of the Khor and Bikin rivers, tributaries to the Ussuri, which in turn empties into the mighty Amur River. Once the forests of these watersheds were contiguous in the Far East, but now they have been isolated by urban development along much of the lower valleys of the Ussuri, Khor, and Bikin river basins.

Khekhtsir Zapovednik contains mixed hardwood-conifer forests. It shares Pleistocene and Tertiary relict plant species and genetic and geologic similarities with the Sikhote-Alin Mountain Range to the east. It encompasses more than 45,000 ha and is bordered by 6-m-wide firebreaks. In this example, then, a greenbelt acts as a viable natural landscape and provides a reserve of ecologic and locally historic importance as part of the broad planning for an urban landscape.

The example of Khekhtsir Zapovednik suggests some lessons on both the limitations and benefits of using greenbelts in urban land management planning. First, the area cannot be expected to maintain itself as a big preserve (in Russia, zapovedniks are preserves that exclude all human intrusion except scientific research), as it lies too near major urban and anthropogenic influences. The ecosystem is, or over time will become, "damaged" and air polluted. Intact primary forests of the area will become increasingly difficult to preserve as urban influence and human encroachment increases. However, the area is an important source of medicinal herbs and serves a positive psychological function for the urban city dwellers. It could act as a good urban forest test site for ascertaining the effects of nearby urbanization on sustaining forest life.

Khekhtsir Zapovednik has remarkable wildlife for being so close to busy Khabarovsk. A pair of Siberian tigers (*Panthera tigris altaica*) has become established there. Also present are Amur forest cats (Felis bengalensis), Himalayan black bears (*Ursus [Selenarctos] thibetanus*), and occasional brown bears (*Ursus arctos*), sables (*Martes zibellina*), Mandarin ducks (*Aix galericulata*) (B. Marcot, pers. obs.), all indicators of locally undisturbed environments and "jewels" of the Far East. It is imperative, if the status of the area is to change from zapovednik to more intensive use, that hunting explicitly not be permitted and that the area be guarded for detecting and prosecuting any poaching or illegal resource-gathering activities. Gathering of edible or medicinal plants and mushrooms for traditional or indigenous purposes, or for other uses, could be managed under a specific program that also monitors effects of such use on the viability of the plant populations. Such a program could serve as a model of sustainable resource use near the urban environment.

To this end, city greenbelts can be established as research centers and parks permitting limited recreation and harvesting, rather than as strict, exclusive preserves. Such sites could be used to test restoration management and ecological recovery, such as evaluating the time and effort needed to recover vegetation and wildlife, and the effects of recreation, hunting, and gathering activities.

Khekhtsir Zapovednik also could serve as a vital habitat corridor link to a similar lowland reserve in China, in the Ussuri River Valley. Some species of wildlife migrate across the border, including black bears and forest cats. Mandarin ducks nest in Russia and feed in China across the Ussuri. However, international peace parks along borders with a history of political conflict may be difficult to establish and even tougher to manage. In such cases, it may be better to establish separately managed "sister" parks so that lines of authority and responsibility remain clear; such sites could still serve as greenbelts for the urban landscape. Eventually, though, joint, cross-border urban greenbelts could become models of international cooperation for conservation of wildlife habitats and populations.

of habitat. Next we describe how this understanding might be achieved.

What Is an Ecosystem Context?

For effective conservation, we advocate studying and managing wildlife in an ecosystem context. What is meant by an ecosystem context? We alluded to an ecosystem context above by adducing the concept of Umwelt, and in our discussions of ecosystem functions, but operationally we propose the following approaches to describing species in an ecosystem context.

An *ecosystem* consists of organisms of various taxonomic designations and levels of biological organization (see table 8.1), along with their interactions among each other and among abiotic conditions and processes. An ecosystem is more than a mere collection of populations (organisms of the same species in a given area), species assemblages (groups of species of particular taxa), or communities (species with their interactions). Understanding wildlife in an ecosystem context entails understanding (1) population dynamics, including demographic and genetic variations; (2) the evolutionary context of organisms, populations, and species, including the contribution of genetic variation to persistence of species lineages, mechanisms of speciation and hybridization, and selection for adaptive traits; (3) interactions among species that affect their persistence and that influence community structure, including obligate mutualisms such as pollination or dispersal vectors, predation, competition, and other interactions; and (4) the influence of the abiotic environment on vitality of organisms (organism health and realized fitness) and populations (viability), including how disturbance mechanisms operate and how organisms respond. Habitat ecology plays a key role in many of these facets of an ecosystem context but itself needs to be subsumed into a broader ecological tapestry.

An ecosystem context also necessitates understanding the role of humans in modifying environments, habitats, and wildlife populations. The question of whether humans are a part of the ecosystem is not a useful question. Rather, evidence clearly shows that humans can greatly affect ecosystems. To "admit" that humans are part of the ecosystem should not be used as rationale that anything goes, that anything we do is "natural." More useful and challenging questions are, To what extent do we want our actions to modify environments, habitats, and wildlife populations? To what extent do we need to change our own resource and land use habits to meet our own goals for conservation? And, perhaps most difficult, how many humans should occupy a specific area? Answers to such questions are critical for successful wildlife management *and* for providing people with sustainable resources; these are two inextricable sides of the same coin. An ecosystem context for studying and managing wildlife should prompt such questions.

A related question that also goes to the heart of management policy and environmental ethics is, What is natural? In areas only relatively recently occupied and changed by human technological presence and high human density, such as much of western North America, the question may seem silly; it is obvious that ancient forests and ungrazed grasslands are natural and that tree farms and croplands are not. However, much of the world lives in ecosystems that have been affected for millennia by human activities, often so much so that the question itself has little relevance.

Take the example of Keoladeo National Park (more popularly known by the name of the adjacent city of Bharatpur) south of Delhi, India, where a rich and productive wetland persists in a landscape long converted to agriculture. Waterbirds by the millions use the wetlands as critical breeding, staging, and wintering habitat. But the park is itself *artificial,* having been created and maintained by people. The park is an integral and essential cornerstone of the local economy, as well, providing many economic and social as well as ecological services. During a major drought in late 2002, most of the basins in the park dried completely, and local people were employed to dig out the silt and deepen the ponds. Similar creation and manicuring of wildlife habitats is common in southern Africa and elsewhere, where pans have been dug or enlarged and filled or supplemented with pumped groundwater.

Is this "natural," in landscapes long deforested and altered by people and where nature is now maintained by intensive human activities? The question is not useful. Instead, we need to ask, To what intensity and extent do we want to help provide for wildlife in landscapes long changed and occupied by humans? To answer this question to the satisfaction of all will require new partnerships between the disciplines of wildlife biology, ecology, landscape architecture, cultural anthropology, environmental planning, and environmental psychology. The environmental engineering sciences can then serve the purpose of helping to design and create patterns of human presence more in concert with wildlife habitats and populations. Such patterns should provide landscapes also designed to meet human social and economic needs, such as infrastructures for transportation, communication, and transmission of goods and services; to meet human cultural needs, such as accommodating social and religious customs and traditional resource use; and to anticipate effects of human activities, such as on soils, slopes, air, water, and esthetics.

Depicting and Managing Key Environmental Correlates

Understanding and managing wildlife in an ecosystem context also raises the question of what characteristics of ecosystems to study and identify for management and planning. The tra-

ditional approach to wildlife habitat management focused on the habitat elements of water, food, and cover. These are essential elements for maintaining the health of organisms and persistence of populations. But they do not adequately describe all facets of ecosystems vital to ensuring realized fitness of individuals and viability of populations and species. In previous chapters we have explored many approaches to empirical studies and modeling of a wide variety of other aspects of species' Umwelts.

A broader approach has been taken in various ecological assessments of wildlife and watersheds in the western United States (Marcot et al. 1997). In this assessment, species (and selected subspecies and plant varieties) were described by their use of *key environmental correlates* (KECs). KECs are biotic and abiotic conditions of a species' environment that proximately influence realized fitness of individuals and viability of populations. KECs can include the biophysical attributes traditionally considered as habitat elements. They can also include other biotic or abiotic factors not traditionally considered as habitat elements, such as use of roads, air quality, hunting or collection pressure, and interspecific interactions.

The purpose of extolling the use of KECs is to extend a focus on environmental factors beyond simple descriptions of vegetation types and their structural or successional stages, so commonly used for wildlife habitat assessment and management. The use of KECs can help shed light on effects (positive or negative) of human activities and other dynamic aspects of ecosystems beyond those affecting just vegetation conditions.

In one assessment in the interior Columbia River Basin in the western United States, Marcot et al. (1997) depicted KECs for a wide variety of selected taxa and species groups of macrofungi, lichens, bryophytes, rare vascular plants, selected soil microorganisms, arthropods, and mollusks, and for all vertebrates. A single, hierarchical classification of KECs was developed for use with all of these organisms. A database of KECs of each wildlife taxon was used to determine which species shared common correlates, how management of some correlates (e.g., large down wood in forests) affect suites of species, and what collective set of correlates should be recognized for managing the full set of species in an area. The KEC classification was developed hierarchically, so that groups of organisms sharing various levels of specificity of correlates could be identified. More recently, use of KECs has been extended to evaluate terrestrial environments as part of watershed assessments within the entire Columbia River Basin under the aegis of the Northwest Power and Conservation Council.

Depicting Species' Key Ecological Functions

Another facet of managing wildlife in an ecosystem context pertains to understanding the ecological roles played by species. The traditional approach to habitat management has assumed simply that wildlife species are a function of their habitat, and that managing wildlife simply entails providing the right kinds of habitats. Instead, the array of key ecological functions (KEFs) of individual species can be explicitly depicted. In this context, *key ecological function (KEF)* refers to the main ecological roles performed by species that influence diversity, productivity, or sustainability of ecosystems (Marcot and Vander Heyden 2001). A given KEF can be shared by many species, and a given species can have several KEFs. These can be depicted, along with KECs, in databases and models of species–environment relations.

Table 11.1 presents a classification of KEFs developed by Marcot et al. (1997) for the same interior Columbia River Basin assessment described above. Main categories of KEFs included trophic relations; herbivory; nutrient cycling; interspecies relations; disease, pathogen, and para-

site relations; soil relations; wood relations; and water relations. Each category was divided into a number of hierarchical subcategories. The KEF classification was used in a wildlife–habitat relationships database to determine which species shared specific ecological functions and the array of functions performed by specified species. Species with the same KEF were called *ecological functional groups* of species. Information on KEFs of species was cross-linked to that on KECs and range distributions of species, as discussed below. The classification of KEFs we present here may be useful for guiding other studies of wildlife biodiversity by helping to focus on ecological roles of species as a complement to the more traditional assessments of biodiversity as species richness.

A species ecological function may often appear in more than one ecosystem. As an example, consider the ecological function of parasite carrier or transmitter (KEF 5.3 in table 11.1). Species with this ecological function in the inland West of the United States include the least bittern (*Ixobrychus exilis*) which is a host to ecto- and endoparasites; the western sage grouse (*Centrocercus urophasianus phaios*), which is a host for protozoan, helminth, and bacterial parasites; and the snowshoe hare (*Lepus americanus*), which supports a variety of ecto- and endoparasites and is a reservoir of several viruses and bacterial pathogens. The collective set of key environmental correlates—including vegetation cover types, structural stages, and other environmental factors—that are occupied by even this simple ecological functional group of three species spans terrestrial forest, shrubland, grassland, and riparian (wetland) ecological systems. In this way, ecological processes for each subsystem can be identified.

Ecological processes, as used here, are those groups of key ecological functions of species that pertain to each part of the ecosystem. For example, ecological processes associated with soil

Table 11.1. A hierarchic classification of key ecological functions of wildlife species

1 Trophic relations
   1.1 Primary producer (chlorophyllous vascular plant)
      1.1.1 Autotrophe (fully independent chlorophyllous plant)
      1.1.2 Hemiparasite (chlorophyllous plant that also partly derives nutrients through attachment to other chlorophyllous plants)
   1.2 Heterotrophic consumer
      1.2.1 Primary consumer (herbivore) (also see below under Herbivory)
         1.2.1.1 Foliovore (leaf eater)
         1.2.1.2 Spermivore (seed eater)
         1.2.1.3 Browser
         1.2.1.4 Grazer
         1.2.1.5 Frugivore (fruit eater)
         1.2.1.6 Sap feeder (sucking insect)
         1.2.1.7 Root feeder (invertebrate)
         1.2.1.8 Sequestration of plant metabolites
      1.2.2 Secondary consumer (primary predator or carnivore)
         1.2.2.1 Consumer or predator of invertebrates, potentially including insects (insectivore)
         1.2.2.2 Consumer or predator of vertebrates (species other than itself)
      1.2.3 Tertiary consumer (secondary predator or carnivore)
         1.2.3.1 Consumer of soil microorganisms
      1.2.4 Largely omnivore (consumer of plants and animals)
      1.2.5 Carrion feeder
      1.2.6 Cannibal
      1.2.7 Coprophagist (consumer of fecal material)
      1.2.8 Aquatic herbivore (invertebrate)
      1.2.9 Consumer of algae, ooze, and plankton in water (invertebrate)
   1.3 Achlorophyllous vascular plant (see 1.9 below for nonvasculars)
      1.3.1 Mycotrophe (indirectly parasitic, nongreen plant that derives nutrients from mycorrhizal fungi that is also associated with a chlorophyllous species that serves as the indirect host)
      1.3.2 Saprophyte (derives nutrients from decaying organic matter through mycorrhizal fungi)
      1.3.3 Parasite (derives nutrients through direct attachment to chlorophyllous plants)
         1.3.3.1 Root parasite
         1.3.3.2 Stem parasite
   1.4 Detritivore (direct consumer of dead organic material)
   1.5 Decomposer (consumer of by-products of decaying organic material)
   1.6 Comminutor (chewing insect, typically feeding on wood or vegetation)
   1.7 Forage or prey relations
      1.7.1 Forage for animals
      1.7.2 Prey for secondary or tertiary consumer (primary or secondary predator or carnivore)
      1.7.3 Carrion source
      1.7.4 Forage for invertebrates
   1.8 Major biomass
   1.9 Achlorophyllous nonvascular plants
      (See 1.3 above for vasculars)
      1.9.1 Mycorrhizal fungus
      1.9.2 Saprophyte
      1.9.3 Parasite
      1.9.4 Decomposer
   1.10 Moss feeder (invertebrate)

Table 11.1. Continued

2 Herbivory
    2.1   Ungulate herbivore (may influence rate or trajectory of vegetation succession and presence of plant species)
        2.1.1 Herbivore of tree or shrub species (browser)
        2.1.2 Herbivore of grasses or forbs (grazer)
    2.2   Insect herbivore (may influence rate of trajectory of vegetation succession or presence of plant species)
        2.2.1 Defoliator
        2.2.2 Dark beetle
        2.2.3 Tree bole feeder
3 Nutrient cycling relationships (see number 6 below for nutrient cycling relationships in soil)
    3.1   Aids in physical transfer of substances for nutrient cycling (C, N, P, other)
    3.2   Nitrogen relationships
        3.2.1 N-fixer
        3.2.2 N-immobilizer
        3.2.3 Source for N mineralization
    3.3   Carbon relationships
        3.3.1 Sequestration of atmospheric carbon
4 Interspecies relationships
    4.1   Insect control
    4.2   Ungulate or other vertebrate population control
    4.3   Pollination vector
    4.4   Transportation of seed, spores, plant or animal disseminules
        4.4.1 Disperses fungi
        4.4.2 Disperses lichens
        4.4.3 Disperses bryophytes, including mosses
        4.4.4 Disperses insects
        4.4.5 Disperses seeds and fruits
        4.4.6 Disperses plants
    4.5   Commensal or mutualist with other species
    4.6   Provides substrates or cover for animals
        4.6.1 Nesting or breeding substrate (e.g., nesting material)
        4.6.2 Thermal, hiding cover, loafing or den site
        4.6.3 Provides microhabitat (as for invertebrates)
            4.6.3.1 Aquatic or riparian environments
            4.6.3.2 Terrestrial environments
            4.6.3.3 Canopy environments
            4.6.3.4 Tree bole environments
        4.6.4 Creates "sap wells" in trees
    4.7   Nest parasite
        4.7.1 Nest parasite species (viz. cowbird)
        4.7.2 Host for next parasitism
    4.8   Primary cavity excavator in snags or live trees
    4.9   Primary burrow excavator (fossorial)
        4.9.1 Creates large burrows (rabbit, badger sized)
        4.9.2 Creates small burrows (smaller than rabbit sized)
    4.10 Competitor
    4.11 Uses burrows dug by other species (secondary burrow user)
    4.12 Uses cavities excavated by other species (secondary cavity user)
    4.13 Endo- or ectoparasite (invertebrate) (also see number 5, Disease, pathogen, and parasite relationships)

Table 11.1.   Continued

5   Disease, pathogen, and parasite relationships
    5.1   Carrier, transmitter, or reservoir of vertebrate diseases (including rabies)
    5.2   Acts as pathogen or disease
    5.3   Parasite carrier or transmitter
    5.4   Carrier, transmitter, or reservoir of plant diseases (invertebrate)
    5.5   Activity increases host susceptibility to plant diseases (invertebrate)
6   Soil relationships
    6.1   Physically affects (improves) soil structure, aeration (typically by digging)
    6.2   Aids general turnover of soil nutrients and layers
    6.3   Aids N retention or uptake in soil
    6.4   Aids soil stabilization
    6.5   Aids rock weathering
    6.6   Detoxifies xenobiotics (invertebrate)
    6.7   Metal accumulator (sequesters heavy metals)
    6.8   Soil (invertebrate) organisms that influence rate or trajectory of vegetation succession and presence of plant species
7   Wood relationships
    7.1   Physically breaks down wood
        7.1.1 Large logs
        7.1.2 Smaller wood pieces
    7.2   Chemically breaks down wood
8   Water relationships
    8.1   Impounds water (e.g., beaver)
    8.2   Bioindicator of water quality
    8.3   Hydrological buffer
    8.4   Improves water quality
    8.5   Contributes to short-term increase in stream flow (invertebrate)
9   Weather, climate, insolation relationships
    9.1   Affects albedo (as of soil, rock, or soil)
10 Vegetation structure and composition relationships
    10.1 Creates canopy gap openings (tree death) (invertebrate)
    10.2 Creates standing dead tree (snags) (invertebrate)

Source: Modified from Marcot et al. 1997.
Note: This classification was developed from taxon-specific information on fungi, lichens, bryophytes, vascular plants, invertebrates, and vertebrates in the inland West of the United States. The numbered codes in this classification are strictly hierarchic (e.g., item 1.1.2 is one element of 1.1, which is one element of the broadest-level category, 1) and can be used in species-environment relations databases.

subsystems include organic matter decomposition, nutrient pooling and cycling, and provision of conditions for mesoinvertebrates and fungi critical to vascular plant productivity. Species ecological functions associated with such processes in soil subsystems include soil aeration, turnover of soil nutrients and layers, nitrogen retention and uptake, and soil stabilization. With a database on KEFs, one can then determine which species play these functional roles that collectively contribute to particular ecological processes.

Further, counting the number of wildlife species with particular KEFs is a crude but useful measure of functional redundancy of that KEF category. The greater the functional redundancy of an ecosystem, the greater the resilience of that ecosystem to perturbations (Lawton and Brown

1994; Mooney et al. 1996; Gunderson et al. 2002). An example of a "functional profile" is shown in figure 11.1 for all forest mammal species in Washington and Oregon (Marcot and Aubry 2003). The profile reveals that functional redundancy among this set of forest mammals is greatest for such functions as serving as prey for predators, consuming invertebrates, and dispersing seeds and fruits. In their analysis of mammals of western U.S. forests, Marcot and Aubry (2003) also determined that the assemblage of forest mammals is not particularly species-rich compared with other taxa, but it is functionally rich in terms of the collective types and numbers of KEFs performed.

Species' key ecological functions and ecological processes all contribute to diversity, sustainability, and productivity over time. For example, the ecological processes and species key ecological functions associated with the soil subsystem of forest ecosystems all contribute to: biodiversity of fungi, lichen, plant, mesoinvertebrate (e.g., soil mites), macroinvertebrate (e.g., earthworms), and fossorial vertebrates (e.g., pocket

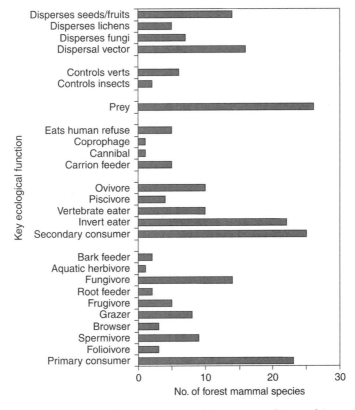

Figure 11.1.   Functional redundancy (number of species) of forest mammals in Washington and Oregon, by selected category of key ecological function (see table 11.1). Such "functional profiles" can be further specified for particular vegetation types or geographic locations (see fig. 11.2). (From Marcot and Aubry 2003, 639 [fig.19.4]; reprinted with permission of Cambridge University Press.)

gophers); to the productivity of plant and animal populations, including tree growth; and to the sustainability of resource growth and use (e.g., sustained timber production and harvest) over the long term.

Collectively, KECs and KEFs describe what has traditionally been referred to as habitat and niche dimensions—respectively, an organism's "address" and "occupation." However, we advocate expanding the address to include facets of the environment not traditionally considered in habitat studies, and expanding the occupation to include the full array of ecological functions that an organism or taxonomic group performs. KECs and KEFs can be described from a conceptual basis of the expected range of addresses and occupations, corresponding to carrying capacity and fundamental niche dimensions; or they can be described from an empirical basis of observed addresses and occupations, corresponding to actual distribution and abundance, and realized niche dimensions.

Beyond functional redundancy, a number of species, community, and geographic patterns of KEFs in wildlife communities can be described (table 11.2; Marcot and Vander Heyden 2001). Marcot and Aubry (2003) used these categories of patterns to assess the functional roles of forest mammals. Marcot (2002) used them to evaluate the ecological roles of vertebrate wildlife associated with decaying wood as an extension of traditional guidelines for managing snags and down wood in public forests. The aim of these evaluations of the functional roles of wildlife was to challenge managers to think functionally and include considerations for functional effects in their management guidelines.

Functional Analysis as Part of Wildlife Assessment and Management

In general wildlife-habitat assessments, we advocate evaluating each aspect of the *triad of habitats, species, and functions*. In many wildlife–habitat relationship models (chapter 10), elements of this triad are presumed to be functions of one another, the most traditional assumption being that species are a function of habitat. However, patterns of species richness or occurrence by community, or of individual species' use of various habitat conditions, do not necessarily correlate with their ecological functions (e.g., fig. 11.2). Thus each member of this triad tells a complementary story.

Various patterns of KEFs among species and within communities have been described (see table 11.2) and used in various projects (e.g., Marcot and Aubry 2003). Further, many of the functional patterns listed in table 11.2 supersede the more general (and vague) concepts of keystone species, key habitats, fully functional ecosystems, and properly functioning conditions.

For example, one of the species-level KEF patterns that is useful for identifying focal species or species potentially at risk is that of a functional specialist. A *functional specialist* is a species that performs very few ecological roles in its ecosystem (that is, a species with very few key ecological functions, as listed in table 11.1), even if other species might perform those roles as well. In western North America, functional specialists include some strictly insect- or invertebrate-feeding herps, such as the Larch Mountain salamander (*Plethodon larselii*) and the sharptail snake (*Contia tenuis*); some insectivorous small mammals, such as the Baird's shrew (*Sorex bairdi*); and some carrion feeders, such as the turkey vulture (*Cathartes aura*). Functional specialists may be very vulnerable to changes in their key resources and thus may suggest special conservation attention. This attention was evidenced by the severe decline in several Old World vulture species whose plummeting populations have been caused by their carrion prey being inadvertently tainted by pharmaceutical chemicals (Milius 2004).

Table 11.2. A taxonomy of patterns of key ecological functions (KEFs) of wildlife species and communities, and how to evaluate them using a wildlife–habitat relationships database. Many of these categories are unstudied in wildlife communities; thus ecological implications should be viewed as working hypotheses. (See table 11.1 for KEF categories.)

Functional pattern	Definition	Ecological implications	How to evaluate
Community patterns			
Functional redundancy	The number of species performing the same ecological function in a community.	High redundancy imparts greater resistance of the community to changes in its overall functional integrity. Low redundancy suggests critical functions to watch.	Tally the number of species by KEF category for specific wildlife habitats, comparing changes over time or among habitats.
Functional richness	The total number of KEF categories in a community.	Functional richness denotes the degree of functional complexity; greater functional richness means more functionally diverse systems. Functional richness also denotes the degree to which the full "functional web" of a community would be provided or conserved.	Tally the number of KEF categories among all species present in a wildlife habitat and, optionally, habitat structure. Compare such tallies resulting from changes in habitats and structures.
Total functional diversity	The total array of KEF categories weighted by their redundancy—i.e, the number of functions times the mean functional redundancy across all functions.	Total functional diversity denotes the total functional capacity of a community. High total functional diversity means many functions and even redundancy among functions; low total functional diversity means few functions and skewed redundancy (some functions with few species).	Tally the number of species by KEF category for a given wildlife habitat; calculate the mean number of species per KEF category. Can assign weights to some KEF categories if they pertain to specific management objectives.
Functional web	The set of all KEFs within a community and their connections among species and thence to habitat elements.	A functional web depicts how habitat elements provide for species, and the array of ecological functions performed by those species. Functions typically extend well beyond the specific habitat elements.	Identify habitat elements of management interest; list all species within a wildlife habitat that are associated with those habitat elements; list all KEF categories associated with those species. Compare changes in habitat elements.
Functional homologies	The functional similarity of communities even if species composition differs.	Two communities are functionally homologous if they have similar functional profiles and patterns of functional redundancy, even if the species performing the functions differ. Functionally homologous communities can be expected to operate in similar ecological ways.	Produce functional profiles across all KEF categories for several communities or for a community over time based on its expected changes in habitat elements, habitat structures, etc. Compare the profiles (e.g., via contingency analysis) and identify statistically similar (functionally homologous) communities.

Table 11.2. Continued

Functional pattern	Definition	Ecological implications	How to evaluate
Geographic patterns			
Functional bottlenecks or cold spots	Geographic locations with very low functional redundancy of an otherwise widely distributed functional category.	Functional bottlenecks denote areas of higher risk of severing functionally connected communities across the landscape. Severing functions might set the stage for degradation of functional ecosystems.	Map wildlife habitats and/or distributional ranges of wildlife species. For a given KEF category, map the number of wildlife species in each habitat or overlay their range maps. Identify locations with the lowest species richness bordering higher richness on each side; these are geographic functional bottlenecks ("cold spots").
Functional linkages or hot spots	Geographic locations with very high functional redundancy.	Functional hot spots denote areas where many species provide a specific ecological function; such communities may be more resilient to changes in environment or habitat for that function.	Map species richness for a particular KEF category as above. Identify locations with the highest species richness; these are functional linkages ("hot spots"). Determine which species occur in a given hot spot and their wildlife habitats, habitat elements, and habitat structures, and how changes might influence the persistence of the species and thus the redundancy of the function.
Species' functional roles			
Functional keystone species, critical functional link species, and critical functions	Functional keystone species are species whose removal would most alter the structure or function of the community. One type of functional keystone species is critical functional links, which are species that are the only ones that perform a specific ecological function in a community. A critical function therefore is the associated functional category represented by only one (or very few) species within a community.	Reduction or extirpation of populations of functional keystone species and critical functional links may have a ripple effect in their ecosystem, causing unexpected or undue changes in biodiversity, biotic processes, and the functional web of a community.	For a given wildlife habitat, tally the number of species for each KEF category (functional redundancy). For KEF categories with only one species, determine which species performs this function. This species is a critical functional link species for this particular function in this habitat.

Functional breadth and functional specialization of species	The number of ecological functions performed by a species.	Species with the narrowest functional breadth (i.e., fewest functions) are functional specialists and may be more vulnerable to extirpation from changes in conditions supporting that function.	For a given wildlife habitat, tally the number of KEF categories for each species. Identify the species with the fewest number of categories. These are functional specialists. Determine their habitat elements and structures and thus their potential vulnerability to changes thereof. Functional specialists that are also functional keystones may be of high priority for conservation attention.
Functional responses of communities			
Functional resilience	The capacity of a community to return to a starting pattern of total functional diversity, richness, and redundancy following a disturbance event.	Functionally resilient communities are better able to maintain their biotic processes in the face of disturbances. Conversely, it is important to know how far a community can be changed by some anthropogenic disturbance event and still be able to return to its starting functional pattern.	Determine the total functional diversity, functional richness, and functional redundancy of a predisturbance community. Then determine the types and rates of recovery of its wildlife habitats, habitat elements, and habitat structures following some disturbance; the wildlife species associated with such recovery stages; and the species' KEF categories and functional diversity, richness, and redundancy for each recovery stage. Compare stages for functional similarity and thus resilience.
Functional resistance	The ability of a community to resist changing its functional diversity, richness, and redundancy, following a disturbance event.	Functionally resistant communities can be counted on to continue to provide specific ecological functions in spite of and during disturbances. They may provide a bastion for a specific desired function in a disturbed or managed landscape.	Analyze as above and determine the degree to which functional diversity, richness, and redundancy do not change for each postdisturbance stage. This determination is a measure of functional resistance.

Table 11.2. Continued

Functional pattern	Definition	Ecological implications	How to evaluate
Functional attenuation	The degree to which the set of ecological functions within a community simplify following a disturbance event.	Functionally attenuated communities provide fewer or lower redundancies of ecological functions. It may be particularly important to know the degree of functional attenuation to be expected following anthropogenic disturbances.	Analyze as above and determine which KEF categories likely drop out and which remain over postdisturbance recovery stages as compared with initial conditions. Calculate functional diversity, richness, and redundancy for each stage to determine the rate of functional attenuation. Compare final-stage to initial conditions to determine overall functional attenuation.
Functional shifting	The degree to which a community changes to a new, stable, functional constitution following a disturbance event.	Communities with low functional resilience or resistance may end up with a new array of functions and a new pattern of functional diversity, richness, and redundancy. It may be particularly important to know how a community might functionally shift following anthropogenic disturbances and thus which functions might be weakened, strengthened, lost, or gained.	Analyze as above and compare KEF categories and functional diversity, richness, and redundancy between predisturbance and final, stable stages. Identify which functions are lost or gained, and which change in redundancy.
Imperiled functions	A function that is represented by very few species (critical functional link species) or by species that are themselves scarce, declining, or moribund, where extirpation of the species would mean loss of the function. Imperiled functions also can be identified geographically.	Loss of imperiled functions serve to degrade overall ecosystem integrity. Even seldom-performed ecological functions might be critical to maintaining ecosystems, such as occasional dispersal of plant seeds in the face of shifting climates.	For a given wildlife habitat, determine KEF categories with the lowest functional redundancy, and the risk level of the associated species. Imperiled functions are those with one or few species that are themselves at risk.

Source: Marcot and Vander Heyden 2001. Reprinted with permission of Oregon State University Press Copyright 2001.

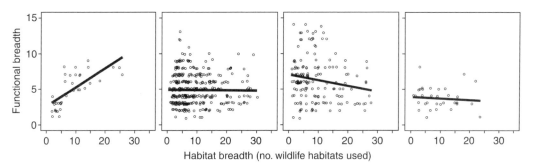

Figure 11.2.   An example of how patterns of the functional aspects of wildlife do not necessarily correlate with patterns of habitat use. These graphs plot functional breadth (number of categories of key ecological functions; see table 11.1) against habitat breadth (number of macrohabitat types used) for all 595 species of nonfish vertebrates regularly occurring in the Columbia River Basin in western North America represented in a wildlife-habitat relationships matrix database (unpublished data from T. O'Neil, pers. comm.). For the most part, there is no statistically significant correlation (P >> 0.05) between functional breadth and habitat breadth. The significant positive correlation with salamanders is due to a set of 16 mostly insectivorous salamanders (most or all species of Dicamptontidae, Rhyacotritonidae, and Plethodontidae) that are tied to ≤5 mesic and humid forest types. The overall conclusion for wildlife assessments is to evaluate habitats, species, *and* functions separately.

Another functional pattern of use in wildlife evaluations and habitat management is that of *critical functional link species,* which are the only species in a community that perform a particular ecological role. Loss of the species means loss of that function. Examples of critical functional link species in various communities and ecosystems are the hippopotamus (*Hippopotamus amphibius*), the American bison (*Bison bison*), and the American alligator (*Alligator mississippiensis*); they each perform the "critical function" of creating or maintaining wetlands or pools by wallowing and are usually the only species in particular communities with this function. Many other species use and depend on the wetlands and ponds they create or maintain. Removal of critical functional link species would remove their function and likely adversely affect many other species (e.g., Campbell et al. 1994; Knapp et al. 1999; Matlack et al. 2002). The many other function-related patterns listed in table 11.2 likewise can be, and are being, used in

wildlife evaluations to help set priorities for wildlife–habitat management.

An Ecography of Key Ecological Functions

The concept of KEFs can be expanded to a practical geographic context. Using geographic information systems, maps of functional redundancy have been produced for a variety or projects in the western United States (e.g., Marcot et al. 2002). Maps of functional redundancy for those species with specific KEFs are made by stacking maps of species' known or expected occurrence (such as from their habitat relations). An example is shown in figure 11.3, which maps the changes in functional redundancy from historic to current time periods for the function of soil digging. Soil digging by vertebrate (as well as invertebrate) organisms can increase soil porosity, increase uptake of organic matter, affect soil bulk density, and otherwise variously influence productivity of soils in an

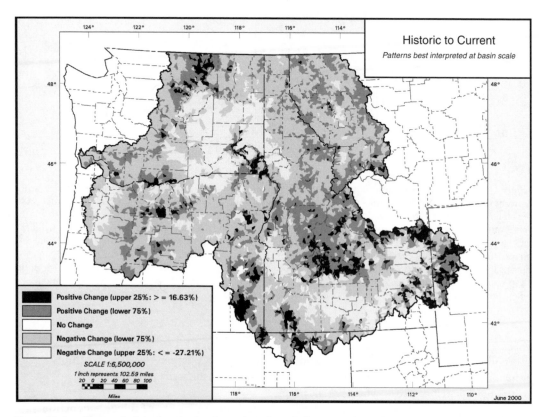

Figure 11.3. Changes in functional redundancy (number of all terrestrial vertebrate wildlife species) from historic (early 1800s) to current (ca. 2000) time periods, for the key ecological function (KEF) of soil digging (see KEF 6.1 in table 11.1). Such maps display locations where functions have likely declined or increased. This map was produced by intersecting maps of wildlife habitat types with information on species occurrence by habitat type and their KEF categories. (From Marcot and Aubry 2003, 642 [fig.19.6]; reprinted with permission of Cambridge University Press.)

area (Scheu et al. 2002; Canals et al. 2003). The map in figure 11.3 shows how redundancy of this function has declined significantly in many major valley and inland lowland areas, such as the Willamette Valley in western Oregon, the Columbia Basin in southeastern Washington, and the Snake River Valley in southern Idaho, largely because of agricultural conversion of na-

tive grasslands, shrublands, steppe, and woodlands in these areas. Restoration of this ecological function may entail restoring some of these native environments along with their native species that perform this function. In this way, geographic patterns of KEFs and their changes can be mapped and used in broad-scale planning and assessment efforts.

Beyond Habitat: Species–Environment Relations

A relational database can list wildlife species along with their KECs, KEFs, and range distributions. Such a database extends beyond the traditional wildlife–habitat relationships approach by considering nonhabitat environmental elements and species' ecological functions. With queries devised to ask questions spanning species and their correlates, functions, and distributions, such a database can be termed a more general *species–environment relations* model.

A species–environment relations model can be used to pose new questions about the roles of species in ecosystems and how managers might provide for species ecological functions through habitat management. New questions to ask can include, Which species provide specific ecological functions, such as nutrient cycling, soil turnover, or insect population control? What are the collective habitat and environmental requirements of such species? Where do they occur? How do such ecological functions affect the productivity and diversity of ecosystems? Such questions are now answerable and are being used in wildlife assessment and habitat management projects, as illustrated with the examples given in the previous section.

In the context of a species–environment relations model, habitats can be viewed as part of ecological subsystems, and species along with their ecological functions can be identified as part of each subsystem. Subsystems include below-ground, surface, and arboreal components of terrestrial, riparian, and aquatic environments. Each subsystem has associated species and ecological processes that contribute to the overall functioning of the ecosystem. The general approach may be (1) to identify a specific habitat, substrate, or KEC of interest and to identify the ecological subsystem in which it pertains; (2) to use a species database to list all species associated with that habitat, substrate, and

KEC, by subsystem; and (3) to list the collective set of key ecological functions of all species by subsystem.

Alternately, one could specify an ecological function and identify all species having this function along with their collective set of habitat requirements and ecological correlates by subsystem. This approach would help identify the set of conditions that habitat managers might provide to help maintain habitat conditions for species with specific ecological functions. Other approaches are also possible, such as overlay-mapping all species associated with a particular ecological function as a way to represent the full geographic distribution of (and potential management risks to) specific functions. In a sense, this map overlay approach would extend the current approaches of spatially explicit modeling of wildlife populations to include species ecological functions.

Classification of land units, including ecoregions or bioregions, existing or potential vegetation communities, and habitat types, are useful and often essential, but they do not by themselves tell us much about the status and trend of associated wildlife species or communities. Instead of striving to map wildlife communities or ecosystems per se, we suggest that abiotic and biotic components of ecosystems and those that most affect wildlife species can be more readily mapped and analyzed.

Beyond Functions: Considering Ecosystem Services

KEFs and KECs of wildlife species, and the environments in which species occur, can be further interpreted in terms of the ecosystem services (or ecological services) they provide. The concept of *ecosystem services* refers to the array of natural resources and processes that are valued by humans and that sustain ecological communities and ecosystems. The literature on ecosystem services broadly describes a very wide array of such resources and processes (e.g., papers in Daily 1997; Daily et al. 1997; Ostfeld and LoGiudice 2003). In reviewing this literature, we developed a set of five categories of ecosystem services that can begin to focus discussion (table 11.3): (1) ecological functions of organisms that maintain ecosystem integrity, such as pollination of desirable plants and food crops; (2) biophysical services mediated by environmental processes and conditions, such as filtering of surface water; (3) ethnobiotic services that provide cultural, social, and religious values of organisms, such as medicinal plants; (4) economic services that provide largely intangible benefits, such as recreational experience; and (5) natural resource services that provide renewable commodities of tangible and direct economic value, such as game animals. Many other examples are listed in table 11.3.

The purpose of classifying ecosystem services is to better identify the many important roles that wildlife species and their ecological functions and habitats play in providing for human needs and ecological systems alike. Valuating ecosystem services of interest to people can help identify economic benefits of managing wildlife species and their habitats (Scott et al. 1998). Rudiments to valuation of ecosystem services have been written of for decades, such as the value of certain bird and bat species consuming insect pests (e.g., Takekawa and Garton 1984) and the economic value of pollinator services (Cane and Tepedino 2001). A fuller classification of ecosystem services can help people recognize and quantify the many economic, ecological, social, and even spiritual benefits of maintaining wildlife.

Beyond Viability: Managing Populations in a Community and Ecosystem Context

Much continues to be written on whether wildlife management should focus on individual species or on ecosystems (e.g., Franklin 1993; Bowen

1999). We do not view this argument as particularly fruitful, as both are necessary for ensuring long-term viability of populations, evolutionary potential of species, provision of habitats, and integrity of ecological communities and systems. An approach that incorporates habitat components and other environmental correlates, as well as that explicitly acknowledges species' functions and ecosystem processes, should strive to encompass several levels of biological organization. Simple coarse-filter approaches to wildlife habitat management may be one useful starting point, but they cannot account for requirements of all ecological entities (see table 3.3).

In a more comprehensive approach, as outlined above, quandaries of defining minimum viable population levels and prioritizing individual species for recovery efforts, as in a triage approach, would be reduced. Ultimately, we are all part of the ecological systems we try to manage, and models that can depict wildlife species can just as well be built for the human species. In this way, species, ecosystems, ecological functions, environmental conditions, ecosystem services, management goals, and management activities can all be merged into one coherent approach for assessment and management. The utility of this comprehensive approach should become more evident for managing wildlife in an evolutionary context, and for use in adaptive management.

Managing Wildlife in an Evolutionary Context

Much of wildlife habitat management focuses on more immediate concerns of habitat mitigation, conservation, and restoration. However, environments will continue to change, climates will vary stochastically, and disturbance events such as major fires will recur; nature is not static. Thus one aspect of management should view wildlife in an evolutionary context. Although specific outcomes of future evolution are unpre-

Table 11.3. Classification of ecosystem services

Class of Ecosystem Service	Definition	Examples
Ecological services of organisms	Key ecological functions and services directly provided or mediated by organisms. *Ecological services of organisms* refers to the strictly ecological roles, played by organisms, that help to keep ecosystems diverse, productive, and fully functional.	Pollination Carbon sequestration Detoxifying xenobiotics Decomposition of wastes Generation and renewal of soil and soil fertility N mineralization Soil protection by biotic crusts Slope stabilization by roots and down wood (CWD) Dispersal of seeds, spores, and disseminules Translocation of nutrients
Biophysical services	Services mediated or provided by macrohabitats in conjunction with biophysical conditions. *Biophysical services* refers more to processes of the physical environment.	Filtering of water by wetlands Filtering of water through hyporheic zones Filtering of air by forest vegetation Providing of O_2 by plants Ameliorating of fire intensity Moderation of floods and droughts Protection from the sun's harmful UV rays Partial stabilization of climate Moderation of temperature extremes and the force of winds and waves
Ethnobiotic services	Cultural values of plants and animals. *Ethnobiotic services* refers to cultural, religious, and other intangible values, as well as native medicinal uses.	Medicinal plants used by native peoples Plants and animals of cultural or religious significance Existence value of organisms Providing of esthetic beauty and intellectual stimulation that enrich the human spirit
Economic services	Services of direct economic significance. These are services of economic significance, but that are not direct tangible commodities per se.	Pollination of crop plants Biological control of agricultural or forestry pests, diseases, or pathogens Recreational services, such as birdwatching, wildlife viewing, and ecotourism Direct use of plant products in pharmaceutical industry
Natural resource services	Direct provision of renewable natural resources and commodities of economic value. These are services that directly produce or support resources to be used or harvested and sold in the marketplace.	Timber (forest growth cycle) Clean water (hydrologic cycle) Game animals and furbearers Nontimber forest products (mosses, mushrooms, ferns, etc.) Healthy rangelands for livestock grazing

dictable—genetic engineering aside—the purpose would be to provide for environmental conditions that permit natural evolutionary processes to proceed, including species interactions and the role of abiotic factors in selection pressures. The ebb and flow of taxonomic lineages—including emergence of subspecies, species, and higher taxonomic groupings—is afforded only by providing a diverse and stochastic array of environmental conditions and gene pools of organisms.

Adaptations and genetic changes in organism lineages can be caused by many factors and can take many forms. Examples are *directional selection,* in which the modal genotype and phenotype of a population shifts consistently in one direction, such as to longer bill length to better exploit some new resource; *neutral selection,* in which no one phenotype is favored, but modes or variations in genomes can still change; *character divergence,* as between different morphs (see table 3.3), which results in two modal phenotypes; *genetic drift,* which may produce new modal phenotypes from random reselection in small, isolated populations; and *inbreeding depression* in small, isolated populations, which shifts modal genotypes and lowers the level of mean fitness.

Identifying the physical and environmental conditions needed to maintain the evolutionary potential of wildlife may seem like an impossible task, as there are so many moving parts and relationships in ecological and evolutionary systems. However, much of this task breaks down to managing for conditions that sustain taxonomic lineages and managing for ecological domains with an eye on evolutionary time frames. Specifically, we can suggest a set of 10 guidelines that together would go far for this goal.

1. *Maintain the full array of native environments and species habitats, prioritizing restoration or protection of those in greatest decline.* An example is the prodigious work to save and restore the last of the North American tallgrass prairie.

2. *Evaluate and provide for long-term viability of all at-risk species.* This helps maintain full species richness and biodiversity. Further, even species (including plants, invertebrates, and vertebrates) that are naturally rare, or made rare by human actions, can serve important functional roles (Walker et al. 1999; Miller et al. 2003).

3. *Maintain environmental conditions and habitats for species along their range peripheries.* It is often on the peripheries where selection pressures are different and new forms can evolve (parapatric speciation). Peripheral conditions also provide for floaters and sink habitats and can contribute to metapopulation stability (Lesica and Allendorf 1995; also see Furlow and Armijo-Prewitt 1995).

4. *Maintain environments at the edge of species' range of tolerance.* This includes such situations as maintaining habitats for montane species at their upper elevational ranges. The benefits are to provide for longer-term movement of population centers under regional climate change, and opportunities for evolution of locally adapted ecotypes and morphs under other changing conditions (see table 3.3).

5. *Maintain locally and regionally endemic species.* Endemic species are the unique signatures of local ecosystems and once lost are gone forever. Together with at-risk (threatened) species, targeting endemic species for conservation can help maintain a significant proportion of biodiversity in an area (Ceballos et al. 1998; Posadas et al. 2001; Bonn et al. 2002).

6. *Maintain locally endemic subspecies and ecotypes.* Endemic subspecies and ecotypes are life forms with unique genomes

particularly adapted to local conditions, and they contribute to overall genetic diversity of species' gene pools.

7. *Identify and maintain cryptic or sibling species complexes.* Cryptic or sibling species complexes are species recently evolved and diverged in some way (often behaviorally) that still appear nearly identical, such as the species complex of red crossbills (*Loxia curvirostra* spp.) in North America. Such complexes contribute to the genetic diversity that underlies adaptation to changing or diverse environmental conditions. In some cases, a sibling species may be endangered and require listing and recovery efforts (e.g., McElroy et al. 1997).

8. *Maintain disjunct populations that are geographically and often genetically isolated.* Disjunct populations sometimes occur in environments with different selective pressures than found in the parent species' range and become centers of new evolutionary forms—morphs, subspecies, and eventually new species. In fact, if the natural situation is a disjunct population, artificially enhancing gene flow to other populations could be detrimental (Storfer 1999; Russell et al. 2002). The argument is the same for maintaining small, isolated patches of vegetation, wetlands, or other environments that can serve as critical refuges and stepping stones for many organisms (e.g., Saunders et al. 1987; Semlitsch and Bodie 1998).

9. *Maintain natural population linkages.* Maintaining natural population linkages helps preserve natural metapopulation structures to ensure that subpopulations do not get isolated and suffer increased risk of local extirpation from environmental disturbances, demographic stochasticity, inbreeding depression, or other threats. Habitat corridors and other link-

ages among populations can take many forms at many scales, depending on the species' requirements and local conditions (Harris 1988; Hudson 1991).

10. *Maintain conditions indicative of particularly genetically diverse populations.* These conditions include polymorphisms (Wiens 1999) and other aspects of what Butler and Mayden (2003) termed "cryptic biodiversity." Such genetically diverse populations have the potential for evolving adaptive behaviors and speciating (Mallet and Joron 1999). Loss of genetic diversity from a population may decrease fitness of a species (Avise 1995), as was documented by Hitchings and Beebee (1998) in common toads (*Bufo bufo*).

Although the above list focuses in part on maintaining naturally segregated conditions (range peripheries, disjunct populations, etc.), there are many cases in which evolution of new species occur without such geographic barriers that isolate populations and prevent gene exchange. Such situations may be indicated by occurrence of ecotypes, sympatric polymorphisms, and sympatric sibling or cryptic species. A number of selection forces can serve to differentiate species sympatrically, such as bacterial agents, mimicry, changes in a single pigment gene or in the song of a bird, or rapid evolution of a sperm surface protein (Morell 1996). Other conditions worthy of attention for maintaining the long-term evolutionary potential of species' lineages include understanding the roles of hybridization, disease and parasitism, effects of commensalisms with and dispersal by humans, adverse effects of exotic species, and other topics we have covered in this book. Many of these factors are left to species-specific evaluations.

Management of species or segments of a species' distribution for evolutionary potential is not a new concept. It has been used, for example,

under the U.S. Endangered Species Act to delineate and manage threatened salmonid stocks.
For this purpose, the formal concept of an "evolutionary significant unit" (ESU) has been defined as a population that (1) is substantially reproductively isolated from other conspecific
population units, and (2) represents an important component in the evolutionary legacy of the
species (Waples 1995; see also Pennock and
Dimmick 1997; DeWeerdt 2002).

Karl and Bowen (1999) suggested a new category for conservation: the *geopolitical species.*
They discovered through mitochondrial DNA
research that the endangered black sea turtle
(*Chelonia agassizii*) is not distinct from the green
sea turtle (*C. mydas*), yet the designation of, and
conservation focus on, the former has persisted
for over a century due to geographical and political considerations. They concluded that the
black sea turtle should continue to be the focus
of endangered species conservation because of
its morphological diversity and the possibility of
its being an incipient species with novel adaptations. This decision corresponds well to our call
for appreciating and conserving novel or endemic subspecies, morphs, and other genetically
unique portions of a species' lineage.

**Managing for Ecological Domains and
Evolutionary Time Frames**

Certainly, managing for evolutionary time
frames can seem a daunting task for managers
concerned with relatively short-term resource
management objectives. We are not suggesting
that evolutionary changes be predicted—this is
not yet possible, given current ecological knowledge (Bennett 1996). Rather, we are suggesting
that the base conditions from which evolution
proceeds be provided, as per the guidelines discussed above. In this way, the array of ecological
domains—the spectra of conditions and variations thereof—can be provided over time, thus

forming the foundation for long-term evolutionary change in organisms.

A final word about evolutionary time frames.
Most biologists view evolution as an immensely
slow process of changes in the morphology,
anatomy, or physiology of organisms that occur
over millennia and eons. This time frame may be
valid for some taxonomic lineages, and higher
taxonomic levels such as families and orders tend
to change more slowly than lower levels such as
genera and species. However, there are startling
new cases of rapid "microevolutionary" shifts
caused by fast changes in environmental conditions due to human activities (e.g., Pergams and
Ashley 1999; Pergams et al. 2003; Sayama et al.
2003). These cases clearly demonstrate that
adaptation of forms and modal genotypes to
changing conditions can occur over centuries,
decades, and, in some cases, just a few years.

This demonstration puts an entirely new perspective on what managers ought to be focusing
on when providing for wildlife habitats and
when projecting effects of management activities. Rice and Emery (2003, 469) cautioned that
"while there may not be any universal rules regarding the adaptive potential of species, an understanding of the various processes involved in
microevolution will increase the short- and
long-term success of conservation and restoration efforts." Microevolutionary processes can
indeed have important consequences for species conservation, management, and restoration
(Ashley et al. 2003). Thus our 10 factors, as listed
above, may pertain to ensuring both long- and
short-term microevolutionary potential in the
face of rapid environmental change.

Adaptive Management:
Learning from Experience

Managing wildlife and habitats in the real world
entails much uncertainty. One approach to deci-

sion making under uncertainty is that of adaptive management.

The Adaptive Management Approach

We have advocated the use of adaptive management as a means of conducting better habitat management over time. The approach entails identifying areas of scientific uncertainty, devising field management activities as real-world experiments to test that uncertainty, learning from the outcome of such experiments, and recrafting management guidelines based on the knowledge gained (Holling 1978; Walters 1986; Irwin and Wigley 1993). Modeling can play a key role in formalizing our current knowledge and identifying important areas of uncertainty (Walters et al. 2000; Lynam et al. 2002). In an ideal situation, management guidelines equate to creation of testable hypotheses (Havens and Aumen 2000); monitoring and adaptive management studies equate to conducting the experiment; and revision of the management guidelines equates to reevaluation and interpretation of the study results in terms of testing the validity of the initial hypothesis.

Adaptive management, however, has seldom been applied successfully or fully in managing wildlife habitat and ecosystems (Gray 2000). Problems are both technical and administrative (Lee and Lawrence 1986). Following are some basic tenets of a successful approach to adaptive management. They can be used as a checklist for specific programs to ensure successful application.

TENETS OF A REAL-WORLD APPROACH TO
ADAPTIVE MANAGEMENT

1. The administration has the political will, and explicit procedures, to accept change. This first tenet summarizes a great deal of experience in the administrative behavior and political reality of resource management organizations and in-

stitutions. Despite good intentions and explicit promises in management plans, if an organization lacks the political motivation or the actual process for accepting new knowledge, then management activities will remain unchanged. What is needed is a clear political mandate to accept change in a timely manner and a formal protocol for weighing and incorporating new information for potential use in updating or reaffirming management decisions.

A corollary to this tenet is that *an explicit risk management framework exists to accept and weigh new information in a timely fashion.* To use new information, resource management organizations must have a basis for acknowledging and incorporating uncertainty into their decision-making processes (Gunderson 1999). Often, however, new information is treated as antithetical to carrying out the management or planning guidelines already chosen, and is used only under duress of litigation or only during intermittent updates to planning in predefined planning cycles (Clark et al. 1996). For federal resource agencies in the United States, such planning cycles run years or decades long.

Another corollary to this tenet is that *performance standards for managers or decision makers explicitly address environmental conservation.* Here is the heart of the problem. If individual managers or decision makers are not held individually responsible for meeting the objectives of adaptive management (and wildlife habitat management), then new information will not be used and management will not change. In this context, environmental conservation refers to wise, sustainable use, which may include strict preservation of habitats or environments, but not necessarily so.

A third corollary is that *what is "at risk" is the condition of the land, not the status of one's career or the political decision space of the management directives.* Wildlife risk analysis and risk managers typically assume the very concept of what is "at risk" is not shared by biologists, managers, decision makers, and politicians alike (e.g., Graham et al. 1991). In reality, biologists explicitly speak of risk in the sense of the likelihood of species extirpation. Managers and decision makers implicitly may use the concept of risk, but in reference more to how a decision may impact their career status or the meeting of overall organizational directives that may extend beyond species or habitat objectives. Upper-management decision makers and politicians may view risk in terms of the political decision space—what is politically acceptable to their peers, to special interest groups, and to funding or voting constituencies. Each of these aspects of risk is authentic, but quite different. Only through clearly specified performance standards for meeting adaptive management goals can specialists, managers, and decision makers be assured to view the concept of risk in the same way. In the absence of such standards, the next best approach is to clearly articulate the bases for risk in a given assessment or decision so it is clear to all how the term and concepts were defined and weighed in any decision affecting public lands and wildlife.

Use of decision modeling techniques can aid managers in choosing an optimal course of action and in articulating their decision criteria. For example, the decision model of McNay et al. (1987) provided a means of prioritizing management for deer populations in coastal British Columbia, Canada, and making explicit the environmental and management conditions that form the basis for the priorities. However, such models are of limited utility if decision makers do not wish to follow such a rigorous procedure or expose their decision-making criteria because of political risk.

2. Options exist to change. This tenet may seem to be an obvious point, but it is often overlooked or deliberately not addressed. In some cases, extensive funding has been provided for research or monitoring while options vanish for changing conditions on the land, such as through protection or restoration of dwindling habitats. One example may be that of the Mt. Graham red squirrel (*Tamiasciurus hudsonicus grahamensis*), a potentially threatened subspecies occurring on a mountaintop in the southwestern United States that is also coveted as a development site for an astronomical observatory (Warshall 1995). In this example, ongoing population studies are being conducted during initial development of the observatory site. The development is designed to retain options for more or less restrictive habitat conservation, depending on the findings, although the degree of risk to the population is still subject to some debate. In other cases, however, monitoring may proceed while scarce habitat continues to be changed, such as monitoring populations of the endangered Lanyu scops owl (*Otus elegans botolensis*) (Severinghaus 1992) while tropical forests on this island of Taiwan are felled or converted to other uses.

An adaptive management approach needs to determine the rate of change in habitats, species, or populations of dire concern, as compared with the pace of the information-gathering and decision-making or decision-changing process, to ensure that critical conservation options are not lost during the information-gathering process. More fundamentally, conducting monitoring and research studies should not be used in place of making difficult decisions about allocation of scarce resources, or as a smokescreen to permit management activities to continue that eliminate options for conservation of a scarce and dwindling resource. Monitoring, research, and funding support, *but do not substitute for,* sound resource management decisions and actions.

A corollary to this tenet is that *irreversible losses of resources or environmental conditions are not incurred during the "testing period."* For example, monitoring population dynamics or demography of an endangered species while its habitat continues to be adversely altered violates this corollary and ensures serious problems for meeting conservation goals. In some cases, adaptive management experiments can deliberately sacrifice or seriously alter some environmental conditions for a habitat or species of interest in the name of quantifying effects of management activities. However, the pace of such experiments and the degree of reversibility of losses must still permit options for changing management activities, such as for in situ protection or recovery of threatened species.

Another corollary to this tenet is that *changes in environmental conditions from human activities together with those from natural disturbances do not outstrip the pace of monitoring and learning and the potential to change management activities.* In other words, adaptive management approaches should attend to the additive effects of both human activities (ongoing management or new experiments) and natural disturbances, and therefrom gauge likelihoods of being able to change activities in time to ensure meeting conservation objectives. For example, if management experiments are testing the effect of various degrees of draining of some rare wetland type over a specified time period, the likelihood of environmental catastrophes such as prolonged drought over the same time period should be factored into the experiment to help ensure that options for change still exist at the end of the experiment.

3. Indicator variables—environmental parameters—can indeed be identified and realistically monitored in a cost-effective way. Many adaptive management studies can be envisioned that cannot be realistically carried out. They may not be viable because indicator variables cannot be readily identified or measured, or because the experiment is so complex or costly that it cannot be completed with adequate sample sizes or intensity of study. These problems are often the case in studies of the response of carnivore populations to management activities, for example, or in studies of ecological processes across entire landscapes that cannot be replicated. If these problems arise, then the directives for management and the focus for associated adaptive management studies need to be more tightly specified and new statistical approaches considered (e.g., Reckhow 1990).

Another corollary to this tenet is that *a statistical sampling frame can be established by which to distinguish effects of human activities from background changes.* One premise of adaptive management is that we can differentiate changes caused by human activities from those caused by natural variation or background noise. By carefully designed manipulation experiments—less often, by using "natural experiments" or passive observation studies—this distinction can be determined. Chapter 4 discusses appropriate study designs. This corollary is vital for determining when we can and cannot affect change through management activities. Adaptive management studies need to pay close attention to sampling design, use of controls and treatments, and ensuring specific confidence and power levels in statistical tests.

Often, conservation questions pertain to conditions that occur across very large landscape areas and thus in contexts that cannot be replicated with adequate controls and in sufficient number to meet the assumptions of traditional statistical techniques. In such cases, we must look to other statistical designs, including use of comparative time series and spectral analysis (de Valpine and Hastings 2002), empirical Bayesian statistics (Oman 2000), and optimization approaches (Hof and Bevers 1999), although some of these approaches are controversial. In some cases, entire landscapes can be devoted to demonstration experiments (Lindenmayer et al. 1999). We advocate that a plurality of approaches be taken to provide the widest possible means of learning when unique conditions violate assumptions of traditional statistics.

A third corollary to this tenet is that *objectives and expected effects should be clearly articulated and quantified.* This corollary may seem an obvious necessity, but it is often overlooked, particularly in observation (nonexperimental) studies that result in post hoc "fishing expeditions" for patterns and management affects, and in demonstration studies designed mostly to justify and conduct a priori desired activities.

Overall, we make the following suggestions to help ensure a successful approach to adaptive management of habitats for wildlife:

First, review the above tenets and their corollaries as a checklist.

Second, clearly separate risk analysis from risk management. Risk analysis should entail estimating (and partitioning) likelihoods of outcomes resulting from potential management actions and from natural conditions and changes ("chance events"). Risk management should entail articulating criteria used for reaching a management decision, including explicating risk attitudes in light of uncertain projections and incomplete information.

Third, in the future we will need to learn from all kinds of information-gathering approaches, including observational or correlative studies, controlled field experiments, laboratory experiments, uncontrolled field trials, and anecdotal experience, as well as theoretical models. All of these approaches can be useful and should be complementary in an adaptive management framework. The challenge is in appropriately combining information gathered from disparate studies and approaches into an overall understanding of the real-world system.

Fourth, technical staffs can help managers interpret and understand ramifications of uncertainties and probabilities of outcomes. Many resource decision makers may not be particularly adept in dealing with scientific uncertainty (Policansky 1993). For example, they may view scientific uncertainty as lack of "proof" of any particular effect, so believe it can be discounted and ignored in, or even touted as supporting evidence for, management decisions. This approach is not an appropriate interpretation of scientific uncertainty. And it is not a correct understanding of the scientific process—that is, to "verify" a hypothesis, we do not seek proof, but rather corroboration or, more accurately, lack of statistical falsification.

Also—and this point is greatly misunderstood by many policymakers—uncertainty is not the same as complete lack of knowledge. An uncertain outcome can be expressed as a low likelihood or as a high variation in potential outcomes, whereas lack of knowledge simply means we do not know and cannot estimate outcome likelihoods. That is, lack of knowledge is not the same as high variance. Ecosystems and their component communities, populations, processes, and disturbance regimes can be highly variable through space and time, and the variance can be precisely measured. In a sense, we can be certain of high variance in some systems! What may be uncertain is a specific future population level or resource productivity level (e.g., the population "standing crop" of some big-game species, in a specific future decade). Even this uncertainty may be expressed as an expected level with some degree of associated variance (such as a mean herd size of harvestable bull elk plus or minus a standard error of prediction). We can do better to express inventories based on samples, and predictions based on projections, as likelihoods and variances, and to help managers and decision makers better understand

and interpret uncertainty and, where appropriate, lack of knowledge.

Real-World Circumstances and Use of Complementary Studies

Ideally, adaptive management would proceed as rigorously defined experiments adhering to all the assumptions and tenets of well-designed scientific studies (chapter 4). In reality, adaptive management studies have to contend with a number of problems, including the following:

- Landscape-scale studies with few or, more usually, no replicates
- "Experiments" with no controls or with controls in vastly different situations (such as higher-elevation wilderness areas)
- Little or no time to collect baseline data
- Losses of selected samples and declining sample sizes over the course of the study due to changes in administrative or management direction
- Unannounced and undirected treatment of controls
- Overall short duration with few truly long-term studies to determine lag and secondary effects
- Changes in management objectives, treatments, and sometimes even land ownership over the course of the study

What can be done in the face of such ruinous circumstances? The answer may be found in using multiple studies of various kinds, and in taking advantage of prior knowledge to establish study objectives, management activities to test, and analysis of results. In some cases, the real-world circumstances listed above degrade an otherwise rigorously defined experiment to the status of an observational or demonstration study. Such studies still have value in incremen-

tally adding evidence to help corroborate or refute management hypotheses. But outcomes of observational or demonstration studies should not be taken as hard evidence of the correctness of assumptions underlying a management approach in the absence of supporting rigorous investigation. It is just too easy to find situations and to craft (inadvertently or otherwise) observational or demonstration studies to provide specific answers regardless of actual effects. An example is locating species thought to be closely associated with specific environments in situations other than what would truly be the modal condition of the population, such as observing a few errant spotted owls in very young-growth landscapes. This is no "proof" (or corroboration) of habitat requirement.

Another kind of investigation that can greatly complement the adaptive management approach is that of retrospective studies. The use of retrospective studies is often of great value for understanding the developmental history and past conditions of habitats and wildlife. Retrospective studies are not experiments; rather, they are hindcast reconstructions of prehistoric or historic events and conditions to help us better understand long-term or recent change. Retrospective studies can borrow from a variety of tools and techniques, including dendrochronology, palynology, paleoecology, archaeology, analysis of historic documents, use of historic photopoints, and other information sources or methods.

Retrospective studies can help inform us on long-term changes in environments, climates, and biota, and complex effects of land management actions (e.g., Caswell 2000; Louda et al. 2003). For example, Crumley (1993, 377) provided examples of retrospective studies to unravel "complex chains of mutual causation in human–environment relations" by tracing past human–environment interactions at global, regional, and local scales (compare this with the discussion on scales of ecological disciplines and management issues presented in chapter 8). Covington et al. (1994) analyzed historic changes in forest ecosystems in the inland West of the United States to help project future changes.

Closing Remarks

How vertebrate-centric is the traditional study of wildlife—including many of the examples in this book! Were humans evolved as poikilotherms, or as invertebrates, or in an aquatic environment, or as clonal organisms, or as canopy-dwelling or flying creatures, our definitions of wildlife for scientific, management, political, legal, ethical, and esthetic interests doubtless would be far different than they are. We urge our community of wildlife scientists and managers to reexamine our species- and life-form biases and to open the doors to all organisms and environments in our work. In this way, we envision a future era of wildlife management that weighs equally, in studies as well as management, the host of mostly unknown creatures, both for their own sake and for their influence on vertebrates. These largely unknown creatures include soil, aquatic, and canopy microbes and mesoinvertebrates, and cryptogams and vascular plants. Collectively, they are critical to the Umwelt and health of vertebrates and to the productivity of the crops, forests, and grasslands we hope to sustain.

Literature Cited

*Emperor Shivaji ruled from 1630-1680. The epigraph appears on a signpost written in the Marathi language in front of Kolkaz Forest Rest House, Melghat Tiger Reserve, central India.

Abate, T. 1992. Which bird is the better indicator species for old-growth forest? *BioScience* 42:8–9.

Alcorn, J. B. 1993. Indigenous peoples and conservation. *Conservation Biology* 7 (2):424–26.

Armstrong, G. W., W. L. Adamowicz, J. A. Beck, S. G. Cumming, and F. K. A. Schmiegelow. 2003. Coarse filter ecosystem management in a non-equilibrating forest. *Forest Science* 49 (2):209–23.

Ashley, M. V., M. F. Willson, O. R. W. Pergams, D. J. O'Dowd, S. M. Gende, and J. S. Brown. 2003. Evolutionarily enlightened management. *Biological Conservation* 111:115–23.

Avise, J. C. 1995. Mitochondrial DNA polymorphism and a connection between genetics and demography of relevance to conservation. *Conservation Biology* 9 (3):686–90.

Bennett, K. D. 1996. *Evolution and ecology: The pace of life*. New York: Cambridge Univ. Press.

Bonn, A., A. S. L. Rodrigues, and K. J. Gaston. 2002. Threatened and endemic species: Are they good indicators of patterns of biodiversity on a national scale? *Ecology Letters* 5:733–41.

Bowen, B. W. 1999. Preserving genes, species, or ecosystems? Healing the fractured foundations of conservation policy. *Molecular Ecology* 8:S5–S10.

Butler, K. F., and T. M. Koontz. 2005. Theory into practice: Implementing ecosystem management objectives in the USDA Forest Service. *Environmental Management* 35 (2):138–50.

Butler, R. S., and R. L. Mayden. 2003. Cryptic biodiversity. *Endangered Species Bulletin* 28 (2):24–26.

Campbell, C., I. D. Campbell, C. B. Blyth, and J. H. McAndrews. 1994. Bison extirpation may have caused aspen expansion in western Canada. *Ecography* 17 (4):360–62.

Canals, R. M., D. J. Herman, and M. K. Firestone. 2003. How disturbance by fossorial mammals alters N cycling in a California annual grassland. *Ecology* 84 (4):875–81.

Cane, J. H., and V. J. Tepedino. 2001. Causes and extent of declines among native North American invertebrate pollinators: Detection, evidence, and consequences. *Conservation Ecology* 5 (1):1 [online]; www.consecol.org/vol5/iss1/art1.

Caswell, H. 2000. Prospective and retrospective perturbation analyses: Their roles in conservation biology. *Ecology* 81 (3):619–27.

Ceballos, G., P. Rodriguez, and R. A. Medellin. 1998. Assessing conservation priorities in megadiverse Mexico: Mammalian diversity, endemicity, and endangerment. *Ecological Applications* 8 (1):8–17.

Clark, T. W., R. P. Reading, and A. L. Clarke. 1996. *Endangered species recovery: Finding the lessons, improving the process*. Washington, DC: Island Press.

Covington, W. W., R. L. Everett, R. Steele, L. L. Irwin, T. A. Daer, and A. N. D. Auclair. 1994. Historical and anticipated changes in forest ecosystems of the inland west of the United States. In *Assessing forest ecosystem health in the inland west,* ed. R. N. Sampson, D. L. Adams, and M. J. Enzer, 13–63. New York: Haworth Press.

Crumley, C. L. 1993. Analyzing historic ecotonal shifts. *Ecological Applications* 3 (3):377–84.

Daily, G. C., ed. 1997. *Nature's services: Societal dependence on natural ecosystems.* Washington, DC: Island Press.

Daily, G. C., S. Alexander, P. R. Ehrlich, L. Goulder, J. Lubchenco, P. A. Matson, H. A. Mooney, S. Postel, S. H. Schneider, D. Tilman, and G. M. Woodwell. 1997. Ecosystem services: Benefits supplied to human societies by natural ecosystems. *Issues in Ecology* (Spring):1–16.

de Valpine, P., and A. Hastings. 2002. Fitting population models incorporating process noise and observation error. *Ecological Monographs* 72 (1):57–76.

DeWeerdt, S. 2002. What really is an evolutionarily significant unit? *Conservation Biology in Practice* 3 (1):10–17.

Franklin, J. F. 1993. Preserving biodiversity: species, ecosystems, or landscapes? *Ecological Applications* 3 (2):202–5.

Fulé, P. Z., W. W. Covington, and M. M. Moore. 1997. Determining reference conditions for ecosystem management of southwestern ponderosa pine forests. *Ecological Applications* 7 (3):895–908.

Furlow, F. B., and T. Armijo-Prewitt. 1995. Peripheral populations and range collapse. *Conservation Biology* 9 (6):1345.

Gehrt, S. D., and J. E. Chelsvig. 2003. Bat activity in an urban landscape: Patterns at the landscape and microhabitat scale. *Ecological Applications* 13 (4):939–50.

Graham, R. L., C. T. Hunsaker, R. V. O'Neill, and B. L. Jackson. 1991. Ecological risk assessment at the regional scale. *Ecological Applications* 1 (2): 196–206.

Gray, A.N. 2000. Adaptive ecosystem management in the Pacific Northwest: A case study from coastal Oregon. *Conservation Ecology* 4 (2):6 [online]; www.consecol.org/vol4/iss2/art6.

Gunderson, L. 1999. Resilience, flexibility and adaptive management: Antidotes for spurious certitude? *Conservation Ecology* 3 (1):7 [online]; www.consecol.org/vol3/iss1/art7.

Gunderson, L. H., C. S. Holling, L. Pritchard Jr., and G. D. Peterson. 2002. Resilience of large-scale resource systems. In *Resilience and the behavior of large-scale systems,* ed. L. H. Gunderson and L. Pritchard Jr., 3–20. Washington, DC: Island Press.

Harris, L. D. 1988. Landscape linkages: The dispersal corridor approach to wildlife conservation. *Transactions of the North American Wildlife and Natural Resources Conference* 53:595–607.

Haufler, J. B., R. K. Baydack, H. Campa III, B. J. Kernohan, C. Miller, L. J. O'Neil, and L. Waits. 2002. *Performance measures for ecosystem management and ecological sustainability.* Wildlife Society Technical Review 02-1. Washington, DC: Wildlife Society.

Havens, K. E., and N. G. Aumen. 2000. Hypothesis-driven experimental research is necessary for natural resource management. *Environmental Management* 25:1–7.

Hitchings, S. P., and T. J. C. Beebee. 1998. Loss of genetic diversity and fitness in common toad (*Bufo bufo*) populations isolated by inimical habitat. *Journal of Environmental Biology* 11 (3):269–83.

Hof, J., and M. Bevers. 1999. Spatial optimization for managed ecosystems. *Forest Science* 45 (4):595.

Holling, C. S. 1978. *Adaptive environmental assessment and management.* New York: Wiley.

Hudson, W. E., ed. 1991. *Landscape linkages and biodiversity.* Washington, DC: Island Press.

Hunter, M. L. 1991. Coping with ignorance: The coarse-filter strategy for maintaining biodiversity. In *Balancing on the brink of extinction,* ed. K. A. Kohm, 266–81. Washington, DC: Island Press.

Hunter, M. L. 2005. A mesofilter conservation strategy to complement fine and coarse filters. *Conservation Biology* 19 (4):1025–29.

Irwin, L. L., and T. B. Wigley. 1993. Toward an experimental basis for protecting forest wildlife. *Ecological Applications* 3 (2):213–17.

James, S. E., and R. T. Coskey. 2002. Patterns of microhabitat use in a sympatric lizard assemblage. *Canadian Journal of Zoology* 80 (12):2226–34.

Karl, S. A., and B. W. Bowen. 1999. Evolutionary significant units versus geopolitical taxonomy: Molecular systematics of an endangered sea turtle (genus *Chelonia*). *Conservation Biology* 13 (5): 990–99.

Keddy, P. A., H. T. Lee, and I. C. Wisheu. 1993. Choosing indicators of ecosystem integrity: Wetlands as a model ecosystem. In *Ecological integrity and the* management of ecosystems, ed. S. Woodley, J. Kay, and G. Francis, 61–79. Delary Beach, FL: St. Lucie Press.

King, B., and B. Warren. 1981. *Endangered birds of the world: The ICBP Bird Red Data Book.* Washington DC: Smithsonian Institution Press and International Council for Bird Preservation.

Knapp, A. K., J. M. Blair, J. M. Briggs, S. L. Collins, D. C. Hartnett, L. C. Johnson, and E. G. Towne. 1999. The keystone role of bison in North American tallgrass prairie. *BioScience* 49 (1):39–50.

Kolasa, J. 1989. Ecological systems in hierarchical perspective: Breaks in community structure and other consequences. *Ecology* 70:36–47.

Lance, G. N., and W. T. Williams. 1967. A general theory of classificatory sorting strategies. I. Hierarchical systems. *Computing Journal* 9:373–80.

Landres, P. B., J. Verner, and J. W. Thomas. 1988. Ecological uses of vertebrate indicator species: A critique. *Conservation Biology* 2:316–28.

Lawton, J. H., and V. K. Brown. 1994. Redundancy in ecosystems. In *Biodiversity and ecosystem function,* ed. E. D. Schulze and H. A. Mooney, 255–270. Berlin: Springer-Verlag.

Lay, D. W. 1938. How valuable are woodland clearings to wildlife? *Wilson Bulletin* 50:254–56.

Lee, K. N., and J. Lawrence. 1986. Adaptive management: Learning from the Columbia River Basin Fish and Wildlife Program. *Environmental Law* 16:431–60.

Leopold, A. 1933. *Game management.* New York: Scribner's.

Lesica, P., and F. W. Allendorf. 1995. When are peripheral populations valuable for conservation? *Conservation Biology* 9 (4):753–60.

Lindenmayer, D. B., R. B. Cunningham, M. L. Pope, and C. F. Donnelly. 1999. A large-scale "experiment" to examine the effects of landscape context and habitat fragmentation on mammals. *Biological Conservation* 88:387–403.

Louda, S. M., R. W. Pemberton, M. T. Johnson, and P. A. Follett. 2003. Nontarget effects: The Achilles' heel of biological control? Retrospective analyses to reduce risk associated with biocontrol introductions. *Annual Review of Entomology* 48:365–96.

Lynam, T., F. Bousquet, C. Le Page, P. d'Aquino, O. Barreteau, F. Chinembiri, and B. Mombeshora. 2002. Adapting science to adaptive managers: Spidergrams, belief models, and multi-agent systems modeling. *Conservation Ecology* 5 (2):24.[online]; www.consecol.org/vol5/iss2/art24.

Maguire, L. A. 1991. Risk analysis for conservation biologists. *Conservation Biology* 5 (1):123–25.

Mallet, J., and M. Joron. 1999. Evolution of diversity in warning color and mimicry: Polymorphisms, shifting balance, and speciation. *Annual Review of Ecology and Systematics* 30:201–34.

Marcot, B. G. 2002. An ecological functional basis for managing decaying wood for wildlife. In *Proceedings of the Symposium on the Ecology and Management of Dead Wood in Western Forests, 2–4 November 1999, Reno, Nevada,* ed. W. F. Laudenslayer Jr., P. J. Shea, B. E. Valentine, C. P. Weatherspoon, and T. E. Lisle, 895–910. General Technical Report PSW-GTR-181. Portland, OR: USDA Forest Service, Pacific Southwest Research Station.

Marcot, B. G., and K. B. Aubry. 2003. The functional diversity of mammals in coniferous forests of western North America. In *Mammal community dynamics: Management and conservation in the coniferous forests of western North America,* ed. C. J. Zabel and R. G. Anthony, 631–64. Cambridge: Cambridge Univ. Press.

Marcot, B. G., T. A. O'Neil, J. B. Nyberg, J. A. MacKinnon, P. C. Paquet, and D. Johnson. 2002. Analyzing key ecological functions as one facet of transboundary subbasin assessment. In *Watershed Management Council Ninth Biennial Conference—Watersheds Across Boundaries: Science, Sustainability, Security, 3–6 November 2002.* Stevenson, WA: Watershed Management Council.

Marcot, B. G., and M. Vander Heyden. 2001. Key ecological functions of wildlife species. In *Wildlife–habitat relationships in Oregon and Washington,* ed. D. H. Johnson and T. A. O'Neil, 168–86. Corvallis: Oregon State Univ. Press.

Marcot, B. G., M. A. Castellano, J. A. Christy, L. K. Croft, J. F. Lehmkuhl, R. H. Naney, K. Nelson, C. G. Niwa, R. E. Rosentreter, R. E. Sandquist, B. C. Wales, and E. Zieroth. 1997. Terrestrial ecology assessment. In *An assessment of ecosystem components in the interior Columbia Basin and portions of the Klamath and Great Basins,* vol. 3, ed. T. M. Quigley and S. J. Arbelbide, 1497–1713. Vol. 3. General Technical Report PNW-GTR-405. Portland, OR: USDA Forest Service, Pacific Northwest Research Station.

Matlack, R. S., D. W. Kaufman, and G. A. Kaufman. 2002. Influence of grazing by bison and cattle on deer mice in burned tallgrass prairie. *American Midlands Naturalist* 146 (2):361–68.

McElroy, D. M., J. A. Shoemaker, and M. E. Douglas.

1997. Discriminating *Gila robusta* and *Gila cypha*: Risk assessment and the Endangered Species Act. *Ecological Applications* 7 (3):958–67.

McNay, R. S., R. E. Page, and A. Campbell. 1987. Application of expert-based decision models to promote integrated management of forests. *Transactions of the North American Wildlife and Natural Resources Conference* 52:82–91.

McNeely, J. A., K. R. Miller, W. V. Reid, R. A. Mittermeier, and T. B. Werner. 1990. *Conserving the world's biological diversity.* Gland, Switz., and Washington, DC: IUCN, World Bank, World Resources Institute, Conservation International, and World Wildlife Fund.

Milius, S. 2004. Vanishing vultures: bird deaths linked to vet-drug residues. *Science News* 165 (5) [online]; www.sciencenews.org/20040131/fob6.asp.

Miller, J. C., P. C. Hammond, and D. N. R. Ross. 2003. Distribution and functional roles of rare and uncommon moths (Lepidoptera: Noctuidae: Plusiinae) across a coniferous forest landscape. *Annals of the Entomological Society of America* 96 (6):847–55.

Mooney, H. A., J. H. Cushman, E. Medina, O. O. E. Sala, and E. D. Schulze. 1996. What we have learned about the ecosystem functioning of biodiversity. In *Functional roles of biodiversity: A global perspective,* ed. H. A. Mooney, J. H. Cushman, E. Medina, O. O. E. Sala, and E. D. Schulze, 475–84. New York: Springer-Verlag.

Morell, V. 1996. Starting species with third parties and sex wars. *Science* 273 (5281):1499–1502.

Morgan, P., G. H. Aplet, J. B. Haufler, H. C. Humphries, M. M. Moore, and W. D. Wilson. 1994. Historical range of variability: A useful tool for evaluating ecosystem change. In *Assessing forest ecosystem health in the inland west,* ed. R. N. Sampson, D. L. Adams, and M. J. Enzer, 87–111. New York: Haworth Press.

Niemi, G. J., J. M. Hanowski, A. R. Lima, T. Nicholls, and N. Weiland. 1997. A critical analysis on the use of indicator species in management. *Journal of Wildlife Management* 61 (4):1240–52.

Oman, S. D. 2000. Minimax hierarchical empirical Bayes estimation in multivariate regression. *Journal of Multivariate Analysis* 80 (2):285–301.

Orians, G. H. 1993. Endangered at what level? *Ecological Applications* 3 (2):206–8.

Ostfeld, R. S., and K. LoGiudice. 2003. Community disassembly, biodiversity loss, and the erosion of an ecosystem service. *Ecology* 84 (6):1421–27.

Pennock, D. S., and W. W. Dimmick. 1997. Critique of the evolutionary significant unit as a definition for "distinct population segments" under the U.S. Endangered Species Act. *Conservation Biology* 11 (3):611–19.

Perault, D. R., and M. V. Lomolino. 2000. Corridors and mammal community structure across a fragmented, old-growth forest landscape. *Ecological Monographs* 70 (3):401–22.

Pergams, O. R. W., and M. V. Ashley. 1999. Rapid morphological change in Channel Island deer mice. *Evolution* 53:1573–81.

Pergams, O. R. W., W. M. Barnes, and D. Nyberg. 2003. Rapid change of mouse mitochondrial DNA. *Nature* 423:397.

Peters, R. H. 1991. *A critique for ecology.* Cambridge: Cambridge Univ. Press.

Peterson, D. L., and V. T. Parker. 1998. *Ecological scale: Theory and applications.* New York: Columbia Univ. Press.

Policansky, D. 1993. Uncertainty, knowledge, and resource management. *Ecological Applications* 3 (4): 583–84.

Posadas, P., D. R. Miranda-Esquivel, and J. V. Crisci. 2001. Using phylogenetic diversity measures to set priorities in conservation: an example from southern South America. *Conservation Biology* 15 (5):1325–34.

Reckhow, K. H. 1990. Bayesian inference in non-replicated ecological studies. *Ecology* 71:2053–59.

Rice, K. J., and N. C. Emery. 2003. Managing microevolution: Restoration in the face of global change. *Frontiers in Ecology and the Environment* 1 (9):469–78.

Rieman, B. E., D. C. Lee, R. F. Thurow, P. F. Hessburg, and J. R. Sedell. 2000. Toward an integrated classification of ecosystems: Defining opportunities for managing fish and forest health. *Environmental Management* 25:425–44.

Russell, K. R., D. C. Guynn Jr., and H. G. Hanlin. 2002. Importance of small isolated wetlands for herpetofaunal diversity in managed, young growth forests in the Coastal Plain of South Carolina. *Forest Ecology and Management* 163 (1–3):43–59.

Saunders, D. A., G. W. Arnold, A. A. Burbidge, and J. M. Hopkins. 1987. The role of remnants of native vegetation in nature conservation: Future directions. In *Nature conservation: The role of remnants of native vegetation,* ed. D. A. Saunders, G. W. Arnold, A. A. Burbidge, and J. M. Hopkins,

387–92. Chipping Norton, NSW, Austral.: Surrey Beatty and Sons Ltd.

Sayama, H., L. Kaufman, and Y. Bar-Yam. 2003. Spontaneous pattern formation and genetic diversity in habitats with irregular geographical features. *Conservation Biology* 17 (3):893–900.

Scheu, S., N. Schlitt, A. V. Tiunov, J. E. Newington, and T. H. Jones. 2002. Effects of the presence and community composition of earthworms on microbial community functioning. *Oecologia* 133 (2):254–60.

Scott, M. J., G. R. Bilyard, S. O. Link, C. A. Ulibarri, H. E. Westerdahl, P. F. Ricci, and H. E. Seely. 1998. Valuation of ecological resources and functions. *Environmental Management* 22:49–68.

Semlitsch, R. D., and J. R. Bodie. 1998. Are small, isolated wetlands expendable? *Conservation Biology* 12 (5):1129–33.

Severinghaus, L. L. 1992. Monitoring the population of the endangered Lanyu scops owl (*Otus elegans botolensis*). In *Wildlife 2001: Populations,* ed. D. McCullough and R. H. Barrett, 790–802. London: Elsevier Applied Science.

Spencer, R., and M. B. Thompson. 2003. The significance of predation in nest site selection of turtles: An experimental consideration of macro- and microhabitat preferences. *Oikos* 102 (3):592–600.

Stone, C. D. 1974. *Should trees have standing?* Los Altos, CA: Kaufmann.

Storfer, A. 1999. Gene flow and endangered species translocations: A topic revisited. *Biological Conservation* 87:173–80.

Takekawa, J. Y., and E. O. Garton. 1984. How much is an evening grosbeak worth? *Journal of Forestry* 82 (7):426–28.

Tracy, C. R., and P. F. Brussard. 1994. Letters to the editor: Preserving biodiversity: Species in landscapes. *Ecological Applications* 4 (2):205–7.

Treitz, P., and P. Howarth. 2000. Integrating spectral, spatial, and terrain variables for forest ecosystem classification. *Photogrammetric Engineering and Remote Sensing* 66 (3):305–18.

Walker, B., A. Kinzig, and J. Langridge. 1999. Plant attribute diversity, resilience, and ecosystem function: The nature and significance of dominant and minor species. *Ecosystems* 2:95–113.

Walters, C. 1986. *Adaptive management of renewable resources.* New York: Macmillan.

Walters, C., J. Korman, L.E. Stevens, and B. Gold. 2000. Ecosystem modeling for evaluation of adaptive management policies in the Grand Canyon. *Conservation Ecology* 4 (2):1 [online]; www.consecol.org/vol4/iss2/art1.

Waples, R. S. 1995. Evolutionarily significant units and the conservation of biological diversity under the Endangered Species Act. *American Fisheries Society Symposium* 17:8–27.

Warshall, P. 1995. The biopolitics of the Mt. Graham red squirrel (*Tamiasciuris hudsonicus grahamensis*). *Conservation Biology* 8 (4):977–88.

Wiens, J. J. 1999. Polymorphism in systematics and comparative biology. *Annual Review of Ecology and Systematics* 30:327–62.

Zorn, P., W. Stephenson, and P. Grigoriev. 2001. An ecosystem management program and assessment process for Ontario national parks. *Conservation Biology* 15 (2):353–62.

12 The Future: New Initiatives and Advancing Education

What is the extinction of the condor to a child who has never known a wren?

ROBERT MICHAEL PYLE,
"INTIMATE RELATIONS AND THE EXTINCTION OF EXPERIENCE"

*For in the end, we will save only what we love, we will love only what we understand,
and will understand only what we are taught.*

LAO TZU

In this book we have reviewed the state of the art of the meaning, measurement, and management of wildlife–habitat relationships, emphasizing the need to conduct all studies within a clear conceptual framework and with the necessary sampling rigor. We live in a time of rapidly increasing technological advancements, including miniaturized monitoring devices, increased ability to study genetics and physiology, fast computers that allow us to construct detailed population models, and so forth. These advances are occurring within the context of a rapidly growing human population that is exerting both direct and indirect pressures on the land base and its inhabitants (see Preface). Concomitant with these increasing human pressures are an increase in extinctions and worries about human-induced changes in the environment.

These changes present us with several closely related challenges. Wildlife scientists are being asked to conduct studies that address large-scale land use problems in relation to anthropogenic changes. As a result, we must study and understand ecological relationships both within and among a variety of scales. Further, we must be able to keep abreast of the latest technological

advances and understand the latest in computer software and modeling capabilities. These requirements necessitate a higher level of education in ecology and in mathematics, statistics, and modeling than previously seen in the wildlife profession.

In this chapter we discuss several problems that will confront wildlife scientists in the coming years and the types of information needed to adequately address them. We will not attempt to solve these problems here; to do so would take another book and much serendipity. Rather, we simply highlight areas that need study, mention some of the pitfalls inherent in their study, and suggest some directions for further work. Because this is a book on habitat relationships, we will focus on this particular area of research, although much of our discussion has broader applicability.

Scale of Conservation

A substantial change in the spatial scale at which field scientists conduct their studies occurred during the 1980s. The historical concentration on single-season, site-specific studies conducted at the microhabitat scale began to give way to multiseason, multisite studies that examine ecological relationships at broader spatial scales. These shifts were driven by both ecological and practical reasons. Ecologically, scientists began to better understand the general factors that drive habitat selection in terrestrial vertebrates—that is, the hierarchical nature of habitat selection as developed by Johnson (1980) and Hutto (1985; see chapter 7). In addition, people began to see the mistakes that had been made in previous studies of habitat, including gross mismatching of spatial scales within a single analysis (Wiens 1989, 227–33; Scott et al. 2002; see chapter 7). At the same time, and from a practical standpoint, land managers were becoming increasingly frustrated at the lack of scientific guidelines that could be applied to the management of properties under their jurisdiction. In particular, they were asking for studies that considered multiple species over broader spatial scales than had been conducted historically (Scott et al. 2002). Pressures from an ever-expanding human population were driving the need to view land management on larger, multispecies scales because of increasingly limited land areas available for conservation.

Changes in the scales at which studies were being conducted also reflected the natural progression of our knowledge. That is, the study of the ecology of animals usually begins with species-specific, detailed studies of life history parameters. Truly, nothing replaces basic field zoology. Remember that scientific studies of animals in North America began in earnest only over the past 100 years or so, although much useful, indigenous knowledge had preceded such scientific inquiry. As our scientific knowledge of animals grows, so does our ability to study and understand interactions between animals and the host of factors influencing their survival and behavior. Eventually it becomes possible to enlarge both the geographic scales and the levels of biological organization in which we view an animal. Stauffer (2002) presented a concise history of habitat studies.

Thus what we see is a merging of both ecological interest and knowledge, with a practical need for solving environmental problems "en masse." Relative to our recent past (i.e., since the 1970s), a plethora of papers has appeared on "landscape-level" analyses. In fact, the need for analyses of large geographic areas was in part responsible for formation of the Society for Conservation Biology and initiation of the journals *Conservation Biology* and *Landscape Ecology*.

Regardless of our need for such analyses, we must ask, Is our knowledge of spatial relationships adequate to meet the conservation challenge? Below we review current thoughts on this question and outline steps we can take to advance our knowledge in this arena of study.

As noted above, increasingly researchers have been turning to studies at the landscape and ecosystem levels. "Landscape management," a reflection of our new interest in expanding our spatial scale of analysis, was discussed in chapter 9. Of particular interest among both scientists and land managers is the controversial area of "ecosystem management." In chapter 11 we presented an approach for studying wildlife in an ecosystem context. Clearly, additional focus on disturbance dynamics and habitat configuration (chapter 9) is needed to augment autecological studies in such an ecosystem context.

Focused Initiatives in Habitat Research

In this section we present some thoughts on how we might want to pursue studies of habitat relationships. There are many approaches to such efforts; we certainly do not present our ideas as a prescription for all wildlife research. Our proposals are primarily a synthesis of the ideas of other thinkers. We hope these thoughts will stimulate further discussion on how the study of habitat relationships may be advanced.

Habitat Selection: An Extensive Concept

Slobodkin (1992) made the critical distinction between extensive and intensive variables, perception of these variables by individual organisms, and the resulting impact on habitat selection theory. *Intensive variables* are those that can be directly perceived by an individual organism at any point in a system, whereas *extensive variables* can only be perceived by considering an overview of an entire system. Systems here could include a study area, a landscape, a community, or an ecosystem. For example, Slobodkin argued that the extensive variable "population size" means less to an individual organism in the population than the intensive variables directly associated with food. In fact, Johnson (1980) implied this relationship when he noted that "habitat usage" studies and studies of feeding were of different orders of complexity and relation. Food is of a lower order than (or is a subset of) habitat and, as such, is a relatively more intensive variable. These comments relate directly to our call for hierarchical studies of habitat use. By separating intensive and extensive variables in wildlife studies of habitat selection, we can better explain the more proximate mechanisms driving the behavior of individual organisms, and the more ultimate context in which such behaviors arise and function (Slobodkin 1992).

Hierarchical habitat selection implies that fitness is affected at each scale. Must an organism, then, be able to perceive the effect that each scale has on fitness for an area to be occupied? or is there some summative process going on? "Macrohabitat" or "landscape" patterns may be only the accumulated picture of many individuals selecting resources at the lowest scale. It does not mean that individual animals recognize dis-

tributions of resources or constraints at population scales or, if they do, that population-scale distributions are of primary importance to the individual. The analogy is the operation of natural selection itself: natural selection operates on individuals or, more specifically, on variation in fitness among individuals, and the accumulated picture of this action is expressed in resultant population parameters.

Developing theories based on assumptions, constraints, and variables that cannot in principle be sensed by individuals perpetuates theoretical incompleteness and thus stymies advancement in understanding and prediction. Witness the approach taken in applied studies of habitat ecology: rather than such studies being framed in terms of testable theory and then designed to understand the process of selection, it has become standard practice to beg the issue and study habitat "use." Here, *use* does not imply any active selection process, any perception by the animal, or any consequences for survival and fitness (e.g., Morse 1980, 89–90; Hutto 1985; Hall et al. 1997). Asking what habitat feature a species uses is neither a "how" nor a "why" question (Gavin 1991). Hobbs and Hanley (1990) showed that measures of habitat use and habitat availability reveal little about the value of the habitat unless the underlying resource distributions are understood mechanistically. Similarly, applied studies of habitat usually equate "selection" with nonrandom use of some item, often vegetation (e.g., see reviews by Johnson 1980; Thomas and Taylor 1990; Alldredge and Ratti 1992). Again, animals can be nonselective under such study designs, whereas statistical evidence of avoidance may really be just an artifact of the study.

We argue, however, that a more consistent and reasonable approach is to view habitat selection as *always* occurring (it is a process; Hall et al. 1997), as a fundamental expression on the scale of the individual, and as something to be viewed in terms of intensive variables. This approach, at least, establishes a more consistent framework because of its more direct relation to fundamental, mechanistic currencies. It also puts great emphasis on ethology (study of behavior) and autecology (study of individual species) within the broader context of landscapes and communities, for us to better understand, explain, and predict patterns of habitat relationships. We need to separate clearly the application of extensive studies of population phenomena (e.g., density dependency, animal–vegetation relationships) from intensive studies of resource acquisition.

Thus we are not arguing against trying to understand the interactions going on within some larger spatial area or an ecosystem. Rather, our point is that different questions are best answered at different scales, and that some questions beg knowledge at multiple scales. In addition, many questions on habitat relationships would likely need to incorporate both intensive and extensive variables. For example, although determining the food requirements of a species (an intensive variable) is often necessary, the availability of food might be determined by intra- or interspecific population densities (an extensive variable). Our argument is that extreme care must be taken when choosing variables, care that includes knowledge of how animals perceive their environment; experimentation will likely be warranted in many cases.

Habitat and Niche

Related and central to our discussion of habitat is the concept of the niche. Like habitat, the concept of the niche has been the subject of much debate and certainly falls under the same criticisms concerning operationalization of terms noted above (Peters 1991). Much of the problem with study of the niche is the lack of an accepted operational definition. Arthur (1987) reviewed the niche concept. He noted the range of definitions stemming from Grinnell's (1917) descrip-

tion of the niche as the ultimate distributional unit, to Elton's (1927) primarily behavioral concept, to Hutchinson's (1978) *n*-dimensional construct (for behavior and distribution). Leibold (1995) expanded on these definitions and the historical development of the niche, concluding that Elton's view and that of MacArthur and Levins (1967) were oriented toward describing how an organism affects the environment by consuming resources and by serving as a resource for higher trophic levels. In contrast, Grinnell's view of the niche emphasized the environmental requirements of the organism.

Arthur (1987) concluded that, to be useful, the accepted concept must be simple and quantifiable. He therefore chose to use MacArthur's (1968) description of the niche, which plots utilization against some quantifiable resource variable—the *resource utilization function* (RUF). Arthur argued that it is better to build complexity as needed—e.g., by using RUFs—than to dissect it, as is necessary when using a hypervolume (Morrison and Hall 2002). (A *hypervolume* can be defined as an "infinitely large set of properties" that cannot be operationalized [Peters 1991, 91].) This rationale appears to meet Peters's, Romesburg's, and Slobodkin's requirements for a testable theory of fundamental causation. In concept, RUFs describe the choice of resources by animals. Choices can be *constrained* by predators, competitors, and various other factors, effects of which can be tested.

From this rationale, then, we should focus our studies on the level of what we classically call a *resource axis* of the niche. Habitat is largely attendant to the underlying reasons why an animal is present at a certain time and place, but it is certainly a population phenomenon under most of our study designs. In the new view outlined here, animals *select habitat* at only the broadest geographical scales, and they *select resources* at the finest scales within those habitats. We need to recognize that, according to our scheme, niche and habitat reside at opposite ends of the intensive–extensive scale, and that we need to cast future studies with these differences in mind.

If this approach is adopted, it is critical to the success of wildlife investigations and management that the distinction between habitat and niche factors, and their relevance to the viability of animals, be recognized. As a review (see chapter 1), recall that by definition "habitat" is a species-specific concept and should not be confused with "habitat type," which is closely related to vegetative associations. *Habitat,* in the view expressed here, is simply a description of the physical attributes of the environment in an observer-defined area around an individual animal. Typical habitat variables include the structure and species composition of vegetation (all layers); the type and coverage of ground cover (e.g., gravel, stones); various special features such as down logs and rocky outcrops; the type and coverage of water (e.g., wet depression, pond); and a host of other factors thought to be relevant to the species of interest. *Habitat factors* thus vary continually as an individual moves through the environment and naturally change substantially over the course of a season. *Niche factors* relate to the behavioral activities of an individual within the habitat and, in fact, are usually determinants of where an animal spends its time. Niche factors include various resources, such as the type and size of food required, and constraints on the acquisition of those resources, such as the activity of predators and competitors. Thus, regardless of the apparent appropriateness of habitat, an animal may be absent because the niche factors are inappropriate. Niche factors are a primary reason why models of the presence or viability of animals based on habitat factors alone often result in poor predictions, and why different habitat models are needed for different locations and times.

Failure to account for niche factors in investigations and management often results in poor success for wildlife diversity and viability. For example, regardless of the physical appropriateness of vegetation, the presence of cowbirds usually results in breeding failure for a host of songbirds, including such focal species as those that are classified as threatened or endangered. Likewise, failing to account for changes in the size distribution and availability of arthropod prey through provision of appropriate plant species across the spring and summer will likely result in inadequate food resources for many species. In addition, the complexity of food webs also needs to be considered in planning. For example, in one study, the patterns of insectivorous bird predation were altered by elk browsing on aspen trees, because the consumption of aspen shoots by elk reduced the quantities of galls produced by sawflies, the presence of which had significant and positive effects on the species richness and abundance of other arthropod species (Bailey and Whitham 2003).

Although a thorough evaluation of niche relationships adds time and effort to the front-end planning of a study or project, such work substantially improves the efficacy of results and thus potentially drives long-term management activities more effectively. For example, if initial analyses indicate that a songbird would experience high nest parasitism from cowbirds and thus poor breeding success, there would be little reason to attempt to restore habitat (i.e., wet meadows) unless intensive cowbird management were an integral part of the restoration plan. Likewise, in the previous example of the web involving birds, elk, aspen, and arthropods, failure to incorporate control of elk in the project area would likely result in poor aspen regeneration and negative responses (e.g., breeding failures) by the bird community.

Given the above rationale, we suggest a focus on the fundamental currency that allows individuals to survive and reproduce. This currency is popularly termed *resource*. Below we will develop some fundamental underpinnings of this theory and proceed with more specific examples. Perhaps this approach will help us to focus our efforts and lead to more rapid advancement of our understanding. Leibold (1995) distinguished between *environmental requirements* and *environmental impacts* of species, and related these to the "habitat" and "functional" aspects of the niche as developed by Grinnell and Elton, respectively. Leibold's functional aspects are reflected in the concept of "key ecological functions," or main ecological roles of organisms, that we presented in chapter 11. Leibold went on to develop a framework for merging these two aspects of the niche into a single concept that could be analyzed through various mechanistic models. Here we are focusing on the "habitat" or "environmental requirements" aspect of the niche, while acknowledging that Leibold presents a comprehensive framework for more advanced study.

Resources and Their Identification

Peters (1991, 91) dismissed the niche as theory because there are likely many resources, which are usually defined relatively and thus are difficult to test. Arthur's (1987) approach, as outlined above, addresses much of this criticism by concentrating on specific RUFs. To be useful in advancing knowledge, the resource currency must have relevance to the fitness of the animal and must be within the animal's perceptive abilities. This requirement helps narrow immense resource dimensionality to a practical set of RUFs.

Abrams (1988) recognized the critical importance of the proper identification of resources and noted the confusion that exists over their definition and enumeration. He identified two central issues: (1) *separation of entities that serve as resources from those that do not;* and (2) *determination of how a set of resource entities can be divided into distinct resources.* Further, he posited that per capita population growth is an increasing function of the rate of resource consumption. Various constraints act to reduce the rate of consumption and thus per capita population growth.

Abrams separates resource "entities" into individual resource types according to the following criteria: (1) the densities of the entities cannot be related by a function that is independent of the consumption rate of any type; and (2) the resource types must be identifiable by a consumer. Thus resources cannot be counted separately from how consumers use them. For example, researchers should realize that a single biological *species* cannot usually serve as the "resource," because a species is composed of individuals with different behaviors (based on sex, age, life histories) and/or phenotypes, and thus a consumer may be selecting a specific subset of the species. Bird diets and foraging behavior vary widely, for example, depending on the availability of various arthropod developmental stages (e.g., Wolda 1990). Likewise, there is no a priori basis for dividing chemical and physical properties into categories—based on size, shape, or content—that is separate from how the consumer uses them. "Resources" can only be documented once consumption rates and subsequent effects on fitness are known. In this way, "resource," like "habitat," is defined by the behaviors of an organism.

What Should We Do?

Most of our recent advances in the study of wildlife–habitat relationships have involved refinements of analytical techniques (e.g., see reviews by Alldredge and Ratti 1986, 1992; Thomas and Taylor 1990), despite clear warnings that our fundamental approaches are flawed (e.g., Van Horne 1983; Hobbs and Hanley 1990). The analytical tool we select is of little importance if the question we ask is trivial or if the approach taken to address a meaningful question is flawed.

There are numerous ways to learn about animals, all of which provide at least some additional insight into their ecology. Here we suggest, however, steps we believe will help to advance our understanding of ecology more substantially, and lead to more effective management actions. Morrison (2002a) reviewed this topic and suggested much of what we provide herein.

We think that researchers should concentrate on identifying and analyzing the separate roles of critical resources and the factor(s) constraining their use. We would rather learn about the animal's behavior along a single resource axis than read yet another habitat study that degenerates into a string of ad hoc suppositions about observed phenomena. Separation and quantification of resources and constraints have clear operational utility. Consumption of resources is a quantifiable occurrence that has consequences for the individual animal and thus the population. The condition of the habitat in which such activity occurs is of interest, but primarily for descriptive purposes, because habitat (as viewed in this approach) is a step removed from the means of survival and reproductive performance. Perry and Andersen (2003), for example, found that the clumping of nest sites by least flycatchers (*Empidonax minimus*) was based on antipredator behavior rather than on differences in habitat condition.

There are several different levels to the study of biology. We must recognize the limitations of purely functional studies of wildlife and avoid unwarranted reliance on their results in basic and applied situations. However, at the same time, it is unreasonable to suppose that a researcher, especially a graduate student, will be able to complete the entire picture of an animal's relation to the environment over the short term. An underlying problem is that studies for graduate degrees are often structured as independent studies that do not facilitate long-term and broad geographic research. We might therefore need to rethink how we structure graduate research. For example, we could manage students by providing opportunities to pursue specific aspects of a long-term research agenda and by narrowing the scope of our questions (Morrison and Hall 2002), although we also acknowledge that students must craft the scientific questions themselves. Extensive studies may be worthwhile and necessary, but chiefly in a screening and descriptive context.

The resources that an animal *encounters* are the parameters in which we should be interested. Models of foraging, for example, view resource selection as a "search-encounter-decide" sequence. In this framework, if we are to properly identify when an encounter occurs—a necessary step in identifying what a resource is to an animal—we must first understand the forager's sensory abilities. It is certainly not enough to simply measure the abundance of some resource, including habitat (Stephens and Krebs 1986, 13; Hobbs and Hanley 1990). In fact, it is actually the rate at which an animal encounters resources or can gain access to them, not the density (or abundance) of the resource, that is of importance. Density is determined by the researcher, who counts items and expresses them by some unit of area. Without relating resource items to the animal's behavior and perception, however, such measures of resource density may or may not be biologically relevant. We should therefore not assume that density of a resource is necessarily an appropriate surrogate to the encounter rate of a resource. Studies of diet, food selection, and foraging behavior have a well-developed methodological base (e.g., see reviews in Morrison et al. 1990) and are useful in the design of studies seeking to determine the fundamental mechanisms of survival and fitness.

The responses of animals (species response curves) along various environmental gradients often differ depending on the number and kind of coexisting species. Thus the *realized niche* may vary (take a different shape), depending on time and location. However, the *fundamental niche* of a species should remain more or less invariant to local conditions. Because environmental conditions vary both within and between localities and over time, identifying the fundamental niche is essential if we are to develop predictions of how species respond to different constraints in different locations and times. Specific research methods will depend on how precise our estimates (of whatever response we are measuring) need to be to meet our objectives. Most studies only examine the "realized habitat" of a species. And studies that do examine a niche parameter seldom address the distinction between the fundamental and realized niche of a species (O'Connor 2002).

Stephens and Krebs (1986,182) recommended that we must first clearly elucidate and separate *resources* and *constraints*. Their approach is as follows: if the interest is in determining the resource, then study systems with well-defined constraints; but if the interest is in constraints, then study systems in which the resources are well known. Obviously, experimental designs will be necessary in many cases (chapter 4; Morrison et al. 2001). Laboratory studies within the broad framework of physiological ecology will allow us to put bounds on the possible behaviors and physiological responses of animals to variations in resources and constraints (Ricklefs and Stark 1998). That is, the limits of resource selection and habitat selection behaviors are defined physiologically.

Using Niche-Related Variables in Habitat Analysis

Compared with purely habitat-based parameters, niche parameters are seldom evaluated in most studies. The focus on habitat-based param-

eters is due to several reasons, including the belief that habitat adequately predicts animal performance, and the fact that data collection for habitat-based parameters is easier than for the more effort-intensive niche factors. As we have reviewed in this book it has long been known that habitat features may be a misleading indicator of habitat quality (Van Horne 1983); recent studies have confirmed this problem (e.g., Wheatley et al. 2002). Thus, although the limitations of habitat alone to develop adequate predictions of animal performance have long been recognized, only recently have more intensive discussions of the need to expand beyond classic habitat variables been initiated (e.g., Scott et al. 2002).

Scattered throughout the literature are good examples of the incorporation of niche-related variables into habitat analysis for wildlife management. For example, Barras et al. (1996) evaluated the role that acorn selection played in habitat selection of wood ducks (*Aix sponsa*). They showed that the physical structure of the acorn, such as top width, shell thickness, and meat content, drove habitat selection. Jamison et al. (2002) showed that invertebrate biomass was greater in habitats used by lesser prairie-chickens (*Tympanuchus pallidicinctus*) than in nonused areas. They concluded that the use of forb cover by prairie-chickens was related to invertebrate biomass, thus providing a mechanistic explanation for the observed habitat use of the species. As reviewed by Wiebe (2001), many descriptive data have been accumulated on characteristics of cavity nests. Nonrandom selection of nest trees on a coarse habitat scale is not surprising given that cavities must be excavated from trees of some minimum size. As Wiebe explained, however, without knowing how reproductive success varies with features of cavity nests, it is impossible to determine whether birds are choosing optimal nest sites or are constrained in their placement of nests (i.e., are they selecting the best of the worst?).

In their study of piping plovers (*Charadrius melodus*) on alkali lakes, Knetter et al. (2002) included measurements of physical structure (i.e., habitat variables) and niche factors in evaluating plover productivity. They especially focused on predator abundance and activity, including measurements of total predator abundance and total predator species; they also evaluated the influence of temperature and precipitation. They standardized point-count outcomes and used four predator indices to assess the predator influence on productivity at each study unit: coefficient of variation of predator sightings, total number of predator sightings, total number of predator species, and an index of predator diversity. Additionally, all precipitation and minimum temperature records were collected from stations found nearest to the breeding sites. They noted that heavy precipitation and cold temperatures can be detrimental to juvenile survival during the first week of life, when chicks need to feed as much as possible and rely on their parents for warmth via brooding. This study clearly indicates the substantial influence that various niche-related factors can have on productivity. Note that in periods when predator activity was minimal or weather conditions favorable, one might conclude that the habitat parameters were indicative of high-quality habitat. Yet, if predator activity was high and weather poor, failure to measure such niche parameters could lead to the conclusion that this same habitat was low-quality.

Guthery et al. (2001) studied the heat hypothesis in determining the cause–effect process governing northern bobwhite (*Colinus virginianus*) abundance. Previous research had found a critical upper temperature (>39°C) at which heat dissipation mechanisms could not neutralize the effect of heat gain. They used a grid system across the landscape to measure wind speed, temperature, and humidity within the vegetation to derive predictions on quail activity and found that the thermal environment was sufficiently intense to suppress bobwhite reproduction. Using their results, one can determine locations where the thermal environment is suitable for bobwhite reproduction. Guthery and colleagues concluded that management measures (e.g., manipulating cover) could lower the heat environment and thus improve bobwhite habitat. It can be seen from these authors' research that they looked for the causal mechanism, and then developed a management recommendation. Had they looked just at bobwhite cover, they would have been forced to speculate as to the cause for differential patterns of bobwhite distribution and reproduction performance. That is, they would not have known whether patterns were due to thermal environment, nest predators, or food availability.

Also interested in the thermal environment, Wiebe (2001) measured the thermal characteristics of northern flicker (*Colaptes auratus*) nest cavities. She included various measurements of temperature and heating, several of which were thought to have biological relevance in how they impacted embryonic development. In addition to measurements on the physical dimensions of cavities, Wiebe recorded measures of the internal nest cavity using data loggers that recorded temperature every 3 to 4 minutes. These data allowed her to create a temperature profile across time and to relate these data to the ambient (external) temperature.

It is not uncommon for avian ecologists to try to relate prey—including prey biomass, abundance, and availability—to some aspect of bird biology (e.g., activity, abundance, productivity) (see Morrison et al. 1990 for review). For example, Keane and Morrison (1999) related within-season, within-day, and sexual differences in foraging behavior of black-throated gray warblers (*Dendroica nigrescens*) to changes in arthropod abundance in pinyon-juniper woodland. They measured prey abundance by placing plastic bags over branches, clipping the branch, and fumigating the contents. Later they sorted, identified, and weighed the arthropods. Changes in arthropod abundance were related to flowering phenology of certain plants, which in turn was related to warbler foraging activity. Changes in plant species composition are extremely slow in this environment, and changes in plant vigor, such as that induced by drought, also occur over years rather than months. Keane and Morrison therefore were able to decipher warbler activity only through measures of prey abundance (although they did not measure other potentially relevant niche factors, such as predator activity and microclimatic variables).

In relating the use and nonuse of locations by lesser prairie-chickens to prey biomass, Jamison et al. (2002) conducted sweepnet sampling of use and associated nonuse areas consecutively on the same day, using a standard 30-cm insect net. They made 100 sweeps through the upper layer of vegetation along two parallel 75-m transects spaced 10 m apart. Each 100 sweeps in an area composed a single biomass sample for that area. They euthanized captured invertebrates, oven-dried them, and determined their biomass.

Quantifying niche relationships is especially relevant in the conservation and management of rare and endangered species. For example, Stevens et al. (2002) studied the endangered Florida snail kite (*Rostrhamus sociabilis plumbeus*), which forages exclusively on the Florida applesnail (*Pomaces paludosa*). They showed that the snails occurred only within a specific range of water temperatures, information that can be used by managers to adjust the timing of water withdrawals so that snail kites have a reliable food source available.

Conceptual Models:
The Envirogram Approach

Models are, among other things, useful means of synthesizing knowledge and identifying potentially useful approaches to ecological studies. We think that wildlife ecologists would benefit from outlining a conceptual flow diagram of the key factors thought to be influencing the animal to be studied. Such a diagram would capture the "causal web" of environmental factors influencing a species or organism, as described in chapter 10. Andrewartha and Birch (1984) developed the *envirogram* approach, in which the main factors influencing species abundance are shown in a diagram. In their scheme, factors that influence the species fall into four categories: *resources, mates, "malentities"* (e.g., weather, competitors), and *predators*. Outside of these factors, a broader web of other biotic and abiotic factors also influences the species indirectly. The research objective is to determine which of these pathways are important. Thus the indirect effects must be explicitly tracked through the factors producing the direct effects, forcing researchers to conceptualize the cause–effect linkage between habitat variables and species abundance. Box 12.1 outlines the development of an envirogram.

The envirograms of Andrewartha and Birch (1984, 19–42) are detailed and complicated. At a minimum, we recommend developing a simplified flow diagram based on the envirogram approach to depict the causal pathway for each variable leading to the critical factor. In current modeling parlance, this simplified flow diagram would constitute an *influence diagram*, which could then serve as a basis for crafting quantified models such as Bayesian belief networks, neural networks, and other such constructs (see chapter 10). Such a flow diagram also effectively constitutes a knowledge-based model, which then serves as the conceptual framework for developing study objectives and testable hypotheses.

This scenario could easily be expanded beyond studies of occurrence to determination of factors driving survival, fecundity, and recruitment. Morrison (2002a) provided two simple examples:

Envirograms: Templates for Modeling Wildlife–Habitat Relationships

J. P. Ward Jr., USDA Forest Service

Knowledge of wildlife–habitat relationships is often provided through quantitative modeling. When quantitative models are developed from observational studies of habitat use or selection, resulting inferences are drawn from correlation, not from cause and effect. Inferences drawn from correlations provide less confidence than those drawn from well-designed and successful experiments. However, a successful experiment is difficult to execute without prior hints about the function or process of the system under study. For example, knowing the types of food taken by a particular organism, its foraging behavior, and the distribution of its existing food makes it easier to design an experiment that examines effects of habitat restoration. This kind of information would aid in determining *what* to manipulate, *when* the manipulation should occur, and *where* it should be conducted. Reliable models are therefore useful for guiding experiments.

Like successful experiments, practical models require thoughtful development. The model presented here represents one method for choosing variables to include in wildlife–habitat models prior to quantitative analysis. It includes an approach designed to identify the reliability of included variables through estimation of model parameters. This example model is drawn from an observational study designed to identify key factors that might be influencing populations of common prey of the threatened Mexican spotted owl (*Strix occidentalis lucida*) (Ward 2001). The approach used in this study was to first narrow the scope of variables to include by developing an *envirogram* for each prey species. Envirograms are conceptual flow charts that show how potential factors influence abundance of a given organism (Andrewartha and Birch 1984). These diagrams depict an organism within the context of a *centrum* and a *web* (box Fig. 12.1).

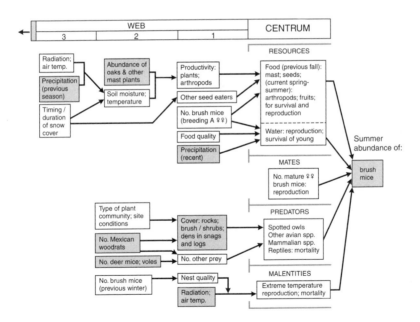

Box Figure 12.1. An envirogram depicting factors hypothesized to influence summer abundance of brush mice (*Peromyscus boylii*) in the Sacramento Mountains of southern New Mexico. Relationships were inferred from published literature (see Ward 2001). Shaded boxes indicate variables included in the a priori regression models used to quantify and rank the relative importance of represented factors. Reprinted with permission of J.P. Ward Jr. Copyright 2001.

(box continued on next page)

Box 12.1   Continued

The *centrum* is comprised of factors that directly affect an organism's abundance. *Direct effects* are categorized as resources, predators, "malentities," or mates. *Resources* are environmental components that enhance an organism's chance of survival and reproduction, and which are either negatively or not influenced by the abundance of the organism (e.g., the organism's food). *Predators* reduce survival and reproduction in the focal population and, in turn, benefit from increases in the organism's abundance. *Malentities* reduce survival and reproduction in a population but are not affected by the organism's abundance (e.g., excessive temperature). *Mates* convey a positive-positive relationship.

The *web* is comprised of *indirect factors,* which include anything that can affect a species by modifying its centrum, including the effect of individuals of the focal species on their own population (e.g., density-dependent effects). Flow in an envirogram tends from distal, indirect influences in the web toward the most proximate, direct effects on the organism's population, as shown in the centrum.

Envirograms are constructed from results of experimental and observational studies reported in the literature. The number of factors and interactions depicted are limited only by knowledge of the ecology of the focal or closely related organism. No envirogram is expected to be comprehensive. However, envirograms developed from Ward's (2001) study included more effects and factors than could be sampled and ultimately modeled. Thus, in addition to providing a template for developing statistical models, the envirograms provided a schematic of possible interactions that can serve as alternative explanations or topics for further investigation. In this sense, envirograms are useful for portraying the relevance of modeled factors within the greater realm of ecological complexity. Because envirograms usually include direct and indirect factors, the resulting models provide a broader perspective on the potential influence of any given "habitat" variable.

Once the envirograms were constructed, Ward (2001) developed a series of models structured from different combinations of variables that represented effects in the envirograms. Parameters were estimated for each model using multiple linear regression and following the information theoretic approach recently advocated by Burnham and Anderson (1998). These procedures allowed a relative weighting of each model according to Akaike's information criterion (AIC), corrected for small sample sizes (AIC_c), and quantification of the relative importance (RI) of each variable (box table 12.1). Following development of envirograms for all five of the most common prey for Mexican spotted owls, Ward (2001) was able to predict which prey species would be more likely to respond to habitat manipulations. Additional examples of how envirograms have been used to model wildlife–habitat relationships can be found in Van Horne and Wiens (1991), James et al. (1997), and Patton (1997).

(box continued on next page)

1. Rainfall → seed production → litter size → recruitment.
2. Water level/hydrology → meadow wetness → flying insect production → fledglings produced → recruitment.

In example 1, the critical factor is rainfall, and the response variable is recruitment. This simple example seems to indicate that rainfall supersedes (drives) all other factors in determining recruitment. Indeed, determining the link between a given amount of rainfall and a given production of seeds would appear to set upper and lower limits on population size through recruitment. A further refinement of this diagram would be to

hypothesize the influence of a specific constraint placed on animals attempting to acquire seeds, such as predators interfering with the activity of a seed-gathering animal. If the basic structure of a flow diagram (envirogram) is confirmed, then a potential management application presents itself, such as manipulating seed availability either directly (e.g., through supplemental feeding) or indirectly (e.g., through predator removal).

In example 2, the critical factor is water level, and the response variable is again recruitment. In this more complicated example, we see that water level likely sets boundaries on flying insects through an influence on meadow condition. However, changing meadow wetness through

Box 12.1   Continued

Box Table 12.1.   Precision and relative importance of ecological factors associated with summer abundance (g/ha) of brush mice in the Sacramento Mountains, New Mexico (1992–1994)

Factor	Original Models			All Models		
	β	CV(%)	RI	β	CV(%)	RI
Number of mature female brush mice in late spring	16.584	6.8	1.000*	16.624	7.3	1.000*
Rock cover (%)	0.229	354.6	1.000*	−0.234	299.6	0.710*
Quantity of shrubs (no./ha)	0.0003	11,095.0	0.868*	0.00002	9,020.6	0.055
Total precipitation (cm) during previous 2 springs and summers	−0.048	595.4	0.145*	0.020	945.1	0.112*
Number of logs and snags (>10 cm diameter)	0.0004	6,168.9	0.132*	0.00002	6,169.1	0.006
Mean grass-forb height (cm)	−0.021	458.8	0.105*	−0.001	460.9	0.005
Quantity of low-growing oaks (no./ha)	0.001	699.3	0.105*	0.005	681.8	0.436*
Quantity of pinyon and juniper trees (no./ha)	0.002	581.9	0.040	0.00008	582.5	0.002
Total precipitation during previous spring	0.011	283.0	0.026	0.0005	283.9	0.001
Number of interspecific competitors (woodrats, deer mice; no./ha)	0.107	103.9	0.014	0.566	98.4	0.080*
Mean temperature during preceding fall and winter months	−0.019	239.8	0.014	−0.0009	240.3	0.001
Amount of alternative prey for Mexican spotted owls (no./ha)[a]	0.[b]	6,420.0	0.000	0.[b]	6,420.0	0.000

Source: Adapted from Ward 2001.

Note: Two sets of regression coefficients (β) are given—one set ("Original Models") averaged over a priori models developed from an envirogram, the other set ("All Models") averaged over both original and exploratory models. Coefficients of variation (CV) of the average coefficients include variation associated with uncertainty of model structure. Relative importance (RI) of each regressor variable is also shown for both model sets and is based on the sums of Akaike weights for each model that included the regressor. Asterisks (*) indicate effects included in the 95% confidence set of models. Factors are ordered from highest to lowest according to RI in the original model set. Additional details on model structure, assumptions, and inferences from these data can be found in Ward (2001).

[a] Mexican spotted owls are a predominant predator of other rodents that co-occur with brush mice. This variable included abundance of Mexican voles, deer mice, and Mexican woodrats.

[b] ≤0.000001.

restoration activities might fail to enhance recruitment unless constraints on the gathering of insects are considered. Although adding insects is an unlikely management alternative, manipulating constraints (e.g., predators, competitors) or adding more meadow might be viable.

In both examples, classical measures of habitat relationships would be unlikely to identify the causal mechanisms involved in driving the response variable. Niche parameters, such as the size distribution of prey, and constraints on acquisition of the prey, such as predator activity, have greater explanatory power. A late frost that kills a substantial portion of the arthropods would not likely be reflected in any habitat parameter. The predator might shift habitat use because of the frost-induced reduction of prey, but the observer would be left with the typical string of ad hoc speculations in lieu of an explanation.

Proposed Research Approaches

The specific research approach will be dependent on the specific question(s) being asked, and could literally entail any laboratory or field methodology. Below we provide three nonmutually exclusive approaches as developed by Morrison (2002a).

THE GRADIENT APPROACH

Most studies examine only a small number of individuals from a narrow part of their overall distributional range. This approach is certainly justifiable if the intent of the study is to gather data in a confined area (e.g., a reserve). However, such studies do not lend much insight into how animals respond across a gradient of resources, nor do they place the particular study in the context of the overall gradient. That is, how does the animal respond when nearing the edge of the gradient of the critical niche parameter(s) identified by the envirogram? The gradient would include a range of a critical resource or, given a constant resource, a range of constraints on the use of that resource. Response variables could include abundance, survivorship, reproductive output, and so forth. A goal is to identify amounts of the resource(s) or constraint(s) that result in thresholds in the response variables. Such considerations can be very useful for explaining the ecological and even evolutionary values of peripheral distributions (see chapter 11).

Van Horne (2002) suggested that we test the importance of processes thought to be important in driving observed habitat relationships and test for the boundary conditions under which these processes may become relatively unimportant. Although we agree with this suggestion, studies of habitat alone, even in the context of boundary conditions, are not likely to elucidate causal processes. Pavlacky and Anderson (2001), for example, studied habitat use of pinyon–juniper (*Pinus–Juniperus*) birds near the limit of their geographic range. They explained the finding that juniper titmice (*Baeolophus griseus*) seemed to select south-to-west aspects that supported productive understory and high herbaceous plant cover by suggesting that such conditions increased the availability of grasshoppers and caterpillars. However, by failing to explicitly explore niche relationships (resources and constraints) as part of their study design, these researchers were forced to speculate post hoc as to the underlying mechanism driving their findings.

THE ACTIVITY/ENERGETIC APPROACH

Laboratory experiments are a classic method for examining the physiological tolerances of animals to varying conditions. As reviewed by Bennett (1987), for example, energy availability and utilization are critical constraints on animal function. Through energetics, ecological physiology links strongly with ecology and behavior (Ricklefs and Stark 1998). Life history strategies and foraging theory (e.g., optimal foraging and giving-up density [GUD]) are governed primarily by the acquisition of energy. Further, much physiological study has identified how complex the thermal environment of an animal is, and the importance of microclimates rather than macroclimates in determining the thermal balance of individual organisms.

Laboratory studies are not sufficient, however, to guide wildlife ecology and applications to management. As noted by Bennett (1987), an important variable often lost in laboratory experiments is behavioral response (although see chapter 4). In fact, behavior is often the primary means by which an animal copes with an environmental challenge (chapter 7). To cope with environmental challenges an animal might (1) conform to the environmental variable; (2) regulate the variable internally (often at great energetic cost); or (3) avoid the challenge altogether, either behaviorally or physiologically (e.g., through torpor or hibernation, migration). Thus many factors must be considered when trying to decipher why an animal occurs where it does, occurs in the numbers it does, and performs in the manner it does. The study of ecological physiology offers great promise in helping to guide studies in wildlife ecology. However, standard measures of habitat alone are not likely to elucidate changes in behavior.

FIELD EXPERIMENTATION

Field experiments on the use of resources, and on the constraints governing use of those resources, offer great promise for further understanding wildlife ecology (chapter 4). Two obvious field experiments are (1) manipulating resource abundance, and (2) manipulating resource availability through changes in constraints. Manipulating resource abundance through food supplementation has been attempted on numerous occasions (e.g., see review by Boutin 1990) and has been shown to alter home range size, population size, and breeding and recruitment. Hall and Morrison (1998) and Morrison and Hall (1998), respectively, linked a field experiment of food supplementation with a longer-term observational study of rodents. The observational study used a naturally occurring drought to examine changes in rodent abundance. Such "natural experiments" are, however, difficult to interpret because of the usual lack of control of numerous confounding factors.

There has been little theoretical work on consumer-resource systems in which consumers actively choose from *two or more* resources. Understanding choice by consumers is a key aspect of understanding the effects of choices on re-

source densities and dynamics. For example, Barras et al. (1996) showed experimentally that wood ducks (*Aix sponsa*) were selecting among the many species of acorns available to them. Another key aspect of consumer-resource systems is the indirect effect on resources of any species that interacts directly with the consumers, but not with their resources. As such, predator-caused changes in prey behavior can affect organisms that are two trophic levels below the level of the predator (see review by Abrams 2000).

Toward a New Understanding of Community Ecology

Wiens (1992) concluded that the study of ecology has been largely *phenomenological*—that is, a pattern is observed and matched with the prediction of a theory that postulates a certain linkage between pattern and process. Here, while the pattern has been empirically determined, the underlying process is still largely inferential—a classic use of abductive logic, or retroduction, in which some cause is inferred from observed outcomes (Niiniluoto 1999). Theories of habitat selection, including questions of scale, certainly fall into this category. Such ad hoc hypotheses have little predictive power and tell us little about *how* the process has occurred. Instead Wiens calls for a turn to *mechanistic ecology,* under which observed phenomena are approached in terms of their underlying causes or mechanisms.

It is our view that we should assume simplicity, search for the underlying dominant factors (resources) driving fitness, and determine the constraints on their use. As concluded by Schultz (1992), "Reductionist studies are needed to prevent obese generalizations. . . . Creative conceptual development is needed to provide appropriate context for mechanistic studies." *Habitat* as a concept is ingrained in both scientific and popular usage. However, the popularity of the concept is not justification for the study of habitat per se. There are different levels of inquiry, but answers to the fundamental "how" questions hold the most promise for scientific understanding and will, in the long run, provide the most generality. We do not, however, advocate purely reductionistic and solely deduction-driven biology. In answering "how" questions, we should use rigorous study designs in mechanistic as well as inductive modes—that is, learning from empirical streams of evidence as well as building generalized theory from such specific examples.

To encourage learning of mechanisms underlying more general patterns, Wiens argued (point 2, 1999, 257–59) that macroparameters such as species richness, diversity, and niche overlap should be deemphasized in favor of studies concentrating on expressing patterns in terms of physiology, behavior, or life history traits. In this vein, we might focus more on attributes of species that influence their use of energy rather than on effects of interactions on their population dynamics. Wiens also argued (point 13, 1989, 262) for development of specific, mechanistically based theories as an approach to advancing understanding of ecological relationships. Wiens developed a list of 18 specific steps that avian ecologists might take to develop a more rigorous understanding of community ecology (table 12.1); his list is certainly generalizable to ecology as a whole. All of Wien's points have been discussed to some extent in our book—in particular, his call for more explicit definitions of terms (point 1) and resources (points 3, 12), consideration of evolutionary as well as ecological constraints (point

Table 12.1.   Desiderata for a more rigorous community ecology

1. Be more explicit about defining the "community" studied and justifying that definition.
2. Deemphasize community macroparameters and focus on individuals, especially aspects relating to energetics, density effects, and habitat selection.
3. Use resource-defined guilds as a framework for intensive comparative studies.
4. Consider both ecological and evolutionary constraints on community patterns.
5. Consider all life stages in community analyses, and evaluate the effects of community openness versus closure.
6. Conduct studies, interpret the results, and generalize from them within the appropriate domains of scales in space and time for the phenomena or biota investigated.
7. Avoid thinking of communities as either equilibrium or nonequilibrium, but examine the dynamics and variability of community measures as features of interest in their own right.
8. Conduct long-term observational and experimental studies.
9. View communities in a landscape context, considering the effects of habitat-mosaic patterns and abandoning notions based on assumptions of spatial homogeneity.
10. Focus on the factors influencing community assembly as a conceptual framework for community studies.
11. Deal with the effects of multiple causes on community patterns.
12. Emphasize the importance of defining and measuring resources and testing the assumption of resource limitation.
13. Develop specific, mechanistically based theory.
14. Frame hypotheses in precise, testable terms whenever possible.
15. Take into account the effects of feedback relationships, indirect interactions, time lags, and nonlinear responses.
16. Avoid extrapolating from particular taxa or habitats to other taxa or habitats, and avoid especially a "north-temperate bias" in thinking about communities.
17. Recognize the importance of replication in both observational and experimental studies.
18. Do not shun or avoid criticism and controversy.

Source: Wiens 1989, 258, table 6.1.

4), framing of studies in the appropriate spatiotemporal context (points 6, 9), and the use of rigorous designs and experimentation (points 8, 14, 15, 17).

Management Implications: Restoration Ecology

As human populations continue to increase in size and distribution over the landscape, there will be fewer opportunities to preserve existing areas in a relatively natural condition. Wildlife conservation will thus depend more and more on the modification of existing reserves, the management of lands between reserves, and the restoration of degraded environments.

Traditionally, wildlife biologists have not dealt with ecological restoration per se. Rather, they have concentrated on modifying vegetation and other environmental features for the benefit of specific species. Conservation biologists, on the other hand, have been especially interested in community-level analyses and landscape-level reserve designs. Restoration projects are usually carried out for highly practical reasons in a nonacademic setting, so there has been a tendency to overlook the scientific value of these projects. Thus, despite attempts to draw attention to its heuristic value (Jordan et al., 1987), restoration has not been recognized as a basic science. Restorationists have only recently started to explicitly consider their profession as a science, as indicated by initiation of the journal *Restoration Ecology* in 1993. The link between restoration and basic ecology, however, remains weak.

We could do a better job of wildlife conservation if we could find a way to integrate more effectively the principles, experiences, and viewpoints of wildlife biology, conservation biology, and restoration ecology. Certainly there are both empirical and academic grounds for such integration. Developing a restoration plan is—or ought to be—the same as developing a plan for an ecological study in wildlife biology or conservation biology (Morrison 2002b). Both wildlife biology and conservation biology offer basic ideas and research techniques, whereas restoration ecology expands the scope to include vegetation and whole systems, along with practical methods. In fact, the goal of restoration ecology transcends the disciplines and may offer a way of integrating them in an adaptive management framework (see chapter 11). In such a framework, restoration activities would be designed as scientific experiments (with the limitations discussed in chapter 4), and then monitored, tested, and readjusted on the basis of results.

Wildlife biologists, however, often engage in academically oriented debates over what constitutes reliable knowledge and "good science," and over how the profession can continue to grow and advance biological understanding (Romesberg, 1981, 1991; Matter and Mannan, 1989; Nudds and Morrison, 1991; Haney and Power 1996). Overall, we are getting better at designing and implementing studies of wildlife and wildlife-resource requirements, demographics, genetics, and so on. Our work usually ends, however, with a list of general management implications. We are good at telling resource managers what the goals should be but seldom include instructions on how these goals can be reached— or on how to determine whether goals *have* been reached. Further, even if the plans are implemented, very little postrestoration monitoring is attempted with any regularity and duration. Frankly, we get the distinct impression that many of our colleagues in wildlife biology feel that implementation of our recommendations is not our responsibility, and that any failure to do so is the fault of "the managers." It is the managers who, in effect, really test ideas by putting them into practice, and who are often in the best position to develop new ideas, to expose the weaknesses of existing ideas, and to redefine research priorities. However, wildlife conservation biologists can play more central roles by advising managers on risk analysis, aiding them with decisions based on risk management, helping to craft integrated applications, and monitoring studies in an adaptive management framework (see discussions on knowledge-based modeling in chapter 10, and on adaptive management in chapter 11).

Clearly, restorationists have a crucial role to play here. The importance of their role becomes especially clear when we consider specific questions that arise in the course of restoration efforts, whether these questions be concerned primarily with wildlife or with whole systems. Morrison (1995, 2002b) listed some of these questions and briefly discussed weaknesses in understanding that could be strengthened by closer collaboration between wildlife conservationists and restorationists.

Directly or indirectly, all restoration projects modify wildlife habitat to some degree. Restoration projects often fail for lack of a clear understanding of how wildlife–habitat relationships are developed. In fact, most reports of restoration projects either do not mention wildlife habitat at all, or they do so only in the collective sense. For example, most of the articles in *Biological Habitat Reconstruction* by Buckley (1989), while describing the reestablishment of wildlife habitat as a primary goal of restoration, do not incorporate species-specific (or other specific) considerations of habitat. Clearly, the tendency to use the term *habitat* loosely, common in wildlife and conservation circles, has been transferred to the restoration field.

Restoration provides unique opportunities to test and refine our ideas about how systems work and the influence of our activities on them. The endangered least Bell's vireo (*Vireo bellii pusillus*), for example, has been induced to breed in several areas of southern California following implementation of a management plan that included restoration of nesting and feeding sites

and reduction of populations of brown-headed cowbirds (*Molothurs ater*), a nest parasite (e.g., Franzreb 1989, 1990). Whether or not successful (sometimes, especially when *not* successful), well-designed restoration projects such as these can teach us much about the critical parameters that drive a particular community or ecosystem. The vireo example incorporates both the resource needs of the species and how a nest parasite can negate the best-designed plans for restoration of the resources. It also serves to illustrate how the study of intensive variables (e.g., nest and food requirements) can be set in the larger-scale context of the influence of extensive variables (e.g., cowbird density) on the species, and how extensive variables alone cannot recover the vireo population.

Wildlife habitats, especially those in regions undergoing development, often occur in a landscape largely dominated by human use. Habitat loss in such a landscape is a complex process that typically separates populations into fragments with different kinds and levels of linkage among them. Since the characteristics of these fragments and linkages have a profound effect on the persistence of wildlife, landscape considerations are crucial to any kind of wildlife habitat restoration in such a setting (see chapters 8 and 9). For example, the influence on wildlife of vegetative patch size and edge-to-interior area is poorly understood, even after decades of research by wildlife biologists (see review in Paton 1994). Here again, restoration provides opportunities to test ideas and to deepen our understanding of these complex interactions—another incentive for wildlife and conservation biologists and restorationists to work together more closely.

The ways in which individual animals adjust their behavior to human influence or manipulation and how these adjustments influence population processes are poorly understood. Wildlife biologists have not worked extensively in this area, and conservation biologists, for their part, have tended to view human influences as a form of disruption that disorganizes systems to the point where studies are unproductive. Neither of these perspectives is appropriate now that virtually all ecosystems are subject to some form of human influence. Though the concept of human influence on wildlife is beginning to receive more attention (see, e.g., Holthuijzen et al. 1990; Griffiths and Van Schaik 1993; Truett et al. 1994), this research still lacks a unified approach or theoretical basis.

Here again, restorationists clearly have an important contribution to make. Restorationists work extensively in human-dominated landscapes and have a wealth of knowledge concerning the development of plant communities under stress. Wildlife biologists, on the other hand, can contribute detailed knowledge of wildlife resource use based on work in relatively natural areas—essential knowledge for the proper design of habitat restoration projects in impacted areas. Collaboration between these two groups can be extremely fruitful. In southern California, for example, wildlife biologists and restorationists assisted land use planners in developing a proposal to enhance the habitat of many animal species in several urban parks (Morrison et al. 1994a, 1994b). These projects, once implemented, will serve as tests of the ideas about wildlife–habitat relationships on which the plans were based.

Jordan et al. (1987) saw the heuristic value of restoration efforts as the basis for an intimate two-way relationship between ecological theory and ecological practice. More recently, Morrison (2000 and papers therein; Morrison 2002b) discussed developing studies of wildlife restoration. We urge restorationists to increase communication with their sibling professional societies (e.g., in North America, the Wildlife Society, the Society for Conservation Biology, the Natural Areas Association, the Society for Ecological Restoration, the Ecological Society of America, and many others). There is no reason why restorationists must be tagged onto a project for the sole purpose of carrying out someone else's recommendations. The results of restoration efforts will be much enhanced by involvement of restorationists in all aspects of a project. They can play a crucial role in developing adaptive management experiments to help managers learn by trial. To participate effectively in such work, however, restorationists will need to apply rigorous approaches to study design, statistics, and sampling.

Translating Wildlife–Habitat Research into Management

Risser (1993) wrote a plea for theoretical ecologists to accept responsibility for clarifying the domain in which results can be used by managers. Arguing that it was unfair to publish papers in the conventional scientific format and then criticize managers for not finding and not using the information correctly, he called for journals such as *Ecological Applications* to include a closing section in each article that describes results in ways that could be used by resource managers. Such a section is, of course, a standard feature in many applied ecological journals (most notably the *Journal of Wildlife Management* and the *Wildlife Society Bulletin*). Even within the resource management community, critics have often called for a closer tie between wildlife researchers and managers (e.g., Baskett 1985).

However, we think that the wildlife biology and management community also bears a responsibility to become educated in the theoretical underpinnings of the ecological systems in which they work. Although we are not calling for all wildlife biologists and managers to become theoreticians per se, it is nonetheless our collective responsibility to review and at least possess a basic understanding of, and appreciation for, the theories on which our management decisions are or could be made. During the initial phases of study design, when the researcher is reviewing the literature and deciding on the variables to measure, thought should be given to how study results might specifically be used in management contexts and who would make such use. As presented in chapter 7, it is standard practice for researchers to divide vegetation measurements into extremely fine categories. But the question seldom asked is, How will my results be translated into management practices? Wildlife researchers are usually consumed by generating publishable R^2 and P values, rather than determining the best result in relation to possible implementation in the field (Anderson et al. 2001). Further, land managers rightfully become frustrated when asked to understand and implement recommendations based on such abstruse variables as log (canopy cover) and (shrub cover)^3, even though the researcher can be complimented for transforming nonlinear data as appropriate (see chapter 7). Researchers, however, should also "transform" their results into practical terms.

Risser (1993), using a paper by Painter and Belsky (1993), further exemplifies our point. The Painter and Belsky work discussed grassland ecosystems but artificially eliminated from consideration the issues of changes in species composition and in nutritive values of the plants because there is little debate on these topics among ecologists. Species composition and nutritive value are, however, central components in determining grassland status by range managers. Risser noted that by arbitrarily dismissing these considerations, ecologists are reinforcing the perception that they are trying to set the conditions of discussion rather than trying to identify the practical considerations used by managers.

Greater emphasis on adaptive resource management (ARM) (Walters 1986) should help researchers, land managers, and administrators, as well as restorationists, to achieve their respective goals. ARM is performed whenever the goal of achieving management objectives and the goal of gaining reliable knowledge are accomplished simultaneously. ARM provides the opportunity to test hypotheses on large geographic scales while allowing some level of management to proceed (Lancia et al. 1993). In addition to C. J. Walters's (1986) text on the subject, students can review the articles published in a special section of the *Transactions of the North American Wildlife and Natural Resources Conference* (1993; see Lancia et al. 1993; also see chapter 11).

Education for the Future

Below we outline our suggestions for enhancing training in the wildlife sciences for university students and professionals.

University Education

"Our current understanding amounts to bungalows. To continue on the present course will only lead to better bungalows . . . it is within our capacity to take a revolutionary course and develop understanding to the state of Taj Mahals" (Romesburg 1991). Romesburg used this analogy to introduce his argument that we are failing to advance our understanding of ecological relationships and thus are failing to advance wildlife management as rapidly as we could. Further, he argued strongly that we are failing to educate our students in a manner that will lead to such advancements in the future. Although not everyone agrees with his conclusions (Knight 1993; but see response by Romesburg [1993]), we think there are certainly many truths in Romesburg's writings that we as professionals would be wise to follow. Romesburg is not alone, nor was he the first, to raise serious questions regarding the rigor with which we educate our wildlife students (e.g., Clark 1986; Gavin 1989; Hunter 1989).

Classically, we fill our students with known facts and figures. As more and more literature is published, and as our technology improves, we are forced to fill their minds with even more facts (Keppie 1990; Romesburg 1991). However, little attention is given to formally training students on *how to think*—that is, how to analyze and synthesize, to be creative. We are not training students to advance the science per se; rather, we are training them to know what we know. Along the way, bright students are able to reorganize and synthesize these facts, challenge established dogmas, and in some manner advance our understanding of what has gone on before us, but many do not.

Romesburg (1991) argued that separate undergraduate and graduate courses in research philosophy are vital to teaching students how to think, and he provided specific topics that such courses should contain. Such courses do indeed teach students how science is structured, how the hypothetico-deductive (H-D) method works, and so on (see chapter 4). Beyond the teaching of the processes of science, however, Romesburg argued persuasively that we should teach students how to invent new ideas. Such courses would introduce students to the diverse structure of theories from many disciplines, and to how such theories can be applied to the natural resources.

In addition, we should educate students on how to identify significant problems (Keppie 1990; Romesberg 1991). Wildlife researchers have been good at studying the same phenomenon in different locations, at different times, and with different species, without ever understanding why the phenomenon is occurring in the first place. The role that a published study might play in advancing science can probably be judged by the firmness with which the author draws conclusions, even if (*particularly if*) conclusions derive from negative results. Typically, our conclusions are nothing more than a list of ad hoc hypotheses about what our study results might mean, and most journal editors reject papers that fail to show statistically significant results. Ad hoc explanations do not advance science, nor do they advance management.

We need to teach our students to push beyond such standard procedures, to identify ways in which we can advance understanding, and then to pursue it. Such an effort might require the learning of a new discipline, be it physiology, genetics, or mathematical modeling. Although the realities of time and money usually prevent students from becoming accomplished physiologists *and* finishing a thesis in a few years, we can at least start the process that will lead them to higher knowledge, broader understanding, greater insight, and more creative research projects in the future. For example, advances in technology have made it possible for a master's-level student to incorporate analyses of reproductive hormones into a field study of small mammals for a few hundred dollars, given that the university has the necessary analytical equipment (which is likely). As developed in chapter 7, for example, the study of time-energy budgets (physiological ecology) should greatly enhance the field of habitat relationships. We also need to encourage more debate on how we approach habitat studies. The argument we made above regarding intensive and extensive variables is an example.

Although it is easy to talk about advances in education, effecting these advances takes a faculty dedicated to this goal. Unfortunately, changes are slow to come in the university setting. Part of this slowness is understandable: it is difficult for faculty who have not been trained in research philosophy, statistics, or modeling to teach such topics. In addition, some simply do not agree that substantial changes are needed (Knight 1993). We think that wildlife science must do a much better job of educating students, and that current wildlife professionals must do a much better job of continuing their education (thus our Dedication and the Preface to this edition).

In addition to more training in specific courses, faculty should increase the rigor with which they teach their courses. For example, in our graduate courses we have found that students resent being challenged to think on a higher order to solve a problem; instead they expect to be told answers as if wildlife ecology and management consist merely of fixed facts. This response is certainly a holdover from their fact-filled undergraduate educations. For example, one of us makes it a practice to start an advanced course by asking students to review an issue—say, ecosystem management or indicator species—and to find ways to solve its inherent problems. This task requires students to explore the literature, synthesize the arguments, understand the theory, and so forth—in other words, to be their own fact finders, synthesizers, and interpreters. The resentment comes from the frustration they feel over their lack of training in independent, critical, and creative thinking.

Specifically, wildlife programs should require all students to receive training in the following areas before leaving a master's program:

- The history of science, and the historic and current links (or gaps) between disciplines
- The general philosophy of knowledge, theory, and creativity
- Research philosophy in the sciences
- How theories and hypotheses in science are developed
- The theory and methods of study design, including impact assessment
- Statistics through nonparametric and multivariate analyses, including Bayesian methods
- Mathematical and computer modeling

In addition, undergraduate wildlife students should be introduced to population genetics, physiology, anatomy, and some topics in organismal-level ecology, including wildlife diseases and parasitology.

Professional Training

Federal land managers in the United States are required to develop inventory and management plans for all land, soil, timber, forage, water, air, fish, wildlife aesthetics, recreational, wilderness, and energy and mineral resources on all areas under their jurisdiction (e.g., USDA Forest Service 1990). Developing such plans is an enormous task that most field personnel, for a variety of reasons, are unable to accomplish fully. Advanced academic training and extensive research experience are usually necessary to design inventories, analyze inventory data, and establish and execute an effective monitoring program (Morrison and Marcot 1995; see also Garcia 1989; Schreuder et al. 1993).

Unfortunately, few field personnel in any federal or state agency have received this training in their university studies, nor are they adequately trained by their agency. There appears to be a genuine reluctance among personnel to advance their education on their own time, including something as simple as keeping current on the literature.

In addition to the changes in the university educational systems discussed above, a substantial change in workplace education is required. This training is ultimately the responsibility of each individual. After all, simply graduating from an accredited institution does not a professional make. However, supervisors and agency policies can create an atmosphere in which staying current, creatively thinking about new ways to approach old problems, establishing programs for auditing advanced courses, and so forth leads to new responsibilities, job advancement, and, most important, sound resource management as a legacy for future generations.

We applaud the many professionals who have indeed taken such burdens upon themselves despite lack of support from their organizations, but far more biologists and managers have failed to keep pace with new knowledge and methods. Managers and other nonresearchers must keep current on the basics of theory, study design, and statistics if they are to make supportable decisions. They must be able to independently evaluate the research that is published.

Toward a Stronger Foundation

Will these proposals be implemented? Romesberg (1993) noted that a profession will instinctively bristle at suggestions that it does not adequately educate, and especially that its intellectual capabilities are inadequate. Further, administrators view creativity in terms of grant dollars and new buildings, and productivity in terms of publications produced, not in terms of restructuring education to emphasize thinking rather than memorization. Professors who challenge themselves to try new ideas and then encourage their students to do the same will be the builders of this foundation. We must change the attitudes of professors, managers, and even ourselves, to see that in the right context, both success *and failure* are the prelude to learning. Such changes should take place in a structured manner, with evaluations, so that modifications can be made as experience is gathered. What we are trying to do is to lay the foundation for future advancements in wildlife science, management, and, ultimately, conservation. The highest possible legacy that any of us can leave is this foundation.

Literature Cited

Abrams, P. A. 1988. How should resources be counted? *Theoretical Population Biology* 33:226–42.

Abrams, P. A. 2000. The impact of habitat selection on the spatial heterogeneity of resources in varying environments. *Ecology* 81:2902–13.

Alldredge, J. R., and J. T. Ratti. 1986. Comparison of some statistical techniques for analysis of resource selection. *Journal of Wildlife Management* 50:157–65.

Alldredge, J. R., and J. T. Ratti. 1992. Further comparison of some statistical techniques for analysis of resource selection. *Journal of Wildlife Management* 56:1–9.

Anderson, D. R., W. A. Link, D. H. Johnson, and K. P. Burnham. 2001. Suggestions for presenting the results of data analyses. *Journal of Wildlife Management* 65:373–78.

Andrewartha, H. G., and L.C. Birch. 1984. *The ecological web: More on the distribution and abundance of animals.* Chicago: Univ. of Chicago Press.

Arthur, W. 1987. *The niche in competition and evolution.* New York: Wiley.

Bailey, J. K., and T. G. Whitman. 2003. Interactions among elk, aspen, galling sawflies and insectivorous birds. *Oikos* 101:127–34.

Barras, S. C., R. M. Kaminski, and L. A. Brennan. 1996. Acorn selection by female wood ducks. *Journal of Wildlife Management* 60:592–602.

Baskett, T. S. 1985. Quality control in wildlife science. *Wildlife Society Bulletin* 13:189–96.

Belovsky, G. E. 1984. Herbivore optimal foraging: A comparative test of three models. *American Naturalist* 124:97–115.

Bennett, A. F. 1987. The accomplishments of ecological physiology. In *New directions in ecological physiology,* ed. M. E. Feder, A. F. Bennett, W. W. Burggren, and R. B. Huey, 1–8. Cambridge: Cambridge Univ. Press.

Boutin, S. 1990. Food supplementation experiments with terrestrial vertebrates: Patterns, problems, and the future. *Canadian Journal of Zoology* 68: 203–20.

Buckley, G. P., ed. 1989. *Biological habitat reconstruction.* New York: Belhaven Press.

Burnham, K. P., and D. R. Anderson. 1998. Model selection and inference: A practical information-theoretic approach. New York: Springer-Verlag.

Clark, R. W. 1986. Case studies in wildlife policy education. *Renewable Resources Journal* 4:11–16.

Elton, C. S. 1927. *Animal ecology.* London: Sidgwick & Jackson.

Franzreb, K. E. 1989. *Ecology and conservation of the endangered least Bell's vireo.* U.S. Fish and Wildlife

Service Biological Report 89(1). Sacramento, CA: USDI USFWS.

Franzreb, K. E. 1990. An analysis of options for reintroducing a migratory native passerine, the endangered least Bell's vireo *Vireo bellii pusillus* in the Central Valley, California. *Biological Conservation* 53:105–23.

Garcia, M. W. 1989. Forest Service experience with interdisciplinary teams developing integrated resource management plans. *Environmental Management* 13:583–92.

Gavin, T. A. 1989. What's wrong with the questions we ask in wildlife research? *Wildlife Society Bulletin* 17:345–50.

Gavin, T. A. 1991. Why ask "why": The importance of evolutionary biology in wildlife science. *Journal of Wildlife Management* 55:760–66.

Griffiths, M., and C. P. Van Schaik. 1993. The impact of human traffic on the abundance and activity periods of Sumatran rain forest wildlife. *Conservation Biology* 7:623–26.

Grinnell, J. 1917. The niche-relationships of the California thrasher. *Auk* 34:427–33.

Guthrey, F. S., C. L. Land, and B. W. Hall. 2001. Heat loads on reproducing bobwhites in the semiarid tropics. *Journal of Wildlife Management* 65:111–17.

Hall, L. S., P. R. Krausman, and M. L. Morrison. 1997. The habitat concept and a plea for standard terminology. *Wildlife Society Bulletin* 25:173–82.

Hall, L. S., and M. L. Morrison. 1998. Responses of mice to fluctuating habitat quality. II. Supplementation experiment. *Southwestern Naturalist* 43:137–46.

Haney, A., and R. L. Power. 1996. Adaptive management for sound ecosystem management. *Environmental Management* 20:879–86.

Hobbs, N. T., and T. A. Hanley. 1990. Habitat evaluation: Do use/availability data reflect carrying capacity? *Journal of Wildlife Management* 54:515–22.

Holthuijzen, A. M. A., W. G. Eastland, A. R. Ansell, M. N. Kochert, R. D. Williams, and L. S. Young. 1990. Effects of blasting on behavior and productivity of nesting prairie falcons. *Wildlife Society Bulletin* 18:270–81.

Hunter, M. D., and P. W. Price. 1992. Playing chutes and ladders: Heterogeneity and the relative roles of bottom-up and top-down forces in natural communities. *Ecology* 73:724–32.

Hutchinson, G. E. 1978. *An introduction to population ecology.* New Haven, CT: Yale Univ. Press.

Hutto, R. L. 1985. Habitat selection by nonbreeding, migratory land birds. In *Habitat selection in birds,* ed. M. L. Cody, 455–76. San Diego, CA: Academic Press.

James, F. C., C. A. Hess, and D. Kufrin. 1997. Species-centered environmental analysis: Effects of fire history on red-cockaded woodpeckers. *Ecological Applications* 7 (1):118–29.

Jamison, B. E., R. J. Robel, J. S. Pontius, and R. D. Applegate. 2002. Invertebrate biomass: Associations with lesser prairie-chicken habitat use and sand sagebrush density in southwestern Kansas. *Wildlife Society Bulletin* 30:517–26.

Johnson, D. H. 1980. The comparison of usage and availability measurements for evaluating resource preference. *Ecology* 61:65–71.

Jordan, W. R., III, M. E. Gilpin, and J. D. Aber, eds. 1987. *Restoration ecology: A synthetic approach to ecological research.* New York: Cambridge Univ. Press.

Keane, J. J., and M. L. Morrison. 1999. Temporal variation in resource use by black-throated gray warblers. *Condor* 101:67–75.

Keppie, D. M. 1990. To improve graduate student research in wildlife education. *Wildlife Society Bulletin* 18:453–58.

Knetter, J. M., R. S. Lutz, J. R. Cary, and R. K. Murphy. 2002. A multi-scale investigation of piping plover productivity on Great Plains alkali lakes, 1994–2000. *Wildlife Society Bulletin* 30:683–94.

Knight, R. L. 1993. On improving the natural resources and environmental sciences: A comment. *Journal of Wildlife Management* 57:182–83.

Lancia, R. A., T. D. Nudds, and M. L. Morrison. 1993. Opening comments: Slaying slippery shibboleths. *Transactions of the North American Wildlife and Natural Resources Conference* 58:505–8.

Leibold, M. A. 1995. The niche concept revisited: Mechanistic models and community context. *Ecology* 76:1371–82.

MacArthur, R. H. 1968. The theory of the niche. In *Population biology and evolution,* ed. R. C. Lewontin, 159–76. Syracuse, NY: Syracuse Univ. Press.

MacArthur, R. H., and R. Levins. 1967. The limiting similarity, convergence, and divergence of coexisting species. *American Naturalist* 101:377–95.

Matter, W. J., and R. W. Mannan. 1989. More on gaining reliable knowledge: A comment. *Journal of Wildlife Management* 53:1172–76.

Morrison, M. L. 1995. Wildlife conservation and restoration ecology. *Restoration and Management Notes* 13:203–8.

Morrison, M. L. 2000. Developing multiple-species conservation reserves and habitat conservation plans. *Environmental Management* 26 (suppl. 1).

Morrison, M. L. 2002a. A proposed research emphasis to overcome the limits of wildlife–habitat relationship studies. *Journal of Wildlife Management* 65:613–23.

Morrison, M. L. 2002b. *Wildlife restoration: Techniques for habitat analysis and animal monitoring*. Washington, DC: Island Press.

Morrison, M. L., and L. S. Hall. 1998. Responses of mice to fluctuating habitat quality. I. Patterns from a long-term observational study. *Southwestern Naturalist* 43:123–36.

Morrison, M. L., and L. S. Hall. 2002. Standard terminology: Toward a common language to advance ecological understanding and applications. In *Predicting species occurrences: Issues of accuracy and scale*, ed. J. M. Scott, P. J. Heglund, M. L. Morrison, J. B. Haufler, M. G. Raphael, W. A. Wall, and F. B. Samson, 43–52. Washington, DC: Island Press.

Morrison, M. L., and B. G. Marcot. 1995. An evaluation of resource inventory and monitoring programs used in National Forest planning. *Environmental Management* 19:147–56.

Morrison, M. L., C. J. Ralph, J. Verner, and J. R. Jehl Jr., eds. 1990. Avian foraging: Theory, methodology, and applications. *Studies in Avian Biology* no. 13.

Morrison, M. L., T. A. Scott, and T. Tennant. 1994a. Wildlife–habitat restoration in an urban park in southern California. *Restoration Ecology* 2:17–30.

Morrison, M. L., T. Tennant, and T. A. Scott. 1994b. Laying the foundation for a comprehensive program of restoration for wildlife habitat in a riparian floodplain. *Environmental Management* 18: 939–55.

Morrison, M. L., W. M. Block, M. D. Strickland, and W. L. Kendall. 2001. *Wildlife study design*. New York: Springer-Verlag.

Morse, D. H. 1980. *Behavioral mechanisms in ecology*. Cambridge, MA: Harvard Univ. Press.

Niiniluoto, I. 1999. Defending abduction. *Philosophy of Science* 66 (3):S436–51.

Nudds, T. D., and M. L. Morrison. 1991. Ten years after "reliable knowledge": Are we gaining? *Journal of Wildlife Management* 55:757–60.

O'Connor, R. J. 2002. The conceptual basis of species distribution modeling: Time for a paradigm shift? In *Predicting species occurrences: Issues of accuracy and scale*, ed. J. M. Scott, P. J. Heglund, M. L. Mor-

rison, J. B. Haufler, M. G. Raphael, W. A. Wall, and F. B. Samson, 25–33. Washington, DC: Island Press.

Painter, E. L., and A. J. Belsky. 1993. Application of herbivore optimization theory to rangelands of the western United States. *Ecological Applications* 3:2–9.

Paton, P. W. C. 1994. The effect of edge on avian nest success: How strong is the evidence? *Conservation Biology* 8:17–26.

Patton, D. R. 1997. *Wildlife habitat relationships in forested ecosystems*. Portland, OR: Timber Press.

Pavlacky, D. C., Jr., and S. H. Anderson. 2001. Habitat preferences of pinyon–juniper specialists near the limit of their geographic range. *Condor* 103:322–31.

Perry, E. F., and D. E. Andersen. 2003. Advantages of clustered nesting for least flycatchers in north-central Minnesota. *Condor* 105:756–70.

Peters, R. H. 1991. *A critique for ecology*. Cambridge: Cambridge Univ. Press.

Ricklefs, R. E., and J. M. Stark, eds. 1998. *Avian growth and development*. New York: Oxford Univ. Press.

Risser, P. G. 1993. Making ecological information practical for resource managers. *Ecological Applications* 3:37–38.

Romesburg, H. C. 1981. Wildlife science: Gaining reliable knowledge. *Journal of Wildlife Management* 45:293–313.

Romesburg, H. C. 1991. On improving the natural resources and environmental sciences. *Journal of Wildlife Management* 55:744–56.

Romesburg, H. C. 1993. On improving the natural resources and environmental sciences: A reply. *Journal of Wildlife Management* 57:184–89.

Schreuder, H. T., T. G. Gregoire, and G. B. Wood. 1993. *Sampling methods for multiresource forest inventory*. New York: Wiley.

Schultz, J. C. 1992. Factoring natural enemies into plant tissue availability to herbivores. In *Effects of resource distribution on animal–plant interactions*, ed. M. D. Hunter, T. Ohgushi, and P. W. Price, 175–97. San Diego, CA: Academic Press.

Scott, J. M., P. J. Heglund, M. L. Morrison, J. B. Haufler, M. G. Raphael, W. A. Wall, and F. B. Samson. 2002. Introduction. In *Predicting species occurrences: Issues of accuracy and scale*, ed. J. M. Scott, P. J. Heglund, M. L. Morrison, J. B. Haufler, M. G. Raphael, W. A. Wall, and F. B. Samson, 1–5. Washington, DC: Island Press.

Slobodkin, L. B. 1992. A summary of the special feature and comments on its theoretical context and importance. *Ecology* 73:1564–66.

Stauffer, D. F. 2002. Linking populations and habitats: Where have we been? Where are we going? In *Predicting species occurrences: Issues of accuracy and scale,* ed. J. M. Scott, P. J. Heglund, M. L. Morrison, J. B. Haufler, M. G. Raphael, W. A. Wall, and F. B. Samson, 53–61. Washington, DC: Island Press.

Stephens, D. W., and J. R. Krebs. 1986. *Foraging theory.* Princeton, NJ: Princeton Univ. Press.

Stevens, A. J., Z. C. Welch, P. C. Darby, and H. F. Percival. 2002. Temperature effects on Florida applesnail activity: Implications for snail kite foraging success and distribution. *Wildlife Society Bulletin* 30:75–81.

Thomas, D. L., and E. J. Taylor. 1990. Study designs and tests for comparing resource use and availability. *Journal of Wildlife Management* 54:322–30.

Truett, J. C., R. G. B. Senner, K. Kertell, R. Rodrigues, and R. H. Pollard. 1994. Wildlife responses to small-scale disturbances in Arctic tundra. *Wildlife Society Bulletin* 22:317–24.

USDA Forest Service. 1990. *Resource inventory handbook.* FSH 1909.14, Amendment no. 1, March 29, 1990. Washington, DC: USDA Forest Service.

Van Horne, B. 1983. Density as a misleading indicator of habit quality. *Journal of Wildlife Management* 47:893–901.

Van Horne, B. 2002. Approaches to habitat modeling: The tensions between pattern and process and between specificity and generality. In *Predicting spe-cies occurrences: Issues of accuracy and scale,* ed. J. M. Scott, P. J. Heglund, M. L. Morrison, J. B. Haufler, M. G. Raphael, W. A. Wall, and F. B. Samson, 63–72. Washington, DC: Island Press.

Van Horne, B. and J. A. Wiens. 1991. *Forest bird habitat suitability models and the development of general habitat models.* USDI Fish and Wildlife Research Report 8. Washington DC: USDI FWS.

Walters, C. J. 1986. *Adaptive management of renewable resources.* New York: Macmillan.

Ward, J. P., Jr. 2001. Responses of Mexican spotted owls to environmental variation in the Sacramento Mountains, New Mexico. PhD dissertation, Colorado State University, Fort Collins.

Wheatley, M., K. W. Larsen, and S. Boutin. 2002. Does density reflect habitat quality for North American red squirrels during a spruce-cone failure? *Journal of Mammalogy* 83:716–27.

Wiebe, K. L. 2001. Microclimate of tree cavity nests: Is it important for reproductive success in northern flickers? *Auk* 118:412–21.

Wiens, J. A. 1989. *The ecology of bird communities.* Vol. 2, *Processes and variations.* Cambridge: Cambridge Univ. Press.

Wiens, J. A. 1992. Ecology 2000: An essay on future directions in ecology. *Bulletin of the Ecological Society of America* 73:165–70.

Wolda, H. 1990. Food availability for an insectivore and how to measure it. *Studies in Avian Biology* 13:38–43.

Afterword

In this book we have tried to review the foundations of wildlife–habitat relationships, describe methods of conducting rigorous studies, and suggest ways in which we can advance our profession. The study of wildlife–habitat relationships encompasses far more than the classic view of habitat as vegetation. It includes consideration of physiology, behavior, ecological history, genetics, evolution, and reciprocal ecological processes. In essence, it forms the basis of wildlife research. As such, it is imperative that we do it correctly. Although we hope that our book will help advance the field, we also hope that it will serve as a basis for more critical debate on how we should proceed.

Much has changed in the realms of wildlife habitat assessment and management since our first edition of this text. If we go out on a limb, we can extrapolate from these trends and foretell the near future for the science and management of wildlife–habitat relationships.

We foresee several general trends in habitat assessment, including continued growth in habitat analysis technologies. Expanding technologies will include more sophisticated computer simulation modeling, improvements to geographically referenced ("spatially explicit") population modeling, and broad-scale analysis of disturbance dynamics aided by new remote-sensing technologies. As we predicted when we wrote our second edition, the then-fledgling Internet and World Wide Web have exploded with content (although also with garbage). However, if used judiciously, these resources provide access to facts, models, tools, maps, data, people, and organizations as never before in the history of humankind. We are really just beginning to explore the best ways to use these astounding new tools to foster communication and collective action for conducting wildlife science and conservation.

We also anticipate continuing development of scientific visualization techniques, including new kinds of remote-sensing data for detecting and displaying change, as well as new analysis methods for discovering and presenting trends in data on wildlife habitats and human habitats alike.

Increasing human population growth, particularly in developing nations, will lead to escalating conflicts in land use allocations and resource conservation. The governments and populace of some developed nations will increasingly resist changing their profligate use of resources, resulting in increased separation from other developed countries that do take on such challenges.

There is much to learn of wildlife conservation plights and successes among all nations of the world, particularly from nations long struggling with high human densities and severe resource scarcities that others of us too often assume will never happen to us. We hope that the trend toward globalization of information will be matched by an earnestness to learn from each other. It is up to us, as professionals, students, and responsible citizens, to talk with one another.

The future of wildlife–habitat relationships assessments and management might broaden to include nonvertebrate species and other taxonomic groups and ecological entities of conservation concern, and to consider the ecosystem context of ecological functions and processes. We hope that broad-scale assessments (e.g., WRI et al. 1992) and policy decisions (e.g., Caldwell et al. 1994) can provide a framework for considering habitat management in an evolutionary context. In the future of ecosystems lies the future of ourselves as well. Management policies that rely less on short-term profit (e.g., maximizing net present value of resources or gross primary production) and at least equally on the needs of future generations would do well to recognize the economic and social benefits—the economic and sociological resiliency and stability—of long-term conservation planning.

The sustainable resource use of the more distant future can be attained by beginning now to articulate an ideal, long-term, resource use scenario. Under such a scenario, humans would learn to adjust their resource use patterns and habits to ensure resource productivity and sustained use in perpetuity. In this idealized future, resource production, such as timber growth and availability of fishery resources and grazing potentials, would be in harmony with rates of use and extraction.

Much has been written on "sustainable futures" over the past several decades (e.g., see papers in Bissonette and Krausman 1995; O'Neill et al. 1996; Sandlund et al. 1992; Holling 2000). We have attempted to outline the primary elements of such a sustainable future in the accompanying box. In addition to changes required in personal resource use habits, attaining environmental sustainability will also entail difficult changes in population centers of growth, resource extraction industries and infrastructures, and even administrative, economic, political, and educational systems. Reaching this goal may entail use of conservation reserves (Franklin 1993). as well as actively managed landscapes (e.g., Everett et al. 1994). However, even if the ideal can never be reached, we feel that articulating its parameters can help determine which components are feasible or desirable ecologically, socially, politically, and economically. Articulating these parameters can also help determine the costs and benefits to future generations of *not* reaching for an environmental ideal.

Finally, we might propose two overall principles for habitat managers. The first borrows from a popular and idealistic modern myth, and may be called the *prime habitat management directive: Insofar as possible, do not interfere with the normal course of development of native ecosystems and populations.* We have learned this lesson from many trials: by suppressing wildfires without considering subsequent adverse changes in forest community structures and increasing susceptibility to catastrophic fire events (Graham 1994); by removing biological diversity from forests, prairies, and grasslands without concern for sources of largely unknown biological elements critical to long-term soil health, re-

Primary elements and assumptions of a resource planning scenario designed for long-term sustainability of habitats for wildlife and humans.

Planning Elements

1. Humans will amend their resource use habits to ensure future, sustained production and availability of resources from forest, grassland, wetland, riparian, and aquatic ecosystems.
2. Design of human occupation of the land is one of the more important facets of this scenario. This includes consideration for urban–forest interfaces, grazing, mining, timber management, locating of urban and rural communities and transportation infrastructures, resource recycling, solid waste management, toxic waste management, air quality control, and changes in resource use habits and lifestyles.
3. Land allocations and patterns will be coordinated among private, state, Federal, and other kinds of ownerships. Each has its own specific—albeit different—contribution to maintaining habitats, wildlife, and biodiversity.
4. Land use activities will focus on the amounts, flows, consumption rates, and renewal rates of material resources. However, additional goals such as recovery of threatened species and recovery of ecological communities at risk would also be addressed.
5. Humans will amend their resource use habits and land occupation patterns according to what will provide for a more aboriginal or natural set of conditions, rather than invest energy and effort into changing natural disturbance regimes and ecosystems to meet predefined desired levels of resource use. The amount and flows of resources used (consumptive or nonconsumptive) will result from the capability of the land to maintain itself.
6. Indigenous peoples will engage in their traditional hunting, gathering, and cultural activities, using desired species, resources, and environmental conditions in perpetuity.
7. Social and economic systems, as well as nonindigenous cultural conditions, resource use habits and patterns, and individual resource-use lifestyles, will be subject to change, to ensure long-term sustained resource use conditions. Administrative boundaries affecting human populations and resource use patterns would be amended along ecological lines.
8. Interregional and international conditions and effects will be explicitly considered in making local resource management decisions.

Assumptions

1. Given present conditions of forests, grasslands, wetlands, and riparian and aquatic systems, it is possible to recover at-risk elements and adjust resource use levels and land allocations to eventually achieve a steady-state of resource use and land occupation. There are no elements that are irreversibly lost at this time or that cannot be recovered; or if some elements are moribund, their loss would not jeopardize overall mission success.
2. Social, cultural, lifestyle, and political institutions would change where necessary to ensure successfully meeting the goals of this scenario.
3. Human populations and their patterns of land use, including placement and kinds of habitations and infrastructures, would change to ensure successfully meeting the goals of this scenario.
4. All legal and aboriginal citizens would participate in, and support, attaining a future that ensures resource availability in perpetuity. Local participation by citizens would aid in transition to a sustainable future.
5. This scenario would provide a conservation benchmark and leadership for considering similar planning for all land administrations, ownerships, regions, and even other nations.

source productivity, and resistance of ecosystems to rampant disease and pathogens (Robinson 1993; Koopowitz et al. 1994); by damming rivers without considering the dispersal blockades they impose to migrating salmon and indigenous peoples (Anderson 1993); by introducing exotic species without first testing or considering their potential for dispersal and ecological havoc on native ecosystems (Coblentz 1990); and by numerous other similar lessons.

The second habitat management principle would pertain where the first proves impossible or infeasible. We can borrow from the medical profession and propose a *habitat management Hippocratic oath: Do no ecological harm.* That is, where "invasive" management is necessary to provide for human resource needs, activities should be designed so as to do no harm to native ecosystems and communities. As an example, manipulation of soils for agriculture or forestry objectives should also strive to avoid wind and water erosion loss and to ensure, at least on a portion of the land base, the full complement of beneficial soil mesoinvertebrates, organic matter, microfungi, invertebrate pollinators, and other components essential to long-term productivity.

For, in the end, in the future of wildlife is the future of us all.

Literature Cited

Anderson, M. 1993. *The Living Landscape.* Vol. 2, *Pacific Salmon and Federal Lands.* Washington, DC: Wilderness Society.

Bissonette, J. A., and P. R. Krausman, eds. 1995. *Integrating people and wildlife for a sustainable future.* Bethesda, MD: Wildlife Society.

Caldwell, L. K., C. F. Wilkinson, and M. A. Shannon. 1994. Making ecosystem policy: Three decades of change. *Journal of Forestry* 92 (4):7–10.

Coblentz, B. E. 1990. Exotic organisms: A dilemma for conservation biology. *Conservation Biology* 4:261–65.

Everett, R. L., P. F. Hessburg, and T. R. Lillybridge. 1994. Emphasis areas as an alternative to buffer zones and reserved areas in the conservation of biodiversity and ecosystem processes. *Journal of Sustainable Forestry* 2 (3/4):283–92.

Franklin, J. F. 1993. Preserving biodiversity: Species, ecosystems, or landscapes? *Ecological Applications* 3:202–5.

Graham, R. T. 1994. Silviculture, fire and ecosystem management. In *Assessing forest ecosystem health in the Inland West,* ed. R. N. Sampson and D. L. Adams, 339–51. Binghampton, NY: Haworth Press.

Holling, C. S. 2000. Theories for sustainable futures. *Conservation Ecology* 4 (2):7 [online]; www.consecol.org/vol4/iss2/art7.

Koopowitz, H., A. D. Thornhill, and M. Andersen. 1994. A general stochastic model for the prediction of biodiversity losses based on habitat conversion. *Conservation Biology* 8:425–38.

O'Neill, R. V., J. R. Kahn, J. R. Duncan, S. Elliott, R. Efroymson, H. Cardwell, and D. W. Jones. 1996. Economic growth and sustainability: A new challenge. *Ecological Applications* 6:23–24.

Robinson, J. G. 1993. The limits to caring: Sustainable living and the loss of biodiversity. *Conservation Biology* 7:20–28.

Sandlund, O. T., K. Hindar, and A. H. D. Brown, eds. 1992. *Conservation of biodiversity for sustainable development.* New York: Oxford Univ. Press.

WRI [World Resources Institute], IUCN [International Union for Conservation of Nature and Natural Resources], and UNEP [United Nations Environmental Program]. 1992. *Global biodiversity strategy: Guidelines for action to save, study, and use Earth's biotic wealth sustainably and equitably.* Washington, DC: World Resources Institute.

Glossary

The following definitions are of standard terms used in studies of wildlife–habitat relationships. Additional terms and definitions are found in other tables and sections throughout the book.

abundance   The number of individuals (cf. *density*).

accuracy   The nearness of a measurement to the actual value of the variable being measured; not synonymous with *precision.*

assemblage   A group of species, often of the same taxonomic class, that co-occur in an area (cf. *community*).

census   A complete enumeration of an entity.

coarse filter   An approach to wildlife and ecological assessment and management that considers broad-scale attributes of species groups and ecosystems rather than species-specific attributes (cf. *fine filter*).

community   The co-occurrence of individuals of all species during a specified time and space and their interactions.

density   The number of individuals per unit area. Crude density includes area of habitat and nonhabitat; ecological density includes only area of habitat.

distribution   The pattern of dispersion of an entity within its range.

ecosystem   The set of all abiotic conditions, and biotic entities and their ecological interactions, in a given area

effective population size   The numbers of reproductive individuals of some population (often referred to as the censused population size) represented as a condition of idealized panmixia.

extent   The area over which observations are made and the duration of those observations.

fine filter   An approach to wildlife and ecological assessment and management that focuses on details of species-specific attributes (cf. *coarse filter*).

fitness   The somatic and reproductive vitality of offspring. *Realized fitness* is the actual fecundity of offspring in an area under particular environmental conditions.

floater   A nonreproductive individual moving among a population that does not hold a territory but that can quickly fill a vacant space.

functional redundancy   The number of species performing the same ecological function in a community. The greater the functional redundancy, the greater the system can resist adverse change in its overall functional integrity.

gap analysis   The evaluation of how concentrations of species overlap with the occurrence of protected areas in a region. Locations where species concentrations lie outside protected areas constitute a "gap" in the conservation protection scheme of the area.

grain   The spatial and temporal resolution of observations; the smallest resolvable unit of study.

guild   A group of species that exploit the same class of environmental resources in a similar way.

habitat   The physical space within which an organism lives, and the abiotic and biotic entities (e.g., resources) it uses and selects in that space. (*Microhabitat* and *mesohabitat* are relative terms referring to the grain size of the area over which habitat is being measured.)

habitat availability   The accessibility and use of physical and biological components in a habitat.

habitat avoidance   An oxymoron that should not be used; wherever an animal occurs defines its habitat.

habitat preference   Used to describe the relative use of different locations (habitats) by an individual or species.

habitat quality   The ability of the area to provide conditions appropriate for individual and population persistence.

habitat selection   A hierarchical process involving a series of innate and learned behavioral decisions made by an animal about what habitat it would use at different scales of the environment.

habitat use   The way an animal uses (or "consumes," in a generic sense) a collection of physical and biological entities in a habitat.

home range   The area traversed by an animal during its activities during a specified period of time (cf. *territory*).

key ecological functions   The major ecological roles played by a species. Examples include herbivory, dispersal of seeds and spores, primary creation of tree cavities and ground burrows, nutrient cycling, and many others. To keep a system "fully func-

tional," one should strive to maintain all categories of naturally occurring functions among all native species.

key environmental correlates   Specific substrates, habitat elements, and attributes of species' environments that are not represented by overall (macro)habitats and vegetation structural conditions. Specific examples of KECs include snags, down wood, type of stream substrate, and many others.

keystone species   Species whose *key ecological functions* are so salient or critical that removal of the species would greatly alter the structure, composition, or function of the community.

landscape   A spatially heterogeneous area used to describe features (e.g., stand type, site, soil) of interest.

landscape feature   Widespread or characteristic features within a landscape (e.g., stand type, site, soil, patch).

metapopulation   A system of populations of a given species in a landscape linked by more or less balanced rates of local extinction and colonization. More loosely, the term is used for groups of populations of a species, some of which go extinct while others are established, but the entire system may not be in equilibrium.

model   Any formal representation of the real world. A model may be conceptual, diagrammatic, mathematical, or computational.

model calibration   The estimation of the values of model parameters from data.

model parameterization   The process of specifying a model structure, including its functions and state variables.

model validation   Comparison of a model's predictions to some user-chosen standard to assess if the model is suitable for its intended purpose.

model verification   The demonstration that a model is formally correct.

niche   Defined variously over time as the position, status, or role of an organism in an ecosystem; also defined as the distribution of an organism resulting from the sum of tolerances along multiple environmental gradients, or the sum of use patterns along multiple resource axes; see text for review.

panmixia Complete and random interbreeding within a population; often an idealized condition against which actual conditions are compared, such as for calculating effective population size.

patch (habitat) A recognizable geographic area that contrasts, in structure or occurrence of resources, with adjacent areas and has definable boundaries.

population Classically, a collection of interbreeding individuals; see text.

precision The closeness to each other of repeated measurements of the same quantity; not synonymous with *accuracy*.

range The geographic limits within which an organism occurs.

resolution The smallest spatial scale at which we portray discontinuities in biotic and abiotic factors in map form.

resource Any biotic and abiotic factor directly used by an organism.

resource abundance The absolute quantity of a resource in a given area.

resource availability A measure of the amount of a resource actually available to an organism (i.e., the amount exploitable).

resource preference The likelihood that a resource will be used if offered on an equal basis with others.

resource selection The process by which an animal chooses a resource.

resource use A measure of the amount of resource taken directly (e.g., consumed, removed) by an organism from a given area.

scale The resolution at which patterns are measured, perceived, or represented. Scale can be broken into several components, including geographic extent, resolution, and other aspects.

scale of observation The spatial and temporal scales at which observations are made.

sensitivity analysis A process in which model parameters or other factors are varied in a controlled fashion to determine their relative influence on model outcome.

sink populations In a landscape, a population or site that attracts colonists or floaters, while not supplying migrants to other sites or populations.

site An area of uniform physical and biological properties and management status (cf. *study area*).

source populations In a landscape, a population or a site that supplies colonists to other patches.

study area An arbitrary spatial extent chosen by the investigator within which to conduct a study (cf. *site* and *scale*).

territory The spatial area defended by an animal or group of animals. Typically a subset of, but may include all of, the overall *home range*.

viability Strictly, the ability to live or grow. In conservation biology, the probability of persistence of a population for an specified period of time.

About the Authors

Michael L. Morrison is Professor and Caesar Kleberg Chair in Wildlife Ecology and Conservation in the Department of Wildlife and Fisheries Sciences, Texas A&M University, College Station. He teaches undergraduate and graduate courses in wildlife management, study design, and restoration ecology. His students and he focus on habitat relationships and the development of restoration plans for wildlife conservation.

Bruce G. Marcot is Research Wildlife Ecologist at the Portland Forestry Sciences Laboratory of the Pacific Northwest Research Station, USDA Forest Service. He works in the general area of environmental relationships of plants and ani-

mals in nature, and depicting them in models and databases. Much of his work focuses on the forests of the Pacific Northwest, but he also conducts substantial work on biodiversity in the forests of India.

R. William Mannan is Professor of Wildlife Ecology in the University of Arizona's School of Natural Resources in Tucson, where he teaches undergraduate and graduate courses. His research interests focus on relationships between animals and their habitats in urban and forest environments, and on animal behavior as it relates to habitat use.

Author Index

Author Index

Author Index

Author Index

Author Index

Author Index

Gibbs, J. P., 109
Gibson, M. J., 113
Giesy, J. P., 342, 348
Gilbert, L. E., 102, 116
Giles, R. H., Jr., 269
Gilliam, J. F., 5
Gilliand, R. L., 178
Gilpin, M. E., 74, 86, 305, 433, 435
Gimblett, H. R., 353, 357
Ginzburg, L. R., 73, 81
Gipson, P. S., 285
Gitzendanner, M. A., 89
Gleason, H. A., 44, 45
Glenn-Lewin, D. C., 44, 45
Glickman, S. E., 145
Glitzenstein, J. S., 186
Glover, T., 223, 233
Godron, M., 256, 269
Godwin, K. D., 208
Goettel, R. G., 109
Goggins, G. C., 287
Golbeck, A. L., 329
Golbert, F. F., 5
Goldberg, D. E., 351
Goldman, D., 31
Goldstein, D. L., 237
Goldstein, M., 182, 193, 198, 200, 208, 209, 216
Gonzalez-Meler, M., 290
Good, D. A., 303
Goodman, D., 343
Goodstein, D., 139
Goodwin, H. T., 22
Gordon, G., 332
Gosz, J. R., 48
Gotfryd, A., 192
Gottman, J. M., 222, 245
Goudie, A., 17, 38
Gould, W. R., 332
Goulder, L., 402
Grabherr, G., 268
Graham J. M., 56, 166
Graham, C. H., 293
Graham, M. H., 214
Graham, R. L., 408
Graham, R. W., 23, 24, 26
Graham, W., 17
Grant, P. R., 274
Grant, W. E., 339
Gravenmier, R. A., 270
Gray, A. N., 285, 407
Graybill, F. A., 136

Green, D. G., 269
Green, D. M., 290
Green, R. H., 152, 155, 168, 190, 194,
Greenberg, C. H., 274
Greene, H.W., 243
Greig-Smith, P., 172
Grier, J. W., 297
Griesemer, J. B., 5
Griffin, P. C., 166
Griffiths, M., 435
Grimm, E. C., 29
Grinder, M. I., 98
Grinnell, J.,
Grinnell, J., 5, 21, 57, 422
Gross, J. E., 240
Gross, P. R., 139
Grubb, G., 110
Grubb, T. G., 342, 348
Gu, W., 264
Guisse, A.W., 357
Gullison, R. E., 309
Gunderson, L. H., 334, 393
Gunderson, L., 407
Gupta, A., 269
Gurevitch, J., 131, 331
Gustafson, E. J., 255
Gutherie, C. G., 93
Guthery, F. S., 30, 48, 135, 136, 138, 284, 425
Gutierrez, R. T., 212
Gutzwiller, K. J., 233

Haas, C. A., 96
Haas, T. C., 344
Haddad, B. M., 273, 290
Haddad, N. M., 96, 263, 307
Haegele, M. A., 110
Hafner, D. J., 26, 27
Hafner, H., 69, 298
Hafner, S. D., 308
Hahn, D. C., 286
Haig, S. M., 303
Haire, S. L., 108
Halaj, J., 33
Hale, A. M., 226
Hall, B. W., 30, 425
Hall, C.A. S., 320
Hall, D. J., 5
Hall, L. S., 166, 170, 222, 419, 420, 431
Hall, S. J., 33, 36, 37
Halley, J., 362
Hallmark, C.T., 17

Author Index

Author Index

Author Index

Author Index

Author Index

Author Index

Author Index

Author Index

Subject Index

473

Subject Index

Subject Index

bobolink, 92
bootstrapping, 362
botfly, 323
Botswana, 292
bottle communities, 144
bottleneck, 30–32, 85–87, 113, 114, 291, 396
boundary effect, 286–288, 289, 306
Breeding Bird Survey, 92–94, 111, 338
Britain, 81, 82
British Columbia, 70, 261, 295, 408
BSD, *see* diversity, bird species,
Bubulcus ibis, 103
budworm, spruce, 273
buffer
    forest, 308
    protected area, 287–288, 309
    riparian, 287, 307, 308
    stream, 286, 287, 392
    wildlife, 287
    zone, 110, 287, 288
Bufo
    bufo, 86, 405
    marinus, 109
Buteo
    buteo, 364
    jamaicensis, 308
    lineatus elegans, 113
    platypterus, 266
    spp., 285
    swainsoni, 160
buzzard, 364

Calidris alba, 233
California, 22, 38, 64, 67, 81, 85, 88, 89, 95, 97, 99, 103, 105,
    108, 263, 308, 363, 434, 435
Callaeas cinerea, 293
Callipepla californica, 30
Canada, 19, 29, 30, 99, 103, 192, 227, 307, 357, 408
Canis lupus, 35, 85
canonical correspondence analysis, 270
canopy
    closure, 162, 163, 169, 170, 171, 192
    cover, 163, 165, 173, 192, 194, 201, 203, 270, 285, 436
    environment, 391
    foraging birds, 110, 164, 205, 272
    forest (woody, vegetative), 56, 72, 169, 200, 274
    gaps, 265, 272–273, 275, 334, 392
    height, 173, 192, 203, 286
    layers, 203
    shrub, 336
    structure, 265

    volume, 165
    wildlife, 265–266, 412
capercaillie, 74
    Capra
        hircus, 105
        ibex, 240
captive
    population, 31–32, 85, 86–88, 90, 233, 271
    propagation, breeding, 86–88, 90, 271, 291, 297
Carduelis
    flammea, 99
    hornemanni, 99
caribou, woodland, 70, 89, 92, 117, 311, 324
carrying capacity, 73, 74, 337, 394
CART, *see* classification and regression tree
Cascade Mountains, 19, 103, 289
cascade, trophic, *see* trophic cascade
Cathartes aura, 394
Catharus ustulatus, 308
cattle, 53, 54
causal web, 324
causation, 162, 221, 224, 320, 411, 420
cave, cavern, 101
cenote, 304
Central America, 21, 304
Centrocerus urophasianus phaios, 389
Certhia americana, 63, 64
Cervus elaphus, 106, 226, 229, 337
Chamaea fasciata, 93, 94
Channel Islands, 81, 85, 105
character
    displacement, 101
    divergence, 404
Charadrius melodus, 425
Chatham Islands, 294
Cheetah, 31
Chelonia
    agassizii, 406
    mydas, 406
cherry, pin, 275
chickadee, 113, 242
    black-capped, 163
    chestnut-backed, 163, 234
    mountain, 285
Chihuahuan Desert, 21
Chile, 200, 202, 301
China, 386
Choristoneura fumiferana, 273
Christmas Island, 274
Chrysemys picta, 286
circular plots, 170, 172, 174, 175, 190

Subject Index

wolverine, 109
woody debris, 32, 56
woodpecker, 133, 196–197, 273
    black–backed, 100
    pileated, 98
    red–cockaded 63, 70, 75, 85, 98, 113, 271, 296
    white–headed 99, 350
wren, rock, 293
wrentit, 93, 94, 113
Wyoming, 273, 285

Xenicus gilviventris, 293

Yellowstone
    (Greater) Ecosystem, 35, 37
    National Park, 271, 272
Yucatan, 304

Zimbabwe, 292
zoogeography, 67, 259, 302
Zosterops lateralis, 294